# Environmental Microbiology
## Principles and Applications

# Environmental Microbiology
## Principles and Applications

**Patrick K. Jjemba**
Biological Sciences Department
University of Cincinnati
Cincinnati, Ohio
USA

**Science Publishers, Inc.**
Enfield (NH), USA          Plymouth, UK

1-06

LOC data available on request.

**SCIENCE PUBLISHERS, Inc.**
Post Office Box 699
Enfield, New Hampshire 03784
United States of America

Internet site: *http://www.scipub.net*

*sales@scipub.net* (marketing department)
*editor@scipub.net* (editorial department)
*info@scipub.net* (for all other enquiries)

ISBN 1-57808-347-8 (HC)
ISBN 1-57808-348-6 (PB)

Published by Science Publishers Inc., Enfield, NH, USA

Printed in India

# Preface

Environmental Microbiology as a subject is relatively young compared to other traditional subjects. It deals with processes in the environment that are directly mediated by microorganisms. However, it is also a rapidly growing field that cuts across various traditional disciplines such as Biochemistry, Ecology, Engineering, Geography, and Microbiology. It is difficult to pinpoint the origin of Environmental Microbiology as a subject but its foundation lies in the initial interest to understand and provide safe drinking water coupled with proper sanitation at the onset of industrialization and associated urbanization. However, during the past three decades the scope of Environmental Microbiology has tremendously expanded to include a variety of other issues, such as the fate of organic pollutants, transformation of metals, and aerobiology.

This book was written for an audience that has a basic understanding of microbiology. Oftentimes, microbiologists tend to overzealously focus on bacteria, inadvertently ignoring other microbes (i.e., algae, fungi, protozoa, and viruses). This discrepancy is redressed herein. Scholars of environmental microbiology come from a variety of disciplines including Microbiology majors, Social Scientists, Engineering, Law (Environmental Law), Agriculturalists, Geography (GIS), Chemists (Environmental Chemists), Toxicologists and so forth. Considering such a diverse audience, not everyone will be content with the depth accorded to all aspects of the topic. However, the reader will find the extensive references rich resources for more in-depth data. The material presented here recognizes the basic foundations and importance of conventional microbiological techniques (which focused greatly on culture-based studies), linking them with information from more recent nonconventional techniques. Various principles are also applied which attest to the undisputable reality that microbes in pure culture may function somewhat differently than in complex multispecies environmental matrices.

This book is unique in that the subject is approached from a **history of microbes** and their place in shaping the environment, rather than a *history of microbiology*. This approach properly introduces the reader to the several different microorganisms and then unveils the role of each (negative or positive) in the environment. That environmental degradation is more prevalent in developing countries is a commonly recognized fact. Quite a number of books address important environmental microbiology issues, such as water treatment, but sad to say orient their presentation exclusively to high-investment treatment systems. This book reaches beyond such economically burdensome schemes by covering the basic concepts of water treatment and modes of application in a variety of backgrounds and economic settings. Basic microbiological concepts such as physiology, genetics, and metabolism are discussed with reference to ecological concepts and biochemical cycling. A chapter on environmental biotechnology is also included.

While writing this book professional advice, editorial opinions, support and encouragement were received from several individuals, including Drs. O. Roger Anderson, Brian K. Kinkle, Mark LeChevallier, Eugene L. Madsen, Lynn Margulis, Sharon Parker, Boakai Robertson, Dorion Sagan, Angela Sessitsch, and Guenther Stotzky, to name but a few. The editorial skills of the copy editor, Margaret Majithia are also greatly appreciated. I am also grateful to the authors and publishers who granted me permission to use their material and to the production staff at Science Publishers Inc. The understanding and patience of my wife Enid, daughter Patricia, and sons Daniel and Eric underpinned the book's realization. It is dedicated to my parents Lakeri (Rachael) and Daniel Kayondo, who sacrificed much to ensure the best education for their family.

Patrick Kayondo Jjemba, Ph.D.
University of Cincinnati, Cincinnati, Ohio

# Contents

# 1

# Microbial Evolution and Diversity

## 1.1  INTRODUCTION

The importance of the environment in our day-to-day activities only became obvious to the general public in the early part of the 1960s when Rachel Carson's book *Silent Spring* was published. Carson (1962) opened her book by describing an imaginary town in the heart of America where all life had at one time lived in harmony with its surroundings. Over a short period of time, mysterious maladies swept the community and all its forms of life. Everywhere in this town was a shadow of death attributed to humans and their activities. Even though Carson clearly indicates in her description of these maladies that this town did not actually exist, the picture she painted brought home humanity's detrimental activities against the environment. Soon after her book was published, the public outcry in the United States prompted Congress to establish the Environmental Protection Agency (US EPA). Various developed countries also created either full ministries or departments within specific ministries to oversee matters pertaining to the environment. In the United Nations, the United Nations Environmental Program (UNEP) was established to help member countries deal with environmental management problems. Today, most countries have, at least in principle, shown a commitment to slow down or even reverse the human-driven devastating effects against the environment.

Microbes, a term we must emphasize, that includes bacteria, protozoa, fungi, algae, and for lack of a better collective term, viruses, play a very significant role in influencing environmental dynamics. Microbes preceded photosynthesis. The subject of environmental microbiology includes microbial ecology which is basically the study of the distribution, activities, and interactions of microorganisms with their habitats (i.e., soil, sediments, freshwater, groundwater, etc.). Such studies normally entail the isolation, identification, and measurement of the activities of organisms in both pure and mixed culture, assessment of their interactions with other living cells, and determination of their response to abiotic environments.

An attempt to understand environmental microbiology requires an examination of microorganisms in the establishment (origin) of life on Earth and their contributions to life through evolution. Based on radiodating measurements, the solar system is estimated to be about 4.6 billion years (Ba) old. It originated when a very hot star (supernova) exploded and generated a new star (the sun) and other components in the galaxy including the Earth. During this early time, the intensity of sunlight was approximately 30% less than it is now (Philander, 1998). We are sure that the Earth's origin was connected with the explosion of a supernova because to date, it is radioactive and made of elements such as iron, silicon, and oxygen. These elements cannot be made in the normal processes of stellar evolution and it takes energy to make them. This radioactivity in turn provides an accurate clock which has been used to establish the time since this explosion occurred and the inception of Earth.

## 1.2 THE ORIGIN OF LIFE

The earliest forms of life on Earth have been estimated to be 3.5-4.0 Ba old, dating from the time when the planet cooled to a point at which liquid water was present. Several scientific theories to explain the origin of life on Earth have been advanced. Most widely subscribed to is the possibility that life started from a primordial soup. It is likely that a rich mixture of gases accumulated on primitive Earth, setting the stage for biological evolution. Ultraviolet light or lightning could have struck the primordial soup of the Archean ocean, causing the fusion of carbon and hydrogen atoms that helped to produce the first life (Margulis and Sagan, 1986). Before the advent of an oxygen atmosphere on Earth, the planet was likely a predominantly reducing environment, the abundant gases being carbon dioxide, methane, ammonia, and nitrogen. Traces of carbon monoxide, hydrogen, and HCN, the latter formed as a result of ammonia reacting with methane, were also present. The different forms of energy that were present even at that early period, including geothermal, ultraviolet (UV) radiation, and radioactivity, could have facilitated the abiotic formation of organic matter, leading to macromolecules that aggregated to form membrane-like interfaces in the surrounding liquid. The absence of an ozone layer enabled fluxes of UV light to reach the Earth's surface. The primordial soup-UV light theory is supported by the fact that some biochemically important molecules such as amino acids, sugars, fatty acids, purines, and pyramidines can be synthesized abiotically when some gases are irradiated with UV or electron discharge. Miller and Urey (1959) bombarded a mixture of water vapor, ammonia,

methane, and hydrogen with a lightning-like discharge for a week and synthesized alanine and glycine, as well as a number of other organic substances. The conditions they provided during this synthesis were somewhat similar to those that existed on prebiotic Earth. Before their experiments, such organic molecules were thought to be produced only by living cells. Since that discovery, more components of complex cellular molecules including ATP, adenine, cytosine, guanine, thiamine and uracil have been synthesized in the laboratory by subjecting various mixtures of gases and mineral solutions to different energy sources such as UV, heat, electric discharge, and electric sparks, reenforcing the possibility of life originating from a primordial soup (Chang et al., 1983; Dickerson, 1978; Robertson and Miller, 1995). When oxygen is included in these synthesis experiments, the synthesis fails since $O_2$ rapidly oxidizes the organic products before they accumulate. In the absence of oxygen and microbial decomposition, these initial life-forming processes may have enabled organic products to accumulate over millions of years, forming the basis of cellular organisms. During this chemical evolution, carbon was abundant and clay minerals, with their surface charge and repeating crystalline structure, could have provided surfaces for polymerization of the more complex organic molecules such as RNA and proteins.

Because the atmosphere on primitive Earth had predominantly reducing conditions with abundant hydrogen and carbon dioxide, the early life forms must have been microbial. Initially anaerobic, they evolved photosynthesis and production of oxygen as a waste product, eventually leading to its accumulation in the atmosphere. People who study evolution call this primitive type of cell which eventually evolved into the present day prokaryotic cells, eugenotes or cenancester. These could in turn have evolved from even more primitive cell-like organisms called progenotes. The progenotes and eugenotes existed in anaerobic environments containing abiotically formed organic matter. As will be discussed in the next chapter, the ability of prokaryotes to spread their genes rapidly must have transformed the planet from a sterile, hostile environment into one abundantly endowed with a variety of species.

Ferrous iron and hydrogen sulfide are thought to have been abundant on primitive earth and the reaction between these two compounds could have been the potential source of energy to the progenitors. The energy generated from these reactions may in turn have formed ATP.

$$FeCO_3 + 2H_2S \rightarrow FeS_2 + H_2 + H_2O + CO_2$$
$$\Delta G^{o\prime} = -61.7 \text{ kJ/reaction} \quad ...(1)$$

$$FeS + H_2S \rightarrow FeS_2 + H_2$$
$$\Delta G^{o\prime} = -61.7 \text{ kJ/reaction} \quad ...(2)$$

Both reactions also yield $H_2$ which could have been used to reduce elemental sulfur ($S^o$) to hydrogen sulfide ($H_2S$), replenishing the supply of $H_2S$ in the environment. Both reactions are fairly simple and require few enzymes, notably hydrogenase and an ATPase to trap the energy released from the reaction. As expected, the early life forms on Earth were biochemically simple, possessing a few enzymes; the more complex forms appeared through mutation and selection. Most hyperthermophilic Archea, which are closest to the earliest organisms on Earth, are also able to reduce $S^o$ and Fe(III) with $H_2$ and form $H_2S$, an observation that offers some validity to the contention that both ferrous iron and $H_2S$ could have initially supplied energy for the earliest progenitors. Thus, iron and sulfate reduction may have been the first forms of microbial respiration.

Isotopic studies of ancient Isua rocks from Greenland and Fig Tree fossils (South Africa) and stromatolites formed in Western Australia (Warrawoona Formations), discussed below, strongly suggest that early life processes were entirely microbial.

## 1.3 MICROBIAL DIVERSITY AND ABUNDANCE

It is important to appreciate how diverse microbes are and the scope of processes with which they are associated. Textbooks devoted solely to the diversity of microorganisms have been published (Goodfellow and O'Donnell, 1993; Priest et al., 1994; Colwell et al., 1996). Bacteria are the most abundant, possibly as a result of their minute size and rapid rate of multiplication. In temperate regions, vis-à-vis the tropics, the diversity of microorganisms and their processes have been more extensively studied. However, the tropics harbor the greatest diversity of organisms compared to other geographic regions on the planet (Croll, 1966; Ehrlich, 1986). Why this is so remains uncertain but the following explanations have been postulated:

(1) productivity (e.g. photosynthesis) is much higher in the tropics than elsewhere;
(2) the tropics are more environmentally stable, making it easier for numerous small populations to exist and less likely to be subjected to accidental extinction; and/or
(3) the tropics are more climatically stable for longer periods of time, allowing species to coevolve and branch more freely.

## 1.4 GEOLOGICAL EVIDENCE OF EARLY MICROBIAL LIFE ON EARTH

The history of life is not a continuum of development but a record punctuated by brief, sometimes geologically instantaneous episodes of mass extinction and subsequent diversifications. Most of the evidence for the existence of early life has, therefore, been based on circumstantial evidence from fossil records. Such records indicate the build-up of oxygen in the atmosphere which, besides improving species diversity, also provided for better aeration that in turn ironically enhanced the rate of decomposition, thus minimizing the preservation of fossil records. Although somewhat better fossils are left by plants and animals compared to microorganisms, plant and animal life is more recent on Earth. The oldest animal and plant macrofossils are about 0.7 Ba old compared to microbial life which has existed for more than 3.5 Ba, a period that only began 1.0 Ba after the Earth was formed and 3.0 Ba before the appearance of plant and animal life. Paleontologists have long struggled to measure evolutionary change during the 2.0 Ba-period before the Cambrian age, known as the Proterozoic, but their analysis has been stymied by the lack of hard-shelled fossilized cells. There are fossils of such groups as the blue-green algae but these are rare and do not provide a complete track through time. Borrowing the words of that famous evolutionary biologist, Charles Darwin, fossil records are oftentimes imperfect and can be equated to a book in which just a few pages are preserved, on which a few lines, a few words, and a few letters still exist. Despite all this, some evidence to document the much earlier existence of microbial life on Earth has been accumulated and is discussed below.

## 1.4.1   The Isua Formation

The oldest rocks found so far on Earth, come from the Isua formation in western Greenland. This 3.75 Ba old sedimentary rock formation offers a record of the cooling and stabilization of the Earth's crust. These strata are too altered by heat and pressure (metamorphosed) to preserve the morphological remains of living creatures. However, the Isua rocks provide a geochemical signature of $^{12}C/^{13}C$ isotope ratios and show the enhanced $^{12}C$ that arises as a product of organic activity, thereby suggesting the existence of a prolific microbial life on the early Earth (Schidlowski, 1988). The proportion of the $^{12}C$ and $^{13}C$ isotopes in carbon of rocks made in the absence of life is recognizably different from the proportion of carbon from rocks that were once living matter. This evidence conveys the possibility that autotrophic C-fixation is

an old process that dates as early as 3.5-3.8 Ba ago. Most carbon is in the stable, light form ($^{12}C$), with less than 1% in the heavy $^{13}C$ and an even smaller amount in the unstable radioactive $^{14}C$. The chemistry of living matter segregates the isotopes such that the lighter $^{12}C$ isotope is preferentially used compared to the heavier $^{13}C$ isotope, a phenomenon called isotope discrimination. Both the organic and inorganic sedimentary C retain the isotopic composition of their progenitor organisms and carbonate rocks respectively. A steady flux of both inorganic C (mostly carbonate) and organic C enter newly formed sediments. Such fluxes have led to the accumulation of organic C. This discrimination raises the ratio of $^{12}C/^{13}C$ above the values that would be measured if all the sedimentary C had an organic source.

The sedimentary rocks in the Isua formation are of further evolutionary interest because their formation suggests that liquid water was present at the time of their formation (i.e., 3.75 Ba ago), a sign that at this early period, the conditions for some form of life existed. Prior to this period, however, the temperatures on Earth were greater than 100°C and thus free water could only have existed as vapor until the Earth subsequently cooled. Microfossils that are 3.75 Ba old which closely resemble cyanobacteria have been found embedded in rocks (Schopf, 1978).

## 1.4.2   Stromatolites and Fig Tree formations

Further evidence of early microbial life on Earth is based on stromatolites and Fig Tree formations in Swaziland and South Africa respectively. In 1977, Dr. Barghoorn hacked off pieces of flintlike rocks from the side of Mt. Barberton which had been formed 3.5 Ba ago when a silica-rich lava repeatedly poured on top of it, forming a chert of organic sediments that appear to have been deposited. Samples from this chert were sliced into thin sections which were placed under the microscope, revealing "microbial fossils". In other samples from a nearby Kromberg formation were found filamentous cyanobacteria-like fossils, evidence that by this early period, photosynthesis already thrived. This evidence is further collaborated by Stanley M. Awramik who, in 1979, discovered well-preserved multicellular filamentous microstructures in the 3.4 Ba-old rocks of the Warrawoona formation in Australia.

At another site in Swaziland, the Earth's oldest non-metamorphosed sediments, actual fossilized cells that date between 3.5-3.6 Ba have been reported (Knoll and Baghoorn, 1977). Stromatolites are fossilized versions of sediment trapped and bound by bacteria and cyanobacteria. Because the Earth was still anoxic during the time of the early formations, these ancient stromatolites were probably made of anoxygenic phototrophs. The anoxygenic phototrophic bacteria in the upper and subsequent layers make the mat opaque. The lower layer contains chemoorganotrophic bacteria (such as sulfur-reducing bacteria). Some of these microfossils have smooth organic walls and, occasionally, internal organic contents. More specifically, some microfossils were preservations in a process of binary fission. It is hypothesized that generation after generation of bacteria in the topmost layers died from exposure to radiation, their remains shielding the lower layers which accumulate sand and sediments to form a sort of living rug. Microbial mats occur in various environments such as hot springs and shallow marine basins where photosynthesis in the uppermost layer of the mat is balanced by decomposition from below (Awramik et al., 1983). Overall, however, such evidence from microbial mats and microscopic sections through chert rocks like those at Fig Tree is rare.

The time of origin of eukaryotes has been estimated by several approaches which include maximum sizes of organic-walled microfossils (1.75-2.0 Ba ago), carbonaceous megafossils (>1.8 Ba ago), modified sterol molecules extracted from Proterozoic rocks (>1.7 Ba ago) and molecular chronometry (1.8±0.4 Ba ago). This period saw a concentrated episode of diversification (the "Cambrian explosion"), an appearance of multicellular animals with hard parts in the fossil record. Han and Runnegar (1992) suggested that the very earliest eukaryotes such as *Grypania* probably lacked mitochondria and were therefore incapable of aerobic respiration until oxygen levels reached at least 1% of the present atmospheric levels. The Burgess fauna, which existed soon after the Cambrian explosion (approximately 530 Ma ago), are the only known major soft-bodied fauna from this primordial time. Considering the fact that the Earth is 4.5 M old (Gould, 1989; Margulis and Sagan, 1986), this Cambrian explosion and subsequent existence of familiar macroscopic life occupies only about 10% of earthly time. A key question by evolution scholars is why plants, animals, and fungi appeared so late and why these complex creatures have no direct simpler precursors in the fossil record of Precambrian times.

## 1.5 ONSET OF PHOTOSYNTHESIS AND RESULTANT DIVERSIFICATION

Early photosynthesis is thought to have been anoxygenic as found in present-day Chlorobiaceae, Rhodospirillaceae, and Chromatiaceae organisms that lack photosystem II and are not able to utilize the hydrogen from water to reduce carbon dioxide. The photosynthesizing organisms possibly used sulfur-reducing compounds such as $H_2S$ and FeS (See Section 1.2), the process probably operating at temperatures close to the boiling point of water. Oxygen was probably not a product from these earlier photosynthetic processes.

About 2 Ba ago, the oxygen levels on Earth started increasing as a by-product of photosynthesis by the photosynthetic ancestors of the present day cyanobacteria (Fig. 1.1). The evolution of photosynthesis is the most important metabolic invention in the history of life on Earth. At its inception, the photosynthesizers used the abundant $CO_2$ and converted it to organic matter and oxygen or its equivalent just

Years before the present
(billions)

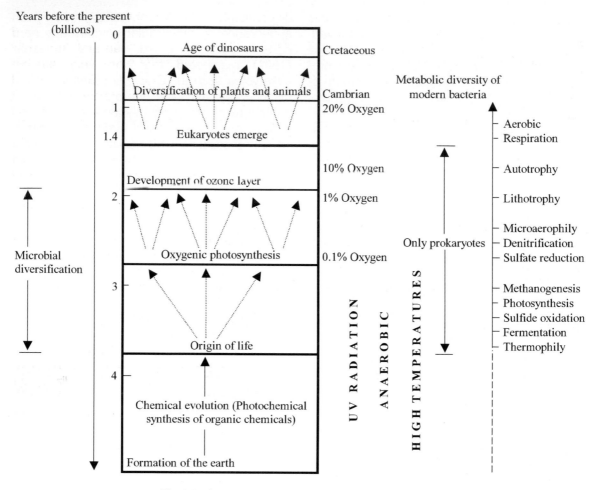

**Fig. 1.1**  Schematic diagram of the evolution of life on Earth.

like plants routinely do today. The $O_2$ generated from this process would have been mopped up immediately by the ubiquitous oxidizable matter in the environment, notably the iron (Fe) and sulfurs. Under these conditions, only microorganisms capable of anaerobic existence, notably the methanogens, prevailed. These photosynthetic microorganisms were also using hydrogen from hydrogen gas and hydrogen sulfide, more abundant then than now. Dissociation of $H_2S$ is a less demanding photochemical reaction than dissociating the O-H bond in water.

Based on 16S rRNA chronometry (see below)

there is sufficient evidence that methanogens also developed early in the ecosystem. The methanogens lived by decomposing organic matter and converting the carbon which had been generated by photosynthesizers, the $CO_2$ and methane thus replenishing these gases in the atmosphere. As photosynthesis continued, the abundance of these gases dwindled. In a frantic search for alternative sources of $H_2$ for photosynthesis, these organisms must have stumbled on water ($H_2O$) which until that time they had not been able to use for photosynthetic purposes because its H-O bond is stronger than the H-H and the H-S bonds in $H_2$ and $H_2S$ respectively.

These cyanobacterial ancestors already had an electron transport system and, through some mutation, constructed a second photosystem center (PS II) enabling the ancestors to reabsorb light at shorter wavelengths and to split the water molecule into hydrogen and oxygen. The split hydrogen was combined with carbon to make organic compounds such as sugars, releasing oxygen into the atmosphere and thus increasing its abundance. Some organisms with a photosystem II would out-compete other phototrophic organisms. Indeed this single metabolic change in bacteria provided for major implications in the history of life on Earth as it enabled life to colonize virtually every spot that had light, water and carbon dioxide. The continued "pollution" of the atmosphere with oxygen by the cyanobacterial ancestors forced the other organisms to acquire the ability to use oxygen too. Many microorganisms that were unable to tolerate this deadly "pollutant" (oxygen) became extinct or found anaerobic zones (such as anoxic water, sediments, and anaerobic soils) where they survive to this day.

The increase in $O_2$ during the Cambrian explosion had significant environmental consequences as it changed the atmosphere from a reducing to an oxidizing one. This change led to the acceleration of other processes. For example, diversity among the algae doubled during this period, further increasing the species turnover rate. About this same period animals, for example, starfish and trilobites, were appearing in the fossil record too, possibly due to the increasing atmospheric oxygen levels and the higher levels of carbon stored in ocean sediments. Fossil evidence also shows a burst in evolution at the advent of increased oxygen levels seemingly associated with an increase in the diversity of multicellular organisms. The increase in species diversity was probably also due, in part, to the more efficient aerobic respiration which breaks down organic molecules, yielding more $CO_2$, water, and energy from the oxidation than could be derived by anaerobes. Thirty-six ATPs are generated from aerobic respiration per sugar molecule broken down through the TCA cycle compared to only 2 ATPs generated from the same amount of sugar under fermentation. This more efficient process led to increased populations and microbial diversity. It is important to note that the appearance of oxygen in the atmosphere did not eliminate the anoxic ecosystems but merely segregated them in the stagnant waters and sediments. The proliferating animals and plants may have found ways to return nutrients such as nitrogen and phosphorus to new parts of water where new forms of algae may have evolved to exploit the watersheds. The nutrient recycling may in turn have helped facilitate the increase in biomass on Earth by allowing the new forms of algae to provide new sources of food for the evolving animals.

Before the oxygenic photosynthesis period, all iron deposits were in the reduced form such as ferrous iron ($Fe^{2+}$). The gradual build-up of oxygen in the atmosphere is also documented based on the layered deposits of oxidized iron (i.e., $Fe_2O_3$) embedded in siliceous rocks. Formation of most of these depositions date from 2.2-1.8 Ba ago. With the advent of $O_2$-producing photosynthesis (eqn 3), the $O_2$ made available could have oxidized $Fe^{2+}$, which existed under anaerobic conditions on the prephotosynthetic Earth, and was released during weathering, depositing the oxidized $Fe_2O_3$ in sediments of the primitive ocean, forming what is called banded iron formations.

$$H_2O + CO_2 \xrightarrow{hv} CH_2O + O_2 \quad ...(3)$$

Thus, the originally reducing atmosphere on Earth changed to an oxidizing one but levels of oxygen in the atmosphere still remained low (0.1 - 1% of the atmosphere) for about a billion years (Table 1.1). A possible explanation for the failure of $O_2$ to build up during that time is that most of it was reacting with reduced $Fe^{2+}$, forming the banded iron formations with other reduced mineral species such as sulfides. However, the build-up of $O_2$ as a sign for the advent of life does not rule out the possibility of the

**Table 1.1** A brief history of life on Earth

| Event | Years before the present time (x10$^6$) | Temperature ($^{\circ}$K) | %O$_2$ (atm) |
|---|---|---|---|
| Big bang singularity | 20-10,000 | >10$^{10}$ | Not applicable |
| Formation of our solar system | Approx. 5,000 | Approx. 10$^3$-10$^4$ | trace |
| Chemical evolution | 4,500-3,800 | >373 (*above boiling*) | " |
| First life forms appear (cells), liquid water (*despite the high temperature*) | 4,200-3,800 | | " |
| Isua formation (Greenland), oldest sedimentary rocks, organic carbon fractionation | 3,750 | " | |
| Fig tree (S. Africa) and Warrawoona formations (Australia), stromatolites (*microbial mass*) containing fossil prokaryotes (eugenotes) | 3,500 | " | " |
| Origin of oxygenic photosynthesis | 2,400 | " | 0.1 |
| Gunflint formation (Canada) containing fossil phototrophs (oxygenic cynobacteria) | 2,000 | " | 1.0 |
| Banded iron formations | 2,000 | | 1.0 |
| Dissolution and oxidation of uranium (Oklo, Gabon) | 1,800 | " | 1.0 |
| First eukaryotes | 1,400 | " | 10.0 |
| Bitter Springs formation (Australia) containing fossil eukaryotes | 1,000 | ±286 | 20.0 |
| First metazoans | 800 | " | " |
| Cambrian Explosion (Paleozoic) (Burgess Shale) | 600 | " | " |
| Dinosaurs flourish (Mesozoic) | 225 | " | " |
| Plants and animals flourish; dinosaurs perish (Cenozoic) | 65 | " | " |
| Homids arise (Holocene) | <1 | " | " |

Compiled from: Gould S.J. (1989); Hawkins S.W. (1988); Lovelock J. (1988); Margulis and Sagan (1986); Schopf J.W. (1978).

presence of abundant anaerobic life prior to this period, including in the subsurface. The ozone layer, a by-product of pure oxygen, was nonexistent at this time and thus the Earth was unprotected from ultraviolet light.

Further evidence of an increase in the levels of oxygen and the early role of microbes in this increase is based on a uranium ore found in the 1970s at Oklo in Gabon. Natural uranium is always in the same isotopic composition of 99.27% $^{238}U$, 0.72% $^{235}U$, and the rest as $^{234}U$. Only the $^{235}U$ isotope is useful in the chain reactions necessary for generating power and in nuclear explosions. During the 1970s, however, a uranium shipment from the Oklo mines in Gabon was found to be depleted in the fissionable $^{235}U$ isotope. This finding by the French Atomic Energy Agency initially caused fears that this fissionable fraction had somehow got in the hands of terrorists who had mysteriously substituted it with spent uranium. Further research and investigation into the mysterious disappearance of the $^{235}U$ fraction led to an interesting discovery, however. Unlike under aerobic conditions, uranium is insoluble in water under anaerobic conditions. It was discovered that in Oklo, when enough oxygen appeared in the Proterozoic, the groundwater was rendered oxidizing and the uranium in the rocks began to dissolve. Traces of the dissolved uranium flowed in the streams into the algal mat that contained microorganisms with a capacity to collect and concentrate this ore, depositing uranium oxide. This event set off a gentle microbe-driven nonexplosive and self-regulating nuclear reaction which ran for millions of years and used up a fair amount of the natural $^{235}U$.

The gradual increase in oxygen in the atmosphere enabled heterotrophic organisms with more energy to oxidize the carbon to carbon dioxide, in the process known as oxygenic photosynthesis, and generally provided for more efficient metabolism, leading to more complex processes. The evolution of porphyrins together with refinements in cell membrane structure could have led to the construction of cytochromes and an electron transport system able to carry on electron phosphorylation. This could have further enabled the formation of bacteriochlorophylls and thus photosynthesis leading to a greater explosion in organisms, due to increased availability of utilizable energy from the sun. Photosynthesis is a major process as it facilitates primary production.

## 1.6 FORMATION OF OZONE AND ITS EFFECTS

When oxygen is subjected to short-wavelength UV radiation, it is converted to ozone ($O_3$), a powerful greenhouse gas which strongly absorbs wavelengths up to 300 nm, thus providing a barrier to the energetic UV rays from the sun reaching the Earth. Formation of ozone further provided protection of the primitive earthly creatures against UV radiation. With such shielding from UV, evolution could have occurred as the different organisms roamed through the different environments on Earth. In the absence of such protection from UV, evolution might have been more limited to isolated environments protected from direct solar radiation such as in water and under rocks. During these changes from predominantly anaerobic to aerobic, various types of multicellular organisms which could not exist without oxygen emerged (Fig. 1.1). The nonnucleated prokaryotes led to the nucleated eukaryotes around 1.4 Ba ago marking an increase in life's complexity. This event is significant in that unlike earlier events which were geologic in nature, it was biological. It is important to note that between the emergence of the first life forms and the appearance of the first eukaryotes more than 2 Ba transpired (Table 1.1; Fig. 1.1). All this time microorganisms were the only forms of life, perfecting their biochemical repertoire, setting the stage for the respective evolution of eukaryotes, which thereafter arose in relatively rapid succession.

## 1.7 A LIVING EARTH

The Gaia hypothesis in its original formulation conceptualizes the Earth as a living organism, the largest in the solar system. In contrast, conventional science depicts the Earth as an inert rock upon which plants and animals happen to live. This hypothesis indicates that the assemblages of organisms on the Earth do not passively adapt to physiological conditions but actively interact with them to suit their existence, affecting the physical and chemical conditions of the biosphere in the process (Lovelock, 1988). Gaia, a Greek word for Goddess of Earth, exercises her powers principally through microbial processes. Such control is hypothesized to have started soon after the first life form appeared. Contrary to this hypothesis is the argument that geological (abiotic) processes produced conditions favorable for life. However, a comparison of Earth with Mars and Venus suggests that the existence of life on Earth affected the physical conditions rather than the physical conditions changing first and then triggering the evolution of life (Table 1.2). Mars and Venus predominantly contain carbon dioxide, whereas the Earth's atmosphere is predominantly nitrogen and oxygen. This difference in atmospheric composition distinctly identifies life on Earth and its absence on the other two planets. Embedded in this observation is the importance of microorganisms in driving the trend of Earth's present-day atmospheric conditions.

**Table 1.2** Composition of the atmosphere of Mars, Venus, and Earth

| Atmosphere | Mars | Venus | Earth | |
|---|---|---|---|---|
| | | | **Without life** | **With life** |
| Carbon dioxide | 95% | 98% | 98% | 0.03% |
| Nitrogen | 2.7% | 1.9% | 1.9% | 79% |
| Oxygen | 0.13% | Trace | Trace | 21% |
| Surface temperature (°C) | −53 | 477 | 290±50 | 13 |

*Source:* Lovelock (1988). With permission of W.W. Norton & Company Inc.

## 1.8 GENETIC MATERIALS IN EVOLUTION

Though complex and fragile, life is quite durable through time as is exemplified by the incredible accuracy with which the inherited information is copied from one generation to the next. Such accuracy is facilitated through genetic replication, the cells undergoing division in eukaryotes through a process called mitosis. Simpson et al. (1957) have estimated mitosis to be more than 500 Ma old. Even with such near perfection, however, there are some changes in genetic material through mutations and genetic recombinations, resulting in changes in the enzymes that are synthesized. Although sometimes harmful, mutations can give more fit genotypes. The environment is also constantly selecting for the best adapted organisms.

Microorganisms have short generation times compared to macroorganisms. So, once genetic changes are introduced into the microbial population, such changes spread more rapidly. Mutations are facilitated by UV and other chemical mutagens. Mutation usually involves minor changes in the base pair sequences. Such minor changes prove efficient or inefficient after being tested by the environment. Understanding mutations requires a knowledge of DNA and RNA. Darnell and Doolittle (1986) speculated that of the two nucleic acids, RNA was formed first. Somewhere in the early stages of evolution, the three-part system (i.e., RNA, DNA, and protein) became fixed in cellular life as the best solution to the flow of biological information. In the early life forms, RNA may have served as an informational and enzymatic agent, preceding proteins as a chief biochemical catalyst. These RNA-organisms were, by necessity, very simple compared to present-day cells. The establishment of DNA as the genome of cells may have resulted from the need to store genetic information in a more chemically stable structure that conserves energy for the cell and thus increases their competitiveness. The RNA-copying system is prone to errors compared to the more precise DNA polymerase. Maintaining the error-prone RNA system for replication could have been less favorable to increasing cellular complexity. That the RNA, DNA, and protein system was an evolutionary success is shown by the fact that to exist, modern cells require all three kinds of informational macromolecules. The onset of meiosis allowed the production of plant and animal sex cells (sperm and egg), a process which enables chromosomes to recombine and generates eukaryotic complex life cycles. Through the evolutionary process, mitosis arose to "ensur" DNA replication in an orderly fashion once the size of the genome increased to a point where replication as a single molecule was no longer possible.

Modern-day eukaryotic cells resulted from endosymbiosis. Under this theory, the cells are hypothesized to have permanently fused together. The first such fusion may have occurred between spirochetes or other motile eubacteria and wall-less archaebacteria such as *Thermophilus acidophila*, then eubacteria, leading to the earliest oxygen users and cyanobacteria, the nucleated cells (protists). Genetic evidence deems it possible that the ancestors of chloroplasts and mitochondria merged genomes with the earliest anaerobic protists in the cell lineage ancestral to algae and plants (Margulis, 1996; Margulis et al., 2000). As will be discussed in Chapter 3, protozoa have the ability to feed by engulfing cells. Available evidence indicates that both mitochondria and chloroplasts began as free-living bacteria but later established residence within the cytoplasm of early eukaryotes that supplied energy in exchange for nutrients and protection from the environment. Mitochondria are the sites of electron transport and oxidative phosphorylation while chloroplasts are the green-pigmented bodies that house photosynthetic reactions. Both chloroplasts and mitochondria, though currently living within eukaryotic cells, have their own DNA and are about the size and shape of many bacteria. It is also possible that endosymbiosis and subsequent coexistence was a result of competition between the early host cells and the endosymbionts for nutrients. Chloroplasts were probably very similar to present-day chloroxybacteria or *Prochloron*. *Prochloron* contains both chlorophyll *a* and *b* which makes it significantly more like plants than are cyanobacteria which contain only chlorophyll *a*. At the very beginning, these cellular mergers were probably not perfect but were eventually perfected into the present-day organisms over evolutional time.

Incorporation of mitochondria and chloroplasts into the eukaryotic cell provided for formation of collective power to assemble. Without mitochondria, nucleated cells and hence the animal or plant, cannot utilize $O_2$ and thus cannot live. We shall discuss the structure of ribosomes below but for now let it be noted that in chloroplasts and mitochondria, they are of the small (70S) form typical of prokaryotes. Phylogeny, which is also detailed below, has likewise established a close relationship between prokaryotes, mitochondria, and chloroplasts.

Formation of eukaryotic cells from unicellular ancestors provided a high concentration of genetic material in the nucleus. Such a concentration of genetic material is advantageous as it effectively facilitates the flow of information along the genetic code. Endosymbiosis and subsequent events facilitated an efficient respiratory (oxygenic) process and favored a burst of organisms on Earth and the ability of these organisms to occupy new and possibly more diverse environments. The prolonged symbiosis of mitochondria and chloroplasts is assumed to have resulted in loss of some genetic material and biosynthetic capabilities, thus limiting their ability to exist independently. Neither mitochondria nor chloroplasts are ever synthesized *de novo* but reach offsprings through cytoplasmic inheritance. Today, mitochondria can no longer survive for more than a few hours outside the eukaryotic cell and cannot be cultured on their own unless provided with a medium that is essentially the same as cytoplasm.

## 1.9  MOLECULAR CHRONOMETRY

Emile Zuckerkandl and Nobel Prize laureate Linus Pauling introduced the idea of calculating chemical bond strength (Zuckerkandl and Pauling, 1965). They indicated that the differences in the nucleic acid or amino acid sequences of homologous molecules can be used to measure evolutionary distances. Expanding this idea, Carl Woese and co-workers were able to sequence DNA and go back in time to construct a phylogenetic tree, a technique that is commonly referred to as molecular chronometry (Woese, 1987; Brown and Doolittle, 1997). This technique, whose data gained wide acceptance, involves indirectly determining ribosomal RNA (rRNA) sequence homology and sequence heterogeneity to calculate the evolutionary distance ($E_D$). Ribosomal RNA is appropriate for this technique because it is universally distributed across living things, is functionally homologous in organisms, and has an identical function in the synthesis of proteins. Furthermore, rRNA is fairly well conserved across species and is readily purified from organisms.

Ribosomes, the basic organelles for protein synthesis, are very intricate, basically comprised of RNA and proteins. The process of protein synthesis is best discussed by basic microbiology and biochemistry textbooks but of relevance in this volume are the respective sizes of ribosomes. Based on mass by volume, ribosomes are of two types depending on whether they are for prokaryotes or eukaryotes, with measurements expressed in sedimentation constant (with Sverdberg units, S). The overall size of ribosomes is either 70S in prokaryotes or 80S in eukaryotes. Each ribosome is comprised of two subunits and in eukaryotes, these have a sedimentation constant of 30S and 50S (Table 1.3). The 30S subunit consists of 21 individual proteins vs. 34 individual proteins in the 50S subunit. Similar subdivisions but of different sizes occur in prokaryotes. The 80S of eukaryotes breaks down into 60S and 40S subunits. Both prokaryotes and eukaryotes also have a small 5S subunit with only 120 nucleotides. In prokaryotes, the initiation always begins with a free 30S ribosome subunit, forming an initiation complex on which is added the 50S subunit to make the active 70S ribosome. At the end of protein synthesis, the two subunits separate into 30S and 50S subunits. The ribosomes of mitochondria and chloroplasts of eukaryotes are similar to those in prokaryotes— a fact that provides additional evidence for the validity of the endosymbiotic theory introduced earlier.

The 16S and 18S rRNA are often used in molecular chronometry. This molecular "clock" is based on applying the known mutation rates of the rRNA to the number of mutations that separate one species from another and calculating backwards to a common ancestor. It can be used to compare both closely and distantly related organisms. The bacterial 16S rRNA and 23S rRNA are large enough and contain several highly variable (rapidly changing) regions, which is good for closely related organisms, whereas

**Table 1.3**  Sizes of ribosomes and their subunits

| Size* | Prokaryotes | | Eukaryotes | |
|---|---|---|---|---|
| Overall size | 70S | | 80S | |
| Subunit sizes (contain RNA and proteins) | 50S | 30S | 60S | 40S |
| Size of RNA | 23S | 16S | 28S | 18S |
| Number of nucleotides | 3,000 | 1,500 | 5,000 | 2,000 |
| Number of proteins | 34 | 21 | 50 | 30 |

*Both prokaryotes and eukaryotes also have a small 5S subunit with 120 nucleotides.

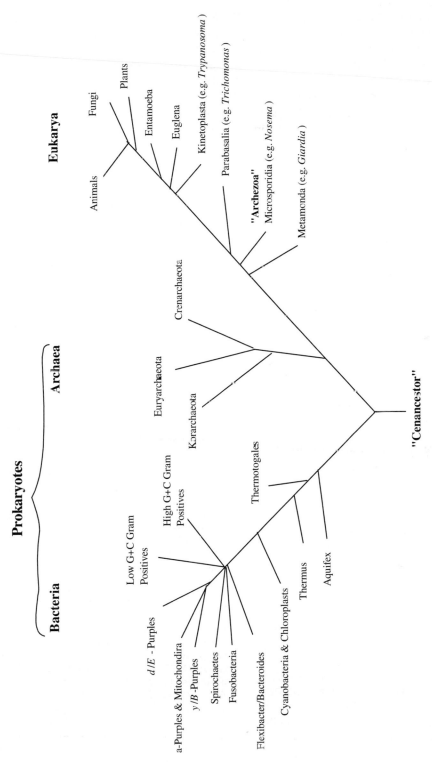

**Fig. 1.2** Universal phylogenetic tree showing the relative positions of evolutionary groups in the three domains Bacteria, Archaea and Eukarya. The evolutionary distance (lines) are not to scale but pertain to the relatedness of base sequences (adapted from Brown and Doolittle, 1997) with permission from the Society for Microbiology.

other regions are more constant and thus good for more distantly related organisms. More widely used in phylogeny, however, is the 16S rRNA because it is more experimentally manageable than the 23S rRNA. Furthermore, 16S rRNA (or 18S rRNA in eukaryotes) is often over 5% of the cell dry weight of growing cells, which makes it relatively easy to isolate. It is 1,500-1,600 nucleotides long (1,800 in eukaryotes), a size that is big enough for good statistics and small enough to sequence. Equally important in the suitability of 16S rRNA as a molecular chronometer is the high unlikelihood of horizontal transfer between unrelated organisms.

Ribosomal RNA can be directly sequenced from crude cell extracts using reverse transcriptase. Researchers have used rRNA to build a family tree which not only reveals the evolution of organisms from one common ancestor, but also a sudden explosion of evolution among the eukaryotes (Fig. 1.2). Based on this phylogenetic tree, all manner of new eukaryotes, including algae and animal ancestors, emerged around the same time (500 Ma ago) thus backing up the somewhat fragmentary evidence accumulated through paleontology. The 16S rRNA phylogeny indicates three domains: Bacteria, Archaea and Eukarya (Fig. 1.2). The Eukarya domain, which includes protozoa, fungi, algae, as well as the plant and animal kingdoms, has a unique 18S compared to a 16S rRNA in Bacteria and Archaea. What was traditionally referred to as prokaryotes are spread among the Bacteria and Archaea. Despite the fact that from our outlook these two domains look alike and have functional properties and organization similar to Bacteria, they differ in lipids, cell wall, and metabolic pathways. Given the information coming out of 16S and 18S rRNA phylogeny, studies of the evolutionary history and relatedness of organisms, this tree will repeatedly be referred to even in subsequent chapters. Significantly shown in the phylogenetic tree is the presence of a universal ancestor of all cells.

Phylogeny, especially if linked with present-day techniques in molecular biology, is greatly increasing the capability of scientists to obtain nucleic acids from natural environments, amplify and sequence them so as to determine what organisms are present (including those that are nonculturable on conventional laboratory media) or existed in the past, and assess their metabolic activity. These exciting possibilities will be discussed in subsequent chapters.

## References

Awramik S.M., J.W. Schopf, and M.R. Walter (1983). Filamentous fossil bacteria from the Archean of Western Australia. *Precambrian Res.* **20**: 357-374.

Brown J.R. and W.F. Doolittle (1997). Archaea and the prokaryote-to-eukaryote transition. *Microbiol. Molec. Biol. Rev.* **61**: 456-502.

Carson R. (1962). Silent Spring. Houghton Mifflin Co., Boston, MA.

Chang S., D. Des Marais, R. Mack, S.L. Miller, and G.E. Strathearn (1983). Prebiotic organic synthases and the origin of life. *In*: J.W. Schopf (ed.). Earth's Earliest Biosphere. Princeton Univ. Press, Princeton, NJ, pp. 53-92.

Colwell R.R., U. Simidu, and K. Ohwada (1996). Microbial Diversity in Time and Space. Plenum Press, New York, NY.

Croll N.A. (1966). Ecology of Parasites. Harvard Univ. Press, Cambridge, MA.

Darnell J.E. and W.F. Doolittle (1986). Speculations on the early course of evolution. *Proc. Natl. Acad. Science (USA)* **83**: 1271-1275.

Dickerson R.E. (1978). Chemical evolution and the origin of life. *Scientific American* **239**: 70-86.

Ehrlich P.R. (1986). The Machinery of Nature: The Living World Around Us and How it Works. Simon and Schuster, New York, NY.

Goodfellow M. and A.G. O'Donnell (1993). Handbook of New Bacterial Systematics. Acad. Press, New York, NY.

Gould S.J. (1989). Wonderful Life: The Burgess Shale and the Nature of History. W.W. Norton & Co., New York, NY.

Han T-M. and B. Runnegar (1992). Megascopic eukaryotic algae from the 2.1-billion-year-old Negaunee iron-formation, Michigan. *Science* **257**: 232-235.

Knoll A.H. and E.S. Barghoorn (1977). Archean microfossils showing cell division from the Swaziland System of South Africa. *Science* **198**: 396-398.

Lovelock J.E. (1988). The Ages of Gaia: A Biography of our Living Earth. W.W Norton and Company, New York, NY.

Margulis L. (1996). Archaeal-eubacterial mergers in the origin of eukarya: phylogenetic classification of life. *Proc. Natl. Acad. Science (USA)* **93**: 1071-1076.

Margulis L. and D. Sagan (1986). Microcosmos: Four Billion Years of Microbial Evolution. Summit Books, New York, NY.

Margulis L., M.F. Dolan, and R. Guerrero (2000). The chimeric eukaryote: origin of the nucleus from the karyomastigont in amitochondriate protists. *Proc. Natl. Acad. Science (USA)* **97**: 6954-6959.

Miller L. and H.C. Urey (1959). Organic compound synthesis on the primitive earth. *Science* **130**: 245-251.

Philander S.G. (1998). Is the Temperature Rising?: The Uncertain Science of Global Warming. Princeton Univ. Press. Princeton, NJ.

Priest F.G., A. Ramos-Carmenzana, and B.J. Tindall (1994) Bacterial Diversity and Systematics. Plenum Press, New York, NY.

Robertson M.P. and S.L. Miller (1995). An efficient prebiotic synthesis of cytosine and uracil. *Nature* **375**: 772-774.

Schidlowski M. (1988). A 3,800-million-year isotopic record of life from carbon in sedimentary rocks. *Nature* **333**: 313-318.

Schopf J.W. (1978). The evolution of the earliest cells. *Sci. Amer.* **239**:110-138.

Simpson G.G., C.S. Pittendrigh, and L.H. Tiffany (1957). Life: An Introduction to Biology. Harcourt, Brace and World. New York, NY.

Woese C.R. (1987). Bacterial evolution. *Microbiol. Rev.* **51**: 221-271.

Zuckerkandl E. and L. Pauling (1965). Molecules as documents of evolutionary history. *J. Theor. Biol.* **8**: 357-366.

# 2

# Prokaryotes

## 2.1 ABUNDANT BUT UNCULTURED

Prokaryotes (Bacteria and Archaea) are probably the most important living things on Earth as they greatly contribute to the cycling of nutrients in soil and marine environments. They have a long evolution and adaptation history which has enabled them to successfully colonize every conceivable environment on this planet, including some environments that, at first sight, may appear inimical to life (see Chapter 7). Dykhuizen (1998) estimated that there are more than a trillion species of prokaryotes in the biosphere. His estimate is based on the premise that comparisons of various communities, for example within a particular forest, coral reef, rhizosphere, every living person, etc., will reveal communities of prokaryotes which differ from each other and don't necessarily overlap. It should be emphasized from the outset that the diversity and breadth of prokaryotes is even now continuing to emerge consequent to exploitation of a whole range of culture independent techniques in the past decade. Since adapting these approaches, the number of divisions within the prokaryotes has increased from 12 in 1987 (Woese, 1987) to more than 36, with some of the phylotypes known only from sequences with no known culturable relatives (Hugenholtz et al., 1998; Dojka et al., 2000; Ward, 2002). Some of these approaches are explored in Chapter 6. About 40% of these divi-

sions have no cultured representatives using conventional plating techniques; they have been identified from 16S and 18S rRNA sequences obtained directly from the environment (Hugenholtz et al., 1998).

Some of the divisions are composed of members that occur only in a limited number of environments but the majority occur in a wide range of environments, indicating that they do have a range of metabolic capabilities. Seemingly abundant in a variety of environments, despite the fact that not many, if any, of their members have been cultured are members of the division Acidobacterium, Verrucomicrobia, green nonsulfur (GNS), OP11 and WS6. Uncultured prokaryotes are also widespread among both of the Archeal kingdoms, Euryarcheaeota and Crenarchaeota (refer to Fig. 1.2). Furthermore, members of the domain Archaea were initially considered to consist only of methanogens that live under anoxic conditions and extremophiles living under inhospitable environments such as hot springs, and other geothermal vents. However, recent evidence shows that Archaea are also fairly widespread in nonextreme environments such as tropical forest soil (Bornemann and Triplett, 1997), temperate soils (Bintrim et al., 1997), and marine environments (DeLong, 1992). On the other hand, members of *Proteobacteria* ($\alpha$, $\beta$, $\gamma$, $\delta$, and $\varepsilon$), *Cytophagales*, (*Bacteroides, Cytophaga,* and *Flexibacter* groups), *Actino-*

*bacteria* (high G+C Gram-positive prokaryotes), and low G+C Gram-positive prokaryotes (firmicutes) account for 90% of the routinely cultured microorganisms (Hugenholtz et al., 1998). Thus, the prokaryotic species that we routinely isolate from the environment are not necessarily the most abundant. Other divisions include *Dictyoglomi, Cyanobacteria, Aquifacae, Thermotoga, Thermodesulfobacteria, Fusobacteria, Chlorobi, Deferribacteres, Chloroflexi, Chrysiogenates, Deinococcus-Thermus, Planctomycetes, Spirochaetes,* and *Fibrobacter.* With this paradox in mind, the data presented here are mostly based on prokaryotes which have been routinely cultured. There is no evidence, however, that the general principles applicable to cultured prokaryotes do not by and large apply to prokaryotes that have not yet been cultured.

## 2.2 PROBLEMS ENCOUNTERED BY PROKARYOTES IN THE ENVIRONMENT

In natural environments, prokaryotes encounter three major problems: availability of essential nutrients, lack of adherence sites, and exposure to noxious chemicals. Each of these problems is elaborated below.

***Nutrient stress:*** Under natural environments, nutrients will be demanded by both the micro- and macroorganisms present. A gradual depletion of some essential nutrients may result in starvation. However, unlike most eukaryotes which readily die off at the onset of starvation, prokaryotes live through a feast-and-famine nature of existence, undergoing processes that ensure low metabolic rates when environmental stresses such as nutrients, and weather extremes set in (Koch, 1971). More specifically, prokaryotes, particularly the Gram-positive bacteria, such as *Clostridium, Bacillus, Desulfotomaculum, Sporosarcina,* and *Thermoactinomyces,* form endospores which protect them against environmental extremes, whereas *Azotobacter* and myxobacteria form thick-walled structures (cysts) which confer resistance to desiccation and UV but unlike endospores, are not resistant to heat. Cyanobacteria can form heterocysts as protective structures. Survival mechanisms under environmental extremes are extensively discussed in Chapter 7. These mechanisms have ecological significance as they contribute to the long-term establishment and persistence of these microorganisms in diverse environments.

***Adherence sites:*** The induction of many metabolic pathways in microorganisms is known to be surface-specific (Lawrence et al., 1995). Thus, the survival and successful reproduction by prokaryotes often require them to be able to colonize a surface and/or integrate into a biofilm community. In most natural environments, prokaryotes do not exist as single cells but rather need to form microcolonies or aggregates to be able to survive. Typically, several species are able to adhere to surfaces such as soil particles, leaves, or rocks, the propensity of attachment depending on the physiological status of the cells. Adherence is attributed to the hydrophobic cell-wall surface (due to the extracellular polysaccharides or polyphenols), and/or the presence of fibriae (fibrils), pili, and flagella (Handley et al., 1991; Stanley, 1983). Adherence by microorganisms has been more widely studied in pathogenesis but is also of relevance in environmental microbiology as it normally provides a direct conduit of the nutrients contained in the attachment site to the attached microorganisms.

Besides participating in adherence directly, flagella also indirectly facilitate movement of the organism in question toward potential aggregates and/or surfaces on which to attach. Thus, to develop a competitive edge, flagella have evolved to occupy different positions on prokaryotic cells (posterior, anterior, lateral, etc.) so as to meet the challenges posed by different surface environments (Lawrence et al., 1995). The aggregates formed may consist of various species with different metabolic capabilities and nutritional requirements, establishing complementary metabolic interdependencies that individual species may not be able to accomplish. As growth continues, some members of the aggregate may be compelled to migrate to another location starting the aggregation process all over again. Thus, they demonstrate some form of density-dependence, a phenomenon referred to as quorum sensing (Miller and Bassler, 2001).

*Noxious chemicals:* Exposure of the microorganism to noxious chemicals produced by themselves or by other species also poses challenges to i.e. prokaryotes in the environment. Such chemicals may include acids from fermentation processes, antibiotics, bacteriocins, and microcins, compounds which kill or immobilize their competitors. Bacteriocins are highly specific and while their role in nature is not known, they probably help in competition among related bacterial strains. Survival under the presence of such chemicals, therefore, requires continuous modification and adaptation to these noxious environments.

Successful adaptation by prokaryotes to both the most accommodative as well as some of the most ubiquitous environments on Earth is attributed to four major properties: (i) metabolic diversity (ii) size, (iii) fairly simple structure, and (iv) genomic plasticity.

## 2.3 SUCCESSFUL ADAPTATION BY PROKARYOTES TO THE ENVIRONMENT

### 2.3.1 Metabolic diversity

All living organisms require energy for maintenance and growth. It is, therefore, a fundamental property for various organisms to obtain such energy through specific means. Metabolic diversity is indeed a hallmark of prokaryotes since they, unlike most other organisms, have three potential sources of energy, namely light (**phototrophy**), reducing inorganic compounds (**lithotrophy**), and organic substrates

(**organotrophy**). From an evolutionary perspective, prokaryotes were the first organisms to develop the ability to use $CO_2$ and sunlight in photosynthesis (see Chapter 1). Cyanobacteria, halobacteria, and phototrophic bacteria are capable of converting energy from solar radiation, a process which involves fixing $CO_2$ as well. Although all three groups fix $CO_2$ through the Calvin cycle, there are some distinct differences to enable them to thrive under specific environments. Unlike cyanobacteria and halobacteria, phototrophic bacteria do not split water to produce oxygen, but rather use $H_2S$, $S^0$, or $S_2O_3^{2-}$ as the external electron donor. While all phototrophs grow anaerobically in light, the nonsulfur-purple bacteria and green bacteria of genus *Chloroflexus* can also thrive anaerobically without light.

The lithotrophs perform aerobic respiration and are capable of utilizing reduced inorganic compounds ($S^{2-}$, $S^0$, $SO_3^{2-}$, $NH_3$, $NO_2^-$, $Fe^{3+}$) as electron donors. By oxidation of such electron donors, lithotrophs generate reducing power and energy for their biosynthetic processes. Some lithotrophs can also use $NO_3^-$, $SO_4^{2-}$ or $CO_2$ as the ultimate electron acceptor under anaerobic conditions. Most lithotrophs are capable of utilizing $CO_2$ as the sole C-source, assimilating it via the Calvin cycle. In such cases, $CO_2$-fixation is catalyzed by ribulose-bis-phosphate carboxylase (RUBISCO).

A variety of bacteria are capable of anaerobic respiration wherein molecular oxygen is replaced by other terminal electron acceptors (Table 2.1). The most important anaerobic respiratory processes are $NO_3^-$ respiration (dissimilatory nitrate reduction), $SO_4^{2-}$ respiration, and carbonate respiration. Reduction of nitrate to nitrite occurs under anaerobic conditions and is carried out by facultative aerobes such as *Pseudomonas* spp. and *Rhizobium* spp., facultative anaerobes (notably enterics), and strict anaerobes (e.g. *Clostridium* spp.). In contrast to the reduction of nitrates by

**Table 2.1**  Terminal electron acceptor in aerobic and anaerobic respiration

| Electron acceptor | Product | Type of respiration | Organisms |
|---|---|---|---|
| Oxygen | $H_2O$ | Aerobic | All strict aerobes and facultative aerobes |
| Nitrate | $NO_2^-$, $N_2O$, $N_2$ | Nitrate reduction | Various aerobes and facultative anaerobes |
| Sulfate | $S^{2-}$ | Sulfate reduction | Various obligate anaerobes |
| Sulfur | $S^{2-}$ | Sulfur respiration | A few facultative and obligate anaerobes |
| $CO_2$, hydrocarbonates | Methane | Carbonate respiration | Methanogens |
| $CO_2$, hydrocarbonates | Acetic acid ($CH_3COOH$) | Carbonate respiration | Acetogenic bacteria, e.g. *Clostridium aceticum* |
| Fumarate | Succinate | Fumarate respiration | Succinogenic bacteria |

Adapted from Stolp (1988) with permission of Cambridge University Press, Cambridge, UK.

nitrate-reducing prokaryotes, sulfate respiration is conducted by obligate anaerobes such as *Desulfovibrio* and *Desulfotomaculum*, generating hydrogen sulfide (Fig. 2.1). *Desulforomonas acetoxidans* uses sulfur as a terminal electron acceptor during acetate oxidation and lives in syntrophic association with phototrophic green sulfur bacteria that oxidize hydrogen sulfide and produce sulfur.

The metabolic diversity of prokaryotes is also displayed by the ability of some members to fix atmospheric nitrogen into ammonium, a form utilizable by other organisms. Thus, genera such as *Azotobacter* sp., *Clostridium* sp., *Rhizobium* sp., some methanogens, and cyanobacteria play a unique role in the cycling of this abundant, but otherwise unavailable nutrient for other organisms on a global scale. Without this activity of fixing $N_2$ directly from the atmosphere, life on earth as we know it would have disappeared long ago as a result of N-starvation.

Generally, nutrients are naturally limiting in most environments. In most instances, prokaryotes compete within themselves and also with other species for various nutrients. However, most prokaryotes often have the abil-

ity to metabolize a variety of substrates, thus diversifying their sources of nutrients. More specifically, some prokaryotes have the ability to concentrate nutrients when the nutrients are present in low concentrations. Other strategies to enhance nutrient acquisition include the production of extracellular hydrolases which break down complex molecules in the environment into simpler ones that can be utilized. As an example, although iron is fairly abundant in soils, it is not readily available for microbial uptake because it mostly exists as ferric hydroxide, a form that is quite insoluble. To enhance its availability to bacteria, some bacteria produce siderophores to chelate the iron, making it more amenable to uptake. Iron is essential in the cytochromes which shuttle energy and in peroxidases.

Similarly, phosphorous concentrations commonly found in the soil solution range from 0.1-10 $\mu M$ (Nye, 1979), the lower concentration range well below that of many micronutrients (Broschart, 1967). Continuous-culture studies using *Rhizobium trifolii*, cowpea rhizobia, and *Bradyrhizobium japonicum* have shown that cells become P-limited at levels of 200 $\mu M$ P (Smart et al., 1984). Below this P concentra-

**Fig. 2.1**  Reduction of sulfate by obligate anaerobes such as *Desulfovibrio* and *Desulfotomaculum* spp.

tion, the alkaline phosphatase enzyme is derepressed, enabling the organism to still acquire some of the limiting nutrient.

## 2.3.2 Size

In culture, prokaryotic cells are abnormally large as they have access to abundant nutrients. In contrast, under natural environments single prokaryotic cells are generally 1-2 $\mu$M and, as highlighted earlier, may occur in aggregates or microcolonies. These assemblages can, depending on species, be described as chains (*Streptococcus*), clusters (*Staphylococcus*), tetrads (*Aerococcus*), pairs (*Neisseria*), packets (*Sarcina*), sheets (*Lampropedia*), filaments (*Leucothrix mucor*), fruiting bodies (myxobacteria), dense mats (*Thiothrix* sp.), or stellate aggregates (*Agrobacterium*). Because of the small size of prokaryotes, their distribution is influenced by the environment. Whether as single cells or in aggregated form, they can attach to soil particles (dust) and be extensively windborne. Through this and other mechanisms, prokaryotes have become widespread throughout the globe. Other mechanisms of spread include being swept by liquid currents (Hamdi, 1971; Madsen and Alexander, 1982), or passively carried in or on animal hosts such as earthworms (Henschke et al., 1989). In addition to these passive mechanisms, some prokaryotes spread actively by motility and chemotaxis, although the distances covered through this means can be insignificant in the short term. Drahos et al. (1986) found that migration of bacteria in the field from the point of application was largely limited to 18 cm horizontally and 30 cm vertically. These may appear short distances to the human eye but spell significant strides for bacteria in relation to their size.

Except where chemotaxis is the mechanism of spread, prokaryotes in most instances find an appropriate site for growth and survival purely by chance. Some prokaryotes have been found to contain magnetic particles (magnetosomes) which enable them to align themselves along the magnetic field lines and swim toward desirable destinations such as the bottom of ponds, lakes, and rivers which are richer in nutrients and provide an anaerobic setting (most magnetotactic bacteria are strict anaerobes or grow at low oxygen levels, i.e., microaerophilic environments). The microbiology of magnetotactic prokaryotes is still in its infancy but it is interesting to note that because of the differences in polarity, those magnetotactic bacteria in the Northern Hemisphere predominantly swim northward while those in the Southern Hemisphere predominantly swim southward.

Anaerobic prokaryotes tend to be smaller in size compared to their aerobic counterparts possibly because growth under the limitation of $O_2$ is limited. Small size is advantageous to microorganisms as the size-to-volume ratio increases with decrease in size. The smaller a cell, the more stable under suspended conditions. Likewise, if food and energy are available, the large surface-to-volume ratio is suitable for fast reproduction (multiplication) without hesitation. This phenomenon partly accounts for the short generation times (minutes to days as opposed to weeks to years for eukaryotes) which provides for rapid population increases (exponential growth) that are typical of prokaryotes. Fast-growing bacteria can double every 15-20 minutes. With such short generation times, even the population of a single mutant can quickly increase in a permissive environment.

## 2.3.3 Structural simplicity

Prokaryotes are single-celled organisms without a distinct nucleus and have a seemingly simple structure. However, structural simplicity is used loosely here as more detailed microscopy shows very intricate complexity. Their cell wall provides rigidity and shape to the cell, is a highly selective barrier, and has an intricate surface chemistry. The cell wall differs distinctly between Gram-positive and negative microorganisms (Fig 2.2). Gram-negative microorganisms have a more complex multilayered cell wall structure than Gram-positive prokaryotes. In Gram-negative organisms, the outer lipopolysaccharide membrane is a bilayer composed of lipids that are endotoxic (Rietschel, 1976). It acts as a very good hydrophobic barrier. Gram-positive organisms, on the other hand, have a thick murein layer as the outermost membrane, a layer that is very thin in Gram-negative organisms (Fig. 2.2). In both the Gram-positive and negative microorganisms, the murein layer is composed of glycoproteins and peptidoglycans that provide structural integrity to the cell and enable it to handle the large osmotic pressure within. Furthermore, the cell wall in Gram-positive microorganisms is composed of muramic acid whereas in Gram-negative microorganisms it is composed of a 2-3 nm thick peptidoglycan layer. Archea has a cell-wall structure that is similar to that of Gram-positive bacteria. While the structure of the cell wall is more extensively discussed in microbial physiology textbooks, its brief discussion here is intended to enable the reader to appreciate the fact that, despite its seemingly simple structure, it is the first line of defense for the cell against a diverse array of environments. It is also noteworthy that *Planctomycetes*, whose members are not readily cultivated during routine isolation but are abundant in the environment (soils, water, and sediments) do not synthesize the universal bacterial cell wall peptidoglycans (Zarda et al., 1997).

Other organelles, such as flagella, ribosomes, and nucleus, though seemingly simple in structure, serve equally complex functions. The prokaryotic cell has a single chromosome that represents the total DNA of the entire cell and unlike eukaryotes is located in the center of the cell and is not enclosed in a membrane (i.e., no nuclear membrane). This unique architecture provides easy accessibility of the DNA in terms of exchanging signals between the outside and inside of the cell. In eukaryotes, the DNA is enclosed within a nuclear membrane and messages have to be "exported" outside the nucleus. In contrast, both transcription and translation occur together in prokaryotes, an arrangement that may have some benefit associated with enhancing adaptation to environmental changes.

### 2.3.4 Genomic plasticity

Our ignorance about what makes prokaryotes successful in a specific niche is quite profound. The enormous success of classical microbiol-

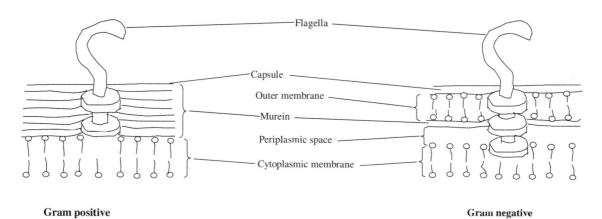

**Gram positive**                                                      **Gram negative**

**Fig. 2.2** Schematic diagram of cell wall structure of a Gram-positive and Gram-negative prokaryote. Notice the presence of a thick murein layer made of both peptidoglycans and glycoproteins in Gram-positive vis-à-vis Gram-negative and the absence of periplasmic space in Gram-positive prokaryotes.

ogy has greatly depended on the selection and utilization of strains of organisms that behave in a reliable and consistent manner on laboratory media. However, nature is quite chaotic and such consistency under laboratory conditions may bear very little resemblance to the natural environment. The conventional microbiology approach may, therefore, be quite misleading if we wish to understand how microorganisms behave outside the confines of a flask, test tube, or petri dish. The reality is that most of the prokaryotes either possess some hidden reserves of genetic information or can activate some genetic rearrangement which can be manipulated under appropriate environmental conditions and thus display genomic plasticity (Terzaghi and O'Hara, 1990). In this regard, prokaryotes are able to adapt to environmental changes much more rapidly than eukaryotes. Such adaptation enables them to be ubiquitous, existing at various extremes, for example high/low pH, temperature extremes, etc., as is discussed in Chapter 7.

Genomic plasticity involves some form of physiological and genetic adaptation. Physiological adaptation in this context refers to the changes in gene expression in the prokaryotic population in response to changes in the environment. Genetic adaptation, on the other hand, is the growth of mutant cells to become the predominant type in the population, enabling the organism to cope with the prevailing environment. It is important to understand how microorganisms acquire new properties. Unlike eukaryotes, prokaryotes are haploid—they have only one copy of their genome. Rare is the case wherein the cell may contain multiple nucleoids during rapid growth simply because cell division has not kept pace with the formation of the nucleus. With a haploid architecture, even the recessive mutations are expressed in the phenotype (the observable characteristics) as opposed to diploids in which recessive alleles are never expressed until after some form of mating. The net result is that, in prokaryotes the outcome of any changes in

genetic composition, due to mutations or otherwise, is phenotypically expressed in the actual cell that is affected rather than waiting to be expressed in the offspring. This unique feature sets the prokaryotes apart, equipping them with enormous genetic plasticity compared to eukaryotes. Thus, any genetic mutation in prokaryotes, whether good or bad, is immediately subjected to the pressure of natural selection, setting a stage for unprecedented diversity among prokaryotes. Besides mutations, plasticity among prokaryotes is also attributed to other mechanisms, such as conjugation, transformation, transduction, and transposons which contribute to genetic exchange. The mutations engendered by each of these exchange mechanisms are discussed in the next section.

The exchange of genetic material by microbes is fairly widespread and can be a programmed or unprogrammed event (Terzaghi and O'Hara, 1990). Programmed rearrangements occur at precisely specified end points and are mediated by site-specific recombination. The general homologous recombination operating upon a pair of separated homologous sequences can, in principle, yield a variety of genetic rearrangements. Typically, a length of 30-150 bp of at least 97% sequence homology is required for normal recombination but some nonhomologous recombinations can also occur within short (5-20 bp) sequences, albeit at a very low frequency.

### 2.3.4.1 Mutations

Mutations enable the microorganisms to access new food sources and survive under more diverse environments. Mutations involve changes in the base sequence of DNA and can be classified either by the nature of the change in DNA (genetic classification) or by the change in the phenotype. In natural environments, mutations are brought about by chemicals, irradiation, replication errors, and insertions of transposons or similar DNA sequences. The rate of mutation is the number of mutations formed per cell doubling and is calculated from the mathematical relationship:

$$a = \frac{m}{\text{Cell generations}} = \frac{m\,(\ln2)}{n - n_0} \quad ...(1)$$

where a is mutation rate and m the number of mutations formed as the number of cells increase from the initial generation ($n_0$) to generation n. Mutations are in themselves rare events. However, through conjugation, transduction, and translation, the resultant mutants do not remain rare for too long in prokaryotes because they are quickly copied and shared, even across species.

In molecular terms, mutations are broadly divided into two classes—microlesions and macrolesions. Microlesions are those mutations in which a single base pair is altered and they include transitions, transversions, and frameshifts (Table 2.2). Macrolesions are more extensive mutations involving deletions, duplications, inversions, and translocations (insertions). Much is known about these processes in *E. coli* K-12 (Gram-negative) and *Salmonella typhimurium* (Gram-positive) but very little, if anything, is known about them in Archea or in the other range of prokaryotes which have not yet been successfully cultured. Deletions involve the loss of segments of DNA and account for approximately 12% of spontaneous mutations in nature. Thus, deletions play a sig-

nificant role in creating genetic diversity in nature. One example of a deletion of ecological significance is the *nifHDK* operon of *Anabaena*. Under sufficient nitrogen, the *nifD* gene is split by an 11.6 kb sequence that is bonded by direct repeat sequences and includes an *xisA* gene encoding a site-specific recombinase recognizing these bracketing direct repeat sequences. Under conditions of nitrogen deficiency, the *xisA* gene is induced, the 11.6-kb block containing the *nifD* genes removed, and the key *nifHDK* operon is expressed (Haselkorn et al, 1986).

Duplication is likely to play an important role in evolution. It results in the formation of additional copies of chromosomal segments (Table 2.2), leading to amplification of the duplicated set of genes. Translocation is the movement of a fragment of DNA from one region of the chromosome to another chromosome. The resultant random alternation in the nucleotide sequences after insertions can change metabolic functions as it can inactivate the gene within which the insertion occurs. Inversions are the reversal of the order of genes and possibly occur spontaneously in microorganisms. Inversions are some of the best-studied rearrangements and involve inverting elements that control the switching between the expression of two alternative genes. Inversions are catalyzed by a site-specific recombinase and

**Table 2.2** Types of mutations

| Class | Type | Molecular change[1] |
|---|---|---|
| Microlesion[2] | Transition | Change in single base, e.g. P → Py or another P |
| | Transversion | Substitution of P → Py or vice versa |
| | Frameshift | AGTCAGT → AGTCAAGT or AGTCGT |
| | | TCAGTCA   TCAGTTCA     TCAGCA |
| Macrolesion | Deletion | abcdefghi → abcghi |
| | Duplication | abcdefghi → abcdefdefghi |
| | Inversions | abcdefghi → abcfedghi |
| | Translocation (Insertions) | abcdefghi → jkldefmno |

[1]Letters represent genes on a double-stranded DNA (for macrolesion mutations) or purines (P) and pyramidines (Py) for microlesion mutations.

[2]Note that bases in mRNA are read in groups of three and these mutations change the code and subsequent amino acids which may, depending on the extent of the mutation change the protein or in the case of frameshift mutations, render the gene product completely inactive.

the inversion process leads to the production of different structural proteins, thus changing the antigenic properties of the organism in question. Among the best studied inversions are the *hin*, *gin* and *cin* systems in *Salmonella typhimurium* and the *pin* gene in *E. coli* (Chater and Hopwood, 1989).

### 2.3.4.2 Mechanisms of genetic exchange in the environment

Unlike the transfer of genes in the clinical environment, gene transfer in the soil and aquatic environment may be hampered by a variety of biotic and abiotic factors (Table 2.3). Simple genetic exchange by mechanisms other than mutation has been demonstrated in natural environments or under laboratory conditions (Table 2.4). They may be regulated by specific environmental signals. However, the transfer of genes in the environment is generally stimulated by conditions that enhance the activity and survival of the host cells. Thus, whereas the transfer of genes is most likely to occur in all organs in a clinical environment, in a natural environment it is most likely to be relatively more favorable in select microcosms such as the rhizosphere, soil invertebrates, sediments, epilithon, and sewage influents as well as similarly polluted water reservoirs (Dröge et al., 1999)—environments that typically have abundant nutrients (Saye et al., 1987).

The mechanisms of gene exchange in microorganisms include conjugation, transduction and transformation, all of which evolved as mechanisms of genetic exchange in prokaryotes during a geological period of intense solar radiation (Margulis and Sagan, 1997). The three mechanisms are unique to prokaryotes as genes donated by one organism can be received by another organism of a different species. This possibility has prompted some scholars to question the validity of classifying prokaryotes into species since there is literally considerable "extraspecies" breeding among them. (The conventional definition of a species is a collection of organisms that share genetic characteristics, a common origin, and the ability to interbreed). The frequency of genetic transfer and its significance in the environment is still unclear but has been reported in legume nodulation by previously nonnodulating rhizobial strains, in tumor induction and pathogenicity by a previously nonvirulent *Agrobacterium radiobacter*, and in enabling microorganisms to acquire novel catabolic activities such as the degradation of specific organic pollutants.

**A. Conjugation** is a process by which DNA is passed from one bacterial cell (donor) to another (recipient) by cell-to-cell contact and is a potential mechanism through which genes may be transferred across species. The trans-

**Table 2.3** Some biotic and abiotic factors affecting the transfer of genes between microorganisms in different environments

| Factor | Environment | | |
|---|---|---|---|
| | Clinical | Aquatic | Soil |
| Temperature | Fairly constant | Fluctuating | Fluctuating |
| pH | Variable; often near 7 | Often near neutral | Variable range |
| Oxygen supply | Aerobic/anaerobic | Anaerobic | Aerobic/anaerobic |
| Nutrient status | Rich | Usually low | Low |
| Moisture | Usually >70% | Usually >50% | Variable (often <50%) |
| Selective pressure | Often high | Low but potentially present | Low but potentially present |
| Cell density | High | Low to high depending on the nutrient status | Low except in specific microenvironments |

**Table 2.4** Examples of *In-situ* transfer of genes in the natural environment and under laboratory conditions

| Mechanism | DNA source | Phenotype | Environment | Organisms | Reference |
|---|---|---|---|---|---|
| Conjugation | Plasmid | N$_2$-fixation | Soil | *Enterobacter* sp. | Klingmüeller (1991) |
| | Plasmid | Catabolism | Fresh water | *Pseudomonas* sp. | Fulthorpe and Wyndham (1991) |
| | Plasmid | Molecular marker | Lab medium | *Rhizobium tropici* | Sessitsch et al. (1997) |
| Transduction | Chromosomal | Antibiotic resistance | Soil | *E. coli* | Germida and Khachatourians (1988) |
| | Plasmid | Antibiotic resistance | Lake water | *P. aeruginosa* | Saye et al. (1987) |
| Transformation | Plasmid | Antibiotic resistance | Sea water | *Vibrio* sp. | Paul et al. (1991) |
| | Chromosomal | Antibiotic resistance | Sea water | *Vibrio* | Paul et al. (1991) |

ferred genetic material may be a portion of the chromosome or a plasmid. During conjugation, the donor produces a pilus to which the recipient becomes attached. Conjugation is almost exclusively determined by plasmids and not by prokaryotic chromosomes. However, chromosomal genes of the donor cell can be transferred to the recipient through integration of a plasmid into it, followed by excision from the donor chromosome. In a few instances, the encoding plasmid assists in the transfer of either chromosomal DNA or other nonencoding plasmids. Conjugation can also be facilitated by the transfer of transposons from donor to recipient cells.

Plasmids are circular with about one-tenth of the circular DNA chromosome and are able to replicate without influence from the chromosomal DNA. While plasmids are not found in every bacterial cell, they are probably present in some individuals of all representatives of both Gram-positive and Gram-negative prokaryotes, in all environments. They may be self-transmissible (conjugative) or nonconjugative. Most conjugative transfer of plasmids occurs within the respective Gram stain but a low frequency of transfer between Gram-positive and Gram-negative organisms has been reported (Veal et al., 1992). Furthermore, genetic transfer has also been observed under laboratory conditions, between bacteria and archea but it is not clear whether such exchange actually occurs in nature. Genetic transfer through plasmid conjugation is normally unidirectional, that is, from the donor to the recipient, although gene transfer in the opposite direction, a phenomenon called retrotransfer, has also been reported (Mergeay et al., 1987). The rate at which plasmids are transferred from donor to recipient is independent of the rate of growth of the recipient cells (Dröge et al., 1999). Therefore, since growth of microorganisms under most environments is low (due to nutrient limitations and other stress factors), plasmid transfer in the environment is probably a rare event.

Once integrated into the chromosome, plasmids can maintain themselves indefinitely. Plasmids in Gram-positive organisms tend to be smaller than those in Gram-negative organisms. They seem to have evolved independently and can transfer themselves from one organism (cell type) to another through conjugation. Plasmids vary in size from 1-200 kbp DNA and are quite widespread in bacteria. Plasmid-mediated genetic rearrangements have been correlated with phenotypic changes (Nano and Kaplan, 1984). Valla et al. (1987) reported ready rearrangement of plasmid and chromosomal DNA in *Acetobacter xylinum*, resulting in loss of cellulose synthesis. Although a single organism may harbor several distinct plasmids, each plasmid with its own replication system, plasmids rarely encode functions that are critical for cell growth or survival. Although they do not contain genes essential for the survival of the host cell, in all environments plasmids possess a remarkable array of factors that may be advantageous in certain environments. Many plasmid-mediated traits are, therefore, cryptic and confer no detectable phenotypes until conditions requiring the expression of these traits emerge (Table 2.5). Though not cardinal for the host organism to survive, plasmids substantially add to the genetic capability of their hosts and thus provide them with a selective advantage over their plasmid-free counterpart.

In the laboratory, the presence of plasmids is determined by subjecting DNA from cells to electrophoresis. Because DNA molecules move through the electrophoresis gel as a function of size and shape, the small circular plasmid DNA migrates faster than the large chromosomal DNA. In bacterial strains freshly isolated from natural environments, it is common to find as many as half a dozen plasmids; after prolonged culture in the laboratory, however, strains tend to lose some of these plasmids, possibly due to the richer nutrient status typical of laboratory media than in the natural environment. This observation not only gives an idea of how important plasmids could be in natural environments, but also signals how some of the work done under laboratory conditions should not be automatically deemed applicable to the natural environment.

The ability of bacteria to conjugate is determined by a number of genes. Studies from the F-plasmid of *E. coli* indicate that the transfer of the plasmid is influenced by the transfer efficiency of the plasmid and the host range of the plasmid. Transfer efficiency is usually expressed as the number of transconjugants per donor cell that will donate its plasmid to a recipient. The efficiency at which the plasmids are transferred depends on environmental factors such as temperature, as well as the possibility of inhibition of plasmid transfer into the recipient by other plasmids already present in the recipient cell. The latter is a phenomenon called fertility inhibition. Another factor that affects plasmid transfer efficiency is incompatibility. Some plasmids are incompatible and cannot coexist in a growing cell as their replication is inhibited. The transfer of a plasmid in the environment is also influenced by the spectrum of species to which it can transfer and maintain itself. For example, the F-plasmid can transfer and replicate between *E. coli* strains but even though it can transfer to other bacteria, it cannot replicate, clearly indicating a narrow host range. Other plasmids may not even be able to transfer across species. In contrast, some

**Table 2.5**  Some phenotypic properties attributable to bacterial plasmids

| Phenotype | Organism |
| --- | --- |
| Toxin production | Some *Vibrio cholerae* and *E. coli* |
| Bacteriocin production | Purple bacteria |
| UV resistance | *Dinoccocus* sp., *E. coli* |
| Nodulation and nitrogen fixation in legumes | *Rhizobium* sp. |
| Induction of plant tumors | *Agrobacterium tumefaciens* |
| Induction of hairy roots in plants | *A. rhizogenes* |
| Galls on plants | *Pseudomonas savastanoi* |
| Antibiotic production | *Streptomyces coelicolor* |
| Insecticidal toxin production | *Bacillus thuringiensis* |
| Adhesins | *E. coli* |
| Catabolism of various aromatic hydrocarbons | *Pseudomonas* spp. |

plasmids have a wide host range as they can transfer across a number of species; hence they are termed "promiscuous".

However, the transfer of plasmids in natural (nonsterile) soils and freshwater systems is rarely detected except under conditions of abundant nutrients (Fulthorpe and Wyndham, 1991; Klingmüeller, 1991; Paul et al., 1991). Studies in soil indicate that nutrient stress greatly reduces the transfer of plasmids by donor cells. The rate at which plasmids are transferred in soil and aquatic environments is, in most cases, maximal at optimal temperatures for the plasmid and/or its host. Moisture content also influences the availability and distribution of nutrients, viability of mating partners, the oxygen status, as well as the motility and thus probability of partners coming into contact with each other.

**B. Transduction** is the transfer of DNA into the host cell mediated by a bacteriophage. In this process, the bacteriophage incorporates a portion of the bacterial genome in the viral capsid and then infects another host upon transferring the bacterial DNA to it. The packaging of DNA in the phage particles physically protects it from being digested by nucleases during its transfer to the new recepient host. Transduction provides an attractive model for transferring genes in the natural environment. Bacteriophages can survive for long periods of time under stress environments in the absence of replicating host cells.

Two types of transduction have been observed—generalized and specialized. Specialized transduction involves lysogenic phages which integrate into the host chromosome, taking some of the adjacent host DNA with them if they excise inaccurately. Through specialized transduction, the genes close to the site of phage integration are incorporated into the chromosome. Thus in this case only the host genes located next to the phage integration site are involved in the genetic exchange process. Generalized transduction, on the other hand, occurs when, due to a packaging error, a random piece of host DNA is encapsulated in a phage particle instead of the phage genome. Although rare, generalized transduction occurs in many bacteria. In this form of transduction, any allele may be incorporated into the phage at by and large equal frequency. The transducing viral particle is unable to lyse the host cell, possibly due to some defective genes. The potential for the exchange of genetic material through transduction is universal within prokaryotes. However, phages often display a narrow host range of infection, such that transduction is most likely to contribute to the exchange of genes among closely related organisms.

Although transduction occurs in the presence of natural microbial populations, its significance under these environmental conditions has not been fully assessed. Despite the prevalence of bacteriophages in the environment, the spectrum of microorganisms that can be transduced depends upon the receptors recognized by the bacteriophages (Ochman et al., 2000). Furthermore, in soils and sediments, bacteriophages may adsorb to clay minerals such as montmorillonite and be protected for extended periods of time, thus serving as reservoirs of bacterial DNA in soil and in other natural habitats. Packaging of genetic material in a transducing bacteriophage probably represents an evolutionary survival strategy for bacterial genes.

**C. Transformation** is the process in which naked DNA is taken up by the cell and incorporated into the genome. In the laboratory, conditions to enable cells to take up naked DNA (i.e., become competent) have been developed. They involve either the electroporation or treating with calcium chloride ($CaCl_2$) and passing through cycles of hot and cold temperatures (a "heat-shock" treatment). The success of transformation is measured in terms of transformation efficiency and reflects the number of cells transformed per g DNA, viz.

Transformants/g DNA =

$$\frac{\text{Number of cells transformed}}{\text{g DNA used}} \times$$

$$\frac{\text{Final volume of cell suspension used}}{\text{Volume of cell suspension plated (mL)}}$$

However, the conditions that induce competence in laboratory studies are unlikely to occur in natural environments. Thus, in nature this process mostly relies on the presence of cells that are able to take up naked DNA as part of their metabolic processes. Such natural cell competence is induced during particular stages of growth, which may vary from one prokaryotic species to another. Studies in *Bacillus subtilis* and *Streptococcus* sp. show that natural competence can be induced by concentrations of a low molecular weight polypeptide which is constitutively synthesized and secreted by bacteria (Dooley et al., 1971; Morrison et al., 1983). Above a critical concentration of the low molecular weight polypeptide, competence is induced. Induction is associated with conditions that favor a state of unbalanced growth. In *Pseudomonas sturtzeri* and *Bacillus subtilis*, natural competence is maximal just before the onset of the stationary phase (Veal et al., 1992). Competence can also be induced in prokaryotic cells by spontaneous cell mutations or by a dramatic onset of cycles of nutritional stress (Paul et al., 1991).

Information about the frequency of competent bacterial cells in natural environments is limited, a factor that limits our ability to fully appreciate the potential significance of transformation in natural environments. When transformation occurs under natural environmental conditions, it consists of four different processes outlined by Stewart (1992):

(1) Development of competence by the prospective recipient cells.
(2) DNA binding onto the potential recipient— a property of the outer surface associated protein that is competence-specific. Binding generally can be inhibited in marine, soil, and sediments by high concentrations of cations.
(3) Uptake of DNA by the competent recipient cells into the cytoplasm.
(4) Recombination of the foreign DNA with the endogenote double-stranded DNA.

Apparently, naked DNA is released into the environment in large quantities from dead or moribund cells. Some viable cells also actively excrete naked DNA during specific growth phases. Such naked DNA could potentially be exchanged across distantly related organisms (horizontal transfer of genes) resulting in organisms that are mosaics of ancestral sequences (Ochman et al., 2000).

However, prokaryotes also release DNase into the environment which rapidly degrades whatever naked DNA may be present. For transformation to occur in a natural setting, therefore, the naked DNA must be able to persist or, somehow, get protected from degradation by the ubiquitous nucleases under these harsh environmental conditions by, for example, binding onto sand, clay, and sediments. DNA adsorbs onto sand particles quite rapidly, irrespective of whether the DNA is linear or supercoiled (Table 2.6). More intense adsorption occurs on clays, particularly the more expansive 2:1 montmorillonite clays (Alvarez et al., 1998). In soils and sediments, the adsorption of DNA is also dependent on the concentration of cations (CEC) and the pH. At a high pH, adsorption of DNA decreases due to an increase in repulsion forces. Genetic transfer through transformation is probably more preva-

**Table 2.6** Kinetics of plasmid DNA adsorption on sand

| Incubation time (min) | %DNA adsorbed | |
|---|---|---|
| | Supercoiled | Linear |
| 1 | 92.9 | 85.9 |
| 2.5 | 97.5 | 93.9 |
| 10 | 98.7 | 98.7 |
| 60 | 99.1 | 99.1 |

*Source:* Romanowski et al. (1991) with permission from the Society for Microbiology

lent under natural aquatic than under terrestrial environments due to the likelihood of DNase being more diluted in aquatic environments.

**D. Transposons and other insertion elements.** Most prokaryotes also have genetic elements called transposons that go from one site to another within the chromosome. First discovered by a plant breeder, Barbara McClintock, who code-named them as "jumping genes", transposons did not gain recognition until the concept of jumping genes was also encountered in antibiotic resistance in humans. In a sense, transposons are, in themselves, not regarded as a mechanism of genetic exchange as their transfer can be mediated by all of the three transfer mechanisms discussed above, i.e., conjugation, transduction, and transformation. Transposons have four significant effects on the recipient organisms:

  (1) inactivation of a gene or operon into which they transpose;
  (2) delivery of novel genes, e.g., antibiotic resistance, resistance to heavy metals, resistance to UV, which are not normally encountered by that organism;
  (3) activation of genes arising from promoter sequences found on some elements; or
  (4) forming structural rearrangements such as duplications, inversions, deletions, and fusions.

Thus, transposons naturally enhance the survival of organisms in which they are found since they are able to facilitate rearrangements of DNA sequences at frequencies higher than those of spontaneous mutations (Pickup, 1992).

Transposons have been widely used through genetic engineering to introduce foreign DNA into organisms. Some of these elements are able to insert anywhere in the genome while others are specific to target sequences. Some transposons carry, in at least one orientation, a good promoter sequence that can switch on transcription of the downstream sequences. Transposable elements may, in some cases, enhance a sequence which in turn can greatly stimulate a nearby, otherwise very weak promoter. It is important to realize that despite the range of mechanisms of genetic exchange between prokaryotes, the introduced genetic material does not guarantee successful gene transfer unless the transferred sequences can be stably maintained in the recipient.

## 2.4 SALIENT FEATURES OF BACTERIAL GENOME

Eukaryotes are diploid and propagate their diversity through the sexual fusion of nuclei. In contrast, prokaryotes are haploid. No nuclei fusion occurs in prokaryotes and there is no sexual mixing of genes. Rather, the mixing of genes in prokaryotes occurs through conjugation, transduction, transformation, and through transposons as outlined above. Although prokaryotes can regenerate their genome very rapidly, and sometimes have multiple copies of the same gene, this is not synonymous to diploidy because the genes are exact copies. The presence of multiple copies of genes is advantageous in that it provides overexpression of some of these genes.

The prokaryotic genome is circular and, unlike in eukaryotic chromosomes, has no introns to separate the different genes. The absence of introns ensures rapid and linear transcription and translation in prokaryotes. In *E. coli* they code for about 1,000 genes. The genes closest to the origin of replication are mostly genes for the synthesis of RNA and DNA. Prokaryotes also have function-specific operons that involve a series of genes acting in sequence to perform a particular function. These specific operons enable prokaryotes to grow rapidly and ensure a rapid physiologic response as a result of various environmental changes. Some genetic functions appear to become activated as a result of a single event that triggers the otherwise nonexpressive (cryptic) genes. When not required, such cryptic genes are normally not expressed by the organism. The cryptic genes are, therefore, at a selective advantage when need arises and can be retained within microbial communities for a long time, contributing much in the process of evolution (Hall et al., 1983). Still, much has to be learned about cryptism and its significance in environmental microbiology but it is apparent that these cryptic genes help bacteria to survive in the natural environment.

Some interesting parallels about the importance of cryptic gene transfer can be drawn from examples in the development of antibiotic resistance in microorganisms. The transfer of antibiotic resistance genes among prokaryotes of medical importance and in the environment is widely documented (Carlson et al., 2001). Furthermore, many of the discoveries in bacterial genetics, including plasmids, have come through studies of the acquisition of antibiotic resistance. Antibiotics are compounds that inhibit the growth of other microorganisms by specifically interfering in their biochemistry. The majority of naturally occurring antibiotics discovered to date are produced by streptomycetes and their close relatives among the actinomycetes. Other major producers of natural antibiotics include myxobacteria, *Bacillus* spp., and mycelial fungi. Production of antibiotics by these microorganisms is not an essential part of the cell differentiation process, however. Many antibiotics could potentially inhibit or even kill their producers. This is averted by a variety of resistance mechanisms that include detoxification, alteration of target sites, cell repair mechanisms, and the cell being impermeable to specific antibody compounds. For example, chlorophenicol-resistant bacteria detoxify the compound by acetylating it using an enzyme, chlorophenicol acetyltransferase (Davies and Smith, 1978). A wide range of $\beta$-lactam antibiotics, such as penicillin and cephalosporin, are also known to be readily detoxified by various $\beta$-lactamases which hydrolyze the $\beta$-lactam ring.

## References

Alvarez A.J., M. Khanna, G.A. Toranzos, and G. Stotzky (1998). Amplification of DNA bound on clay minerals. *Molec. Ecol.* **7**: 775-778.

Bintrim S.B., T.J. Donahue, J. Handelsman, G.P. Roberts, and R.M. Goodman (1997). Molecular phylogeny of archaea from soil. *Proc. Natl. Acad. Sci. (USA)* **94**: 277-282.

Bornemann J. and E.W. Triplett (1997). Molecular microbial diversity in soils from eastern Amazonia: evidence of unusual microorganisms and microbial population shifts associated with deforestation. *Appl. Environ. Microbiu.* **63**: 2647-2653.

Broschart F.M.H. (1967). The Soil-plant System in Relation to Inorganic Mineral Nutrition. Acad. Press, New York, NY.

Carlson S.A., T.S. Frana, and R.W. Griffith (2001). Antibiotic resistance in *Salmonella enterica* serovar *typhimurium* exposed to microcin-producing *Escherichia coli*. *Appl. Environ. Microbiol.* **67**: 3763-3766

Chater K.F. and D.A. Hopwood (1989). Diversity of bacterial genetics. *In*: D.A. Hopwood and K.F. Chater (eds.). Genetics of Bacterial Diversity. Acad. Press, New York, NY.

Davies J. and D.I. Smith (1978). Plasmid determined resistance to antimicrobial agents. *Ann. Rev. Microbio.* **32**: 469-518.

DeLong E.F. (1992). Archea in coastal marine environments. *Proc. Natl. Acad. Sci. (USA)* **89**: 5685-5689.

Dojka M.A., J.K. Harris, and N.R. Pace (2000). Expanding the known diversity and environmental distribution of an uncultured phylogenetic division of bacteria. *Appl. Environ. Microbiol.* **66**: 1617-1621.

Dooley D.C., C.T. Hadden, and E.W. Nester (1971). Macromolecular synthesis in *Bacillus subtilus* during development of the competent state. *J. Bacteriol.* **108**: 668-679.

Drahos D.J., B.C. Hemming, and S. McPherson (1986). Tracking recombinant organisms in the environment: -galactosidase as a selectable non-antibiotic marker for fluorescent pseudomonads. *Bio/Tech.* **4**: 43-48.

Dröge M., A. Pühler, and W. Selbitschka (1999). Horizontal gene transfer among bacteria in terrestrial and aquatic habitats as assessed by microcosm and field studies. *Biol. Fert. Soils* **29**: 221-245.

Dykhuizen D.E. (1998). Santa Rosa revisited: Why are there so many species of bacteria? *Antonie van Leeuwenhoek* **73**: 25-33.

Fulthorpe R.R. and R.C. Wyndham (1991). Transfer and expression of the catabolic plasmid pBRC60 in wild bacterial recipients in a freshwater ecosystem. *App. Environ. Microbio.* **57**: 1546-1553.

Germida J.J. and G.G. Khachatourians (1988). Transduction of *Escherichia coli* in soil. *Can. J. Microbiol.* **34**: 190-193.

Hall B.G., S. Yokoyama, and D.H. Calhourn (1983). Role of cryptic genes in microbial evolution. *Molec. Biol. Evol.* **1**: 109-124.

Hamdi Y.A. (1971). Soil-water tension and the movement of rhizobia. *Soil Biol. Biochem.* **3**: 121-126.

Handley P.S., L.M. Hesketh, and R.A. Moumena (1991). Charged and hydrophobic groups are localized in the short and long tuft fibrils on *Streptococcus sanguis* strains. *Biofouling* **4**: 105-111.

Haselkorn R., J.W. Golden, P.J. Lammers, and M.E. Mulligan (1986). Developmental rearrangement of cyanobacterial nitrogen-fixation genes. *Trends Genet.* **2**: 255-259.

Henscke R.B., E. Nücken, and F.R. Schmidt (1989). Fate and dispersal of recombinant bacteria in a soil microcosm containing the earthworm *Lumbricus terrestris*. *Bio. Fert. Soils* **7**: 374-376.

Hugenholtz P., B.M. Goebel, and N.R. Pace (1998). Impact of culture-independent studies on the emerging phylogenetic view of bacterial diversity. *J. Bacteriol.* **180**: 4765-4774.

Klingmüeller W. (1991). Plasmid transfer in natural soil: A case by case study with nitrogen-fixing *Enterobacter*. *FEMS Microbiol. Ecol.* **85**: 107-116.

Koch A.L. (1971). The adaptive responses of *Escherichia coli* to a feast and famine existence. *Adv. Microb. Physiol.* **6**: 147-217.

Lawrence J.R., D.R. Korber, G.M. Wolfaardt, and D.E. Caldwell (1995). Behavioral strategies of surface-colonizing bacteria. *Adv. Microb. Ecol.* **14**: 1-75.

Madsen E.L. and M. Alexander (1982). Transport of *Rhizobium* and *Pseudomonas* through soil. *Soil Sci. Soc. Amer. J.* **46**: 557-560.

Margulis L. and D. Sagan (1997). What is Sex? Simon and Schuster, New York, NY.

Mergeay M., P. Lejeune, A. Sadouk, J. Gerits, and L. Fabry (1987). Shuttle transfer (or retrotransfer) of chromosomal markers mediated by plasmid pULB113. *Molec. Gen. Genet.* **209**: 61-70.

Miller M.B. and B.L. Bassler (2001). Quorum sensing in bacteria. *Ann. Rev. Microbiol.* **55**: 165-199.

Morrison D.A., S.Lacks, W.R.Guild, and J.M. Hageman (1983). Isolation and characterization of three new classes of transformation-deficient mutants of *Streptococcus pneumoniae* that are defective in DNA transport and genetic recombination. *J. Bacteriol.* **156**: 281-290.

Nano F.E. and S. Kaplan (1984). Plasmid rearrangements in the photosynthetic bacterium *Rhodopseudomonas sphaeroides*. *J. Bacteriol.* **158**: 1094-1103.

Nye P.H. (1979). Soil properties controlling the supply of nutrients to root surfaces. *In*: J.L. Harley and R.S. Russel (eds.). The Soil-Plant Interface. Acad., Press, New York, NY, pp 21-38.

Ochman H., J.G. Lawrence, and E.A. Groisman (2000). Lateral gene transfer and the nature of bacterial innovation. Nature **405**: 299-304.

Paul J.H., M.E. Frischer, and J.M. Thurmond (1991). Gene transfer in marine water column and sediment microcosms by natural plasmid transformation. *Appl. Environ. Microbiol.* **57**: 1509-1515.

Pickup R.W. (1992). Detection of gene transfer in aquatic environments. *In*: E.M.H. Wellington and J.D. van Elsas (eds.). Genetic Interactions among Microorganisms in the Natural Environment. Pergamon Press, New York, NY, pp. 145-164.

Rietschel E. (1976). Absolute configuration of 3-hydroxy fatty acids present in lipopolysaccharides from various bacterial groups. *Europ. J. Biochem.* **64**: 423-428.

Romanowski G., M.G. Lorenz, and W. Wackernagel (1991). Adsorption of plasmid DNA to mineral surfaces and protection against DNase I. *Appl. Environ. Microbiol.* **57**: 1057-1061.

Saye D.J., O. Ogunseitan, G.S. Sayler and R.V. Miller (1987). Potential for transduction of plasmids in a natural freshwater environment: Effect of plasmid donor concentration and a natural microbial community on transformation in *Pseudomonas aeruginosa*. *Appl. Environ. Microbiol.* **53**: 987-995.

Sessitsch A., P.K. Jjemba, G. Hardarson, A.D.L. Akkermans, and K.J. Wilson (1997). Measurement of the competitive index of *Rhizobium tropici* strain CIAT899 derivatives marked with the *gusA* gene. *Soil Biol. Biochem.* **29**: 1099-1100.

Smart J.B., A.D. Robson, and M.J. Dilworth (1984) A continuos culture study of the phosphorus nutrition of *Rhizobium trifolii* WU95, *Rhizobium* NGR234 and *Bradyrhizobium* CB756. *Arch. Microbiol.* **140**: 276-280.

Stanley P.M. (1983). Factors affecting the irreversible attachment of *Pseudomonas aeruginosa* to stainless steel. *Can. J. Microbiol.* **29**:1493-1499.

Stewart G.J. (1992). Transformation in natural environments. *In*: E.M.H. Wellington and J.D. van Elsas (eds.) Genetic Interactions among Microorganisms in the Natural Environment. Pergamon Press, New York, NY, pp. 216-234.

Stolp H. (1988). Microbial Ecology. Organisms, Habitats, Activities. Cambridge Univ. Press, Cambridge, UK.

Terzaghi E. and M. O'Hara (1990). Microbial plasticity. Relevance to microbial ecology. *Adv. Microbial Ecol.* **11**: 431-460.

Valla S., D.H. Coucheron, and J. Kjosbakken (1987). The plasmids of *Acetobacter xylinum* and their interaction with the host chromosome. *Molec. Gen. Genet.* **208**: 76-83.

Veal D.A., H.W. Stokes, and G. Daggard (1992). Genetic exchange in natural microbial communities. *Adv. Microb. Ecol.* **12**: 383-430.

Ward B.B. (2002). How many species of prokaryotes are there? *Proc. Natl. Acad. Sci. (USA)* **99**: 10234-10236.

Woese C.R. (1987). Bacterial evolution. *Microbiol. Rev.* **51**: 221-271.

Zarda B., D. Hahn, A. Chatzinotas, W. Schönhuber, A. Neef, R.I. Amann, and J. Zeyer (1997). Analysis of bacterial community structure in bulk soil by in-situ hybridization. *Arch. Microbiol.* **168**: 185-192.

# 3

# Protozoa

It is always difficult to capture a definition of protozoa in one sentence. Generally, protozoa are eukaryotes that possess a nuclear envelope, have an animal-like form of nutrition (heterotrophic), and depend on external sources of food. Motile forms move either by hairlike vibrating cilia, whiplike flagella or extension of amoeboid pseudopodia. Most are unicellular although their unicellular nature is not always accompanied by the presence of a single nucleus. Generalizations about protozoa are difficult to make because some of them are multicellular and live together as colonies. Furthermore, some large unicellular protozoa, e.g. large amoeba, ciliates, and foraminifera, are multinucleate, the nuclei providing surfaces through which enough mRNA can be transported so as to supply the large cell. Multinucleate protozoa are distinct from metazoans because protozoa do not form tissues of specialized cells. Protozoa range in size from 2 $\mu$m to almost 8 cm (Fig. 3.1), a range that is almost 5 log units and corresponds to approximately one-third of the logarithmic range of all organisms of biological interest. The size of the smallest protozoa overlaps that of the largest prokaryotes such as cyanobacteria whereas that of the largest protozoa almost overlaps the smallest terrestrial mammal, i.e., the pygmy shrew. In terms of shape, protozoa can, depending on the species, assume a variety of shapes such as amorphous, oval, wormlike,

flattened, etc. Considering the variation in size and shape, it is not surprising that protozoa play an enormously diverse role in nature.

## 3.1 EVOLUTION OF PROTOZOA

Protozoa are the most abundant phagotrophs in the biosphere, with sea water containing about 1,000 flagellates and freshwater sediments containing about 10,000 ciliates per milliliter (Finley, 1990). Some organically rich habitats such as activated sludge and wastewater harbor more than $10^5$ protozoa per milliliter (Curds, 1973). Protozoa are believed to have evolved from prokaryotes about 1.5 billion years ago and could have led the way into modern-day eukaryotic life style (see Chapter 1). To understand this evolutionary relationship, it is important to make comparisons with their nearest relatives among the large group called protista, such as algae, fungi, and slime molds, with which the characteristics of protozoa overlap. For example, slime molds are unicellular and coenocytic but have both fungal and protozoan characteristics. Just like protozoa of the group Sacordina, slime molds are amoeba-like but form fruiting bodies containing spores, a trait typical of fungi. They are therefore discussed in Chapter 4 among fungi under which they fit best. For comparative purposes, the mode of nutrition, motility and structure within protista is outlined in Table 3.1. Algal protists are photosynthetic (and therefore primary pro-

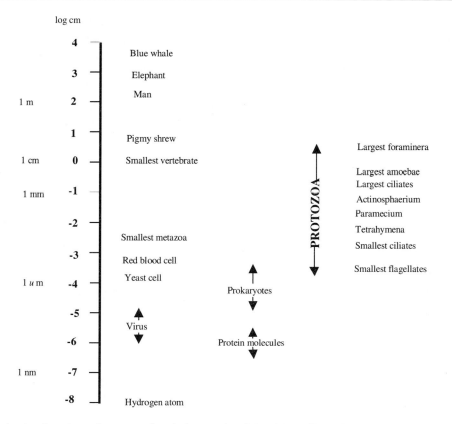

**Fig. 3.1** Scale showing sizes of protozoa in relation to other living things (From Fenchel, 1987 with permission from Springer-Verlag GmbH & Co. KG).

ducers) whereas most protozoa are phagotrophic heterotrophs whose primary food source is bacteria. However, some protozoa, for example dinoflagellates, are also photosynthetic, thus providing the biggest overlap with other protists through their mode of nutrition. Algae and fungi are filamentous and unicellular and some algae have amoeboid movement just like some protozoa. Algae and fungi are discussed jointly in Chapter 4.

Protozoa show much more diversity than plants, animals, and fungi combined, as is evidenced from the rRNA sequence of the Eukarya domain (see Fig. 1.2) The phylogenetic tree shows major branches of Entamoeba, Microsporidia, Metamonda (e.g. *Giardia*), Euglena, Kinetoplasta (e.g. *Trypanosoma*), and Parabasalia (e.g. *Trichomonas*)—all of which are protozoa. It should be noted that unlike all the other mem-

**Table 3.1** Nutritional, motility and structural characteristics of protista

| Group | Structure | Nutrition | Motility |
|-------|-----------|-----------|----------|
| Algae | Filamentous and unicellular | Phototrophic | Flagellar (amoeboid) |
| Fungi | Filamentous and unicellular | Osmotrophic | Nonmotile (except some spores) |
| Protozoa | Unicellular (some are multicellular) | Phago-/osmotrophic | Cilia, flagella and/or amoeboid |
| Slime molds | Coenocytic and unicellular | Phagotrophic | Amoeboid |

bers of the eukarya domain which have 18S rRNA in their cytoplasm, *Giardia*, which branched off earliest among the eukarya, has 16S rRNA just like the archaea. Thus, despite striking similarities among the various members of protozoa, there are some distinct differences which make it harder to draw the line on what comprises protozoa.

## 3.2 MAJOR GROUPS OF PROTOZOA

Four phyla of protozoa are recognized based on motility, feeding structures, cell surface and presence of symbiotic bacteria (Table 3.2). These include the Sarcomastigophora (amoeba and flagellates), Ciliophora (ciliates), Microspora, and the wholly parasitic Sporozoa (Apicomplexa). Generally, the most ubiquitous species are small (<10 μm diameter), feed on bacteria or on dissolved inorganic nutrients, and are able to move freely through the soil. They include the small ciliates, flagellates, and amoebae. Although not all protozoa are associated with diseases, those that have a potential to cause diseases are spread across all the four taxonomic groups (Table 3.2; Fig. 3.2). Of major economic importance are members of the phylum Apicomplexa, some of which are discussed in Chapter 16. Members of the phylum Microspora are quite distinct from other protozoa as they have a membrane-bound nucleus, multiply by binary fission, and lack both mitochondria and Golgi membranes.

Whereas the numbers and biomass of protozoa vary greatly between climates and types of vegetation, in terrestrial ecosystems, amoebae appear to be more predominant in bulk soil, comprising 50-90% of the protozoan population (Clarholm, 1981; Bischoff and Anderson, 1998; Anderson, 2000; Bass and Bischoff, 2001). Amoebae are categorized into four distinct morphotypes (morphotypes 1, 2, 3, and 4). Most abundant in soil are members of morphotype 4, for example *Vanella* sp. and

*Platyamoeba* sp., which are flattened or discoid amoebae characterized by circular or fan-shaped pseudopodia. They can exploit diverse types of soil. Morphotype 1, such as *Acanthamoeba* sp., *Chaos* sp., *Echinamoeba* sp., *Amoeba proteus*, and *Mayorella* sp. have extended lobose fine pseudopodia. Morphotype 2 are limax non eruptive amoeba whereas morphotype 3 are limax eruptive amoebae that exhibit sporadic, bulgelike locomotion. Examples of the former include *Glaeseria* sp., *Saccamoeba* sp., and *Hartmanella* sp. and the latter *Naegleria* sp., *Vahlkampfia* sp., and *Heteramoeba* sp. However, Stout (1980) contends that enumeration methods tend to detect fewer ciliates and that ciliates are quite prevalent in wetter soil habitats. In the rhizosphere, flagellates tend to be more predominant than amoeba and ciliates (Elliot and Coleman, 1977; Table 3.3).

## 3.3 ENVIRONMENTAL ADAPTABILITY, SURVIVAL, AND DISPERSAL OF PROTOZOA

Protozoa are very adaptable and diverse organisms, existing in soil (including hypersaline soils), sediments (including sulfur-rich sediments), acid mine drainages, surface water, and deep waters. However, unlike some prokaryotes that can exist at temperatures as high as 110°C in thermovents and at high pressure in deep seas, protozoa, just like other eukaryotes, do not exist above 50°C. Because protozoa, and other protists for that matter, are structurally more complex, the membrane integrity at high temperatures is distorted and some organelles, particularly mitochondria and the Golgi apparatus can be lost. At these high temperatures, protists cannot grow but may exist as cysts. However, the cysts can only remain viable for shorter periods at temperatures as high as 120°C or as low as −70°C. Protozoa are also more resistant to a variety of antibiotics than are bacteria although they are

**Table 3.2** Characteristics of the four protozoa phyla

| Phylum | Common name | Distinguishing characteristics | Representatives | Habitats |
|---|---|---|---|---|
| Ciliophora | Ciliates | Have projections called cilia that are similar to flagella in structure but much shorter. Almost all ciliates have a cytostome through which they feed and possess two sizes of nuclei, i.e., macronucleus and micronucleus. The duality of nuclei is a major distinguishing feature of this phylum | *Paramecium* sp., *Balantidium coli* | Rumen, marine and fresh water, animal parasites, rhizosphere |
| Sarcomastigophora | Flagellates (Mastigophora) | Possess flagella that move in a whiplike manner. Flagella are for movement toward food | *Giardia* sp., *Trypanosoma* sp., *Leishmania* sp., *Tricomonas vaginalis* | Fresh water, animals (as parasites) |
| | Sarcodina (Amoebae) | Have characteristic pseudopodia used for movement. | *Entamoeba histolytica, Acanthamoeba* sp., *Naegleria* sp. | Animals (as parasites), marine and fresh water |
| Apicomplexa | Sporozoans | Have a complex of special organelles at the tips of thier cells which contain enzymes used to penetrate host tissues. They have complex life cycles and need several hosts to complete those cycles. | *Plasmcdium* sp., *Cryptosporidium* sp., *Toxoplasma gondii* | Insect vectors, Animals (as parasites) |
| Microspora | Microsporida | Lack mitochondria and microtubules | *Microsporidium* spp., *Nosema* spp. | Insects, e.g. silkworms |

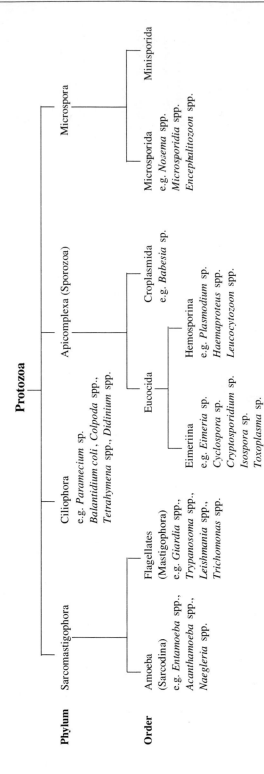

**Fig. 3.2** Taxonomic relationships within protozoa.

**Table 3.3** Shift of protozoa densities in a soybean rhizosphere over time

| Type | % occurrence of protozoa at | |
|---|---|---|
| | 3 days | 14 days |
| Flagellates | 69 | 70 |
| Amoeba | 9 | 29 |
| Ciliates | 22 | 1 |

*Source:* Jjemba (2001) with permission from the Society of Protozoologists.

sensitive to some, notably bacitracin, actidione, chloromycetin, and neomycin (Stout and Heal, 1967).

Most prevalent in the environment are protozoan species that have rapid excysting mechanisms that enable them to exploit their bacterial prey. The stability of predominant protozoan types also greatly depends upon the presence of adequate levels of moisture. The majority of protozoa are aerobic heterotrophs. However, facultative and obligate anaerobes do exist in places such as the hind gut of some insects, the rumen, and in sediments. Facultative anaerobic protozoa, such as *Loxodes* sp., can exist at the oxic-anoxic interface (Finley, 1985) whereas anaerobic protozoa, such as *Pelomyxa* sp., one of the largest protozoa, may have many mineral inclusions which seem to, among other purposes, help retain them in sediments and thus protected from oxygen. Protozoa found in deep water tend to be very different from those predominant near the surface as they are well adapted to survive under anaerobic conditions. Anaerobic protozoa also tend to have bacterial endosymbionts living inside them. For example, endosymbiotic protozoa inside termite guts and in the rumen have chains of methanogenic bacteria on their surface. Many endosymbiotic protozoa also have membrane vesicles called hydrogenosomes which carry phosphoclastic reactions that generate ATP (see Section 3.4). Protozoa typically eat bacteria and enclose them in digestive vacuoles. The consumed bacteria are then broken down by numerous lysosomes. However, methanogenic bacteria are not broken down because they lack peptidoglycans. The methanogens therefore end up in specialized vesicles similar to mitochondria and chloroplasts and can have special ecological and clinical significance (see Section 3.4).

The availability of food under most environmental conditions is patchy both in time and space. The protozoa present have to adapt to this reality of life by changing their physiological state. If the conditions are conducive for the prey bacteria to exist in large numbers, the population of the predators also tends to explode until it eventually reaches a constant level that corresponds to the new condition. Under unfavorable conditions such as depleted food supplies, food shortages, moisture stress, oxygen deficiency, unfavorable temperatures, toxic chemicals, etc., the existing protozoa form a protective coat (cyst) and become dormant. Under starvation, both the cell volume and the rate of respiration are reduced (Table 3.4). The reduction in some instances is as much as ten times or higher. At the onset of starvation, cell division continues for a generation or two, producing progeny that are smaller in size than the normal cells. The mitochondria formed under such conditions of starvation are also diminished in size. In *Acanthamoeba* sp., encystment is also followed by a depletion of the adenosine phosphates to about 85% because of the reduction in ATP (Stout, 1980). Changes also occur in the cytoplasmic membrane and in the activity of a variety of enzymes such as alkaline phosphatase. Such adaptive changes are definitely not unique to *Acanthamoeba* sp.; they are widespread among protozoa. The cyst structure also enables parasitic protozoan species to survive outside a host, an important attribute because parasitic protozoa tend to have life cycles that involve more than one host.

The cysts formed also facilitate dispersal of the organisms through the air as dust particles, just like fungal spores. The cysts can stay viable in a desiccated state for many years and

**Table 3.4** Cell volumes and respiratory rates of growing and starving cells of four species of protozoa

| Species | Cell volume ($\mu m^3$) | | Respiration ($nl\ O_2\ cell^{-1}\ h^{-1}$) | | Weight specific respiration ($\times 10^{-6}\ nl\ O_2\ \mu m^{-3}$) | |
|---|---|---|---|---|---|---|
| | Growth | Starvation | Growth | Starvation | Growth | Starvation |
| *Chaos carolinensis* | $5 \times 10^7$ | $1.25 \times 10^7$ | 16.9 | 4.2 | 0.34 | 0.34 |
| *Paramecium caudatum* | $9.5 \times 10^5$ | $1.7 \times 10^5$ | 1.8 | 0.19 | 1.89 | 1.12 |
| *Tetrahymena pyroformis* | $1.5 \times 10^4$ | $1.7 \times 10^3$ | 0.15 | $3.8 \times 10^{-3}$ | 9.7 | 2.24 |
| *Ochromonas* sp. | 240 | 50 | $3.2 \times 10^{-3}$ | $7.1 \times 10^{-5}$ | 13.3 | 1.4 |

*Source:* Fenchel and Finlay (1983) with permission from Springer-Verlag GmbH & Co. KG.

may be widely distributed globally by wind. Formation of cysts is an adaptation for survival and enables growth to resume rapidly once food resources become available again. There is every indication that cysts for protozoan pathogens, notably *Giardia* and *Cryptosporidium* also remain viable and, therefore, pathogenic for long periods of time. Despite all this, protozoa pass quickly between active and latent metabolic phases. The population of small flagellates and ciliates doubles in 2-3 hours whereas large ciliates may double in 10-20 hours. Large amoebae may take several days to double in population.

## 3.4 PROTOZOA AS SYMBIONTS AND PARASITES OF METAZOANS

Various types of protozoa are infected with obligate bacterial endosymbionts and parasites, but the environmental significance of these relationships is only beginning to be appreciated. For example, free-living amoebae that have endosymbiotic and endoparasitic bacteria are well adapted to hostile environments, resistant to desiccation, and to various disinfection processes. These amoebae have also been found to protect endoparasitic bacteria such as *Legionella* and *Mycobacterium* spp. against disinfection in water systems (Fritsche et al., 1998; Winiecka-Krusnell and Linder, 1999). Obviously, these relationships are of environmental, epidemiological, and clinical interest in particular, in situations wherein one

or both of the components are pathogenic. It is not clear whether the presence of the bacterial species as endosymbionts amplifies the bacterial or protozoan parasite infections. However, in the case of *Legionella* spp., the bacteria cannot multiply extracellularly in the environment (Abu Kwaik et al., 1998). Aspects of *Legionella* spp. as they pertain to its parasitic relationship with protozoa are discussed in Chapter 11.

However, not every protozoan symbiont signifies undesirable relationships with its host species. Symbiotic protozoa that permanently live in anaerobic habitats, such as the rumen of mammals, rely on anaerobic metabolic pathways for the release of energy, facilitating the digestion of food in their host species. The fluids in the rumen contain approximately $10^6$ ciliate protozoa and about $10^{10}$ bacteria per milliliter. Rumen protozoa do not possess mitochondria, possessing instead intracellular bacteria and hydrogenosomes. Hydrogenosomes function like mitochondria in that they oxidize pyruvate and can respire oxygen but do not contain cytochromes. Through this process, hydrogenosomes generate $H_2$ gas which would otherwise jeopardize the reoxidation of reduced NAD in the cytosol (i.e., $NADH + H^+ \rightarrow NAD^+ + H_2$) if it builds up. Hydrogenosomes also generate $CO_2$ and acetate. To preclude failure of reoxidation of the reduced NAD, ruminant ciliates and other protozoa rely on their endosymbiotic methanogens, e.g. *Methanobacterium formicicum* and *Methanoplanus endosymbiosus* to remove the excess $H_2$ gas wastes generated, the methanogens using the gas as a substrate. This symbiosis is strengthened by the close association between hydrogenosomes and methanogens in the protozoan cell. Other bacteria (probably also methanogens) can also use the acetate produced by hydrogenosomes, i.e.

$$CH_3COO^- + H_2O \rightarrow CH_4 + HCO_3^{3-}$$

This symbiotic relationship succeeds and is thermodynamically feasible if the methanogens keep the $H_2$ pressure low enough ($\leq 10^{-4}$ atm) for oxidation of the acetate.

Considerable research on protozoa as parasites of animals and terrestrial plants has been conducted. Most renown are parasites in humans and domestic livestock, of which the least sophisticated are protozoa that gain entry into the body through the oral canal and exit the body through feces. Through this mode of entry and spread, the respective protozoa do not face the animal's immune system and, to successfully maintain themselves and spread even further, they just have to avoid destruction by the digestive processes of the host species. In the alimentary canal, they derive most of their nutrition by absorbing food digested by the host. Other parasites inhabit specific tissues or the bloodstream. Some protozoan parasites of major economic significance are discussed in Chapter 15.

## 3.5 MOTILITY, TAXES, AND OTHER MODES OF POSITIONING

All protozoa show some motility at least during part of their life cycle. Motility enables them to move to new and more suitable environs and to physically go to the food (prey) or to generate currents to bring the food to them. Movement of protozoa can occur in liquids in the form of swimming. Swimming in mastigophora and ciliates is facilitated by flagella and cilia respectively. The flagella and cilia move back and forth at a frequency of about 50 Hz, the movement between neighboring cilia occurring out of

phase relative to each other, thus generating a wave across the surface and a propulsion that allows the organisms to move. Other mechanisms used by protozoa for locomotion include propulsion using tentacles, fibrilla and, in the case of amoebae, pseudopodia. In all cases, both the internal concentration and the flux of $Ca^{2+}$ across the cell membrane control motility. Increased levels of $Ca^{2+}$ reduce the beat of cilia and flagella whereas in pseudopodium-based movement, an increase in $Ca^{2+}$ induces the production of pseudopodia (Naitoh and Eckert, 1974; Naitoh and Sugino, 1984). Calcium levels also intracellulary regulate the direction of ciliary beating, thus controlling the forward and backward swimming movements. A number of protozoa also have the ability to creep along solid surfaces. Movement, coupled with body size and shape, enables protozoa to access small pore spaces and to move over air-liquid interfaces.

## 3.6 PHYSIOLOGICAL ECOLOGY OF FREE-LIVING PROTOZOA AND THEIR IMPACT ON THE ENVIRONMENT

Under aerobic conditions, protozoa will mainly feed on bacteria although some of them also feed on small algae, other protists, organic matter, and other small particulate nutrients. Of the 258 protozoan species (63 flagellates, 98 amoeba, and 97 ciliates) from the rhizosphere of plants tested by Ekelund and Rønn (1994), 40% were strict bacterial feeders, 36% had bacteria as a main feed source, while 20% lived completely or partly on other protozoa. A smaller number of protozoa feed on fungi, detritus, and algae. Over a short period of time, protozoa can survive some extreme environmental conditions if they can generate ATP to sustain basic physiological functions. Over the long term, however, they need some mechanism for capturing sufficient food and essential nutrients. All protozoa need water to move and feed. Thus, fluctuations in the supply of water and food are the fundamental problem facing protozoa in terrestrial environments. During feeding, protozoa may excrete low molecular weight compounds and growth factors necessary for bacterial growth. Nutrition and the role of protozoa in mineralizing nutrients are crucial in considering the physiological ecology and impact of protozoa on the environment.

## 3.6.1 Nutrition

Four major modes of nutrition are recognized for protozoa: diffusion, permeation, phagocytosis, and pinocytosis. Diffusion involves the passage of some small organic molecules across the membrane passively. Permeation is that process whereby the dissolved substances penetrate in the cell across the plasma membrane. The uptake process is selective and is an active, rather than passive process that expends energy. It is restricted to specific points ("pores") on the cell envelope. Pinocytosis denotes the abscission of vesicles from the outer membrane into the cell cytoplasm. It is most prevalent in amoebae but also occurs in other types of protozoa. Most widespread among protozoa, however, is phagocytosis, a process by which whole particles are engulfed in a membrane-covered vacuole. In amoebae, phagocytosis involves the protozoa flowing around the solid food particle and enclosing it in a "food vacuole" inside which lysosomal enzymes are discharged, lysing the ingested particle. Phagocytosis is induced by chemicals such as hyaluronidase, trypsin, cytochrome C, peptones and lectin as well as a variety of SH-bearing substances, such as glutathione, cysteine and phospholipids (Nisbet, 1984). Phagocytosis occurs anywhere on the cell surface. The ingested particles are then digested in the vacuole. Studies show that no vacuoles are formed in the absence of food or foodlike particles. Inert materials such as latex beads are ingested if their size is comparable to that of food particles (Mueller et al., 1965). Pinocytosis occurs in *Paramecium* and other ciliates and involves taking in food by waving the cilia

toward a mouthlike opening (cytosome). The swallowed food is digested in the food vacuole. The food vacuole changes in pH from alkaline to acidic, the acidity killing the bacterial prey. Subsequently, the vacuole contents become alkaline once again before they are excreted. In ciliates and flagellates (mastigophora), the "mouth" through which ingestion takes place is at a specific position on the cell, surrounded by cilia that concentrate and retain the food particles.

Under marine conditions, protozoa are able to concentrate food particles from the environment through various mechanisms, prior to phagocytosis, permeation, or pinocytosis. Fenchel (1987) estimated a protozoan to concentrate and phagocytize particles from a volume of water which is $10^5$ times the respective protozoan cell volume. Most studies of feeding in protozoa have been done in water systems, with ciliates, where three mechanisms of concentrating food apply:

(1) Filter feeding, i.e., direct transportation of water through a "filter mat" formed by pseudopodial tentacles, flagella, and cilia.

(2) Direct interception of the particles as they are carried along the flow, the particles ending up attached to the sticky surface of the protozoa until they are phagocytized, or taken up either through permeation or pinocytosis. This mechanism is also referred to as raptorial feeding.

(3) Diffusion of food particles to the protozoan predator.

The first two mechanisms depend on motility of protozoa whereas the third mechanism requires motility of prey. In liquid medium, the predominant mechanism in individual cases depends on the ratio of the radius of the prey and the predator, the shorter prey being consumed by the shorter types of predators, such as *Bodo* sp., *Cyclidium* sp., and *Colpoda* sp. Likewise, the larger prey are mostly consumed by larger predators such as *Bursaria* sp., *Didinium* sp., and *Stentor* sp. (Fig. 3.3). For ciliates, if the prey-to-predator size ratio is

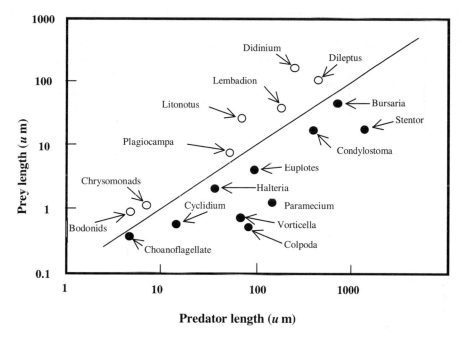

**Fig. 3.3** Average food particle size as a function of cell size for several raptorial (open circles) and filter feeding (closed circles) protozoa. The line corresponds to a 1:10 food particle size: predator cell size ratio (from Fenchel, 1987 with permission from Springer-Verlag GmbH & Co. KG).

greater than 0.1, raptorial (interception) feeding predominates. On the other hand, if the ratio is less than 0.1, filter feeding predominates.

In natural ecosystems, protozoa are present together with other microorganisms which mediate the food cycle. A diversity of potential prey species enables them to choose between suitable and unsuitable prey. However, protozoan predators do not reduce their prey to extinction because the bacterial prey continues to multiply even in the presence of actively grazing protozoa. This phenomenon is demonstrated clearly, for example, when the replication of bacteria is arrested by an antibiotic (Fig. 3.4). Under arrested replication conditions, continued predation of the bacteria by protozoa reduces the bacterial population to extinction, which in turn diminishes the protozoan density. Protozoa generally require a large number of bacteria per protozoa cell division, ranging between $10^3$ and $10^4$ bacteria cells for ciliates (Seto and Tazaki, 1971; Ramirez and Alexan-

der, 1981). Generally, the rate of growth of protozoa is dependent on the number of bacteria, such that the more bacteria there are above a certain threshold (about $10^6$ bacterial cells ml$^{-1}$), the faster the rate of growth of protozoa and its predatory activity (Ramirez and Alexander, 1981).

Factors controlling food selection by protozoa remain largely unknown but some prey possess morphological features such as extracellular capsules, inconveniently large size, or projections that discourage predation. Among the filter-feeding protozoa, inert latex beads are accepted as readily as bacteria of comparable size (Nisbet, 1984) suggesting that size and not quality could be the basis for selection. Chemoreception as well as the surface chemistry of the predator and/or the prey are also believed to play a key role in enabling some protozoa to select suitable prey. Preston et al. (1982) indicated that receptors on the plasma membrane of *Acanthamoeba* spp.

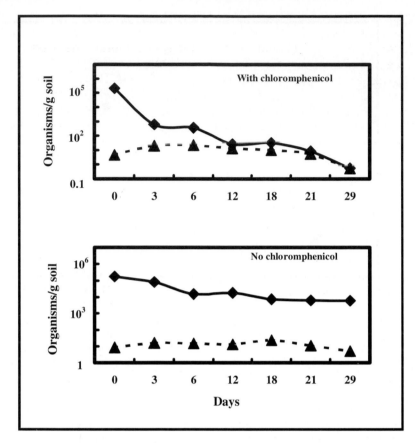

**Fig. 3.4** Survival of *Rhizobium* in nonsterile soil in the presence and absence of chloramphenicol. Triangles represent the indigenous soil protozoa density whereas the diamonds represent the density of bacteria (*Rhizobium* KO4SR) (based on data from Habte and Alexander, 1978 with permission from Elsevier).

respond to motile *Pseudomonas* sp. by binding them to its surface in large numbers through the polar flagella of the *Pseudomonas* prey. *Acanthamoeba* spp. only engulf food at the posterior and bacteria must be moved toward this end. This action is achieved by aggregates of bacteria being moved as a cap posteriorly followed by endocytosis. If the plasma membrane is pretreated with a plant lectin, such as Concanavalin, which blocks the carbohydrate binding sites on the plasma membrane glycoproteins, the bacterial prey will not bind. Binding of the bacterial prey onto some protozoa might also be facilitated by some kind of electrostatic attraction (Lee, 1980).

Some bacteria as well as bacterial and fungal spores are able to resist digestion and can be ejected live after being engulfed by protozoa (Zwart and Darbyshire, 1992; Pussard et al., 1994). This possibility has interesting ramifications, especially with pathogenic endosymbiotic bacteria as was pointed out earlier in Section 3.4. Most models of predation regard the prey to be in solution and it is not certain whether protozoa can consume bacteria attached to soil aggregate surfaces. It is likely, however, that amoebae are capable of engulfing attached bacteria as they move across the surface. For example, studies by Clarholm (1981) showed that amoebae decreased their bacterial prey in soil. Other types of protozoa had little impact on the bacteria. Such dominance by amoebae is probably attributable to their gliding/sliding motion on surfaces which may enable them to feed on the soil particles, sites on which most of the bacteria grow. The amorphous cell morphology of amoebae is also highly flexible and possibly well adapted for grazing activities within the thin water films surrounding the soil particles. On the other hand, ciliates select their prey based on ease of capture. A potential food organism which is small enough or slow enough as to be easily trapped is usually preferred to a larger vigorous prey. The amount of energy required to pursue a larger and more vigorous prey may not be equatable to its additional food content.

Under laboratory conditions, Caron (1987) showed preferential grazing by marine heterotrophs *Cryptobia sp.* and *Monas sp.* on *Pseudomonas halodurans* unattached to chitin particles compared to *Bodo sp.* and *Rhynchomonas nasuta* which preferred grazing on attached and aggregated bacteria. Both *Bodo sp.* and *R. nasuta* had very limited ability for feeding on unattached bacteria. No sign of preferential feeding or susceptibility of specific bacteria to predation was observed in the soybean rhizosphere (Jjemba, 2001).

The most extensive studies of the relationship between soil protozoa and bacterial flora were done by Singh (1941; 1942; 1945; 1948; Singh *et al.*, 1958; Anscombe and Singh, 1948). Based on many bacterial strains and a wide range of soil flagellates, amoebae, and the ciliate *Colpoda steinii,* Singh (1964) concluded that bacteria can be classified as either (i) eaten readily and completely, (ii) inedible strains, (iii) strains slowly but completely eaten, and (iv) strains that are only partly eaten. He found no correlation between edibility and Gram reaction, motility, slime production, and presence of proteolytic ferment, but found some relationship between edibility of bacteria by amoebae and *C. steinii* and pigment production by the prey. The pigments of some of the

inedible strains are toxic to protozoa (Singh, 1945). The dynamics of protozoa grazing on bacteria are further detailed in Chapter 8.

A few types of protozoa, e.g. *Tetrahymena* sp. and *Colpoda* sp., can grow axenically in solution of defined media (Rasmussen and Orias, 1975; Curds and Cockburn, 1968) but the growth of protozoa on a defined medium is generally less efficient compared to growing on bacteria prey. No protozoan groups have evolved free-living forms which do not ingest particulate material, probably because their size in relation to bacteria does not enable them to compete effectively with bacteria in the environment for dissolved nutrients. As a matter of fact, many protozoa are obligate predators of other organisms, notably bacteria, diatoms, cyanobacteria, smaller protozoa and, in some cases, smaller metazoans such as rotifers and crustaceans. The foregoing observations emphasize how the modes of nutrition for protozoa play a key role in ensuring the successful existence of protozoa in the environment.

### 3.6.2 Enhancement of mineralization of nutrients

All naturally occurring organisms in an ecosystem contribute to the flow of energy and the recycling of nutrients within that system. The rate at which nutrients are mineralized depends on the concentration of the nutrients in question inside the cells of the mineralizing organisms, and the rate at which the nutrients are transported from the source into the mineralizing cells/organisms. Under terrestrial and aquatic systems, most nutrients are in plant biomass. Through litterfall, the decomposing litter is attacked by fungi and bacteria. Thus bacteria collectively contain much more N and P than other cells. When protozoa feed on these bacteria, the nutrients are released from the bacterial biomass. The rate of growth and yield from organisms are, in nature, mostly limited by N and P. Thus, protozoan predators directly ingest the N and P reservoir (i.e., bacteria), a mechanism that efficiently supplies these nutrients for assimilation, subsequently making them available for other organisms. All in all, grazing by protozoa changes the physiology of the bacteria, promoting a more rapid assimilation and a faster circulation of nutrients through the ecosystem.

In sewage sludge, studies by Curds (1973) showed that the presence of protozoa reduces the half-life of *E. coli* from 16.1 h to 1.8 h, indicating a dramatic increase in the rate at which nutrients are turned over. Decomposition in soils and under marine conditions occurs more rapidly in the presence of bacteria and protozoa than in the presence of bacteria alone. Using $^{32}$P, Barsdate et al. (1974) determined the rate of P transfer between ponds in the presence of *Carex* detritus from a tundra pond seeded with natural bacterial microflora with or without protozoan predators. The rates of transfer of P were greater when the protozoan predators were present than when they were absent (Table 3.5). Furthermore, the rate of uptake of P by the bacterial cells was fourfold greater in

and protozoa compared to when only bacteria are present (Elliot et al., 1979). Amending the soil nitrogen to either 95 or 245 g $NH_4^+$-N $g^{-1}$ soil significantly increased the percent shoot N in grass seedlings when both bacteria (*Pseudomonas* sp.) and protozoa (*Amoeba* sp.) were present than when only bacteria were present (Fig. 3.5). Clarholm (1985) found that net mineralization to produce inorganic N from soil organic matter (SOM) was strongly enhanced in the presence of protozoa. Mineralization is enhanced even further in the presence of living roots. Apparently, root exudates stimulate bacterial growth which in turn increases the nitrogenous SOM degradation rate and subsequent assimilation of nitrogen into bacterial biomass. Eventually, protozoan populations increase due to increased bacterial numbers and as the bacteria are consumed, their cell N is remineralized more rapidly by the protozoa and taken up by the plant roots. It is safe to assume that mineralization of other nutrients besides N and P is also enhanced by protozoa.

**Table 3.5**  Rate of P uptake by bacteria in grazed and ungrazed systems with *Carex equatilis* litter

| Transfer velocity | Bacteria alone | Bacteria and ciliates |
|---|---|---|
| Water to microorganism | | |
| µg P/L/h | 18.9 | 147.3 |
| µg P/bacterium/h (x10$^{-7}$) | 16.7 | 191.3 |
| *Carex* litter to water and resin bags in water | | |
| µg P/L/h | 0.028 | 0.84 |
| µg P/bacterium/h (x10$^{-7}$) | 0.025 | 1.09 |

*Source:* Barsdate et al. (1974) with permission from Blackwell Publishers, Oxford, United Kingdom.

the presence than in the absence of predators due to the high proportion of young rapidly dividing bacterial cells present in the grazed microcosm.

Protozoa also play an important role in the mineralization of nitrogen and other nutrients. Amoebae feeding on bacteria consistently increased the mineralization of nitrogen (Table 3.6). Plants have also been reported to take up more nitrogen in the presence of both bacteria

**Table 3.6**  Effects of amoebae and nematodes feeding on bacteria on the concentration of $NH_4^+$-N in soil after 14 days of incubation

| Biological treatment | µg $NH_4^+$-N $g^{-1}$ soil |
|---|---|
| Bacteria alone | 20.9 |
| Bacteria with amoebae | 32.7* |
| Bacteria with nematodes | 18.3 |
| Bacteria, amoebae and nematodes | 19.9 |

*Significantly different at the 0.01 level of significance.
*Source:* Woods et al. (1982)  with permission from Elsevier.

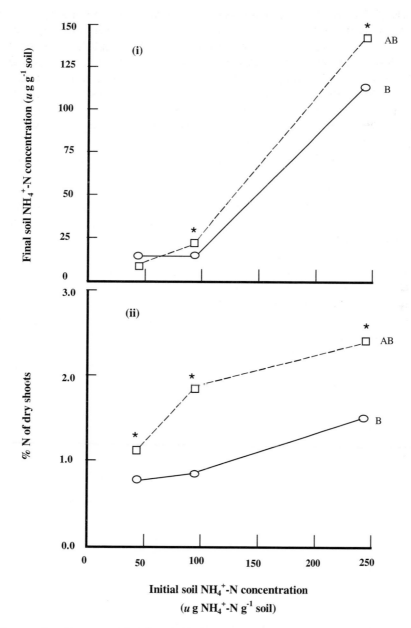

**Fig. 3.5** Final soil ammonium-N concentration (i) and %N of dry shoots (ii) in the presence of only bacteria (B) and bacteria with amoebae (AB) 60 days after growing blue grama (*Bouteloura gracilis*) in clay loam soil microcosm that initially contained 45 μg $NH_4^+$-N $g^{-1}$. The soil was amended with either 50 or 200 μg $NH_4^+$-N $g^{-1}$ soil as ammonium sulfate to provide initial concentrations of 45, 95, and 245 μg $NH_4^+$-N $g^{-1}$. Asterisk (*) represents significant differences at $p<0.05$ (from Elliot et al., 1979 with permission from the authors).

The role of protozoa in regenerating nutrients in marine environments has also been widely demonstrated (Johannes, 1965, 1968). Under marine conditions, most of the nutrients are in algal bodies. The bacteria accumulate nutrients through decomposition. The bacteria are in turn preyed on by protozoa, releasing the nutrients directly into a soluble form for uptake by algae. Johannes (1965) compared the regeneration of $^{32}$P from *Spartina* in the presence of either only bacteria or in the presence of bacteria and a ciliate, *Euplotes vannus*. The dissolved inorganic phosphorus (DIP) was significantly higher, 7 days and thereafter, in the system that had both *E. vannus* and bacteria compared to the system that had only bacteria (Fig. 3.6). With an additional supply of *Spartina* 21 days after the initial supply an even more substantial release of DIP in the presence of both the bacteria and the ciliates was noted compared to when only the bacteria were present, despite the fact that the density of bacteria was always high in the absence of the ciliates. The grazing of bacteria by protozoa maintains the bacterial population in a prolonged state of "physiological youth", thus greatly increasing the rate at which bacteria

assimilate organic material. Some of the regenerated dissolved inorganic phosphorous is attributable to autolysis of the detrital material by the enzymes generated by bacteria and possibly due to the direct excretion of phosphorus by bacteria (Fig. 3.7).

## 3.7 REPRODUCTION IN THE SUCCESSFUL EXISTENCE OF PROTOZOA IN THE ENVIRONMENT

Both sexual and asexual reproduction occur in some protozoa species, while other species reproduce only sexually or asexually. Sexual processes among protozoa are quite diverse as they occur either by conjugation or autogamy. A common factor during sexual reproduction is the production of haploid gametic nuclei, two of which fuse during fertilization. Sexual reproduction is credited for maintaining genetic variation through recombination, thus accelerating the fixation of more favorable alleles. Diversity in sexual reproduction in protozoa and a variety

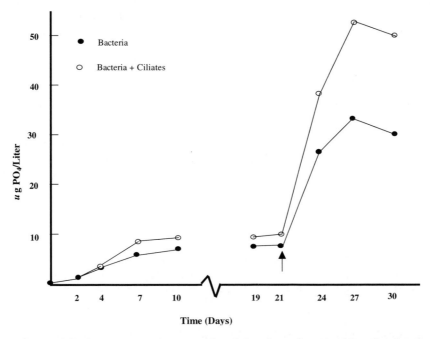

**Fig. 3.6**  Influence of ciliates, *Euplotes vannus,* on regeneration of phosphorus from dead *Spartina. Spartina* addition was at 21 days (from Johannes, 1965 with permission from the American Society of Limnology and Oceanography).

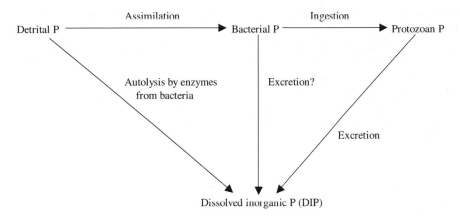

**Fig. 3.7** Pathways for regenerating detrital P in marine environments. Detrital P is assimilated by bacteria which are in turn grazed by protozoa. The protozoa subsequently excrete the P which ends up in the dissolved inorganic fraction (adapted from Johannes, 1965 with permission from the American Society of Limnology and Oceanography).

of other protists is reflected in the fact that the gametes may differ in size and structure between species, and in the fusion of gametic nuclei before or after the gametes differentiate. Sexual reproduction in protozoa can be almost as plastic as that in prokaryotes. Most protozoa are haploid just like prokaryotes but a few, especially the ciliates such as *Tetrahymena*, *Paramecium,* and *Euplotes* sp., are diploid. Conjugation in ciliates is highly specialized and involves the pairing of two protozoan cells, a haploid micronucleus from each cell migrating to the other cell. The haploid micronucleus fuses with the host cell's micronucleus, forming a zygote nucleus. Each of the now fertilized cells divides into two daughter cells with recombined DNA. However, the two exconjugants that are generated are diploid and genotypically identical. Thus, despite the same name, conjugation in protozoa is quite different from conjugation in prokaryotes. Autogamy is the process whereby the gametes are all produced by one parent cell and therefore have the same genetic make-up. Unlike conjugation, autogamy leads to homozygosity at all loci (Dini, 1984).

Ciliates, for example *Paramecium aurelia*, can reproduce sexually through conjugation or autogamy (Sleigh, 1989). This may not be unique to *P. aurelia* alone and whatever determines which type of sexual reproduction occurs is not known for certain but environmental conditions such as starvation, pH extremes, heavy metals, higher temperatures, age of the members of the clone, and chemical characteristics of the surface membranes may play a role (Dini, 1981; Sleigh, 1989). Sexual processes in protozoa are known to occur under specific circumstances. For example, flagellates in the hind guts of roaches and termites are predominantly haploid and divide mitotically. However, they can develop sexual stages in molting insects, the sexual stages being stimulated by the host insect's molting hormone (ecdysome) (Sleigh, 1989). Some ciliates and flagellates excrete water-soluble "gamones" which attract mating types (van Houten et al., 1981). Diploidy is common among binucleate ciliates.

Many groups of protozoa, e.g. most of the lobose amoebae and several flagellates such as choanoflagellate, euglenids, and chrysomonads and the kinetoplastids lack sexual processes and reproduce by binary fission, multiple fission (schizogony), or budding. Binary fission occurs when a constriction ring gradually bisects the protozoan cell, producing a cleavage furrow, eventually yielding two daughter cells. The nucleus undergoes multiple divisions before the cell divides into multiple progeny. Budding, a specialized type of fission, occurs when nuclear division produces daughter nuclei. It occurs when a smaller cell is released from the surface of the parent cell, culminating into a progeny nucleus. Each of the daughter nuclei migrates into a cytoplasmic bud that is eventually released. Whether sexual or asexual, reproduction initially involves some division of the nucleus in mitosis and, in the case of sexual reproduction, mitosis is then followed by meiosis. The division results into a doubling of the amount of DNA. The extensive possibilities that occur under sexual and asexual, as well as the fact that the same protozoan species can perpetuate its progeny sexually, asexually or through both means, greatly contribute to the diversity and success of protozoa as a group in the environment.

# References

Abu Kwaik Y., L-Y. Gao, B.J. Stone, C. Venkataramna, and O.S. Harb (1998). Invasion of protozoa by *Legionella pneumophila* and its role in bacterial ecology and pathogenesis. *Appl. Environ. Microbiol.* **64:** 3127-3133.

Anderson O.R. (2000). Abundance of terrestrial gymnoamoebae at a northeastern US site: A four-year study, including the El Niño winter of 1997-1998. *J. Eukaryotic Microbiol.* **47:** 148-155.

Anscombe F.J. and Singh B.N. (1948). Limitation of bacteria by micro-predators in soil. *Nature* **161:** 140-141.

Barsdate R.J., R.T. Prentki, and T. Fenchel (1974). Phosphorus cycle of model ecosystems: significance for decomposer food chains and effects on bacterial grazers. *Oikos* **25:** 239-251.

Bass P. and P.J. Bischoff (2001). Seasonal variability in abundance and diversity of soil gymnoamoebae along a short transect in southwestern USA. *J. Eukaryotic Microbiol.* **48:** 475-479.

Bischoff P.J. and O.R. Anderson (1998). Abundance and diversity of gymnoamoebae at varying soil sites in northern U.S.A. *Acta Protozologica* **37:** 17-21.

Caron D.A. (1987). Grazing of attached bacteria by heterotrophic microflagellates. *Microbial Ecol.* **13:** 203-218.

Clarholm M. (1981). Protozoan grazing of bacteria in soil: impact and importance. *Microb. Ecol.* **7:** 343-350.

Clarholm M. (1985). Possible roles for roots, bacteria, protozoa and fungi in supplying nitrogen to plants. *In*: A.F. Fitter (ed.). Ecological Interactions in Soil. Blackwell, Oxford, UK, pp. 355-365.

Curds C.R. (1973). The role of protozoa in the activated sludge process. *Ameri. Zool.* **13:** 161-169.

Curds C.R. and A. Cockburn (1968). Studies of the growth and feeding of *Tetrahymena pyroformis* in axenic and nonaxenic culture. *J. Gen. Microbiol.* **54:** 343-358.

Dini F. (1981). Relationship between breeding systems and resistance to mercury in *Euplotes crassus* (Ciliophora: Hypotrichida). *Mar. Ecol. Prog. Ser.* **4:** 195-202.

Dini F. (1984). On the evolutionary significance of autogamy in the marine *Euplotes* (Ciliophora: Hypotrichida). *Amer. Natur.* **123:** 151-162.

Ekelund, F. and R. Rønn (1994). Notes on protozoa in agricultural soil with emphasis on heterotrophic flagellates and naked amoebae and their ecology. *FEMS Microbiol. Revi.* **15:** 321-353.

Elliot E.T. and D.C. Coleman (1977). Soil protozoan dynamics in a shortgrass prairie. *Soil Biol. Biochem.* **9:** 113-118.

Elliot E.T., D.C. Coleman, and C.V. Cole (1979). The influence of amoebae on the uptake of nitrogen by plants in gnotobiotic soil. *In*: J. Harley and R.S. Russell (eds.). The Soil-Root Interface. Acad. Press, New York, NY, pp. 226-229.

Fenchel T. (1987). Ecology of Protozoa: The Biology of Free-living Phagotrophic Protists. Science Tech. Publishers, Madison, WI.

Fenchel T. and B.J. Finlay (1983). Respiration rates in heterotrophic, free-living protozoa. *Micro. Ecol.* **9:** 99-122.

Finley B.J. (1985). Nitrate respiration by protozoa (*Loxodes* spp.) in the hypolimenetic nitrite maximum of a reproductive freshwater pond. *Freshwater Biol.* **15:** 333-346.

Finley B.J. (1990). Physiological ecology of free-living protozoa. *Adv. Micro. Ecol.* **11:** 1-35.

Fritsche T.R., D. Sobek, and R.K. Gautom (1998). Enhancement of in vitro cytopathogenicity by *Acanthamoeba* spp. following acquisition of bacteria endosymbionts. *FEMS Microbiol. Lett.* **166:** 231-236.

Habte M. and M. Alexander (1978). Mechanisms of persistence of low numbers of bacteria preyed upon by protozoa. *Soil Biol. Biochem.* **10:** 1-6.

Jjemba P.K. (2001). The interaction of protozoa with its potential prey bacteria in the rhizosphere. *J. Eukaryotic Microbiol.* **48:** 320-324.

Johannes R.E. (1965). Influence of marine protozoa in nutrient regeneration. *Limnol. Oceanog.* **10:** 434-442.

Johannes R.E. (1968). Nutrient regeneration in lakes and oceans. *In*: M.R. Droop and E.J. Ferguson Wood (eds.). Advances in Microbiology of the Sea. Acad. Press, New York, NY, Vol.1, pp. 203-213.

Lee J.J. (1980). Nutrition and physiology of the Foraminera. *In*: M. Levandowsky and S.H. Hutner (eds.). The Biochemistry and Physiology of Protozoa. Acad. Press, New York, NY, Vol. 3, pp. 43-66. (2nd ed.).

Mueller M., P. Röhlich, and I. Törö (1965). Studies on feeding and digestion in protozoa, VII. Ingestion of polystyrene latex particles and its early effect on acid phosphatase in

*Paramecium multimicronucleatum* and *Tetrahymena pyriformis. J. Protozool.* **12:** 27-34.

Naitoh Y. and R. Eckert (1974). The control of ciliary activity in protozoa. *In:* A.M. Sleigh (ed.) Cilia and Flagella. Acad. Press, New York, NY, pp. 305-352.

Naitoh Y. and K. Sugino (1984). Ciliary movement and its control in *Paramecium. J. Protozool.* **31:** 31-40.

Nisbet B. (1984). Nutrition and Feeding Strategies in Protozoa. Croom Helm Ltd., London, UK.

Preston T.M., Davies D.H., and King C.A. (1982). Surface binding and subsequent capping of certain bacteria by *Acanthamoeba. J. Protozool.* **29:** 318 (abstract).

Pussard M., C. Alabouvette, and P. Levrat (1994). Protozoan interactions with the soil microflora and possibilities for biocontrol of plant pathogens. *In:* J.F. Darbyshire, (ed.). Soil Protozoa. CAB International, Wallingford, UK, pp. 123-146.

Ramirez C. and Alexander M. (1981). Increased bacterial colonization of the rhizosphere by controlling the soil protozoa in the *Rhizobium*-legume system. *In:* P.B. Vose and A.P. Ruschel, (eds.). Associative $N_2$-Fixation. CRC Press, Boca Raton, FL (USA). Vol. I, pp. 69-86.

Rasmussen L. and E. Orias (1975). *Tetrahymena*: growth without phagocytosis. *Science* **190:** 464-465.

Seto M. and Tazaki R. (1971). Carbon dynamics in the food chain system of glucose-*Escherichia coli-Tetrahyemena vorax. Jpn. J. Ecol.* **21:** 179-188.

Singh B.N. (1941). Selectivity in bacterial food by soil amoebae in pure mixed culture and in sterilized soil. *Ann. Appl. Biol.* **28:** 52-64.

Singh B.N. (1942). Selection of bacterial food by soil flagellates and amoebae. *Ann. Appl. Biol.* **29:** 18-22.

Singh B.N. (1945). The selection of bacterial food by soil amoebae, and the toxic effects of bacterial pigments and other products on soil protozoa. *Brit. J. Exper. Path.* **26:** 316-325.

Singh B.N. (1948). Studies on giant amoeboid organisms,1.The distribution of *Leptomyxa reticulata* Gooday in soils of Great Britain and the effect of bacterial food on growth and cyst formation. *J. Gen. Microbiol.* **2:** 8-14.

Singh B.N. (1964). Soil protozoa and their probable role in soil fertility. *Bull. Natl. Inst. Sci. India* **26:** 238-244.

Singh B.N., Mathew S., and Anand N. (1958). The role of *Aerobacter* sp., *Escherichia coli* and certain amino acids in the encystment of *Schizopyrenus russelli. J. Gen. Microbiol.* **19:** 104-111.

Sleigh M.A. (1989). Protozoa and other Protists. Arnold, Publ., New York, NY.

Stout J.D. (1980). The role of protozoa in nutrient cycling and energy flow. *Adv. Microb. Ecol.* **4:** 1-50.

Stout J.D. and O.W. Heal (1967). Protozoa. *In:* A. Burges and F. Raw (eds.). Soil Biology. Academic Press, New York, pp. 149-195.

Van Houten J., D.C.R. Hauser, and M. Levandowsky (1981). Chemosensory behaviour in protozoa. *In:* M. Levandowsky and S.H. Hutner (eds.). Biochemistry and Physiology of Protozoa Acad. Press, New York, NY, Vol. 4, pp. 67-124 (2nd ed.).

Winiecka-Krusnell, J. and E. Linder (1999). Free-living amoebae protecting legionella in water: The tip of an iceberg? *Scand. J. Inf. Dis.* **31:** 383-385.

Woods L.E., C.V. Cole, E.T. Elliot, R.V. Anderson, and D.C. Coleman (1982). Nitrogen transformation in coil as affected by bacterial-microfaunal interactions. *Soil Biol. Biochem.* **14:** 93-98.

Zwart K.B. and Darbyshire J.F. (1992). Growth and nitrogenous excretion of a common soil flagellate, *Spumella* sp., —a laboratory experiment. *J. Soil Sci.* **43:** 143-151

# 4

# Fungi, Algae and Their Association

Up to the end of the eighteenth century, algae, fungi and lichens were crudely classified together. Combining these three in the same chapter herein should not be viewed as a step back in time and create the misconception that they are taxonomically related. Rather, they are combined in a single chapter to serve as a basis for discussing their similarities and the culminating association—the lichens. Slime molds have properties that are similar to those of fungi and have always been treated as such (since they are more widely studied by mycologists). They are multicellular filamentous organisms such as rusts, mildews, and smuts. However, based on rRNA phylogeny, they appear not to be closely related to fungi nor to either of the other two kingdoms, i.e., animals or plants (Hasegawa et al., 1993). Slime molds live in a manner similar to fungi and protozoa as they form amoebalike structures (see Section 3.1). Two types of slime molds are known, cellular and acellular. Cellular slime molds resemble amoeba and feed by ingesting fungi and bacteria by phagocytosis. They reproduce asexually by initially aggregating to form a slug, a yellowish amorphous fruiting body which in turn forms spores. The spores in the fruiting body are then released to germinate into amoebae-like structures. Acellular slime molds are multinucleated amoebalike organisms which engulf organic debris and bacteria. They reproduce both sexually and asexually. Slime molds

are not discussed further in this chapter. Some slime molds are discussed, however, in Chapter 12 because of their economic importance in aerobiology and airborne plant diseases.

## 4.1 FUNGI

### 4.1.1 Characteristics of fungi

Fungi are tubelike eukaryotic osmotrophs which lack chlorophyll, have a mycelium (pl. mycelia) and hyphae. Mycologists separate them into higher and lower fungi. The higher fungi are mostly found in terrestrial environments and many tend to have a symbiotic or parasitic relationship with plants. Hence it appears that terrestrial plants coevolved with higher fungi. The lower fungi, on the other hand, are mostly found in water and have fewer hyphae with a septum (pl. septa), the cross-walls within the hyphae (Fig. 4.1a). A few classes of fungi have hyphae with no septa, which thus appear like long continuous cells with many nuclei. These are called coenocyctic hyphae (Fig. 4.1b). However, even when a septum is present, it has a pore which allows for continuous mixing of fungal cytoplasm between fungal "cells". Thus, technically, the presence of a pore in the septum leaves open the question as to whether the individual septate compartments represent cells since it enables the cytoplasm and nuclei from different compartments to migrate and mix. Microscopic and biochemical studies show that the mixing of

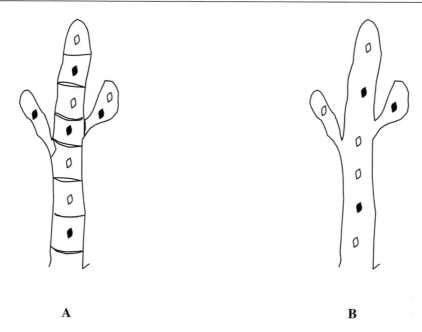

**A**                    **B**

**Fig. 4.1** Types of fungal hyphae. (A) Septate with cross-walls dividing the hypha into cell-like units. (B) Coenocytic hypha without cross-walls.

organelles and fluids is highly regulated, just as in cells (Moore, 1998). Some fungi have cytoplasm that is interconnected (coenocytic), a feature very similar to slime molds. Their multinucleate nature is not known to occur naturally in any other organisms except a few algal groups. Many of these nuclei within the same cytoplasm are of different genetic composition. The septa have pores which facilitate the flow of cytoplasm within the hyphae, an attribute which provides interconnectedness and continuity throughout the thallus. This continuity provides for many nuclei in a common cytoplasm. Fungi with septa normally produce nonmotile airborne spores. This multinucleate state allows for nuclear-to-nuclear and nuclear-to-cytoplasm interactions which are largely unique, providing unprecedented variability and diversity in fungi (Ross, 1979). The intraorganismal nuclear and cytoplasmic interactions have enabled fungi to develop a multiplicity of forms and developmental patterns and to invade and survive on nearly every substrate in a wide range of earthly environments.

Fungi are radial in growth and the hypha extends at the tip, actively colonizing the solid or semisolid substrate. Furthermore, the surface-area-to-volume ratio of a cylinder increases with decreasing diameter. Thus, the smaller and finer the hyphae, the more effective the fungus remains in contact with the substrate. Fungi also have a high surface-area-to-mass ratio. Also, the hyphae branch continuously, thus colonizing the substrate with minimal biomass accumulation (Park, 1985a, b). In terms of total mycelial length, growth in fungus is exponential, just as in prokaryotes. Hyphae branching is initiated when the mean volume of the cytoplasm per hyphal tip exceeds a particular critical value (Trinci, 1974). The materials necessary for the extension (growth) of the hyphae apex are produced throughout the mycelium and translocated toward the growing tip. The rate at which the peripheral growth zone (Kr) expands is estimated based on the mathematical formula

$$Kr = \mu w \qquad \qquad ...(1)$$

where $\mu$ is the specific growth rate in liquid culture and w the width of the center of the colony in which growth has ceased either due to depletion of nutrients or negative autotrophism since hyphae tend to avoid their neighbors. A hypha apex approaching another apex will turn away, possibly to avoid limitation by oxygen concentrations.

Fungi are identified by the type of spores. The spores are either sexual or asexual and they are formed from aerial mycelia in a variety of ways, depending on the species. Asexual species are formed by the myxocellum of one organism, whereas asexual spores result from the fusion of nuclei, two mating strains of the same species. Several types of asexual spores are known: sporangiospores, arthrospores, and conidiospores. Sporangiospores develop inside a structure, the sporangium, that is usually a differentiated apex of some specialized hyphae. They may be self-motile (zoospores) through water by flagella or nonmotile, thus relying on outside help for dissemination. Conidiospores are asexual spores which are unicellular or multicellular and are not enclosed in a sac. They are differentiated at the tips of specialized hyphae and are always nonmotile, relying on external means such as wind, water (rainsplash), animals, or built-in explosive mechanisms for dissemination. Sexual spores (ascospores) are often produced only under special circumstances, for example, changes in seasons, temperature, moisture content and nutrient availability. Many fungi have both a sexual and an asexual phase, usually separate in time and space but often occurring simultaneously.

Fungi are classified into five phyla: Ascomycota, Basidiomycota, Deuteromycota (Fungi Imperfecti), Chyridiomycota, and Zygomycota (Table 4.1). Classification of Chyridiomycota has only been resolved fairly recently using ribosomal RNA phylogenetic analysis (Bowman et al., 1992). Because of their flagellated nature, they had wrongly been characterized as protists.

Formerly, fungi were repeatedly referred to as lower plants and were implicitly assumed to conform to the behavior of plants. This concept turned out to be quite misleading and actually slowed our understanding of this important group. Whereas fungi possess some feature of both animals and plants, they also have many features that are unique to themselves. Five characteristics delimit what constitutes a fungus, three of which, namely lack of chlorophyll, eukaryotic composition, and heterotrophism, are also found in all animals and in some plants. The fourth characteristic prevalent in most fungi is the possession of more than one nucleus (multinucleate) in a common cytoplasm. A fifth characteristic in fungi is the ability to obtain food by absorbing soluble nutrients from their environment, a characteristic fungi share with prokaryotes and some other eukaryotes (algae and some protozoa). In this respect, both fungi and prokaryotes are unique since they must digest their substrates externally prior to absorbing the smaller molecules. This mode of nutrition exposes the substrate to competition by other organisms in the vicinity until it is internalized. Considering how successful and widespread both prokaryotes and fungi are in the environment, the circumstances to which their substrates are exposed to uptake by other organisms in the immediate vicinity may have compelled them to evolve extracellular mucilages, membranes, extracellular enzymes and cell walls to improve their competitive ability in nature.

Before the advent of 16S and 18S rRNA chronometry, the chronological fossil record of fungi was skimpy. Currently, the major criteria used to establish relationships within fungi mostly depend on the few stable characteristics of contemporary species. This approach caused mycologists considerable difficulty because fungi are inherently quite variable. This variability is complicated by the fact that many morphological features of fungi are affected by particular environments. It is common for the same fungal species growing on several host plants,

**Table 4.1** Fungal Classification

| Phylum | Characteristics | Examples |
|---|---|---|
| Ascomycota | Possess ascospores which are located inside a specialized fruiting body (ascus). Most of them are filamentous with perforated septa in the hyphae. The perforation is not centrally located. Each hyphal compartment complimentary haploid nuclei (dikaryotic cell). The walls are multi-layered. Their sexual spores are usually conidiophores produced in long chains otherwise called club fungi. A = ascospores (sexual) and C = conidiophores (asexual). | *Neurospora crassa, Eupeni cilium, Aspergillus* sp. |
| Basidiomycota | Produce basidiospores which are usually non-septate. Where septate, the sept are perforated. Most are filamentous but can reproduce sexually to form mushrooms as fruiting bodies with a basidia that suspends the basidiospores. Similar to ascomycota, each hyphal compartment contains two complimentary haploid nuclei. | *Saccharomyces cerevisiae, Cryptococcus* sp. |
| Zygomycota | The conjugation fungi. Saprophytic with a coenocytic hypha. Filamentous Produce resting spores called zygospores (large spores enclosed in a thick wall) as a sexual state. Lack perforations in the septa. S = sporangium and Z = zygospore. | *Rhizopus nigricans* |
| Chytridiomycota | Possess flagella just like protists. | *Blastocladiella emersonii, Chytridium confervae, Spizellomyces acuminatus* |
| Deuteromycota (Fungi imperfecti) | Produce asexual or sexual conidiophores, chlamydospores, and orthospores. Have perforated chlamydospores and orthospores. Have perforated septa. This phylum is described as a holding category for fungi to which no sexual spores have been observed. Thus, called fungi imperfecti. | *Cladosporium* sp., *Spirothrix* sp., *Pneamocystis carnii* |

for example, to possess different forms and structures on each host—a difference which may mislead one to classify this same species under a different family or genus. Under different environmental or nutritional conditions, many fungi can also pass through dissimilar developmental pathways but end up as identical mature forms.

From an evolutionary standpoint, the similarity between bacteria and fungi is even more profound for bacteria of the actinomycetes group. Actinomycetes, like fungi, have hyphae. However, the hyphae of fungi and actinomycetes differ distinctly. Those of actinomycetes are much finer (1-2 μm) in diameter and nonnucleated as opposed to the 5-10 μm nucleated fungal hyphae. Hence fungi have evolved while actinomycetes and other bacteria are still evolving from their simple model. Fungi are very diverse. As a matter of fact, they are more diverse than plants and animals. Analysis based on the 16S and 18S rRNA phylogeny shows that these three eukaryotic kingdoms—animals, fungi and plants—are very closely related (see Fig. 1.2). There are similarities between these three kingdoms in complex biosynthetic pathways including syntheses of hydroxyproline, chitin, ferritin, and cellulose (Wainright et al., 1993). The evolutionary relationship is especially tighter between fungi and animals which branched off from a common ancestor that is more recent than either did with the plant lineage (Fig 1.2).

Fungi characteristically interact with other organisms, particularly plants. Almost all of the terrestrial plant species form some mutualistic association with fungi. Most notable of these associations is mycorrhiza which contributes to the plants' mineral nutrition. Fossilized fungal-plant associations have been encountered in geographically widespread regions (Moore, 1998). The nutritional diversity of fungi may have eased the nutritional problems encountered by the first terrestrial plants during the course of evolution.

## 4.1.2  Fungal biology

Fungal biology is used mostly for grouping purposes. The basic vegetative body of the fungus (*thallus*) is essentially a multinucleate protoplast enclosed in a rigid wall. In fungi, the cell wall is composed of chitin, glucan, and cellulose-long chain polymers of *N*-acetylglucosamine or glucose. Under the electron microscope, the walls typically have a fibrous appearance. At the molecular level, the walls are composed of adhering microfibril long chain molecules, either branched or unbranched. Initially, the wall is extensible to accommodate the increasing size of the protoplast growing inside but becomes rigid and inextensible at maturity. Fungi germinate from a single uninucleate spore which develops into a multinucleate mycelium. Many fungal species produce clones, mainly by asexual sporulation, which may become geographically widespread.

All nutrients that enter the fungus pass through the cell wall. Most developed for absorbing and assimilating food materials is the thallus. The thallus is a mycelium with some hyphae and includes fruiting bodies. It also functions in reproduction by either becoming a reproductive structure itself or providing a foundation from which reproductive structures develop. Growth of the thallus is usually by expansion in all directions at once or by apical extension of branching cylindrical filaments, the hyphae. The hyphae usually form a foraging network called mycelium. In some fungi, notably the Basidiomycotina, Ascomycotina, and Deuteromycotina (Imperfecti Fungi), the hyphae have localized cross walls (septa) which are rarely complete but rather are perforated so that cytoplasmic continuity from one segment to the other still occurs (see Table 4.1). Most of these perforations are still large enough to allow the passage of cell organelles such as mitochondria and possibly the nuclei (Ross, 1979). Although the higher fungi, particularly the higher Basidiomycotina, possess highly regulated hyphae with only two nuclei per septation ("cell"), in most fungi the number of nuclei in any one septation is often nonuniform even though the size of the septa may be quite uniform.

In higher fungi such as ascomycetes and basidiomycetes, adjacent hyphae of the same species have a strong propensity to fuse. As a result, many mycologists considered the thalli of these fungi as physiologically cooperative mosaics regardless of whether the component hyphae are genetically similar or different. More recent reports indicate that genetically different mycelia are often mutually antagonistic and hyphal fusions attempted between them usually break down (Brasier, 1992).

### 4.1.3 Ecological importance of fungi

The physiological ability of fungi makes them extremely important in the environment in four major ways, namely:

1. Decomposition of organic materials which, together with bacteria and protozoa, recycle C, N, and a whole range of other elements including metals and metalloids. Thus, they play an important role in the mineralization of C and other biogeochemical cycles. Through the use of extracellular enzymes, such as celluloses, hemicelluloses and pectinases, fungi are the primary decomposers of the robust plant materials.
2. Causing diseases and destruction to plants (crops) and animals, thereby impacting the social and economic life of populations through deaths, food shortages, and loss of products in storage. Virtually every important crop is attacked by one or more types of fungi.
3. Production of a wide range of products due to the numerous peculiar metabolic pathways unique in fungi. Products attributed to fungi include antibiotics, alcohol, enzymes, organic acids, bread-making, and cheese-ripening to name but a few.
4. Use of fungi in experimental biology to study important processes such as morphogenesis, gene regulation, nucleic

acid and protein synthesis, thus providing an avenue for further understanding these processes in eukaryotes, particularly in comparison with what is learnt from prokaryotes, so as to obtain a sense of whether some of those findings apply in "higher organisms".

Although fungi have acquired notoriety as agents of disease and producers of psychedelic toadstools, they play an important role in nutrient cycling through decomposition of organic matter and the uptake of nutrients into plants via mycorrhizae. Fungi are generally considered to lack sufficient metabolic diversity to allow them to play an important role in cycling nutrients in the environment. However, nutrient cycling should not be viewed only from the viewpoint of the element being transformed in relation to its increased availability for plant growth. Rather, it should also be viewed in terms of the benefits to the organisms involved in cycling the nutrient. A most obvious benefit—often ignored out of hindsight—is the supply of energy to the organisms involved in the cycling of nutrients.

Mycorrhizae associate with roots and provide a mechanism by which plants enhance their ability to access nutrient from soil. Mycorrhizae and their attributes have been documented worldwide (Lapeyrie et al., 1991; Read, 1991; Bagyaraj and Varma, 1995). Mostly phosphorus and water are channeled through the mycorrhizae hyphae. The concentration of soluble P in soils is usually in the range of only 0-1.2 mg$^{-1}$L (Barber, 1995). Phosphorus is also one of the most immobile nutrients in soil because it often combines with aluminum and iron oxides, especially in highly weathered tropical and subtropical soils, to form insoluble complexes (Sanchez, 1976). Most plants have a high affinity uptake system for P, thus taking it up more readily than it diffuses into solution, a situation that often creates a P depletion zone around the roots. Thus, the sustained P uptake requires continuous growth of the roots into new areas that are not P-depleted in the immediate vicinity of the root zone. To maximize P uptake, plants associate with mycorrhizae. The mycorrhizae are ideal for this purpose because they are fast growing and small in diameter, which provides for a high surface area-to-volume ratio that is favorable for absorption. Root hairs also provide a high surface area-to-volume ratio but can only grow to a certain extent compared to mycorrhizae. Thus, to continuously explore new ground, plants rely on mycorrhizal fungi so as to ensure a high rate of acquisition of P, water, and other nutrients such as $NH_4^+$, which is also relatively immobile in soil.

In acid soils where P is even more unavailable due to fixation, ericoid fungi directly mineralize P by releasing phosphatases which are directly absorbed by both the fungus and the plant roots (Read, 1983). This is why ericoid mycorrhizae are also particularly prevalent in acid soils. The contribution of mycorrhizae is significant in the growth and survival of conifers, orchids, etc., which produce millions of very tiny seeds without much stored nutrients (see Chapter 8). In order to survive, these seeds have to have mycorrhizae early in their germination to provide them nutrients. This natural process is not always perfect and farmers have worked around this by almost always avoiding propagating some agricultural plants with tiny seeds (e.g. orchids) from seed. Most studies of mycorrhizal fungal effects have been done on plants singularly grown in pots. In the field, however, the significance is more complex as a single mycorrhizal network might infect more than one plant type, thereby facilitating the flow of photosynthates between plants as well. Other types of fungi, such as *Penicillium bilaii*, have also been implicated in solubility phosphates in soil (Cunningham and Kuiack, 1992).

Besides P, other nutrients which are held and availed of in the fungal biomass include Ca, K, N, Mg, and sulfate (Lodge, 1993; Gharieb et al., 1998). The fungal biomass, at least in tropical and subtropical environments, is directly related to soil moisture, under which nutrients are more liable to leaching. Immobilization of nutrients in fungal and other microbial biomass may provide a reduction in loss of nutrients through leaching.

## 4.1.4  Distribution

The ecology of fungi is important to the functioning of the whole ecosystem as they are dominant microorganisms in all terrestrial ecosystems. They are major saprophytes, symbionts and pathogens. The potential for fungus to spread and establish where conditions are favorable is enormous. It is possible to have about 10 fruiting bodies in a 5 $m^2$ area, with each fruiting body distributing millions of spores per hour. Fungi occur in both aquatic and terrestrial environments but their distribution is limited by temperature. Compared to prokaryotes, very few types of fungi inhabit areas on the high end of the temperature scale. However, fungal spores can exist at fairly high temperatures, a mechanism that enables them to overcome both wet and dry hot climatic cycles. Determination of reproductive success for any fungal individual is difficult because of the difficulty in tracking the dispersal and establishment of spores. Spores are readily distributed by wind, water, and animals. Their ability to spread extensively by wind has important implications for the quality of air, as discussed in Chapter 11 dealing with bioaerosols.

Although fungi are perceived as spatially discontinuous and ephemeral, some basidiomycetes clones can exceed 500 m in diameter. In one particular case, an individual of the fungus *Armillaria bulbosa* was found to occupy a minimum of 15 hectares and weighed more than 10 tons. At this size and weight, it is one of the largest living organisms, rivaling the blue whale and the giant redwood (*Sequoiadendron giganteum*). Using the estimate that its rhizomorphs grow at a rate of about 0.2 m per year, with a length of 635 m across the area it covers, this fungus has been estimated to be at least 1,500 years old (Smith and Bruhn, 1992). Despite its enormous size, this particular clone has surprisingly remained genetically conserved as shown by random amplified polymorphic DNA (RAPD) analysis. Fungal succession is unique due to the indeterminate growth and the spread of fungal mycelia. Unlike plant communities in which succession is predictable and tends, under a stable environment, to build up climax communities and have a very tight and conservative nutrient circulation with minor leakages, increase in fungi species diversity during succession leads to an increased release of nutrients from the system.

Among fungi, two types of succession are recognized, namely substratum (or resource) succession and seral succession (Södeström, 1996). Substratum succession refers to the succession of species that occurs on any decomposable material, for example cellulose, wood, or leaves. This type of succession is usually accompanied by degradation of the resource, leaving only the more slowly mineralizable material. Seral succession, on the other hand, occurs in a developing ecosystem such as a forest in different phases of its establishment. In such a system, succession of the fungal community will follow the development climax of the developing plant community. Needless to say, seral succession spells long-term studies.

Fungi may interact either between themselves or with other hosts. Such interactions may be mutualistic (where the partners in the relationship may be advantaged), unaffected (neutralistic or commensal) or antagonistic (parasitic) in nature. Some fungal associations are discussed in Section 4.2.4 and interactions are widely explored in Chapter 8. Numerous fungi are parasites of animals, plants or both, a characteristic which makes them economically very important. *Phytophora infestans,* a fungus responsible for the great potato blight in Ireland in the early 1800s, was one of the first microorganisms to be associated with disease. Fungal infections are classified as subcutaneous, cutaneous, systemic, opportunistic, or superficial, depending on the degree of tissue involved and the mode of entry into the host. A few interesting cases of fungal parasitism against humans include athlete's foot (*Tinea pedis*) caused by various fungal agents, in particular *Trichophyton rubrum* and *Epidermophyton floccosum*. The causative agents have a niche on our skin and are always present. However, if one does not take the necessary routine sanitary care, like showering, the chances of the fungal agents developing into a clinical or subclinical but nonlethal infection are increased. A number of fungal diseases such as *Candida albicans, Aspergillus fumigatus, Cryptococcus neoformans, Fusarium* spp., *Histoplasma cepsulatum,* and *Coccidioides immitis* are of increasing concern, particularly among immunocompromised individuals. The infections emerge in altered hosts under permissive environmental conditions and selective antifungal pressure. Several fungi, such as *Pythium* sp., *Xanthomonas* sp., *Ceratomystis ulmi*, are also known to cause diseases in plants. Overall, this brief discussion demonstrates how fungi and fungal activity are intertwined with our daily life.

### 4.1.5 Environmental adaptability

To survive unfavorable periods, for example winter, dry seasons, wildfires, or a low density of host populations, fungi adapt inactive resting stages, namely rhizomorphs, conidia, and spores. *Armillaria* rhizomorphs are known to withstand aboveground fires of extremely high temperatures (Smith and Bruhn, 1992). The resting spores tend to have a double wall that is thought to enhance their resistance to environmental extremes. Research in this area is aimed at understanding what breaks off such dormancy. In many instances, when the conditions favoring germination of the resting spores are restored, a few spores initially germinate, this percentage only increasing with time. Unlike bacterial endospores which solely function for survival in adverse conditions, fungal spores also serve for reproductive purposes. The germinating asexual spore produces an organism that is genetically identical to the parent.

A variety of fungal species are either tolerant or resistant to high levels of heavy metals such as mercury, cadmium, zinc, and copper and may become dominant organisms in some polluted sites (Gadd, 1986). Tolerance to heavy metals is reportedly exhibited by fungi from polluted and unpolluted habitats, suggesting that in many cases survival at high heavy metal levels is an intrinsically based property rather than an adaptive change. Uptake of metals by fungi is effected by surface binding and/or intracellular influx mechanisms, the former being independent of metabolism while the latter is energy dependent. The uptake of heavy metals by fungi has a potential for industrial applications such as transforming metals, metalloids, and organometalloids. Some transformation processes such as methylation, reduction, and dealkylation, are of major environmental significance as they modify the toxicity and/or mobility of some metals (Table 4.2). For example, mercury (II) can be reduced to metallic mercury (0) by *Candida albicans* and *Saccharomyces*

*cerevisiae*. The methylated products of metalloids are often volatilized. This mechanism has particular potential in remediating sites contaminated with specific metals. Several types of fungi also have the propensity to sorb radioactive substances. *Aspergillus* sp., *Alternaria tenulis, Chaetomium distortium, Fusarium* sp., *Mucor meihei, Penicillium* sp., *Rhizopus arrhizus, Trichiderma herzianum,* and *Zybgorenchus macrocarpus* are some of the many fungi used at the industrial level to treat waste water from radionuclide plants (Lloyd and Macaskie, 2000). The chitin and chitosan components of the fungal wall are responsible for the biosorption of these metals.

Soil provides the matrix for maintaining a natural reservoir of many fungi. For many fungal species, the soil environment also provides a medium for growth and potential dispersal. Both biotic and abiotic conditions in the soil influence the survival and activity of fungi. These conditions can be used to control pathogenic fungal infections. For example, high tempera-

**Table 4.2**   Some metal transformations attributed to fungi in nature

| Mechanism | Metal transformation | Responsible fungi |
|---|---|---|
| Reduction | Mercury (II) to metallic mercury (0) | *Candida albicans, Sacchromyces cerevisiane* |
| | Silver (I) to metallic silver(0) | *Sacchromyces cerevisiane* |
| | Copper (II) to Copper (I) | *Debaromyces hansenii* |
| | Selenium (IV) and (VI) to metallic selenium (0) | Several species of *Fusarium* |
| | Tellurite ($TeO_3^{2-}$) to metallic tellurite [Te(0)] | *Schizosaccharomyces pombe* |
| Dealkylation | Trybutyltin oxide to Sn(II) | |
| | Trimethyl lead | *Phaeolus schweintzii* |
| | Organomercury to Hg(II) which in turn reduces to Hg(0) | *Penicillium* sp. |
| Methylation | Monomethylasenic acid to trimethylarsine | *Candida humicola, Penicillium* sp., *Gliocladium roseum* |
| | Selenite ($SeO_3^{2-}$) and selenate ($SeO_4^{2-}$) to dimethylselenide [$(CH_3)_2Se$] and dimethyldiselenide [$(CH_3)_2Se_2$] | *Penicillium* sp., *Alternaria alternata* |
| | Tellurite ($TeO_3^{2-}$) and Tellurate ($TeO_4^{2-}$) to dimethyltelluride [$(CH_3)_2Te$] and dimethylditelluride [$(CH_3)_2Te_2$] | *Penicillium* sp. |

Compiled from Orsler and Holland (1982), Gadd and Sayer (2000)

**Table 4.3** Deuteromycotina fungi with potential or in use as microbial control agents of insects

| Entomopathogenic fungus | Hosts |
| --- | --- |
| *Metarhizium anisopliae* | Spittlebugs, planthoppers, beetles, mosquitoes, and termites |
| *Hirsutella thompsonii* | Planthoppers and mites |
| *Beauveria bassiana* | Caterpillars, beetles, mites, grasshoppers, true bugs, mosquitoes, termites, planthoppers, and ants |
| *Verticillium lecanii* | Aphids, whiteflies, thrips, mites, and grasshoppers |

Compiled from Roberts and Hajek (1992) and Hajek (1997).

tures and high moisture levels adversely affect the survival of *Beauveria bassiana*, the fungal pathogen that causes mortality in silkworms and a wide range of other insects (Krueger et al., 1991). Some fungi, particularly the pathogenic species, exhibit what is called dimorphism, i.e., two forms of growth, depending on environmental conditions. Under lower temperatures (25°C), they grow as a mold whereas at higher temperatures (37°C), the same fungus has a yeastlike growth. Dimorphism has also been shown to be induced by $CO_2$ concentrations and nutrient status, i.e., starvation.

Fungi have also been used in biological control processes that are of economic importance in the environment. For example, the entamopathogenic fungus *Entomophthora muscae*, infects and kills domestic flies (*Musca domestica*) (Hajek, 1997). There is growing interest and focus on the use of such entomopathogenic fungi for insect control rather than relying on synthetic chemical pesticides. Since they live with their hosts, entomopathogenic fungi have to disperse, locate and recognize new hosts, and successfully cause infection in order to perpetuate themselves over time. The fungus produces pheromones which attract other flies, particularly male flies, to the dead infected flies, which on contact with the carcass also become infected and die. The use of entomopathogenic fungi to control insect pests emphasizes a radical change in the approach to insect pest control and could provide protection against some pests. When using chemical insecticides, the strategy is often to blanket-apply the specific

environment so that each insect (targeted or otherwise) that comes into contact with the lethal dose is affected. At present, mass production and application of several entomopathogenic fungi are being extensively investigated (Table 4.3).

### 4.1.6 Physiological ecology

Fungi produce spores that are readily transported. Dispersal of ascospores (asexual spores) of ascomycetes is frequently discharged when asci dehisce. Within this phylum, some fungal species, for example *Cordyceps* sp., produce long stromata extending up to many centimeters so that ascospores are discharged from perithecia above the soil surface, thus increasing exposure to free air and to passing insect vectors and hosts. Dispersal of fungal spores is also greatly facilitated by rainsplash. The individual or aggregate spores dispersed by rainsplash are often surrounded by mucus (Gregory et al., 1959) or slime (Roberts and Humber, 1981). The coating either serves as a means of adhering to the host or, when dissolved by water, suspends the spores in the rain droplets. Dispersal of fungal fruiting bodies is mostly by wind or by active discharge through hydrostatic pressure. If the primary conidia or spores do not land on a suitable host, they germinate to produce secondary conidia or spores which are further dispersed, a process repeated again and again until a suitable host is encountered. While hydrostatic ejection may not carry the conidia or spores to greater distances, it may expose them to locations where they can become airborne. This way, some spores get deposited in marine environments. Fungi are also quite adapted to dry environments.

The fruiting bodies and spores subsequently formed mainly occur in that part of the mycelia where vegetative growth and hyphal elongation have ceased. Both hyphal elongation and vegetative growth can be terminated due to lack of nutrients, specific changes in irradiance, temperature changes, and/or $CO_2$ or $O_2$ concentrations. Generally, fruiting and sporulations in fungi are associated with light intensities lying between UV (320-400 nm) and the blue (400-520 nm) wavelength but the duration of exposure to specific light intensities also plays a part (Moore, 1998). A downward shift in temperature and/or an increase in $CO_2$ concentration also induces sporulation in fungi. The spores formed get spread to a variety of environments, e.g., marine, dry, etc.

Fungi have evolved with external biodegradative enzymes which are able to degrade cellulose and hemicellulose and this makes them the major decomposers in forest environments. Oftentimes fungi do the initial degradation work and the bacteria take over. Some basidiomycetes can degrade lignin using extracellular enzymes (lignases) which are non-specific. Biotechnology has recognized the use of basidiomycetes, in particular *Pharnechaete chrysosporium* and they are currently used in degrading other types of recalcitrant pollutants such as coal tar ( a highly insoluble molecular weight compound) and effluents from the textile industry (Kunz et al., 2001). The fungus expresses extracellular enzymes, especially lignin peroxidase and manganese peroxidase that degrade a number of recalcitrant compounds.

Fungi also produce low-molecular weight iron-chelating siderophores, e.g. ferrichromes which solubilize Fe(III). Most common of the siderophores is ferrichrome. Siderophores solubilize the otherwise insoluble ferric oxides, hydroxides and oxyhydroxides, making the iron available for assimilation by biological systems. Fungi also solubilize a variety of other metals. A

rapid screening mechanism to detect solubilization of insoluble compounds was developed by Sayer et al. (1995). The method involves observing clear zones of solubilization around fungal colonies growing on solid medium amended with the desired insoluble metal compound.

All fungi are heterotrophic and thus require preformed organic compounds as a source of energy for growth. This necessity led to the evolution of fungi as saprotrophs, deriving nutrients from dead or living detritus, and/or symbionts deriving nutrients from living hosts. Fungal metabolism is primarily aerobic and thus they tend to grow on surfaces of their substrate rather than throughout the substrate. However, some fungi, for example yeasts are facultative anaerobes. Fungi have a wide range of organic carbon/energy sources which include pectins, cellulose, lignin, monosaccharides, ethanol, hydrocarbons, acetate, propanol, butane, phenolic compounds, humic acids as well as fulvic acids. Some gases such as methane, $CO_2$ and carbon monoxide can also be utilized as C sources by fungi. Most of fungi are resistant to osmotic pressure and have the ability to grow in high salt or sugar concentrations. They are also able to grow at low N and moisture levels compared to prokaryotes. All these traits enable fungi to grow on unlikely surfaces, such as refrigerators at low temperatures, paper, painted walls, etc.

Most fungi produce antibiotics and toxins. This ability can give the antibiotic and toxin producers, together with the fungal-like antibiotic-producing bacteria (actinomycetes) a competitive advantage in the environment. However, some of these organisms produce antibiotics only under laboratory conditions and not in the environment. When produced in soil, the antibiotic amounts are unlikely to be produced in substantial quantities because their production expends large amounts of valuable energy. Such organisms have the potential to produce these antibiotics, therefore, but in mixed community, this trait is not greatly expressed.

### 4.1.7 Reproduction in fungi

In nearly all organisms, sex is the fundamental process of two nuclei coming together to form an offspring and is generally referred to as sexual reproduction. In many organisms, this is the only means of self-propagation. A bewildering array of sexual interactions is found in fungi. However, for most fungi, a second way, i.e., asexual or vegetative propagation and often a more productive alternative route exists. Some fungi, particularly the Deuteromycotina (Fungi Imperfecti), which lack sexual reproduction, propagate themselves only through asexual means. When they reproduce vegetatively, often not only two or three progenies are generated but rather millions or even billions. This provides a fantastic potential for explosive growth and widespread infection of the environment by fungi. While vegetative propagation involves no organized fusion of nuclei and is often assumed to be just a means of increasing numbers and not genetic variability, this is not entirely the case with fungi since the peculiarities of nucleus-cytoplasm interactions do permit vegetatively produced progeny to differ genetically from each other and from the parents. During asexual reproduction, only the genetic constitution of an individual fungus is multiplied by mitotic nuclear divisions. It is not clear why so many fungi invest their energy and resources in the more complex sexual reproduction phase, considering the fact that they have the ability to propagate asexually. In cases where sexual reproduction occurs, it appears to be initiated by changes in moisture, temperature, or the availability of nutrients.

Sexually reproducing fungi have fairly typical genetics which involve Mendelian segregation and chromosomal recombination. However, there are some distinct differences between the genetics of most fungi and the majority of other eukaryotes. For one, fungi have a smaller genome size than most eukaryotes. They also possess small nuclei and chromosomes that vary in shape and are difficult to study using conventional microscopy. Thus, progress in studying sex in fungi has been slow until recently. A more recent understanding of the karyotypes in fungi has greatly relied on electron microscopy and molecular biology techniques. Most fungi are haploid during a greater part of their life cycle, diploidy occurring only briefly just before meiosis commences. However, some basidiomycetes are diploid throughout their life cycle. Most haploid fungi have evolved to accommodate two haploids, enabling genetically different nuclei to coexist in the same cytoplasm. Furthermore, hyphae from different haploid parental mycelia can fuse, enabling both cytoplasm and nuclei to exchange, forming a heterokaryotic mycelium which grows vegetatively until conditions are ripe for sexual reproduction to occur. When it occurs, sexual reproduction involves the heterokryon undergoing karyogamy followed by a meiotic division, to generate spores. Meiosis occurs in the meiotic cells and the chromosomes segregate, assort independently and cross over just like other eukaryotes. However, unlike in other eukaryotes, the nuclear membrane in fungi remains intact through prophase I. It should be noted that the number of nuclei and hence genome size, differs even within the same fungal strain. This trait could be a blessing in disguise as it may provide additional sequences in fungi in which mutations may occur, allowing the adaptation of various fungal species to new or broader environments with little risk. Fungal genetics has recently been described by Moore (1998). It is important to note that for most fungi, the mycelium has a number of alternative developmental pathways. It could continue growing as a hypha, produce asexual spores, or progress into the sexual cycle.

It is not clear how gene regulation works in multinucleate cytoplasm. Most of the fungal genetics work has been done with yeast, an important player in major industrial processes such as brewing and baking. Yeast reproduces by budding from single cells. However, the importance of yeast in natural environmental processes like mineralization and composting is largely unknown. Genetic observations in laboratories and industrial processes with yeast may not entirely hold true for other fungi in the natural environment, emphasizing the need to investigate fungal genetics in nature. For example, during normal mycelial growth vegetative hyphae are negatively autotrophic and usually avoid each other, to promote the exploration and exploitation of the substrate. However, fusion of nuclei from different fungi during sex and reproduction requires that hyphae grow toward each other (positive autotrophism). It is not clear as to how and why the negative autotrophism is reversed to enable hyphal cell fusion in nature. Much as it is desirable for the fusing nuclei to be as genetically different as possible, the security of the fusion cell requires that if cytoplasms are to mingle, they must be as similar as possible or else the heterokryon will abort, killing the hyphal compartments involved in incompatible anastomoses. It is not clear how autotrophism and incompatibility are resolved in the natural environment.

In most fungi, the dominant nuclear condition is haploidy, a diploid nucleus appearing by karyogamy (fusion of a nucleus) just before meiosis. Because of this condition, mutations will show up immediately in haploid organisms making fungi, just like bacteria, a tremendous resource for genetics research. This haploid condition may exist even after plasmogamy (fusion of the cytoplasms containing the nuclei) has brought together the nuclei in a classic sexual act. The nuclei may not fuse but continue to multiply in conjunction for a long period of time before any karyogamy. Such behavior is unique to fungi and is quite dominant in Basidiomycotina and some Ascomycotina.

## 4.2 ALGAE

Algae are large, morphologically and physiologically diverse organisms with chlorophyll and the ability to conduct $O_2$-evolving photosynthesis. Algae range from single-celled forms to aggregations of cells or filaments. Many of the unicellular algal forms are motile and can be easily mistaken for protozoa. There is good fossil evidence about the existence of algae billions of years ago. Moldowan and Talyzina (1998) found microfossils from lower Cambrian formations in Tallinn (Estonia) and determined the dinoflagellates (single-celled protists) to be as old as 5.2 Ba. Fossil records show that algae were the first photosynthetic cellular organisms. All crytogenic groups of plants and ultimately flowering plants may have arisen from algae. Once $O_2$ became abundant, much more energy became available from carbon and this could have enhanced the colonization (and possibly evolution) of photosynthetic organisms (see Chapter 1; Section 1.5).

Algae are able to grow in areas low in carbon but must have light and water. Thus, algae are found throughout the photic (light) zone of bodies of water. Algae are physiologically and biochemically very similar to other plants as they possess the same basic biochemical pathways, chlorophyll *a*, as well as carbohydrate and protein end products comparable to those of higher plants. Algae are major producers of $O_2$. At least half of the world's primary production is estimated to occur under marine environments mostly from photosynthesis by algae. Furthermore, three-quarters of the Earth's surface is covered by water and it is estimated that algal activity in the water generates approximately 80% of the Earth's global $O_2$.

The type of chlorophyll pigments, together with the common algal C reserve materials, are a major basis for classifying algae (Table 4.4). Chlorophyllic pigments are also responsible for the distinctive colors of many algae. Brown algae are macroscopic and occur in coastal waters up to a depth of 0-75 cm. Dinoflagellates are unicellular free-floating organisms and are commonly referred to as plankton. Some dinoflagellates, e.g. *Gonyaulax* sp. and *Gambierdiscus toxicus* are reknowned for their ability to produce neurotoxins. The production of toxins, popularly referred to as red tides, has ecological implications in large fish. Fish may

**Table 4.4** Properties of algae

| Algae | Pigments | C reserve materials | Cell wall | Habitats |
|---|---|---|---|---|
| Brown algae (Phaenophyta) e.g. *Laminaria* sp. | Chlorophylls *a,c*; Xanthophylls | Mannitol, Laminarin (β-1,3-glucan) | Cellulose, alginic acid | Marine environments that receive light with red to orange wavelength, i.e., ≈0-75 m depth |
| Dinoflagellates (Pyrrophyta) e.g. *Gonyaulax* sp., *Gambierdiscus toxicus* | Chlorophylls *a,c*; Xanthophylls | Starch (α-1,4-glucan) | Cellulose | Marine, freshwater environments with red wavelength |
| Euglenids (Euglenophyta) e.g. *Euglena* sp. | Chlorophylls *a,b*; Xanthophylls | Paramylon (β-1,2-glucan) | None present | Fresh water |
| Golden-brown (Diatoms) e.g. *Navicula* sp. | Chlorophylls *a,c,e*; Xanthophylls | Lipids | Silica; Cellulose; CaCO₃ | Soil, freshwater and marine environments that receive light in the red wavelength region |
| Green algae (Chlorophyta) e.g. *Chlamydomonas* sp. | Chlorophylls *a,b* | Sucrose and Starch (α-1,4-glucan) | Cellulose | Moist soil, freshwater and marine environments that receive light in the red wavelength region |
| Red algae (Rhodophyta) e.g. *Polisiphonia* sp. | Chlorophylls *a,d*, phyccerythrin, phycocyanin | Starch (α-1,4- and α-1,6-glucan), glycerol-galactoside | Cellulose; Xylans | Marine environments that receive light with violet to blue wavelength (≈150-275 m depth) |

experience a cumulative effect of toxins from algae as they feed on smaller fish and other creatures that in turn have fed on toxin-containing algae. Euglenoids are unicellular algae with a single anterior flagellum. Most of them have a red eyespot at the anterior which senses light and directs the cell in the right direction. Golden-brown algae (diatoms) are unicellular or filamentous and consist of two parts of the wall that fit together like a petri dish fitted with a cover glass, a unique morphological pattern that is used in identifying them. Green algae are either unicellular or multicellular and are believed to have given rise to present-day terrestrial plants. Red algae tend to live at greater ocean depths than other types of algae. Chlorophylls *a* and *b* are also found in cyanobacteria, an indication of some evolutionary linkage between algae and cyanobacteria (a prokaryote).

## 4.2.1 Ecological distribution

Algae are ubiquitous, occurring in practically every habitable environment on Earth where photosynthesis can be conducted at a rate sufficient to generate net production. They exist in hot and cold deserts, soils, air, and of course, all kinds of aquatic habitats where they are the primary producers in the food chain. They can survive temperatures higher than $100°C$ (Trainor, 1983), although the survival mechanisms at these high temperatures are not well known. Algae are more predominant in marine environments at various depths. Their abundance in this environment is very significant, considering the fact that more than 70% of the Earth's surface is covered by water. However, their distribution is limited by low temperature extremes, possibly because of the limitations low temperatures impose on photosynthetic processes.

Soil also contains algae which sometimes bloom under the right moisture and light conditions. Soil algae are principally from members of the Chlorophyceae, Cynophyceae, Bacillariophyceae, and Xanthophyceae. Identifying and quantifying soil algae requires isola-

tion and culturing and is, therefore, likely to be biased by the medium used. The abundance of algae in soil can alternatively be determined by using chlorophyll levels. Cynophyceae appear to be more abundant in soils of pH higher than 5 whereas Chlorophyceae are often found in acid soils, albeit moisture content ought also to be taken into consideration. Many soil algae species produce thick-walled akinetes or hypnozygotes, an adaptation to dry conditions. While many algae species occur on the soil surface, others exist below the soil surface, where they are able to maintain at least minimal metabolic levels in a resting stage.

Using the energy produced in photophosphorylation, algae convert $CO_2$ in the atmosphere into carbohydrates and generate $O_2$ as a by-product. The fixation process is facilitated by ribulose 1,5-bisphosphate carboxylase (RUBISCO) and this enzyme merits recognition as the most important on Earth as it is the primary $CO_2$-fixing enzyme in plants, cyanobacteria, and algae. The enzyme is highly conserved and probably has a very stable genetic background. Algae are very nutritionally versatile as they are not always photosynthetic and may display heterotrophy and mixotrophy. However, these other types of nutrition are not the rule. Since photophosphorylation is a light-driven process, seasonal changes do affect algal populations, especially in temperate and semitemperate zones. The availability of nutrients also causes fluctuations in algal populations, causing algal blooms.

Most algal growth is not limited by $CO_2$ but rather by both nitrogen and phosphorus. These two nutrients can enter the waterways from domestic uses and farm and industrial waste disposal processes. Such disposal may cause algal blooms, a condition called eutrophication. In the long run, the excessive algal bloom is detrimental because when the algae die, the decomposing algal cells deplete some of the dissolved oxygen in the water. The undecomposed organic (algal) material settles at the bottom, hastening the death of the bloomed water life. Trace metals and vitamins are also required by algae. These nutrients are supplied by other microorganisms living together with the algae in clumps. In essence, therefore, algal blooms are indicators of nutrient enrichment (Caliceti et al., 2002). Such enrichment in lakes, rivers and oceans can also occur due to contamination with raw sewage, runoff from fertilized agricultural fields or from the mineralization of dead organisms. On top of N and P, diatoms also require substantial amounts of silicates to flourish. In marine environments, nitrogen seems to be most limiting (Ryther and Dunstan, 1971) whereas P is most limiting in fresh water (Schindler, 1977).

Except for Dinophyceae which are poisonous, very rarely do algae cause serious human diseases. However, some members of Cyanophyceae and Chlorophyceae routinely encountered in house dust are suspected of causing some allergies in susceptible individuals (see Chapter 11). Sunlight provides infrared wavelengths which are absorbed by water and have a heating effect whose magnitude depends on the size of the body of water and its ambient temperature. Continued surface heating decreases the water density and stratifies the water into two layers - a warmer upper and a cooler lower layer. This creates a region (thermocline) where the temperature and water density change abruptly within the water system. Because there will be no exchange of water between the two layers, a differential in nutrients available to the algae, which mostly reside in the upper layer, occurs. Nutrients are thus more plentiful in the deep ocean than within the surface water (Fig. 4.2). In the temperate oceanic environments growth and reproduction in diatoms rely upon the seasonal availability of nutrients rising to the surface from deeper waters in zones of upwelling. When the nutrients are depleted, diatoms can change their form to a resting spore, essentially shutting down metabolism, and sink to deeper waters. Nutrient accumulation in deep waters is further facilitated by the fact that the phytoplankton and its grazers, the zooplanktons, on death, sink to

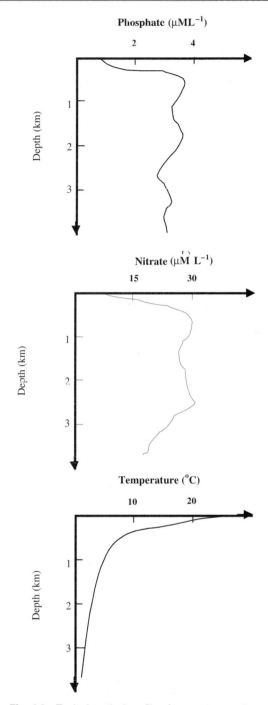

**Fig. 4.2** Typical vertical profile of temperature, nitrate and phosphate with increasing depth in the ocean. Differences between surface and deep ocean layers are quite apparent for all three parameters (redrawn from Philander, 1998 with permission from Princeton University Press) .

the abssyal ocean and decompose into constituent elements. In the tropics, this stratification occurs all year round but in the temperate regions, this is more common in summer. The only exception to this nutrient differential between surface and deepwater nutrient concentration is along the equator where the equatorial undercurrent causes upwelling of cold nutrient-rich water from below. The surface ocean water along the equatorial line thus characteristically registers higher chlorophyll values than its surroundings because of a more thriving phytoplankton population. It is one of the most biologically productive regions of the oceans (Philander, 1998).

## 4.2.2 Physiological ecology

There are two major ecological groups of algae, namely phytoplanktons and seaweeds.

### 4.2.2.1 Phytoplanktonic algae

Phytoplanktonic algae are mainly unicellular. However, filamentous and colonial forms also occur, especially in fresh waters. Filamentous and colonial forms also occur in virtually all bodies of water where they float freely. Their ecology has been comparatively well studied. Their size ranges from <5 µm to large colonial shapes. For example, *Volvox* sp. have a diameter of 500 µm. Their cell size and the surface area-to-volume ratio have important ramifications in the uptake of nutrients, buoyancy, and resisting predation by grazers. Buoyancy is important in maintaining the phytoplankton populations in positions where solar rays are available to support photosynthesis, and to move or sink the planktons when need arises. With the exception of gas vesicle-forming Cyanophyceae, all algae have densities greater than that of water and have to sink or actively swim towards light. In some cases, flotation in phytoplanktons is also facilitated by ionic regulation (South and Whittick, 1987). The heavier ions such as $SO_4^{2-}$, $Ca^{2+}$ and $Mg^{2+}$ are replaced by lighter ones such as $K^+$ and $Na^+$. In Cynophyceae, gas vacuoles which consist of a number of closely packed vesicles are formed. When these vesicles are exposed to high external pressure, which in some instances can be as high as 37 atm, they collapse and the cell loses buoyancy. However, gas vacuole-forming species are uncommon in more turbulent marine environments.

In water, the rate at which phytoplankatonic cells sink directly depends on size, the small cells tending to sink more slowly than the larger. The physiological status of the cells also affects their rate of sinking, with dead or senescent cells sinking faster than viable cells. The surface to volume (S/V) ratio decreases with increasing cell volume. This has important physiological implications as the exchange of energy and nutrients occurs through the surface of phytoplanktonic cells. The S/V ratio and phytoplankton growth correlate positively (Sournia, 1981).

Phytoplankton populations change with seasonal changes, particularly in temperate regions where an increase in irradiance in spring brings about a surge in numbers. This population surge is further enhanced in the surface layer by the development of a thermocline and continues until predation and depletion of nutrients brings about a decline in abundance in the summer. Spring phytoplanktonic blooms are dominated by diatoms but these are consumed by grazers and over time replaced by larger dinoflagellates and green flagellates. Diatom blooms may redominate during fall when the thermocline disappears and allows the nutrient-rich deeper waters to mix with the surface waters. In contrast, tropical waters show less seasonal variation than temperate regions due to the rapid mineralization that occurs in tropical waters at higher temperatures. Furthermore, tropical lakes generally support a high algal biomass but display less species diversity (South and Whittick, 1987).

Phytoplanktons are mainly grazed by zooplanktons. But under marine conditions, protozoa (mainly ciliates) and in fresh waters, crustaceans and rotifers play a role in reducing phytoplanktons. The smaller cells are more susceptible to predation. Grazing is actually quite beneficial to the natural environment since it is a first step in the remineralization of inorganic nutrients to support growth of other algae. In some cases, more than 50 algal species may coexist within uniform water bodies. Such coexistence would, at first sight, seem to contradict the competitive exclusion principle which states that populations of two or more dissimilar organisms occupying the same niche but differing in rate of growth will not coexist for long. Under this principle, the fast-growing species tend to displace the other (slow-growing species). However, the water environment is so heterogeneous due to spatial and temporal fluctuations that diverse species can coexist because of the provision of intermediate nutrient ratios. Various species may be limited by different nutrients.

### 4.2.2.2 Seaweeds

Seaweeds occur on solid substrates in habitats such as shorelines and riverbeds where water movement prevents the accumulation of sediments. They comprise three major taxonomic groups: Chlorophyceae, Phaeophyceae, and Rhodophyceae, although other divisions (Bacillariophyceae; Chrysophyceae; Cyanophyceae; Xanthophyceae) may be locally important. In evolutionary terms, seaweeds represent distinct lines of development which resulted from growing in a common habitat. In contrast to phytoplanktons, very little information is known about the importance of nutrients in controlling the growth and diversity of seaweed communities. However, a report by Darley (1982) indicates that both ambient nutrient levels and the capacity of the seaweed to internally store nutrients are crucial. During the time when nutrients are readily available, growth of seaweeds is rapid. As the nutrient supply slumps due to changes in seasons, depletion of source, or slight changes in water levels, the stored nutrients, depending on their abundance, can still sustain growth (Fig. 4.3). Seaweeds are mostly grazed on by fish (especially in the tropics), mollusks, sea urchins, and crustaceans. Selection by grazers mainly depends on nutritional quality, size, and presence or absence of toxins.

### 4.2.3 Environmental adaptability

Algae can position themselves in a way to ensure most efficient capture of the right wavelength for photosynthesis. If they happen to be at low ambient irradiance, some algal cells increase the concentration of photosynthetic pigments by increasing the number of thylakiods per cell, reducing the level of light at which photosynthesis becomes saturated. Under such light-limiting conditions, the concentration of carotenoids also increases and enhances the light harvested. At even deeper water levels where light is not available, some phytoplanktons may survive by dramatically dropping their metabolism to basal maintenance levels. These adaptations reduce competition between the different types of algae.

Reproduction in algae can be asexual, sexual, or vegetative in nature, but sexual reproduction is noticeably absent among the Cyanophycota, Prochlorophycota and some Euglenophycota. Other forms of adaptation to specific environments include the ability to form dispersal and survival forms such as spores, cysts and akinetes, as well as the ability to form large colonial growths to resist grazing by the predators.

### 4.2.4 Symbiotic algal associations

Symbiosis between algae and invertebrates, fungi, protozoa, ferns or mosses, and vascular plants is common (Table 4.5). The relationship may involve two or more individual organisms. The association may occur at varying degrees and may lead to the formation of morphological structures totally different from those of any of the components growing by themselves. Whatever shape or morphology adapted, the algal symbiont tends to strive toward displaying the maximal area of its photobiont cells to light, thus improving its fitness and ability to spread into otherwise hostile environments. Of outstanding

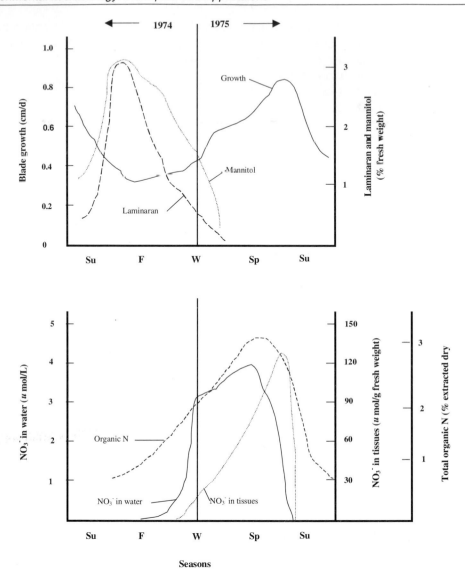

**Fig. 4.3** Relationship between rate of growth of a seaweed (*Laminaria longicruris*), its internal C and N reserves, and $NO_3^-$ concentration in sea water. Levels of ammonium ($NH_4^+$) were highly variable with no clear trend. Su = summer; F = fall (autumn); W = winter; and Sp = spring (redrawn from Darley, 1982 with permission from Blackwell Publishers, Oxford, United Kingdom).

importance in the environment are lichens and the *Azolla-Anabaena* relationship.

### 4.2.4.1 Lichens

Lichens have conventionally been studied in botany and are not micro in nature. However, their coverage in environmental microbiology is justified as they develop from a symbiotic partnership of one or more fungi with one or more algae and/or cyanobacteria. The coexistence between the two partners that comprise a lichen is mutualistic, with both partners benefiting. Because of this unique partnership, lichens cannot be neatly fitted into any conventional classi-

**Table 4.5** Examples of algal associations

| Association with | Algae | Associate/resultant | Notes |
| --- | --- | --- | --- |
| Invertebrates | *Codium* sp., *Caulerpa* sp. | Saccoglossan mollusks | Mollusks suck the cytoplasm from green algae accumulating the chloroplasts. The chloroplasts continue to photosynthesize but are unable to divide, synthesize chlorophyll or RuBP carboxylase. Photosynthates metabolized by host or chloroplasts digested to provide nutrients. |
| Protozoa | *Cyanophora korschikoffiana* | *Cyanophora paradoxa* | Synthesis of photosynthetic pigments. |
| Fungi | Wide range of algal species | Lichens | Algae photosynthates transferred to the associated fungus. It is not clear whether the algae get any benefit from the relationship (except perhaps protection against predators). |
| Bryophytes and vascular plants | Heterocytous members of Nostocaceae | Hortwort (*Anthoceros*) and Liverwort (*Blasia*) | *Anthoceros* and *Blasia* have cavities in their gametophyte thalli containing Nostoc. N is excreted as ammonia and taken up by the bryophyte host which in turn supplies fixed C to the alga. |
| | *Anabaena azollae* | *Azolla* | *Azolla* contains *Anabaena azollae* in cavities in the upper lobes of its floating leaves. |
| | *Cephaleurus* sp. | A wide range of vascular plants e.g. tea, coffee and citrus | Algae grow beneath the cuticle of the host, on the stems, leaves, flowers and fruits, and may produce orange-red hematochrome pigments giving a fungallike red-rust appearance. One of the few examples of parasitic algae because they draw photosynthates from the host plants. |
| | Several macro- and microalgae, e.g. *Polysiphonia lanosa* | Epiphytes on plants | Most show no specificity or relationship with host plants. Appear to be a result of competition for light, nutrients or space. |

fication and were referred to by the father of taxonomy, Linnaeus, as *rustia pauperrimi* which is loosely translated as the "poor trash of vegetation". Lichens have traditionally been mostly ignored and their importance in ecological studies often grossly underrated. Lichens occupy about 8% (i.e., $1.2 \times 10^7$ km$^2$) of the earth's terrestrial surface (Seaward, 1996). They are mostly abundant in the tundra region but do exist in temperate, desert, and tropical forests as well.

In most lichens, the fungal component (mycobiont) constitutes more than 90% of the biomass. All but the simplest lichens display a distinct internal structure, and the alga is restricted to zones in the surface layers of the fungal thallus (Fig. 4.4). The fungal/algae relationship is not an obligate one since both the algal component (phycobiont) and the mycobiont are capable of independent growth in axenic culture (Ahmadjian, 1981). Although there is only a fragmentary fossil record of lichens prior to the pre-Cambrian period (Seaward, 1996), it is very likely lichens were some of the earliest colonizers of terrestrial habitats, coping with the relatively low oxygen levels prevailing at that time. This likelihood is especially valid considering the fact that the resultant symbiont-lichen-is better equipped to cope with extreme environments. As a matter of fact, lichens are some of the most successful symbioses. They can comprise a wide range of symbiotic interactions which manifest in many morphological and physiological adaptations to various environments, often inhabiting areas in which neither algae nor fungi can survive alone, such as rocks. For example, under conditions of extended desiccation, lichens devise physical means of concealment which may include mimicry or camouflage. Some lichens are remarkably long lived and can persist for hundreds of years.

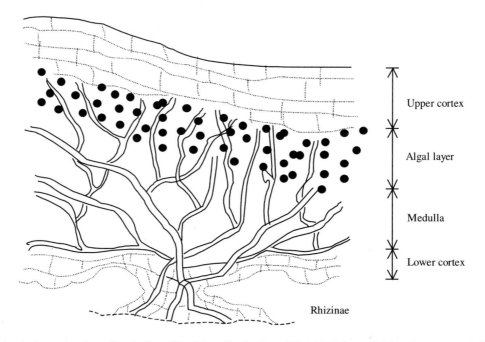

Upper cortex

Algal layer

Medulla

Lower cortex

Rhizinae

**Fig. 4.4**  Anatomical cross section of the thallus of the lichen *Xanthoria parietina*. Medulla predominantly composed of fungal hyphae and accounts for approximately 40% of the lichen volume. Algae and air spaces comprise approximately 7% and 8% respectively. Rest of the volume occupied by extracellular matrix (redrawn from South and Whittick, 1987 with permission from Blackwell Publishers, Oxford, United Kingdom).

Because lichens grow very slowly, they require a stable substratum and some illumination. They can grow on rocks, undisturbed soil, tree barks, and on man-made structures such as concrete. They can tolerate desiccation and temperature extremes. The association of the phycobiont (algal partner) and the mycobiont (fungal partner) into a complete lichen is only possible under carefully controlled stressful conditions. As a matter of fact, removing a lichen from a stressful environment leads to outgrowth of one of the partners in the association, breaking up the lichen as an entity. Ahmadjian (1967) observed that the key feature that has to be controlled in the establishment of lichens is moisture (i.e., wetting and drying cycles).

The flow of nutrients in lichens is one way in the sense that nutrients flow from the alga to the fungus. In some cases, up to 90% of the total carbon fixed by the alga during photosynthesis is transferred to the fungus. The alga, while giving up its valuable nutrients, receives both protection from desiccation and anchorage by the fungus. Cyanobacteria synthesize glucose whereas green algae synthesize polyols, erythritol, sorbitol or ribitol which are converted to mannitol in the fungal thallus (Fig 4.5). The ability of lichens to fix atmospheric nitrogen is dependent on the possession of cyanobacteria as one of the lichen components. Nitrogen fixation in cyanobacteria occurs in heterocysts and the frequency of these heterocystic cells in lichenized cyanobacteria greatly increases compared to free-living and nonlichenized cyanobacteria. Unlike the nonlichenized cyanobacteria, the N-fixing cyanobacteria may ironically show symptoms of nitrogen deficiency (chlorosis) due to the substantial transfer of the fixed nitrogen to the mycobiont.

A major contribution by lichens is their role as agents of biological weathering, their pedogenic action being both chemical and physical. They can secrete organic acids that chemically weather rocks, a process that provides nutrients to support growth. Lichens have an enormous capacity to accumulate elements such as nitrogen, phosphorus, and sulfur by directly absorbing these nutrients from the surfaces on which the lichens are growing or the dust particles that settle on them, thereby increasing the availability of these elements to secondary colonizers. Nitrogen is also accumulated by lichens through direct fixation from the atmosphere, by the cyanobacteria present in the association. Irrespective of the means of nutrient accumulation by lichens, the nutrients are recycled

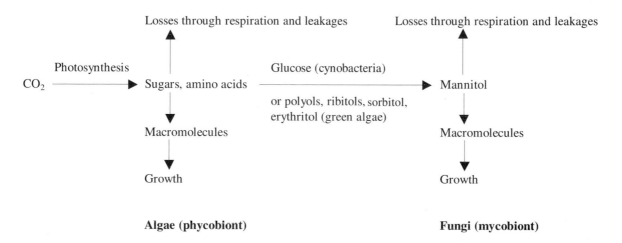

**Fig. 4.5** Carbon flow through lichens (modified from South and Whittick, 1987 with permission from Blackwell Publishers, Oxford, United Kingdom).

through the soil-formation process. In studies by Boucher and Nash (1990), the standing biomass of the epiphyte *Ramalina menziesii* was estimated at 706 kg$^{-1}$ha and the litter fall of this epiphyte in a blue oak (*Querus douglasii*) woodland contributed 13%N, 4%P, 7%K, 1%Ca, 3%Mg, and 8%Na of the annual canopy nutrient turnover. The decomposition rate of lichens is fairly rapid but is generally less than that of vascular plants. The common saprophytic agents rarely appear to be involved in this decomposition process, perhaps due to the lichen substances present. The ability of lichens to facilitate the biodeterioration of rocks and cement-based monuments is due to the metal oxalates (e.g. calcium oxalate) and oxalic acid which chemically disrupt these structures. As the lichens decompose, together with the detached rock substratum, they subsequently form primitive soil.

Many species of lichens reproduce vegetatively by involving structures that ensure the codispersion of both the phycobiont and the mycobiont. The vegetative dispores formed, the simplest of which is soredium, consist of 25-100 µm diameter balls of a few phycobiont cells wrapped in a mycobiont hypae. If the fungal spores on the lichen reproduce sexually, the spores must locate an algal partner to enter the lichenized state. Ahmadjian and Jacobs (1981) successfully synthesized the lichen *Cladonia cristatella* in axenic culture by mixing clumps of algae and the mycobiont hyphae on freshly cleaned strips of mica which had been soaked in the algal culture medium and incubated at a relative humidity of 95%.

Many lichens are quite sensitive to rapid environmental disturbances and human activities, particularly agricultural practices, urbanization, recreation, deforestation as well as soil, water, and air pollution. In fact lichens are increasingly used in environmental impact assessment studies in order to monitor environmental perturbations. In such studies, interpretations are based on the presence and/or absence of particular species of lichens. For example, the diversity in lichen species has been used in monitoring sulfur dioxide pollution (Hawksworth and Rose, 1970), and in pollution by lead as indicated by the distribution of *Stereocaulon pileatum* which corresponds to urban and highway developments (Seaward, 1996). There is a negative relationship between the diversity of lichen species and sulfur dioxide levels (Fig 4.6). Whereas the use of bioindicators to detect pollution is sensitive, when the pollutant concentration falls, the lichen species diversity does not necessarily reestablish itself at the same rate that the concentration in pollutants falls. Bioindicators such as lichens cannot completely replace the direct physical and chemical measurements of air and soil pollutant concentrations; both direct measurements and bioindicators are necessary for making detailed or large-scale surveys of the distribution of pollutants. Thus, if used jointly, they supplement each other. Using bioidicators in environmental monitoring is particularly advantageous where on-site instrumentation would be expensive to install and maintain. However, in cases where lichens are used as the bioindicators, a clear understanding of the identification of individual species and taxonomy in general is cardinal.

### 4.2.4.2 Azolla-Anabaena association

*Azolla* are floating aquatic heterosporous ferns. They form a symbiotic relationship with filamentous oxygenic phototrophic cyanobacteria (*Anabaena*) that fix nitrogen. Nitrogen fixation occurs in specialized cells, called heterocysts. *Azolla* are found worldwide and are almost exclusively in association with their symbiotic endophyte. The relationship is very supplementary as *Azolla* contains chlorophylls *a*, *b*, as well as carotenoids whereas *Anabaena* filaments contain chlorophyll *a*, carotenoids and phycobiloproteins, an arrangement that provides a light-harvesting combination within the 400-700 nm wavelength range. Whereas the endophytic *Anaebaena* contributes only about 6-10% of the total photosynthetic capability of

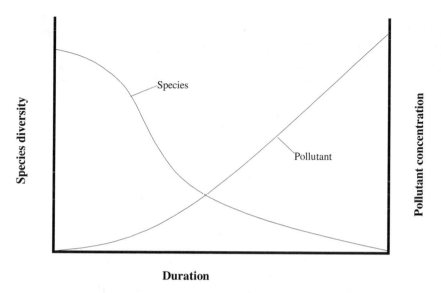

**Fig. 4.6**   Logarithmic relationship between the diversity of species and pollutant levels. Duration could be days, weeks, months, or years, depending on the potency of the pollutant. Notice that small increases in the pollutant lead to sizable decreases in the diversity of species.

the symbiotic association, it greatly contributes to the nitrogen requirement of the *Azolla* (Peters and Calvert, 1983).

There is considerable synchrony in the development of the *Azolla-Anabaena* association. As a matter of fact, the rate of growth of the *Anabaena* is coordinated with that of its *Azolla* host. The *Anabaena* filaments come into contact with the leaf priomordia in the apical meristem of the *Azolla*. A pocket develops on the undersurface of the leaf and closes to form a cavity, thus trapping the alga. The alga then divides and a heterocyst is formed within the cavity. The *Anabaena* microsymbiont accounts for 15% of the biomass of the association and is capable of providing all the nitrogen required by the *Azolla*. The fixed nitrogen (as ammonium) is catalyzed by the nitrogenase enzyme, a molybdenum-iron rich protein (see Chapter 9; section 9.3.1). N-fixation in *Azolla-Anabaena* fields is estimated at about 30-50 kg N ha$^{-1}$y$^{-1}$. The rate of nitrogen fixation by the *Anabaena* symbiont is dependent on the growth stage of the *Azolla* leaf host in the sense that increased cell differentiation at the leaf apex is also accompanied by an increase in heterocysts differentiation (Peters and Calvert, 1983). The *Azolla* in turn supplies the bulk of the carbohydrates to the N-fixing endophyte. However, the fixation of N is usually limited due to:

(i)   limitation from other nutrients particularly P;

(ii)   competition between green plants and algae; and

(iii)   predation of the algae by phytoplanktons.

The symbiotic relationship offers *Anabaena* a competitive advantage against other free-living cyanobacteria, "lifting" it above potential competitors to tap solar energy.

## References

Ahmadjian V. (1967). The Lichen Symbiosis. Blaisdell Publ. Co. Waltham, MA.

Ahmadjian V. (1981). Algal/fungal symbioses. *Prog. Phycol. Res.* **1:** 179-233.

Ahmadjian V. and J.B. Jacobs (1981). Relationships between fungus and alga in the lichen *Cladonia cristella* Tuck. *Nature* **289:** 169-172.

Bagyaraj D.J. and A. Varma (1995). Interaction between arbuscular mycorrhizal fungi and plants: Their importance in sustainable agriculture in arid and semiarid tropics. *Adv. Microb. Ecol.* **14:** 119-142.

Barber S.A. (1995). Soil Nutrient Bioavailability. A Mechanistic Approach. John Wiley and Sons, New York, NY.

Boucher V.L. and T.H. Nash (1990). The role of the fruticose lichen *Ramalina menziesii* in the annual turnover of biomass and macronutrients in a blue oak woodland. *Bot. Gaz.* **151:** 114-118.

Bowman B.H., J.W. Taylor, A.G. Brownlee, J. Lee, S.-D. Lu, and T.J. White (1992). Molecular evolution of the fungi: Relationship of the Basidiomycetes, Ascomycetes, and Chytridiomycetes. *Molec. Biol. and Evol.* **9:** 285-296.

Brasier C. (1992). A champion thallus. *Nature.* **356:** 382-383.

Caliceti M., E. Argese, A. Sfriso, and B. Pavoni (2002). Heavy metal contamination in the seaweeds of the Venice lagoon. *Chemosphere* **47:** 443-454.

Cunningham J.E. and C. Kuiack (1992). Production of citric and oxalic acids and solubilization of calcium phosphates by *Penicillium bilaii*. *Appl. Environ. Microbiol.* **58:** 1451-1458.

Darley W.M. (1982). Algal Biology: A Physiological Approach. Blackwell Scientific Publ. Boston, MA.

Gadd G.M. (1986). Fungal responses towards heavy metals. *In:* R.A. Herbert and G.A. Gadd (eds.). Microbes in Extreme Environments. Acad. Press, New York, NY, pp. 83-110.

Gadd G.M. and J.A. Sayer (2000). Influence of fungi on the environmental mobility of metals and metalloids. *In:* D.R. Lovely (ed.). Environmental Microbe-Metal Interactions. ASM Press, Washington DC pp.237-256.

Gharieb M.M., J.A. Sayer, and G.M. Gadd (1998). Solubilization of natural gypsum ($CaSO_4.2H_2O$) and the formation of calcium oxalate by *Aspergillus niger* and *Serpula himantiodes*. *Mycol. Res.* **102:** 825-830.

Gregory P.H., E.J. Guthrie, and M.E. Bunce (1959). Experiments on splash dispersal of fungus spores. *J. Gen. Microbiol.* **20:** 328-354.

Hajek A.E. (1997). Ecology of terrestrial fungal entomopathogens. *Adv. Microb. Ecol.* **15:** 193-249.

Hasegawa M., T. Hashimoto, J. Adachi, N. Iwaba, and T. Miyata (1993). Early branchings in the evolution of eukaryotes-an ancient divergence of *Entamoeba* that lacks mitochondria revealed by sequence data. *J. Molec. Evol.* **36:** 380-388.

Hawksworth D.L. and F. Rose (1970). Qualitative scale for estimating sulfur dioxide air pollution in England and Wales using epiphytic lichens. *Nature* **227:** 145-148.

Krueger S.R., J.R. Nechols, and W.A. Ramoska (1991). Infection of chinch bug, *Blissus leucopterus* (Hemiptera: Lygaeidae), adults from *Beauveria bassiana* (Deuteromycotina: Hyphomycetes) conidia in soil under controlled temperature and moisture conditions. *J. Invert. Path.* **58:** 19-26.

Kunz A., V. Reginatto, and N. Durán (2001). Combined treatment of textile effluent using the sequence *Phanerochaete chrysosporium*-ozone. *Chemosphere* **44:** 281-287.

Lapeyrie F., J. Ranger, and D. Vairelles (1991). Phosphate solubilizing activity of ectomycorrhizal fungi in vitro. *Can. J. Bot.* **69:** 342-346.

Lloyd J.R. and L.E. Macaskie (2000). Bioremediation of radionuclide-containing wastewaters. *In:* Lovely D.R. (ed.). Environmental Microbe-Metal Interactions. ASM Press, Washington DC, pp. 277-327.

Lodge D.J. (1993). Nutrient cycling by fungi in wet tropical forests. *In:* S. Isaac, J.C. Frankland, R. Watling, and A.J.S. Whalley (eds) Aspects of Tropical Mycology. Cambridge Univ. Press, Cambridge, UK, pp. 37-57.

Moldowan J.M. and N.M. Talyzina (1998). Biogeochemical evidence for dinoflagellate ancestors in the early Cambrian. *Science* **281:** 1168-1170.

Moore D. (1998) Fungal Morphogenesis. Cambridge Univ. Press, Cambridge, UK.

Orsler R.J. and G.E. Holland (1982). Degradation of tributyltin oxide by fungal culture filtrates. *Int. Biodeter. Bull.* **18:** 95-98.

Park D. (1985a). Application of Horton's first and second laws of branching to fungi. *Trans. British Mycol. Soc.* **84:** 577-584.

Park D. (1985b). Does Horton's law of branching length apply to open branching systems? *J. Theor. Biol.* **112:** 299-313.

Peters G.A. and H.E. Calvert (1983). The *Azolla-Anabaena azollae* symbiosis. *In:* L.J. Goff (ed.). Algal Symbiosis: A Continuum of Interaction Strategies. Cambridge Univ. Press, Cambridge, UK, pp. 109-146.

Philander S.G. (1998). Is the Temperature Rising?: The Uncertain Science of Global Warming. Princeton Univ. Press, Princeton, NJ.

Ray T.B., B.C. Mayne, R.E. Toia Jr., and G.A. Peters (1979). *Azolla-Anabaena* relationship. III. Photosynthetic characterization of the association and individual partners. *Plant Physiol.* **64:** 791-795.

Read D.J. (1983). The Biology of Mycorrhiza in the Ericales. *Can. J. Bot.* **61:** 985-1004.

Read D.J. (1991). Mycorrhizas in ecosystems. *Experimentia* **47:** 376-391.

Roberts D.W. and R.A. Humber (1981). Entomogenous Fungi. *In:* G.T. Cole and B. Kendrick (eds.). Biology of Conidial Fungi. Acad. Press, New York, NY, vol. 2, pp. 201-236.

Read D.J., D.H. Lewis, A.H. Fitter, and I.J. Alexander (eds)(1992). Mycorrhizas in Ecosystems. CAB International, Wallingford, UK.

Roberts D.W. and A.E. Hajek (1992). Entomopathogenic fungi as bioinsecticides. *In:* G.F. Leatham (ed.). Frontiers in Industrial Mycology. Chapman and Hall, New York, NY, pp. 144-159.

Ross I.K. (1979). Biology of the Fungi. McGraw-Hill, New York, NY.

Ryther J.H. and W.M. Dunstan (1971). Nitrogen, phosphorus and eutrophication in the coastal marine environment. *Science* **171:** 1008-1013.

Sanchez P. (1976). Properties and Management of Soils in the Tropics. John Wiley and Sons, New York, NY.

Sayer J.A., S.L. Raggett, and G.M. Gadd (1995). Solubilization of insoluble metal compounds by soil fungi: development of a screening method for solubilizing ability and metal tolerance. *Mycol. Res.* **99:** 987-993.

Schindler D.W. (1977). Evolution of phosphorus limitation in lakes. *Science* **195:** 260-262.

Seaward M.R.D. (1996). Lichens and the Environment. *In:* B.C. Sutton (ed.) A Century of Mycology. Cambridge Univ. Press. Cambridge, UK.

Smith M.L. and J.N. Bruhn (1992). The fungus *Armillaria bulbosa* is among the largest and oldest living organisms. *Nature* **356:** 428-431.

Södeström B. (1996). Some aspects of fungal ecology. *In:* D.F. Jensen, H.-B. Jansson, and A. Tronsmo (eds.). Monitoring Antagonistic Fungi Deliberately Released into the Environment. Kluwer Acad. Publ., Norwell, MA, pp. 11-16.

Sournia A. (1981). Morphological bases of competition and succession. *Can. Bull. Fish. Aquatic Sci.* **210:** 339-346.

South G.R. and A. Whittick (1987). Introduction to Phycology. Blackwell, Boston, MA.

Trainor F.R. (1983). Survival of algae in soil after high temperature treatment. *Phycologia* **22:** 201-202.

Trinci A.P.J. (1974). A study of the kinetics of hyphal extension and branching initiation of fungal mycelia. *J. Gen. Microbiol.* **81:** 225-236.

Wainright P.O., G. Hinkle, M.L. Sogin, and S.K. Stickel (1993). Monophyletic origins of the metazoa: An evolutionary link with Fungi. *Science* **260:** 340-342.

# 5

# Viruses and Related Particles

## 5.1 WHAT ARE VIRUSES?

Viruses are small (0.02-0.3 $\mu$m) acellular genomic entities containing four possible types of nucleic acids: single- or double-stranded DNA and single- or double-stranded RNA, surrounded by a protein (or lipid) coat. The evolutional origin of viruses is certainly not clear but they probably originated as pieces of DNA and RNA that broke off from a cell and, through the course of evolution, "learnt" to spread to new cells. Viruses were first discovered by Dimitri Ivanovsky in 1890, during the time when a tobacco mosaic epidemic raged across plantations in Crimea. However, their physical observation was not possible until the advent of electron microscopy and X-ray diffraction. Prior to these techniques, viruses were only described as filtrable agents because they could pass through filter columns of micropore size (Chamberland candle) normally used to retain bacterial cells. Their discovery occurred when Ivanovsky passed sap from mosaic-infected tobacco plants through the Chamberland candle and found that the resultant purified liquid could still cause the disease in healthy plants. Bacteria were ruled out as a possible causative agent in the filtrate because they are unable to pass through the filter column but, more importantly, because the filtrate could not elicit microbial growth on media. Subsequent investigations by Martinus Beijerinck in 1898 ruled out the possibility of the infective agent being a toxin since it

caused a chain of infections in a series of plants, the sap of one plant infecting the next one in sequence, something that an inert toxin cannot do. Beijerinck proposed the term filterable virus (virus, Latin word for poison). Based on these experiments, Beijerinck further found that to cause disease, the contagious fluid has to be incorporated in the host cells.

Viruses differ from cellular organisms in size and their existence as obligate intracellular parasites. They perpetuate their progeny by exporting themselves to other living entities. Thus viruses occur in any environment that sustains life and infect all life forms (i.e., plants, animals, insects, bacteria, fungi, and algae). Those most commonly found in the environment are the single-stranded RNA and the double-stranded DNA types. Their nucleic acid is surrounded by a protein coat (capsid), itself enclosed in an envelope containing lipids and lipoproteins. The protein coat confers specificity on the virion and protects it against adverse environmental conditions. Besides the capsid, a typical virion may also possess a sheath (tail) and tail fibers which facilitate adsorption of the virion on the host cell. Some virions are devoid of an envelope and have been termed naked virions. General structures of some viruses of concern in the environment are shown in Fig. 5.1.

Outside the host, viruses occur as virions, a survival stage which neither grows or divides.

| Family | General structure | Characteristics | Virus of concern | Disease/syndrome | Transmission and vectors |
|---|---|---|---|---|---|
| Arenaviridae | | Enveloped particles of 50-300 nm with ssRNA and ribosomes. The ribosomes are of unknown function | Arenaviruses | Lassa fever | Rodents as reservoirs; person-to-person contact & consuming contaminated food/water |
| Bunyaviridae | | ssRNA enveloped particles of 100 nm diameter with spikes. Has internal ribonucleoprotein filaments of 2 nm width. | Rift valley virus | Rift valley fever | Mosquitoes; flooding |
| | | | Hantaviruses | Hantavirus pulmonary syndrome | Transmitted by rodents e.g. Deer mouse (*Peremyscus maniculatus*) |
| Calicivirus | | ssRNA icosahedral particles of 37 nm diameter | Norwalk and Norwalk-like viruses | Gastroenteritis | Shellfish; contaminated water |
| Filoviridae | | Long filamentous particles of 800 - 14000 x 80 nm with helical nucleocapsid of 50 nm diameter. | Ebola virus | Ebola | Bats? Person-to-person contact. |
| | | | Machupo virus | Bolivian hemorrhagic fever (BHF) | Rodents e.g. *Calomys callosus* are the vectors. Person-to person contact. |
| | | | Marburg virus | Muburg hemorrhagic fever (MHF) | Vectors include monkeys, mice & bats. Person-to-person contact. |
| Flaviviridae | | Enveloped particle with 40-60 nm diameter. ssRNA enveloped particle | Yellow fever virus | Yellow fever | Mosquitoes (*Aedes aegypti*) are the main vectors |
| | | | West Nile virus | West Nile virus | Mosquito vectors |
| Orthomyxoviridae | | Has several segments of ssRNA-enveloped particles of 100 nm diameter with 5-10 nm spikes and 9 nm helical nucleocapsids. | Influenza virus | Influenza | Aerosol |
| Paramyxovirus | | Enveloped ssRNA particle of 150 nm diameter with spikes. Particle contains a 12-17 nm diameter nucleocapsid. | Measles virus | Measles | Aerosol |

(Fig. 5.1 Contd.)

(Fig. 5.1 Contd.)

| Family | General structure | Characteristics | Virus of concern | Disease/syndrome | Transmission and vectors |
|---|---|---|---|---|---|
| Piconaviridae |  | Icosahedral particles of 30 nm diameter. Have ssRNA. | Coxsackieviruses (several strains) | Meningitis | Contaminated water/food; some are airborne |
| | | | Hepatitis A virus | Hepatitis A | Contaminated water/food |
| | | | Poliovirus | Polio | Contaminated water/food; person-to-person contact. |
| Reoviridae |  | dsRNA with icosahedron particles of 60-80 nm diameter. | Rotavirus (i.e. Reoviruses, Caliciviruses, and Astroviruses) | Rotaviral gastroenteritis | Contaminated water/food/toys/utensils; person-to-person contact. |
| Retroviridae |  | ssRNA enveloped in 100 nm diameter particles with a core containing a helical nucleoprotein. | Human immunodeficiency virus (HIV) | Acquired immunodeficiency syndrome (AIDS) | Tainted bodily fluids; intimate sexual contact |
| Rhabdoviridae | 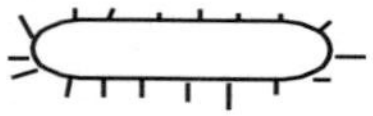 | Bullet shaped ssRNA enveloped particle of 100-430 x 70 nm with 5-10 nm spikes. | Rabies virus | Rabies (also in canines) | Transmitted through bites from infected canines. |
| Togaviridae |  | Particles have ssRNA and are enveloped. The envelope has spikes and a diameter of 60-70 nm. | Equine encephalitis virus | Equine encephalomyelitis | Vectors include rodents and mosquitoes. |

**Fig. 5.1** Some of the viral-related diseases or syndromes of environmental consequence which affect humans[1]

[1]Compiled from Dimmock and Primrose (1994) and Peters and Olshaker (1997), ds=double stranded; ss=single stranded

As a matter of fact, viruses do not reproduce outside their host and, therefore, are obligate intracellular parasites. Three basic types of virus reproduction have been established: lytic, chronic, and lysogenic. During lytic infection the virus attaches to a host cell and injects its genetic material (DNA or RNA) into the cell, directing the host cell to subsequently manufacture more copies of the virus. When the cell bursts, viral progenies are released and the cycle repeated with other host cells. Chronic infection is somewhat similar to lytic infection except that the viral progenies are nonlethal and can be released by extrusion or budding (instead of fatal bursting) over several generations. Under lysogenic infection, the viral genetic material becomes part of the genome of the host cell and reproduces as part of the normal genetic material of the host. Its release from the host cell can be triggered by some event, such as stress, in a lytic fashion. Thus, irrespective of the type of reproduction, there is some form of adsorption of the virus by the host cell, penetration of the viral nucleic acid into the host cell, replication of the nucleic acid within the host cell, maturation, and subsequent release of the virions. Typically, during the lytic phase the phages destroy their hosts upon release of the virus. In the latent phase, the viral genome may become integrated into the host genome where it is maintained from generation to generation. Viruses fill specific receptors on the susceptible host and then, like a hypodermic syringe, inject their DNA into it. Once inside the host cell, the injected DNA replicates itself, forming a coat protein. Inside the host, the virus can take over its host's DNA replication machinery, replicating itself and releasing its genome from the coat, giving various virions. Subsequently, the host cell which has various virions attached on it becomes weak and bursts, releasing the virions to other cells. However, viroids are host-specific and have to find the right host for this process to be successful. Such specificity dictates that a compatible cell may not be present in less than $10^6$

host cells per unit volume. This has ecological implications as the distribution of a particular type of virus may not be readily tracked except in specific environments favorable to the specific host cell. For example, some bacteriophages are only encountered in waste water or other water contaminated with mammalian fecal matter. On the other hand, since every cellular organism is susceptible to viruses, viruses are probably the most abundant and diverse biological entities on Earth.

Unlike DNA-based viruses, RNA-based viruses tend to be more error-prone during replication, a trait that creates even more diversity within the RNA viruses. For example, each time the influenza virus enters a host, the resultant viral progenies turn out to be substantially different in terms of antigenicity, virulence, and other biological properties. This has implications in vaccination since having the flu does not immunize one from the next flu strain as the new strain would most likely be antigenically different from that of the previous infection to which the body had developed immunity (Earn et al., 2002).

In the past, viruses were named according to the diseases or symptoms they cause, their hosts, or the geographic location where they were first noticed. This nomenclature is ambiguous and has created some confusion, for the same virus may have more than one host or cause different symptoms. Efforts are currently underway to harmonize and rationalize virus nomenclature by adapting binomial nomenclature based on particle morphology, presence absence of an envelope, or type of DNA and RNA genome, i.e., single or double-stranded, and types of hosts preferentially infected (Ackerman, 1992; Dimmock and Primrose, 1994; Cooper, 1995).

## 5.2 OCCURRENCE OF VIRUSES IN THE ENVIRONMENT

Viruses are obligate parasites with the single goal of replicating themselves by attachment to a host cell and penetration of its nucleic acid, taking over its reproductive machinery in order to produce massive copies of themselves. Since viruses are always reproducing on some other cells' nucleic acid and machinery, any mutation that might make them more capable in a particular environment is selected, allowing the viruses to adapt more quickly than their hosts can. Distribution of viruses depends on the density of susceptible host population. Thus inevitably viruses are discussed conjunctively with their hosts. Hosts include plants, animals, bacteria and protists. Since the existence of viruses is so intertwined with the host, normally they do not kill it but rather live with it. The hosts may not be susceptible to a uniform extent and, in some cases, only particular stages in the life cycle are infective. For example, in actinomycetes only the germ tubes and mycelia tips are susceptible to viral infection (Williams et al., 1989). Furthermore, the growth rates of the hosts vary greatly. The rate of growth of some hosts, in particular the unicellular organisms, is quite high, with populations that rapidly increase with an increment in supply of resources. On the other hand, the duration it takes multicellular hosts to develop can be significantly long thus affecting the relative abundance of hosts, limiting the distribution of some viruses. However, the spread and distribution of viruses that infect humans have been greatly accelerated by improvements in global travel and trade.

Transmission of viruses from one host to another can be vertical or horizontal. Vertical transmission is hereditary and is facilitated from parents to offspring through eggs, sperms, pollen, etc. Vertical transmission of viral diseases is common in seedborne diseases, the frequency of transmission markedly varying with the host genotype. Vertically transmitted viral diseases can therefore persist in plants, animals, algae, protozoa, and bacteria because the viral genome is capable of integrating within the genome of these hosts. Vertical transmission of viruses is not greatly impacted by envi-

ronmental factors and is of little consequence in environmental microbiology as opposed to horizontal transmission. Horizontal transfer of viruses usually involves a vector. Typical vectors for viruses include nematodes, thrips, and a variety of other insects which transmit viruses to the uninfected contemporaries on which they routinely feed. The greatest exposure of the human population to viruses is through contact with sewage, waste water and related animal refuse. The degree of contamination of the environment by sewage varies greatly across the globe but the pollution of lakes, wells, and rivers by viruses due to the discharge of untreated or inadequately treated sewage has been documented around the world. Such pollution has implications not only for drinking water, but also for water used for recreational purposes. Bitton (1980) estimated that as many as 63% of river water in many parts of the world may be positive for viruses. Once discharged into waterways or on land, the viruses may survive for several months if the environmental conditions are conducive. The microbiological aspects of treating sewage and waste water are extensively discussed in Chapter 13.

To facilitate dispersal, viruses are composed of morphologically distinct particles (virions) with two main components—a nucleic acid genome and a proteinaceous coat (capsid). Most virions are spherical in shape though a few are cylindrical or pleomorphic. Viruses are released from infected host cells in an aggregated state and probably remain so upon entering natural waters, waste water, or soil. Once outside the host cell environment, viruses are exposed to a variety of adverse environmental stresses of abiotic (pH, temperature, oxygen content, cations, light, etc.) and biotic (polysaccharides, enzymes, predators, etc.) nature. Their stability under these stresses is important to their successful transmission from one host to another. The packaging of DNA in the phage is possibly a useful means of preserving extracellular DNA in natural environments since the phage can replicate and retain its stability for a significant period of time under most environmental conditions. However, some varions may individually have a short half-life in the environment although their collective persistence can be the result of continuous discharge of contaminated wastes in specific localities.

A great deal about the ecology of viruses in the environment has been learned from bacteriophages, viruses that attack bacteria. Bacteriophages are quite widespread in terrestrial and aquatic environments. As a matter of fact, these environments have been used as convenient sources of phages for studies in taxonomy, genetics, molecular biology, and a whole range of other disciplines. There has been great interest in the ecological role of bacteriophages in order to determine their impact on bacterial populations through lysis and the potential possibility of lysed cells transferring genetic material to others and thereby contributing to biogeochemical cycling of nutrients. It is not clear whether bacteriophages greatly impact the microbial population in soils for example, and for that matter, whether they may exert some control on the many important chemical transformations in the terrestrial environment either directly by lysing the key microbial players, or indirectly as agents of genetic exchange between and within microbial populations. On the contrary, viruses are known to influence particle matter and energy transfer in aquatic systems (i.e., the food web), affect the community species composition, and impact biogeochemical cycles (Fuhrman, 1999). Thus, at least in aquatic environments, viruses are not inert particles but actively effect ecological changes. This aspect is elaborated in Chapters 8 and 9.

Since they are obligate parasites, however, bacteriophages must survive until they encounter a suitable host cell in which to multiply. Studies show that they are density dependent, increasing with an increment in number of host bacterial species. Bacteriophages that infect *E.*

*coli*, coliphages, are widespread in waste water and other water bodies contaminated with sewage. Since the isolation and assay of coliphages is easy and inexpensive, they have been widely used as indicators of fecal pollution in particular and viruses in general. However, there is increasing evidence that the prevalence of coliphages does not correlate with other enteric viruses of concern in water and biosolids. Where coliphages are present in substantial amounts in marine environments, growth of their host bacteria (*E. coli*) in sufficiently high numbers is a precondition. In marine environments, the concentration of coliphages is usually very low since they have to cope with the prevailing sea-salt conditions and the associated ionic environment which can affect lysis, adsorption onto the host cells (and therefore infection), phage survival as well as replication. Depending on the strain however, *E. coli* can withstand a good amount of salinity. This discrepancy between host and coliphage may, at least in part, explain why the abundance of coliphages is not always a good predictor of other waterborne viruses in particular and other pathogens in general.

Viruses are mostly associated with a wide range of diseases in humans (Table 5.1), as well as other animals and plants. Viruses also act on prokaryotes (in that instance, they are referred to as bacteriophages), eukaryotic algae, protozoa and cyanobacteria, possibly sickening them too. As can be seen from Table 5.1, viral infections to humans are facilitated through a variety of means, some of which relate to the environment. Some of the connections may not be very apparent initially but even where vectors such as bats, mice, and mosquitoes are involved, the thriving and subsequent contact of these vectors with humans can be greatly influenced by environmental conditions (such as flooding) and how we maintain our surroundings. Thus, there have been numerous outbreaks of a wide range of fatal viral diseases, especially in Asia, Africa, and

**Table 5.1**   Some of the viral-related diseases that affect humans

| Virus | Disease/syndrome | Reservoirs/carrier/transmission |
|---|---|---|
| Human immunodeficiency virus (HIV) | Acquired immunodeficiency syndrome (AIDS) | Bodily fluids; intimate contact |
| Poliovirus | Poliomyelitis | Water; person-to-person |
| Measles virus (a paramyxovirus) | Measles | Aerosol |
| Hepatitis A virus | Hepatitis A | Contaminated water |
| Hantaviruses, e.g. Sin Nombre virus | Hantavirus pulmonary syndrome (HPS) | Rodents, e.g. deer mouse (*Peremyscus maniculatus*) |
| Machupo virus | Bolivian hemorrhagic fever (BHF) | Rodents, e.g. *Calomys callosus* |
| Arenoviruses | Lassa fever | Rodents as reservoirs; person-to-person |
| Rubella virus | Rubella | Vertical transmission |
| Marburg virus | African hemorrhagic fevers | Vectors include monkeys, bats, mice |
| Ebola virus | African hemorrhagic fevers | Bats? |
| Rabies virus | Rabies (also in canines) | Bites from infected canine. |
| Influenza virus | Influenza | Aerosol |
| Rift valley virus | Rift valley fever | Mosquitoes; flooding |
| Monkey B virus (*Herpesvirus simiae*) | Encephalitis | Can be latent in host over a long period. |
| Yellow fever virus | Yellow fever | Mosquitoes (*Aedes aegypti*) |
| Junìn (an arenovirus) | Venezuelan equine encephalitis (same disease in horses, donkeys, mules, and burros) | Vectors include rodents and mosquitoes |
| Norwalk and Norwalk-like viruses | Gastroenteritis | Shellfish; contaminated water |
| Coxsackie viruses (several strains) | Meningitis, respiratory tract infections | Contaminated waterborne (some airborne) |
| Lymphocytic choriomeningitis virus (LCV) | Lymphocytic choriomeningitis (fever, malaise, vomiting, etc.) | Rodents, e.g. mice, hamsters, etc. |

Central and South America within the past three decades. One of the leading causes of mortality in developing countries is infant diarrhea. Viruses such as rotaviruses (i.e. Astroviruses, Caliciviruses, and Reoviruses) and Norwalk, are a major cause of this syndrome. Thus, the associated economic importance emanating from viral diseases arising from the environment can be significant. However, the occurrence of viruses in a particular environment does not always lead to a disease in each potentially susceptible individual. The human imunodeficiency virus (HIV) has been found in feces and waste water and it was initially feared that it could potentially cause contamination problems (Ansari et al., 1992; van der Hoek et al., 1995). The disease process for any viral infection depends on the infectious agents, reservoirs and carriers, escape routes, means of entry, modes of transmission, as well as susceptibility of the host. Thus, both host and environmental factors influence the disease process. Epidemiology has conventionally considered studying the spread of diseases in animals and plants but not the ecological aspects. However, recent events that centrally involve viral infections have increasingly changed that approach. Currently, an efficient and effective approach to conducting epidemiological studies involves drawing blood samples and determining which antibody can best combat the virus of interest (Wilson et al., 1994; Mills et al., 1997).

As briefly indicated earlier, the major sources of viruses that contaminate the environment are sewage and animal waste when these products are disposed, ending up in some land-based system, such as agricultural land, forested areas, recreational areas (golf courses), and pastures. Following these applications, some viruses find their way into surface water, groundwater and soils (Derbyshire and Brown, 1978; Vaugh et al., 1978). During treatment of sewage and/or during agricultural operations such as sprinkler irrigation using tainted waste water, some viruses are also introduced in the atmosphere as aerosols, traveling considerably long distances (Baylor et al., 1977a, b; Katzenelson et al., 1976). A majority of the world population depends on on-site toilet facilities such as pit latrines and septic tanks to dispose of domestic fecal matter. Some of these leak, discharging fecal matter into groundwater. More than 100 different viruses of human origin are known, with infected individuals shedding in excess of $10^6$ viral particles per gram feces, irrespective of whether the individuals show any signs of disease or not (Slade and Ford, 1983). Because of the complexity of detection, data about the prevalence of viruses in the environment, for example in water or biosolids, reveal more information about the distribution of testing laboratories rather than the distribution of viruses per se. Testing for viruses in the environment is almost unheard of in developing countries but their presence is believed to be high. In a number of these countries, sewage treatment and/or disposal is less efficient and more likely to introduce viruses into the environment. However, even in countries where testing for viruses is routinely done, the information obtained through conventional methods, i.e., cell culture, takes a long time to obtain. A typical test for enteric viruses in biosolids and water routinely takes a month or longer to obtain the final results. By that time the results obtained would be of diminished value. Furthermore, no cell line can produce a cytopathic effect with all the diverse array of viruses occurring in the environment.

Viral infections have been implicated in a number of disease outbreaks, even in developed countries (Scandura and Sobsey, 1997; Dowd et al., 1998). Some viruses of significant concern in the environment, notably Hepatitis A, rotaviruses, Norwalk, as well as Norwalk-like viruses, are not readily detectable by conventional cell culture techniques. For example, Reoviruses are readily and frequently detected in environmental samples showing fecal contamination but have a high survival rate in the environment. However, in cell culture, they can interfere with the replication of enteroviruses, giving false negative assays. Hepatitis A virus annually causes more than 25,000 cases of hepatitis in the United States alone (LeChevallier et al., 1999). With the current advancements in molecular biology, these viruses can be detected in environmental samples, even when they are present in fairly low quantities (Atmar et al, 1996; Schwab et al., 2000; Le Guyader et al., 2000). However, when detected by molecular techniques alone, it still remains questionable as to which of the viruses detected are viable and therefore potentially infectious. Combining both molecular techniques and conventional cell culture techniques (say, by using a combination of multiline cell culture) can greatly reduce the time needed to detect these viruses, detect virus types that are hard to quantify solely through cell culture, and resolve issues of viability in a shorter timeframe. These approaches are still under development and validation.

## 5.3 SURVIVAL OF VIRUSES IN THE ENVIRONMENT

Unlike bacteria which are independent living organisms, viruses cannot be fought by targeting their life processes because they have no processes of their own. Everything that viruses do is tied to the host cells' mechanics. Our bodies fight viral infections using the immune system. The immunity response occurs as a reaction between a foreign body (antigen) and a product of the host (antibody) produced in response to the antigen. Each antibody is specific for the antigen that stimulates its production. Once the antigen and antibody combine, the antigen is inactivated and rendered harmless. To continue attacking other antigens of the same type, the antibodies produced following the initial stimulation are released into the bloodstream, circulating with the serum to neutralize antigens that are identical to that which initially stimulated the production in the first place, thus showing extreme specificity and conferring immunity to the host (Fig. 5.2). Such acquired immunity is for life in some cases and this concept has been widely utilized to control viral diseases through vaccination. From a clinical perspective, vaccines are designed to stimulate the body's own immune system to develop antibodies to a virus by infecting it with some of the virus vaccine. There are two types of vaccines, viz. killed vaccines which use dead viral material, and live, attenuated vaccines which use a small amount of a weakened live virus. The acquired immunity may be natural or artificial depending on whether the antigen-antibody reaction occurs in the host itself or not. More specifically, the target hosts can be vaccinated with a weakened (attenuated) or a small dose of the disease causative agent in response to which the subject forms antibodies and becomes protected thereafter from the infection (naturally acquired immunity). It is important to note that immunity is not only relevant to viral infections, but also applicable to several other sources of infection.

The ultimate survival of viruses in the environment depends on both biotic and abiotic factors (Table 5.2). Their inactivation is encouraged by low humidity, high temperatures, pH

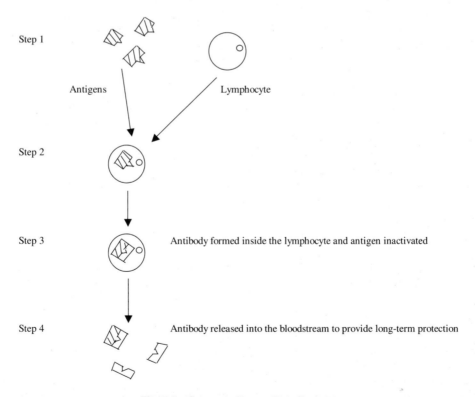

Step 1

Antigens     Lymphocyte

Step 2

Step 3       Antibody formed inside the lymphocyte and antigen inactivated

Step 4       Antibody released into the bloodstream to provide long-term protection

**Fig. 5.2** Concept of immunity to the host.

**Table 5.2** Some biotic and abiotic factors that affect the survival of viruses in the environment

| Factor | Remarks |
|---|---|
| **Abiotic** | |
| Chemical | |
|    pH | Increasing the pH to 11-12 rapidly inactivates most viruses as the capsid configuration is deformed, which in turn reduces infectivity and adsorption onto solids. |
|    Dissolved oxygen | Some viruses prefer anaerobic or microaerophilic environments |
|    Cations | Some di- and trivalent cations, such as $Cu^{2+}$, $Mg^{2+}$, and $Al^{3+}$, increase the adsorption capacity of viruses on clay particles. However, most heavy metals (e.g. $Pb^{2+}$, $Hg^{2+}$, etc.) are virucidal. |
|    Organic pollutants | Some may be virucidal. |
| Physical | |
|    Temperature | High temperatures are detrimental but repeated freezing and thawing also inactivate viruses. |
|    Pressure | Inactivated under high osmotic pressure (i.e. under saline conditions). On the contrary, most viruses are not adversely affected by hydrostatic pressure. |
|    Aggregation | May inactivate some viruses but can also protect some types of viruses against inactivation. |
|    Irradiation | UV from natural solar radiation inactivates some viruses either directly, or indirectly by actively affecting the virus hosts (e.g. bacteria and algae). |
| **Biotic** | |
|    Type of virus | Differences in resistance between viruses and viral strains to inactivation. |
|    Predation | Some types of viruses may be less susceptible to predation. |
|    Host activity | Some bacteria enzymes can inactivate viruses, e.g. *Pseudomonas aeruginosa* enzymes inactivate Coxsackie A9 virus and Hapatitis A virus in soil (Nasser et al. 2002). |

extremes, and high solar irradiation, to name but a few. The availability of water, for example through rainfall, modifies the number and diversity of potential hosts for viruses, thus indirectly influencing the occurrence of viruses and their vectors. A clear example of the complex influence of rainfall on viruses and their transmission is the hantavirus pulmonary syndrome outbreak in 1993 at Four Corners in the USA. Unusually heavy rainfall attributed to the El Niño weather phenomenon increased the production of piñon nuts, an important food for deer mice, the vectors for the virus. On the other hand, for some viruses, excessive rain which results in flooding tends to minimize numbers and movement of most common aerial virus vectors but probably does not greatly affect soil-inhabiting vectors (e.g. fungi and nematodes) directly. Arming ourselves with such environmental information can increase our ability to predict the potential or increase in risks to some area-specific viral disease outbreaks and devise control measures. Such strategies would, understandably, require multidisciplinary collaboration.

Temperature affects the activity of passive vectors and hence viral transmission, making it less effective at 24°C or above. At 55 - 65°C, most viruses are inactivated, though some thermotolerant variants can survive temperatures as high as 75°C and extremely acidic environments (Rice et al., 2001; Rachel et al., 2002). Significantly greater tolerance for high temperatures is shown by phages for thermophile bacteria such as *Micropolyspora, Thermoactinomyces*, and *Bacillus stearothermophillus* (Williams et. al., 1989). There is growing interest in studying high temperature-tolerant viruses and how they interact with the environment. Understanding how they are able to survive under extreme environments would give insights into how life exists at high temperatures. When inactivation at high temperatures occurs, it is due to the denaturation of the protein coat and possibly the nucleic acids. Once damaged, the coat protein and related capsid are unable to attach to the host cell while damage to nucleic acids disables transcription. At very low temperatures, viruses are not inactivated and, as a matter of fact, are routinely stored at $-70°C$ in protein solutions. However, their infectivity is negatively affected by repeated freezing and thawing (Table 5.3), indicating that they can be adversely affected by diurnal and seasonal cycles of temperature. Temperatures in temperate regions can oscillate between below and above freezing over short periods of time. Temperature possibly explains, in part, why most viral diseases are more widespread in the tropics where temperatures are more stable compared to temperate zones.

Most enteric viruses are sensitive to a variety of organic compounds such as the hetercyclic dyes, methylene blue or neutral red. Thus, some of these compounds can be used in specific situations to sanitize environments contaminated by viruses. However, some chemicals, such as formaldehyde, cause viruses to lose their infectivity without destruction of their antigenic properties while others, such as phenol, destroy the capsid without altering the nucleic acid. The presence of $O_2$ can also inactivate some viruses, particularly those of enteric origin. Ultraviolet light also affects viral nucleic acid, the lethal radiation affecting the pyrimidine rings, subsequently forming thymine dimers or causing hydration of acid residues. Sunlight plays a role in inactivating some viruses in natural waters. As a matter of fact, ultraviolet light is currently used to disinfect water by some treatment plants. Treatment of water, waste water and sludges is discussed in Chapter 13.

Viruses are negatively charged biocolloid particles (Block, 1983; Dowd et al., 1998). This characteristic enables some of them to be adsorbed quite effectively on inanimate surfaces such as glass, benches, dust particles, soil, clothing, surviving for long periods of time. The survival of viruses on such surfaces has implications in their transmission within hospital environments. Studies by Mahl and Sadler (1975) showed the persistence of viruses on glass slides at low (7%) and at high (96%) relative humidity (Table 5.4). Adenovirus 2 and Poliovirus 2 survived on glass at desiccation levels even after 8 weeks, though the population of the latter virus had greatly diminished by this time. The survival of Coxsackie virus was greatly reduced to below detection after 8 weeks irrespective of the moisture content. Herpes simplex virus, on the other hand, survived for long durations only on inanimate surfaces and only at low relative moisture content (Table 5.4). On another note, results indicate that the survival of either free or cell-associated HIV is rapidly inactivated in the

**Table 5.3**    Inactivation of poliovirus (Sabin 3) after freezing and thawing (Jjemba, unpubl.).[a]

| Buffalo Green monkey cell passage | Number of days to attain 70% cytopathic effect with different initial concentrations of the virus stock | | | |
|---|---|---|---|---|
| | 200 PFU[b] | 20 PFU | 2 PFU | 0.2 PFU |
| 145 | 3/3[c] | 3/3 | 3/3 | 6/- |
| 146 | 4/4 | 4/4 | 4/- | —[d] |
| 147 | 3/3 | 5/5 | — | — |

[a]A fresh virus stock was prepared and applied to a fully confluent Buffalo Green monkey cell culture (passage 145) in duplicate T-25 cm flasks. The flasks were incubated at 37°C. The remaining virus stock for each concentration was frozen at $-70°C$ and then thawed to inoculate another set of cell monolayer (passage 146). The stock was frozen once more, thawed, and used to inoculate cells in passage 147. In each case, the flasks, and a negative control, were incubated for 14 days with the maintenance medium replenished after 7 days.
[b]Plaque-forming units.
[c]Days until 70% cytopathic effect was visually detected in the duplicate flasks.
[d]No cytopathic effect detected after 14 days of incubation at 37°C.

**Table 5.4**  Survival of various viruses on glass slide at two relative humidity levels (25°C)

| Type of virus | Relative humidity (%) | Density of viruses recovered per slide (in CCID$_{50}$/ml) or (PFU/ml)[a] | | | |
|---|---|---|---|---|---|
| | | 1 day | 1 week | 2 weeks | 8 weeks |
| Poliovirus 2 | 96 | $1.2 \times 10^8$ | $4.1 \times 10^7$ | $6.0 \times 10^6$ | NVD |
| | 7 | $1.1 \times 10^6$ | $7.5 \times 10^5$ | $3.2 \times 10^5$ | $2.0 \times 10^2$ |
| Herpes simplex virus | 96 | $4.5 \times 10^3$ | NVD[b] | NVD[b] | –[c] |
| | 7 | $1.2 \times 10^2$ | $2.0 \times 10^3$ | $3.9 \times 10^3$ | $9.2 \times 10^2$ |
| Adenovirus 2 | 96 | $10^{8.25}$ | $10^{6.75}$ | $10^{6.0}$ | $10^{2.75}$ |
| | 7 | $10^{7.5}$ | $10^{6.25}$ | $10^{7.0}$ | $10^{5.25}$ |
| Coxsackie virus | 96 | $10^{5.25}$ | $10^{4.0}$ | $10^{2.5}$ | NVD[d] |
| | 7 | $10^{4.5}$ | $10^{2.25}$ | $10^{2.5}$ | NVD[d] |

[a]Adenovirus 2 and Coxsackie virus were enumerated per slide (in CCID$_{50}$ml$^{-1}$) whereas Poliovirus 2 and Herpes simplex virus were enumerated as PFUml$^{-1}$. PFU: Plaque-forming units; CCID$_{50}$: Cell culture infection dose and refers to the highest viral dilution (i.e., the smallest amount of viruses) capable of producing morphological changes under the microscope in 50% of the culture.
[b]No virus detected at 1 week for Herpes simplex virus.
[c]-Not determined.
[d]No virus detected at 3 weeks for Coxsackie virus.
*Source*: Mahl and Sadler, 1975.

environment (Moore, 1993). The jury is still out but no transmission of HIV and subsequent development into AIDS has been associated to water or biosolids, strongly indicating that this is an unlikely route of transmission

The presence of organic matter may also influence the number of viruses by influencing the metabolic rate of the host bacterium. Aggregation helps some viruses persist in the environment. In soil, viruses can adsorb on clay surfaces and this may in fact provide some protection to them or, in cases where the soil has a high cation exchange capacity, inactivate them. Subsequent survival of viruses in the environment may also be, at least in part, indirectly enhanced due to better survival of their host bacteria as the bacteria attach to sediments, clay minerals, and biofilms, becoming better protected from disinfectants and attack by predators (Quignon et al., 1997). Various researchers have also used the poliovirus to show that following the application of biosolids or waste water, most of the enteroviruses that may be present in these materials are adsorbed in the top few centimeters of the soil horizon (Landry et al., 1980; Wekerle, 1986). However, under certain conditions, such as rainfall,

favorable soil pH (near neutral), soil texture, changes in the water table, and degree of topographic slope, some of the adsorbed viruses are desorbed and migrate farther down into the soil profile, eventually reaching groundwater (Scandura and Sobsey, 1997). The problem is compounded by the fact that once in deeper soil layers, viruses appear to survive for longer periods of time compared to those adsorbed on surface soils (Lance et al., 1976). The adsorption of viruses on soils is both virus-strain and soil dependent, with some strains only poorly adsorbing to soils (Landry et al., 1979). Both the size and the isoelectric point of viral particles influence adsorption (Dowd et al., 1998). Thus, the larger the viral particle, the higher its overall number of charges available to interact with the environmental matrix (e.g. soil or sediments). Likewise, the isoelectric point, that is the pH at which the viral particle shows a net neutral charge, correlates negatively with the level of viral particle adsorption on soil. A low isoelectric point is usually associated with a high net negative charge on the surface of the viral particle (Dowd et al., 1998). Viral strains that adsorb poorly on soils, therefore, tend to be frequently encountered in sewage-recharged aquifers (Vaugh et al., 1978; Marzouk et al., 1979). Viruses can also be inactivated at pH extremes below 4.0 or above 8.0 (Table 5.5) though some of them can remain relatively stable at pH conditions that fall outside this range.

Biological factors that contribute to virus inactivation in sea water include the direct predation of the host bacteria by protozoa and other predators (such as *Bdellovibrio bacteriovorus*) and by lysis of some marine bacteria such as *Vibrio marinus*. Inactivation of some viruses has also been attributed to enzymes such as proteases generated by bacteria (Block, 1983; Deng and Cliver, 1995). Prophages sometimes modify the polysaccharide surface antigens of their host bacteria, rendering the latter resistant to attack from extraneous viruses. Furthermore, phage genes can directly impact the pathogenicity of bacteria that infect vertebrates. An example of this phenomenon is the increase in pathogenicity of *Neisseria/Clostridia* which cause diphtheria, botulism or toxic shock in vertebrates (Betley et al., 1992). Some viruses can also be inactivated when they adsorb on surfaces and colloids but the major contributor to their inactivation appears to be ultraviolet light, particularly UV-B (Noble and Fuhrman, 1997; Fuhrman, 1999).

It is important to emphasize that the potential inactivation of viruses by the factors outlined above does not entirely remove the threat of viruses from the environment since they are sometimes found a great distance away from the point of discharge. For example, enteroviruses (poliovirus, echovirus, Coxsackie virus) were detected as far away as 13 km from the point of sewage discharge, indicating that in freshwater conditions, viruses can survive for longer periods (Bitton, 1980). Dahling and Safforman (1979) also encountered substantial concentrations of human enteric viruses 300 km downstream from the source of viral contamination after 7 days of transport of the contaminated water in the middle of winter in Tanana River (Alaska). Thus, viral inactivation in the environment is an unpredictable and slow process.

## 5.4 OTHER VIRUS-LIKE PARTICLES

The best known of these are viroids (RNA) and prions (protein). Viroids are very tiny and composed of single-stranded pieces of RNA. The RNA is circular, pairing internally as it coils, thus resembling a double-stranded structure. However, because they are single stranded, viroid RNA does not encode for any protein. They replicate in the nucleus of the host cell. They are only transmitted by vector (since naked RNA cannot exist and remain viable in the environment) or maternally through vegetative material. If not maternally transmitted, their sig-

**Table 5.5** Survival of a streptomycete phage in soils of different pH

| Soil type | Horizon | Soil pH | Phage recovered (PFU g$^{-1}$ soil)[a] | |
| --- | --- | --- | --- | --- |
| | | | 0 h | 200 h |
| Podzol | F1 | 3.7 | 0 | 0 |
| | F2/H layer | 3.9 | 0 | 0 |
| | A2 | 4.0 | 0 | 0 |
| | A1 | 4.2 | $1 \times 10^2$ | 0 |
| Clay | | 6.1 | $3 \times 10^3$ | $3 \times 10$ |
| Garden soil | | 7.0 | $2 \times 10^3$ | $2 \times 10^2$ |
| Saline soil | | 8.0 | $6 \times 10^4$ | $3 \times 10^3$ |

[a]The initial concentration of $1 \times 10^6$ PFU was added to g$^{-1}$ dry soil.
*Source*: Sykes et al., 1981 reprinted with permission from the Society for General Microbiology.

nificance in the environment relates to conditions that favor the existence of their vectors, such as aphids and other insects which feed by sucking. They are of consequence to plants and seem to be directly unimportant to human health and well-being.

Prions are microscopic infective protein particles (33-35 kD) which are similar to viruses but lack DNA or RNA. They are able to somehow replicate and modify the activities of enzymes rather than nucleic acids without the help of autonomous nucleic acid genes (Prusiner, 1992). They are not destroyed or modified by chemical and physical processes that affect nucleic acids. Thus, chemicals such as nucleases and hydroxyamines do not adversely affect them, possibly because they are devoid of nucleic acids. However, they replicate in permissive cells. Although the mechanism by which prions replicate is unclear, some investigations suggest that it is driven by a specific polynucleotide. They transmit their identity to their progeny by an epigenetic process that includes the refolding of a host of proteins. Prions were largely unheard of until 1986 when they first caused bovine spongiform encephalopathy (BSE; also known as mad cow disease) in England, disappearing and eventually reappearing in the same country a decade later. The disease is characterized by the prions inducing changes in molecules which destroy neurons in the brain. The cow disease is similar to the one in sheep called scrapie (so named because infected sheep would scrape their hide until it was raw), known since the 1940s. However, a critical search from history indicates that prion-related diseases, notably scrapie, date back as far as 1732 although at that time efforts to deliberately transmit this disease to healthy sheep remained unsuccessful; the goal was achieved in 1936 (Prusiner, 1992). The currently more frequent mad cow disease is also similar to kuru and Creutzfeldt-Jacob disease in humans both of which are 100% fatal and attributed to infectious agents resistant to procedures that inactivate nucleic acids.

Prion diseases are collectively referred to as transmissible spongiform encephalopathies (TSE) because of their effect on the nuerons, and ultimately, the brain. Other diseases attributable to prions include TME, CWD, and GSS (Table 5.6). In all TSE cases, a normal host-prion protein ($PrP^C$) is converted, through unknown mechanisms, to a disease-specific prion protein ($PrP^{SC}$ or $PrP^{TSE}$). Some TSEs such as GSS possibly arose due to a point mutation in cases with genetically-based CJD whereas Kuru may have been associated with ritualistic cannibalism by some tribe in New Guinea. Kuru has become increasingly rare since this practice stopped. As to how the agent moves from one host to another is still not known but it is believed that the agent that causes scrapie jumped to cattle through feeding them infected sheep parts in their feed (Reisner, 2001). This theory has reinforced the

**Table 5.6** Examples of transmissible spongiform encephalopathies (TSEs)

| Disease | Natural host | Distribution and occurrence |
|---|---|---|
| Bovine spongiform encephalopathy (BSE) | Cattle | Western Europe; increasing frequency |
| Chronic wasting disease (CWD) | Wild ruminants e.g. Mule, deer, and elk | North America |
| Creutzfeldt-Jacob disease (CJD) | Humans | Worldwide; 1:100,000 y |
| Gerstmann-Stäussler-Sheinker syndrome (GSS) | Humans | Genetic; rare |
| Kuru | Humans | Papua New Guinea |
| Fatal familial insomnia (FFI) | Cats | Genetic; very rare |
| Scrapie | Sheep and goats | Mostly western Europe |
| Transmissible mink encephalopathy (TME) | Mink (*Mustela vison*) | North America and northern Europe |

indications that kuru was indeed transmitted as outlined above and certainly justifies peoples' fears about eating beef products in countries where BSE outbreaks have been reported. In the livestock industry, it was a common practice to supplement the basic meal of grains with offal leftovers from butchered animals, but this practice has reportedly stopped since recent outbreaks of BSE in Europe. An increased occurrence of CJD in dairy farmers in Western Europe suggests a connection of this disease with BSE cases in their herds (Almond et al., 1995).

Except for kuru, the occurrence of TSEs in developing countries is not well documented possibly because it mostly goes undetected. The proper detection of prion proteins requires very refined methods by immunoblotting and immunochemistry. Approximately $10^5$ prion protein molecules are needed for one infection unit (Reisner, 2001). The infectivity of prion-related diseases is resistant to UV and ionizing radiation. However, several other normal disinfection procedures which are effective against viruses are not readily effective against prions (Table 5.7). This observation may have important implications about the survival of prions in the environment. Seemingly effective in inactivating prions is treatment with sodium hypochlorite (Chlorox) alone or treatment with sodium hydroxide (1-2M) combined with either autoclaving or boiling. Some prions have been reported to survive after ashing at 360°C (Brown et al., 1990). Taylor (2001) observed that merely increasing the autoclaving time and/or temperature can, in fact, increase the thermostability of some prions, indicating the hardiness of these materials. Most simplistic, but reportedly effective against prions is boiling the agents in 1M sodium hydroxide for one minute (Table 5.7). The lack of effective therapies for prion diseases, all of which are fatal, still poses a significant challenge, although treatment with polyene antibiotics has shown

**Table 5.7** Inactivation of TSE agents by various procedures

| Procedure | Little titre reduction | Significant titre reduction | No infection detected |
|---|---|---|---|
| **Chemical methods** | | | |
| Sodium hypochlorite (1.65% available chlorine) | | | ✓ |
| Sodium hydroxide (1-2 M for 2 hours); Sodium dichoroisocyanurate (1.65% available chlorine); Chaotropes e.g. guanidine thiocynate; 95% Formic acid; | | ✓ | |
| Formaldehyde; Organic solvents; Hydrogen peroxide; Chlorine dioxide; Iodine and iodates; Peracetic acid; Phenolic disinfectants; Proteolytic enzymes; | ✓ | | |
| **Physical methods** | | | |
| UV irradiation; gamma irradiation; microwave irradiation | ✓ | | |
| Autoclaving (134-138°C for 18 minutes or 132°C for 1 hour); or dry heat at over 200°C | | ✓ | |
| **Combined chemical and physical methods** | | | |
| Autoclaving (121°C for 0.5 to 1.5 hours) in/or after 1M sodium hydroxide treatment; or boiling in 1M sodium hydroxide for 1 minute | | | ✓ |
| Autoclaving (121°C) in 5% sodium dodecyl sulfate; or heating in 1M HCl | | ✓ | |
| Autoclaving after aldehyde, alcohol or dry heat treatment | ✓ | | |

Compiled from Brown et al., 1990 and Taylor, 2001

some promise by prolonging onset of the clinical signs in animal models (Demaimay et al., 1997). Much information about prions is still lacking but they have been noted to possess considerable structural plasticity. Their survival over time in the environment is also still largely not known. One wonders whether specific environmental factors may be favorable in converting normal PrP$^C$ protein into disease-specific protein (PrP$^{TSE}$).

# References

Ackermann H.W. (1992). Frequency of morphological phage descriptions. *Arch. Virol.* **124:** 201-209.

Almond J.W., P. Brown, S.M. Gore, A. Hofman, P.W. Wientjens, R.M. Ridley, H.F. Baker, G.W. Roberts, and K.L. Tyler (1995). Greutzfeldt-Jakob disease and bovine spongiform encephalopathy: Any connection? *Brit. Med. J.* **11:** 1415-1421.

Ansari S.A., S.R. Farrah, and G.R. Chaudhry (1992). Presence of human immunodeficiency virus nucleic acids in wastewater and their detection in polymerase chain reaction. *Appl. Environ. Microbiol.* **58:** 3984-3990.

Atmar R.L., F.H. Neill, C.M. Woodley, R. Manger, G.S. Fout, W. Burkhart, L. Leja, E.R. McGovern, F. Le Guyader, T.G. Metcalf, and M.K. Estes (1996). Collaborative evaluation of a method for the detection of Norwalk virus in shellfish tissues by PCR. *Appl. Environ. Microbiol.* **62:** 254-258.

Baylor E.R., M.B. Baylor, D.C. Blanchard, L.D. Syzdek, and C. Appel (1977a). Virus transfer from surf to wind. *Science* **198:** 575-580.

Baylor E.R., V. Peters, and M.B. Baylor (1977b). Water-to-air transfer of virus. *Science* **197:** 763-764.

Betley M.J., D.W. Borst, and L.B. Regassa (1992). Staphylococcal enterotoxins, toxic shock syndrome toxin and streptococcal pyrogenic exotoxins—a comparative study of the molecular biology. *Chem. Immunol.* **55:** 1-35.

Bitton G. (1980). Introduction to Environmental Virology. Wiley-Interscience, New York, NY.

Block J.-C. (1983). Viruses in environmental waters. *In*: G. Berg (ed.). Viral Pollution of the Environment. CRC Press, Boca Raton, FL, pp. 117-145.

Brown P., P.R. Liberski, A. Wolff, and D.C. Gajdusek (1990). Resistance of scrapie agent to steam autoclaving after formaldehyde and limited survival after ashing at 360°C: Practical and theoretical implication. *J. Infec. Dis.* **161:** 467-472.

Cooper J.I. (1995). Virus and the Environment. Chapman and Hall, New York, NY.

Dahling D.R. and R.S. Safferman (1979). Survival of enteric viruses under natural conditions in subarctic river. *Appl. Environ. Microbiol.* **38:** 1103-1110.

Demaimay R., K.T. Adjou, V. Beringue, S. Demart, C.I. Lasmézas, J.-P. Deslys, M. Seman, and D. Dormont

(1997). Late treatment with polyene antibiotics can prolong the survival time of scrapie-infected animals. *J. Virol.* **71:** 9685-9689.

Deng M.Y. and D.O. Cliver (1995). Antiviral effects of bacteria isolated from manure. *Microb. Ecol.* **30:** 43-

Derbyshire J.B. and E.G. Brown (1978). Isolation of animal viruses from farm livestock waste, soil and water. *J. Hyg.* **81:** 295-302

Dimmock N.J. and S.B. Primrose (1994). Introduction to Modern Virology. Blackwell Science, Cambridge, MA.

Dowd S.E., S.D. Pillai, S. Wang and M.Y. Corapcioglu (1998). Delineating the specific influence of virus isoelectric point and size on virus adsorption and transport through sandy soils. *Appl. and Environ. Microbiol.* **64:** 405-410.

Earn D.J.D., J. Dushoff, and S.A. Levin (2002). Ecology and evolution of the flu. *Trends Ecol. Evolu.* **17:** 334-340.

Fuhrman J. (1999). Marine viruses and their biogeochemical and ecological effects. *Nature* **399:** 541-548.

Katzenelson E., I. Buium and H.I. Shuval (1976). Risk of communicable disease infection associated with wastewater irrigation in agricultural settlements. *Science* **194:** 944-946.

Lance J.C., C.P. Gerba, and J.L. Melnick (1976). Virus movement in soil columns flooded with secondary sewage effluent. *Appl. Environ. Microbiol.* **32:** 520-526.

Landry E.F., J.M. Vaughn, and W.F. Penello (1980). Poliovirus retention in 75-cm soil cores after sewage and rainwater application. *Appl. Environ. Microbiol.* **40:** 1032-1038.

Landry E.F., J.M. Vaughn, M.Z. Thomas, and C.A. Beckwith (1979). Adsorption of enteroviruses to soil cores and their subsequent elution by artificial rainwater. *Appl. Environ. Microbiol.* **36:** 680-687.

LeChevallier M.W., M. Abbaszadegan, A.K. Campor, C.J. Hurst, G. Izaguirre, M.M. Marshall, D. Naumoritz, P. Payment, E.W. Rice, J. Rose, S. Schaub, T.R. Slifko, B.D. Smith, H.W. Smith, C.R. Sterling and M. Stewart (1999). Emerging pathogens-viruses, protozoa and algal toxins. *J. AWWA* **91:** 110-121.

Le Guyader F., L. Haugarreau, L. Miossec, E. Dubois, and M. Pommepuy (2000). Three-year study to assess human enteric viruses in shellfish. *Appl. Environ. Microbiol.* **66:** 3241-3248.

Mahl M.C. and C. Sadler (1975). Virus survival on inanimate surfaces. *Can. J. Microbiol.* **21:** 819-823.

Marzouk Y., S.M. Goyal, and C.P. Gerba (1979). Prevalence of enteroviruses in groundwater of Israel. *Ground Water* **17:** 487-491.

Mills J.N., T.G. Ksiazek, B.A. Ellis, P.E. Rollin, S.T. Nichol, T.L. Yates, W.L. Gannon, C.E. Levy, D.M. Engelthaler, T. Davis, D.T. Tanda, J.W. Frampton, C.R. Nichols, C.J. Peters, and J.E. Childs (1997). Patterns of association with host and habitat: Antibody reactive with Sin Nombre virus in small mammals in the major biota communities of the South Western United States. *Amer. J. Trop. Med. Hyg.* **56:** 273-284.

Moore B.E. (1993). Survival of human immunodeficiency virus (HIV), HIV infected lymphocytes, and poliovirus in water. *Appl. Environ. Microbiol.* **59:** 1437-1443.

Nasser A.M., R. Glozman and Y. Nitzan (2002). Contribution of microbial activity to virus reduction in saturated soil. *Water Res.* **36:**2589-2595.

Noble R.T. and J. Fuhrman (1997). Virus decay and its causes in coastal waters. *Appl. Environ. Microbiol.* **63:** 77-83.

Peters C.J. and M. Olshaker (1997). Virus Hunter: Thirty Years of Battling Hot Viruses around the World. Anchor Books, Doubleday, New York, NY.

Prusiner S.B. (1992). Molecular biology and genetics of neurogenerative diseases caused by prions. *Adv. Virus Res.* **41:** 241-280.

Quignon F., L. Kiene, Y. Levi, M. Sardin, and L. Schwartzbrod (1997). Virus behaviour within a distribution system. *Water Sci. Tech.* **35:** 311-318.

Rachel R., M. Bettstetter, B.P. Hedlund, M. Häring, A. Kessler, K.O. Stetter, and D. Prangishvili (2002). Remarkable morphological diversity of viruses and virus-like particles in hot terrestrial environments. *Archi. Virol.* **147:** 2419-2429.

Reisner D. (2001). The Prion Theory. Background and Basic Information. *In*: H.F. Rabenau, J. Cinatl, and H.W. Doerr (eds.). Prions. A Challenge for Science, Medicine and Public Health System. Karger, New York, NY, pp. 7-20.

Rice G., K. Stedman, J. Snyder, B. Wiedenheft, D. Willits, S. Brumfield, T. McDermott, and M.J. Young (2001). Viruses from extreme thermal environments. *Proc. Natl. Acad. Sci. (USA)* **98:** 13341-13345.

Scandura J.E. and M.D. Sobsey (1997). Viral and bacterial contamination of groundwater from on-site sewage treatment systems. *Water Sci. Tech.* **35:** 141-146.

Schwab K.J., F.H. Neill, R.L. Fankhauser, N.A. Daniels, S.S. Monroe, D.A. Bergmire-Sweat, M.K. Estes, and R.L. Atmar (2000). Development of methods to detect "Norwalk-like viruses" (NLVs) and Heptatis A virus in delicatessen foods: Application to a food-borne NLV outbreak. *Appl. Environ. Microbiol.* **66:** 213-218.

Slade J.S. and B.J. Ford (1983). Discharge to the environment of viruses in wastewater, sludges, and aerosols. *In*: G. Berg (ed.). Viral Pollution of the Environment. CRC Press Inc., Boca Raton, FL (USA).

Sykes I.K., S. Lanning, and S.T. Williams (1981). The effect of pH on soil actinophage. *J. Gen. Microbiol.* **122:** 271-280.

Taylor D.M. (2001). Resistance of transmissible spongiform encephalopathy agents to decontamination. *In*: H.F. Rabenau, J. Cinatl, and H.W. Doerr (eds.). Prions. A Challenge for Science, Medicine and Public Health System. Karger, New York, NY, pp. 58-67.

Van der I loek L., R. Boom, J. Goudemit, F. Snijders, and C.J. Sol (1995). Isolation of human immunodeficiency virus type 1 (HIV-1) RNA from feces by a simple method and difference between HIV-1 subpopulations in feces and serum. *J. Clin. Microbiol.* **33:** 581-588.

Vaugh J.M., E.F. Landry, L.J. Baranosky, C.A. Beckwith, M.C. Dahl, and N.C. Delihas (1978). Survey of human virus occurrence in wastewater recharged groundwater on Long Island. *Appl. Environ. Microbiol.* **36:** 47-51.

Vaugh J.M., E.F. Landry, L.J. Baranosky, C.A. Beckwith, M.C. Dahl, and N.C. Delihas (1978). Survey of human virus occurrence in wastewater recharged groundwater on Long Island. *Appl. Environ. Microbiol.* **36:** 47-51.

Wekerle J. (1986). Agricultural use of sewage sludge as a vector for transmission of viral diseases. *In*: J.C. Block, A.H. Havelaar, and P. L. Hermite (eds.). Epidemiological Studies of Risks Associated with the Agricultural Use of Sewage Sludge: Knowledge and Needs. Elsevier Appl. Sci. Publ., New York, NY, pp. 106-122.

Wilson M.L., L.E. Chapman, D.B. Hall, E.A. Dykstra, K. Ba, H.G. Zeller, L.M. Traorelazimana, J.P. Hervy, K.J. Linthicum, and C.J. Peters (1994). Rift Valley fever in rural northern Senegal: Human risk factors and potential vectors. *Amer. J. Trop. Med. Hyg.* **50:** 663-675.

Williams S.T., M. Goodfellow and G. Alderson (1989). Genus Streptomyces Waksman and Henrici 1943, 339AL. In: S.T. Williams, M.E. Sharpe, and J.G. Holt (eds) Bergey's Manual of Systematic Bacteriology. Volume 4 pp. 2452-2492. Williams and Wilkins, Baltimore, MD.

# 6

# Methods in Environmental Microbiology

Environmental microbiology focuses on synecological approaches which emphasize the distribution, abundance, and behavior of microorganisms in their natural habitats and how these are influenced by both biotic and abiotic factors. The facets of microbiology, including environmental, food, industrial, medical, and soil microbiology, virtually rely on a reductionistic approach to advance our knowledge. Implicit in this approach is the common practice of using small samples as representative of what is actually happening in a particular environment. The precision of such reductionistic approaches is ideally aimed at reflecting the behavior of microorganisms as populations in their natural environment. However, when we begin with a field-derived environmental sample and make microscopic examinations, we are more likely to impose artifacts which may ultimately affect the results obtained. During sampling and sample-handling, therefore, an effort should be made to minimize artifacts in order to clearly understand the structure and activity of microorganisms in their natural habitats. In some instances, this still requires the development of new methodologies and improvement of those that currently exist. In this chapter, we broadly analyze the general methods used in environmental microbiology and assess their pros and cons.

Our present knowledge of the biology of prokaryotes, protozoa, fungi and algae is mostly based on investigations in pure (axenic) cultures. Key questions such as which microorganisms are present in a particular environment and what are their functions, often get only partially answered. For example, if only culturing techniques are applied to assess prokaryotes, it has been repeatedly shown that only a portion of the diversity is retrieved compared to what is observed when direct enumerations are made in soil, sediments, fresh and marine water, sewage, and a variety of other environments (Bohlool and Schmidt, 1980; Ward et al., 1990; Amann et al., 1995). This reality is becoming increasingly recognized for eukaryotes as well (Tuomi et al., 2000; Moreira and López-García, 2002). It so happens that at any one time culturing microorganisms from the natural environment will always leave some organisms uncultured either because they are nonculturable or because they are culturable but the medium used does not favor their growth (Fig. 6.1). We examine these possibilities in the next section but in highlighting them should not underestimate the importance of autecological approaches, which are helpful in characterizing morphology and biochemical and physiological properties. Instead the importance of carrying out studies beyond autecology is emphasized in achieving

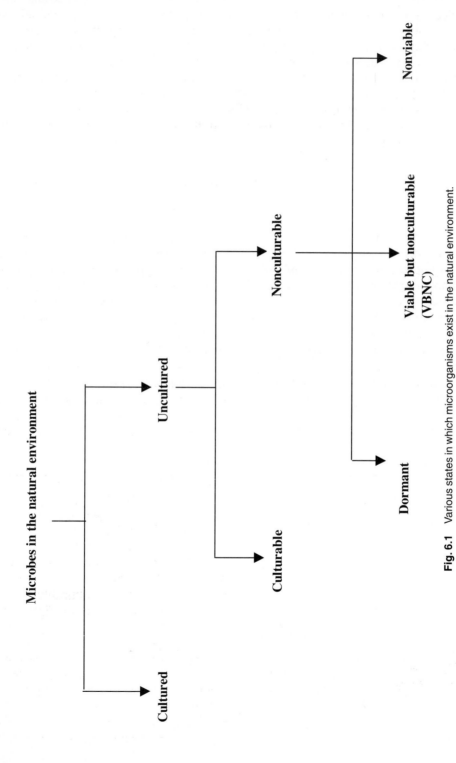

**Fig. 6.1** Various states in which microorganisms exist in the natural environment.

major ecological relevance. Studying microorganisms in natural habitats is often aimed at determining the identity of existing microorganisms, their distribution, their densities (enumeration), as well as their metabolic (e.g. growth rates) and physiologic potential or state. The methodologies in environmental microbiology are therefore discussed here keeping these broad objectives in mind.

## 6.1 WHAT MICROORGANISMS ARE PRESENT?

To accurately assess the microorganisms occuring in a particular environment, the obstacles likely to be encountered should be anticipated. First of all, sampling is a crucial step for all subsequent analyses as it can alter the complexity of the sample compared to the situation in the field. However, the sensitivity of the microbial population in a sample to such perturbations greatly depends on the prevailing environmental conditions. It is also important to recognize that each assay or detection method has its own limitations (Table 6.1) which must be kept in context while interpreting the results. Some of the major obstacles confronted in satisfactorily establishing the microbial diversity of any particular environment include:

(1) difficulty in distinguishing between dormant and active cells;

(2) types and rates of biogeochemical metabolic processes expressed by active cells *in situ* can differ greatly.; and

(3) the physiology of microbial communities within a particular environment can change rapidly consequent to taking samples from it.

These obstacles are particularly expressed by the fact that when direct counts of prokaryotes for example in environmental samples are undertaken, they commonly depict populations that are 100 to 1,000 times greater than viable populations obtained by culturing (Roszak et al., 1984; Bhupathiraju et al., 1999). This discrepancy cannot be totally accounted for by the lack of distinction between viable and nonviable bacteria because if one uses different types of culture media, distinctly different profiles of the species that grow are obtained. For example, Sorheim et al. (1989), using different media to isolate bacteria from the same soil, consistently recovered phenotypically different populations. A substantial proportion of

**Table 6.1** Limitations associated with various census methods

| Method | Limitation |
| --- | --- |
| Viable counts | All growth media are selective. |
| Microscopy (general stains) | May not distinguish between viable and nonviable. |
| | The sample size is small compared to the entire community. |
| | Cell morphology is not insightful. |
| Microscopy (immunostains) | Since the antibody-producing cells have to be cultured prior to infecting say a rabbit, the cells must be culturable. |
| | Antibody specificity is uncertain. |
| Microscopy (nucleic acid probes) | Must penetrate cells. |
| | Have to ensure that it is specific (must not cross-react with nontarget genomes). |
| | If ribosome numbers are low, they may not be detected. |
| Biomarker extraction (e.g. G + C DNA-based typing, rRNA sequencing, etc.) | Studies show that ca.10% of the cells may not lyse (10% of $10^9$ is a lot of cells!). |
| | Targeted cell material may bind to environmental matrices such as soil colliods and remain unavailable. |
| | Various steps may be selective. |
| | Impossible to extract 100% of the marker. |

the uncultured organisms are culturable but their failure to grow on a particular medium can be due to the fact that every growth medium selects for and against a particular set of populations. Likewise, incubation conditions at any particular time do not meet the physiological requirements for all the organisms that may be present and are therefore selective. Thus, in essence all growth-based enumerations of microorganisms are biased one way or another and it is impossible to define conditions suitable for the activation and growth of all types of microorganisms present in a particular environment.

Furthermore, among the uncultured is a nonculturable proportion simply because they are dormant, viable but nonculturable (VBNC), or nonviable (Fig. 6.1). Nonviable organisms are unable to grow regardless of growth conditions or resuscitation treatments whereas dormant cells require some form of resuscitation, e.g. subjection to temperature shock (Kamprelyants et al., 1993), to stimulate their ability to grow on a "proper" medium. By definition, VBNC microorganisms fail to grow on any known type of media, possibly as a result of an imbalance in metabolism when some microorganisms are transferred from their natural environment where they are starved, producing an instantaneous production of superoxide and free radicals (Bloomfield et al., 1998). Superoxides and free radicals accumulate in lethal doses as such stressed cells lack the ability to adapt and detoxify. In that sense, therefore, potentially viable microorganisms are killed by the very conditions intended to recover (culture) them. These viable but nonculturable organisms have aroused much interest in microbiologists to determine their species and function in the environment (Madsen, 1998; Ward, 2002), leading to efforts to devise culture-independent techniques for identifying them. VBNC species in the environment spell many unique characteristics of untold potential. For example, despite their abundance in the environment, only a few

*Planctomyces* and members of other groups such as *Acidovorans*, and *WS6* members have ever been cultivated in culture (Dojka et al., 2000; Hugenholtz et al., 1998). Zarda et al. (1997) also encountered abundant *Planctomyces* in densities of $10^9$ cells $g^{-1}$ soil, coequal with those of $\alpha$- and $\delta$-proteobacteria in a pristine temperate forest soil. The fact that *Planctomycetes*, a group that does not synthesize the universal bacterial cell wall peptidoglycans is abundant in a hostile environment (soil) is intriguing. Although the focus on uncultivated microorganisms has centered on prokaryotes, they too have been detected among the Eukarya (Moreira and López-García, 2002).

Studying the types and distribution of microorganisms takes both direct and indirect methods. Direct methods include microscopy after staining with a nucleic acid-specific stain such as 4',6'-diaminido-2-phenylindole (DAPI), acridine orange, epifluorescence using antibodies that are specific for a target population or by labeling with nucleic acid probes (e.g. 16S rDNA) that will hybridize specifically with a target organism and immunofluorescence. Indirect methods involve isolating, culturing and identification based on colony characteristics/ morphology as well as biochemical tests. Enumeration on growth medium usually follows a procedure of dilution to extinction referred to as the most probable number (MPN) technique with the general assumption that each colony originated from a single cell. Whereas growth medium-based techniques greatly underestimate prokaryotes, they often overestimate fungal populations in environmental samples due to heavy sporulation in the latter and also fragmentation of fungal hyphae during treatment of the samples to determine colony-forming units (CFUs).

Caution should be taken to ensure that our techniques for assessing microbial populations in natural ecosystems are adequate as the characteristics of an organism determined under laboratory conditions may dramatically differ compared to those displayed by the same organism under field conditions. During sampling of a natural habitat, the physiology of the cells is oftentimes changed due to the structural disturbance of the environment. Subsequent handling and treatment of the sample also introduces further physiological changes. For example, on placing the sample in a container, some of the nutrients in the sample may attach onto the surface of the container and disfavor those populations that are poor scavengers outright. The mode of transport and storage before assaying may create conditions that differ from the ones in the field. Rochelle et al. (1994) reported a significant shift in microbial diversity in samples from anaerobic deep marine sediments. The samples they stored under aerobic conditions for 24 hours before freezing contained mainly β and γ proteobacteria, those stored at 16°C mainly contained α proteobacteria, whereas the ones they stored under anaerobic conditions and froze within 2 hours maintained a more diverse microbial composition. Most assays in environmental microbiology involve drying, mixing, sieving before the actual measurements of interest are taken. There is evidence that even for samples from the same source, performing any of these processes differently can affect our measurements.

In reality, it is easier to get more meaningful information about what microorganisms are present in a particular environment and what they are doing by comparing their number between communities or niches than by conducting a census of the entire microbial population within a particular community (i.e., determine relative diversity rather than species composition). Various models have been proposed to estimate the diversity of microorganisms in the environment. The respective models generally assume that microbial species are either log-normal or nonparametrically distributed. Under log-normal distribution, most species are assumed to have an intermediate number of individuals and a few species have

very large or very small populations (Ward, 2002). A nonparametric distribution on the other hand, relies on information about the proportion of species observed more than once in a particular environment relative to those observed only once (Hughes et al. , 2001). Whatever approach is adapted, a full description of the structure of the microbial community should reflect the diversity of that community (i.e., the relative number of species present), the number of individual cells of each species, and an assessment of their physiological role in relation to other species present in the environment.

## 6.2 PHYSIOLOGICAL STATUS AND ACTIVITY OF MICROBES IN THE ENVIRONMENT

Determining the physiological status and activity of cells is of interest because it enables assessing whether the microbes in a particular environment are providing any service. In determining the physiological status of microorganisms in their environment, two dilemmas become apparent. One is based on the premise of the Heisenburg uncertainty principle which states that the closer we examine a given microbiological niche or process, the more likely artifacts will be imposed on the measurements. Secondly, as the degree of influence from artifacts increases, confidence in the validity of these measurements decreases (Fig. 6.2). The physiological status and activity of naturally occurring microbial communities change rapidly with slight changes in their environment. Thus, a sufficiently long duration of disturbance to an environmental sample will be accompanied by changes in the anaerobic, catabolic, cellular, enzymatic and gene regulatory processes, which in turn results in changes in population and community structures. Therefore, sampling from a site to subsequently conduct laboratory studies introduces artifacts whose impact on the laboratory-based experiments compared to the *in-situ* situation greatly depends on three factors:

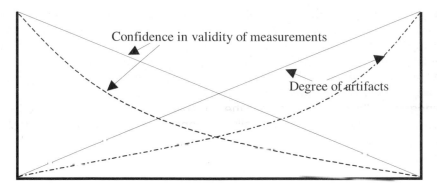

**Time and/or space**

**Fig. 6.2**  Relationship between the degree of influence of artifacts and the level of confidence in the data generated extrapolated over time and space in environmental microbiological studies. Notice that this relationship is not always linear.

- degree to which environmental conditions have changed,
- duration of the change, and
- specific responses of the microbial population that carry out the process of interest.

To minimize the impact of artifacts, a sampling strategy that aims at capturing, immediately fixing, and then analyzing the sample is more favorable if the information obtained after analysis of the fixed samples can be directly related to the composition and metabolic characteristics of the community at the time of fixation. Cells are typically fixed with (para)formaldehyde, alcohol, or by rapidly freezing in dry ice or liquid nitrogen (Table 6.2). The method of capturing and fixing has to be compatible with the intended analytical proce-

**Table 6.2**  Methods of fixing naturally occurring microbial communities

| Information sought | Fixation procedure |
| --- | --- |
| Total viable counts | Cool on ice |
| Total microscopic counts | Add alcohol, formaldehyde, or paraformaldehyde to the sample |
| Extraction of biomarkers such as RNA, DNA, chlorophyll, phospholipids, ATP, etc. | Freeze in liquid nitrogen or dry ice |

dure and depends on the information desired after fixation. To that effect, the choice of method or technique adapted and its validity can only be successful when its limitations are thoroughly understood. Thus, fixation by adding formaldehyde ties all the key macromolecules and is appropriate for total microscopic counts and, just like freezing, cannot be applied if viable counts are needed since both processes kill the cells.

The physiological role and activity of microorganisms in the environment can also be assessed by isolation of the organisms following enrichment. Thus, we can monitor growth of the organism(s) of interest within an environment to demonstrate Koch's postulate and show that the organism is responsible for the process under study. Although intended for medical microbiology, a modification of this postulate in the context of environmental microbiology should, as outlined by Waksman (1927) include:

(1) determining whether the organism is present in active form in the environment that demonstrates the metabolic activity of interest;

(2) determining whether it is present in large numbers in many samples exhibiting the activity;

(3) isolating or using equivalent modern nonculture-based techniques to study the organism in pure culture;

(4) inoculating the organism into a sterile environment (e.g. sterilized soil) or finding an environment lacking the activity and then unequivocally establishing that the organism causes the activity; and finally

(5) reisolating the organism from the inoculated environment.

Thus, the presence of the organism in an active form as stipulated in the first postulate alone does not in itself equate to *in-situ* activity in environmental microbiology. Combined with the presence in large numbers (postulate 2) of the organism in active form, this attributes to *in-situ* growth. Postulates 3 to 5 are fully obtained from medical microbiology and may not always be wholly adhered to in environmental microbiology. They assume that the organism is capable of growing in pure culture, can be isolated in a sterile environment, and is readily reisolated from the environment once introduced.

It is possible, however, for organisms of consequence in environmental microbiology to resist growth on laboratory medium or fail to act when introduced in a sterile environment, let alone be successfully reisolated from the environment to which they have been introduced. Examples of these anomalies to the postulate include most of the endomycorrhizae which have not been successfully cultured (see Chapter 8) and some methanogens such as *Methanobacterium bryantii* and *Syntrophomonas wolfei* which are obligate syntrophs as well as most members of *Planctomyces*, and *Acidovorans*. Thus, relying on growth is especially useful in situations wherein organisms of consequence can be readily isolated but will not tap the metabolic potential of the uncultured (i.e., both the culturable and nonculturable; see Fig. 6.1) microorganisms, which as indicated earlier, are present in the environment in surprisingly large numbers.

Laboratory experiments can be substituted with *in-situ* studies to determine the physiological status of microorganisms using various techniques. Some of these techniques and their limitations are outlined in Table 6.3 and usually aim at assessing aspects such as growth, viability, starvation, heat evolution, and dormancy. Thus, from such *in-situ* studies, we are able to observe whether at the time of sampling or fixation, the cells are:

- dividing (i.e., whether there is any growth *in situ*),
- small in size, usually indicative of a limitation in nutrients,
- have abundant storage granules such as poly-β-hydroxyalkanoate (PHA) granules, indicative of a sufficient C-supply but a deficiency in other nutrients, or
- increasing in number over time, also indicative of growth.

To detect and measure *in-situ* microbial processes, methods should emphasize field approaches thus reducing laboratory artifacts, and strike a balance between closeness to the field site and interpretability of laboratory-based results. *In-situ* studies should also aim at distinguishing between abiotic (e.g. volatility, sorption, transport) and biotic factors at sites that differ in geochemistry. To describe microbial processes *in situ*, we can take some lessons from other disciplines in, particular geology and ecology. In geology, the processes of building mountains, formation of crystals, and understanding plate tectonics are not based on recreations from laboratory, flask-oriented studies, but rather are deduced from field observations. Likewise in ecology, mass balances from watersheds provide a chance for ecologists to check on the accuracy of ecological process measurements. A variety of biochemical measurements which can be done *in situ* to assess the physiological status of microorganisms are outlined in Table 6.4.

**Table 6.3** Measures of physiological status and their limitations

| Technique | Information | Limitations or uncertainties of application |
|---|---|---|
| Microscopic examination of Fixed stained cells | Cell size, frequency of cell division, RNA:DNA based on acridine orange (cell color changing from green to orange). | Frequency of cell division presumes that the rates of cell constriction and separation are constant. Does not necessarily represent cells that were viable prior to fixation. |
| Fixed cells probed with 16S rRNA | Protein synthesis-based activity of ribosome content. | RNA, DNA, protein, and cell growth relationships are species specific. In some environments, most cells are in an inactive state. |
| Sample incubated with respiratory stains e.g. CTC, TTC, INT, etc | Dark red pigment indicates active respiration. | Substrate specific; causes changes in community composition. |
| Sample incubated with cell division inhibitor, e.g. nalidixic acid, piromidic acid, ciprofloxacin, etc. | Elongated cells indicate ability to grow. | Inhibitors not universal; causes changes in community composition. |
| Glass slide inserted into habitat, then retrieved | *In-situ* colonization and growth of native population. | Growth on glass may not be identical to that in natural habitat (*in-situ* growth). Some may not grow. |
| Retrieved slide placed in microcolorimeter | Measures relative heat attained from different communities. | Microcolorimeter assay is not synonymous to the field and its conditions differ. |
| Extraction of biomarkers and then sequencing or constructing bacteria artificial chromosomes (BACs) | Leads to creation of enormously large libraries. | Efficiency of extraction is uncertain. Sifting through the enormous libraries could be mind-boggling. |
| Phospholipid fatty acid | An increase in trans versus cis isomers indicates starvation. | Transfer from cis to trans is unlikely to be universal. |
| Incubate sample with radioactive pollutant | Biodegradation potential. | Biodegradation potential is not equivalent to field activity. |
| Storage bodies | Accumulation of poly-$\beta$-hydroxyalkanoates (PHA) granules under high carbon conditions. Accumulation of polyphosphate granules under high P conditions. | C insufficiency or limitation says little; PHAs are not universal. Rare |
| ATP | Physiological readiness indicated by ATP energy change. | Energy change calculations are based on uncertain preculture assumptions; efficiency of extraction is sample-dependent. ATP extraction from soil usually inefficient. |
| Incubate sample with radioactive thymine | DNA replication indicates growth. | Cannot be done *in situ*; thymidine may be catabolized instead of anabolized. |
| Incubate sample with other catabolic or anabolic building blocks | Metabolic uptake indicates "readiness". | Which blocks are appropriate and at what concentrations |

Modified from Madsen (1995) with permission from Marcel Dekker, New York.

**Table 6.4** Summary of techniques of measuring *in-situ* biogeochemical activity and associated limitations

| Technique | Information | Limitations |
|---|---|---|
| Field chambers with periodic monitoring of head space gases | $CO_2$ evolution for respiration<br><br>$O_2$ uptake for respiration<br><br>$CH_4$ evolution or uptake for methanogenesis or methanotrophs respectively<br><br>$N_2O$ evolution for denitrification | Only addresses processes that affect volatile compounds. Flux rates of these compounds have to be relatively rapid to minimize alterations in environmental conditions such as temperature, light, etc. Many physiological processes serve as reactants for other processes which are often overlooked. |
| Add acetylene and measure ethylene | Nitrogenase activity as an indicator of nitrogen fixation | Indirect and has a conversion problem since for every mole of ethylene produced 1/3 mole of nitrogen is fixed. |
| Erect barriers around site; collect water samples and measure changes in water chemistry | Various elements, e.g. if ammonia is added and $NO_3$ measured to determine nitrification | Mass balances are difficult to achieve. Depth of profile can be limited, causing an underestimation of nitrification. |
| Deploy crop residues in nylon mesh bags; periodically retrieve and weigh/analyze | Decomposition | Placement of residue in bags may reduce the contact between soil and the organisms. Retrieval of residues is likely to be biased toward large fragments. |
| Application of stable isotopes (with or without field chambers). | Conversion of labeled compound from one to another, e.g. $^{15}NO_3^-$ to $^{15}N_2$ and $^{15}N_2O$ as measurements of denitrification.<br><br>Comparison of $^{15}N$ in leguminous and nonleguminous plants after supplying $^{15}N$-fertilizers to quantify nitrogen fixation.<br><br>Conversion of $^{15}N_2$ to $^{15}N$ in plant biomass to determine N fixation | Stable compounds including fertilizers are expensive. Ambiguous qualitative rates of biochemical processes often obtained due to dubious assumptions about the experimental controls. Use of chambers often imposes some changes on the *in-situ* conditions. |
| Fractionation of naturally occurring stable isotopes. | Detecting signature ratios of $^{13}C/^{12}C$ in soil $CO_2$ produced from crude oil of the same signature ratio to study biodegradation of oil.<br><br>Preferential conversion of C in organic matter to $CO_2$ along flow path in aquifer by drilling holes at different points in the flow path to study respiration, N- and S-transformation. | Relies on knowledge of analytical chemistry and plant as well as microbial physiology but the data suffer from uncertaintie associated with variabilities in field sites and/or biochemical principles used in interpreting the stable isotopic fractionation data. |
| Chemical analysis of field samples (no chamber deployment required) | $N_2O$ profiles above soil surface or in soil waters to indicate denitrification or nitrogen fluxes. | $N_2O$ may also be produced during nitrification. |

*Table 6.4 contd.*

*Table 6.4 contd.*

| | | |
|---|---|---|
| | Total soil mass balances for nitrogenous substances to assess $N_2$ fixation and denitrification. | Requires highly sensitive instrumentation. Fluxes to $N_2$ are marked by high ambient $N_2$ in the atmosphere and analytical errors in total N determination are likely to mask small changes caused by denitrification. |
| | Differences in nitrogen content of $N_2$-fixing and nonfixing (nonnodulating) legumes to quantify nitrogen fixation. | Plants have to be identical in root patterns except for nodulation. |
| | Ratios of degradable to degradation-resistant crude oil components to determine biodegradation. Degradation-resistant hydrocarbons serve as an internal tracer over time. | The internal tracers may undergo abiotic changes and thus differ from the degradable. They may also become more degradable with environmental changes, e.g. temperature. |
| | Detection of unique intermediary metabolites in field samples (ethylene, dihydrodiols, etc.) to study biodegradation. | Requires rapid handling of samples and generation of a unique (rare) metabolite. |
| | Release of radioactive pollutants to soil, water; subsequent detection of $^{14}CO_2$ in the field as an indicator of biodegradation. | Even though just a trace of radiolabel, it is oftentimes very unethical. Recovery rate is low. |
| | Replicated field plots with and without addition of nutrients and/or microbial inocula to monitor biodegradation. | Requires a uniform site. |
| Extraction of nucleic acids | Detection of messenger RNA (mRNA) shows gene expression to determine $N_2$-fixation, photosynthesis and biodegradation. | Messenger RNA shows gene expression but falls short of documenting true biogeochemical change. |
| Microelectrodes | Vertical profiles of $O_2$, $H_2S$, $NO_3^-$. Coupled with calculations from diffusion models as it normally shows a concentration gradient to assay respiration, sulfate reduction or denitrification. | Biotic versus abiotic and dynamic causes of concentration gradient may be difficult to discern. Calculations of metabolic rates based on diffusion models may or may not be accurate. |
| Laboratory incubation of "intact fields cores". | Results of headspace analyses to assay methanogenesis, methanotrphy, denitrification, or respiration. | Samples are disturbed and thus not artifact-free. |

Modified from Madsen (1995) with permission from Marcel Dekker, New York.

## 6.3 GENERAL APPROACHES TO DETERMINING MICROBIAL BIOMASS AND ACTIVITY

The size of microbial biomass reflects the potential influence of microorganisms to conduct biochemical transformations in the environment. Several methods have been developed to quantify the microbial biomass and activity in complex substrates. Most notable of these are the ATP assay, respiration, and the chloroform fumigation technique. With the advent and continuous improvements in molecular biology, biomarker techniques to determine biomass have also been explored. The principles underlying these techniques are briefly discussed below.

### 6.3.1 Fumigation

The chloroform fumigation technique is the most widely accepted method of quantifying the actual microbial biomass and is generally regarded as the standard approach for assaying microbial biomass in some environmental matrices, particularly soils and sediments. It is based on the fact that chloroform vapor kills microorganisms and in soil, the killed microorganisms represent dead microbial biomass that is subsequently metabolized (mineralized) by soil microorganisms introduced into the treated soil by addition of untreated soil (Jenkinson and Powlson, 1976; Jenkinson et al., 1979). The $CO_2$ evolved during fumigation can be measured by gas chromatography and the C-flush converted to biomass C using a $k_c$ factor of 0.45 (Jenkinson et al., 1979). $CO_2$ production is followed for 10 days and directly correlated with the microbial biomass (Sparling et al., 1986). This approach has been further modified by assaying for the amount of carbon released from fumigated versus nonfumigated samples (Wu et al., 1990).

### 6.3.2 ATP concentration

The ATP assay is based on the premise that ATP is an indicator of life and occurs in all living cells. Its content changes rapidly with changes in the physiological status of cells and is rapidly decomposed in dead cells (Jenkinson and Oades, 1979; Balkwill et al., 1988). The underlying assumption with this technique, therefore, is that the ATP in cellular materials is more or less proportional to the biomass. ATP measurements have been used successfully under marine environments to quantify microorganisms (Lee et al., 1971; Strickland et al., 1969). In quantifying microbial biomass in soils and sediments, however, the use of ATP is limited by problems of

- efficiently extracting the ATP from the soil; and
- relating the amounts of ATP obtained to the actual biomass in the respective environmental matrix.

The first problem relates to how the sample is handled prior to assaying for ATP. ATP, like other phosphates, can adsorb on positively charged soil colloids (Jenkinson and Oades, 1979). Dry soil tends to have lower ATP per unit weight compared to moist soil. Air-drying the soil results in a marked drop in amount of ATP extracted from the same soil when it is moist (Ahmed et al., 1982; Fairbanks et al., 1984). Once dry, however, the concentration of extractable ATP remains quite stable. The decrease in ATP during drying can be somewhat ameliorated by freeze-drying the soil in liquid nitrogen because this fixation process increases dispersion of the soil, resulting in a better extraction of organic materials. Freeze-drying may also cause cellular changes that in turn greatly improve the efficiency of ATP extraction per se. To correct for incomplete recovery of ATP from environmental matrices, Contin et al. (2002) proposed including control treatments spiked with ATP and then simultaneously extracting the ATP from these treatments side by side with unspiked samples. Alternatively, the ATP can be extracted sequentially. A comparison of these two strategies gave comparatively similar amounts of ATP in soils from five locations (Table 6.5). The

**Table 6.5** Amount of ATP recovered from soils after either extracting ATP sequentially or spiking the soil and correcting for incomplete recovery[a] (Contin et al., 2002 with permission from Elsevier).

| Site | Recovered after spiking | | Total ATP recovered (nmol g$^{-1}$) after six sequential extractions |
|------|------------------------|---|---|
| | Percent ATP recovered | Total ATP recovered (nmol g$^{-1}$) after correction for incomplete recovery[b] | |
| Highland arable | 69.8 | 3.45 (0.24) | 3.25 (0.22) |
| Highland grassland | 75.3 | 9.75 (3.12) | 9.82 (2.36) |
| Manor garden | 74.2 | 11.13 (2.04) | 11.82 (1.59) |
| Great Knott | 77.8 | 3.82 (0.38) | 4.11 (0.30) |

[a]Numbers in parentheses are standard errors.
[b]Percent recovered was used to correct for the ATP amount.

total biomass computed based on these ATP extraction approaches was also comparably similar (Contin et al., 2002).

Generally, the concentration of ATP depends on the growth stage of the organisms present in the environment (Fairbanks et al., 1984). Thus, ATP concentrations change with addition of nutrients, predation, and changes in the physiological state of the cells. Overall, the method is extremely sensitive. For example, if a crude enzyme is used to assay for ATP, higher ATP values are recorded than when a purified enzyme is used (Tate and Jenkinson, 1982). The approach is also less time consuming compared to plate or direct counts and gives very valuable information about the intensity of biochemical microbial processes. It is particularly useful in following changes in microbial biomass during short periods (a few minutes) or when the quantities of environmental sample are too small to be incubated. It should, preferably, be used in conjunction with other methods such as respiration, fumigation, or direct counts to assess the activity of microorganisms.

## 6.3.3 Respirometry

The overall microbial activity, including growth, reproduction, and metabolism in complex systems such as soils can also be evaluated by measuring the uptake of $O_2$ (aerobic respiration) (Orchard and Cook, 1983) or $CO_2$ evolu-

tion (Anderson and Domsch, 1978). Measurements of $O_2$ uptake in soil electrolytic respirometry are based on replacement of $O_2$ at a concurrent absorption of $CO_2$ in a closed system containing a substrate with the respiring organisms. For example, a substrate such as glucose is added to the environmental sample such as soil in well-sealed bottles and the rate of $CO_2$-respiration at a known temperature determined using chromatography. The respiratory rate is directly correlated with the rate of electrolysis, which can be monitored. The method is applicable to several unamended and amended soils (Pažout and Vančura, 1981). Alternatively, the $CO_2$ evolved can be trapped in sodium hydroxide and determined by titration (Lodge, 1993). By combining the biomass assay using the chloroform fumigation technique with respiratory values, it was found that at 22°C a substrate-induced maximum respiratory rate of 1 ml $CO_2$h$^{-1}$ corresponds to about 40 mg microbial biomass carbon present in the soil (Stolp, 1988).

Respirometry has been widely used with radioactive tracers (i.e., radiorespirometry) in monitoring the metabolism of $^{14}CO_2$-labeled compounds by microorganisms *in situ*. The soil is exposed to $^{14}CO_2$ and the $^{14}C$ then incorporated into the cellular organic fraction determined by heating the exposed soil by pyrolysis at high temperatures (600°C) while flushing the soil with helium containing $^{14}CO_2$. Alternatively, metabolism is assayed by the wet oxidation of the soil with potassium dichromate and determining the radioactivity of the evolved $^{14}CO_2$ using a liquid scintillation counter. This method is rapid, highly sensitive and allows for quantifying primary productivity *in situ* with minimal perturbation to the natural soil ecosystem (Fig. 6.3). Specific examples wherein this method was used to determine the degradability of organic compounds in the environment are discussed in Chapter 9.

### 6.3.4 Calorimetry

All microbial processes are accompanied by either the evolution or absorption of heat which can be analyzed calorimetrically. The heat may originate from the biochemical activity of living cells in the soil matrix, metabolic activity of living organisms, or from nonbiological processes in the system. To distinguish these sources, microbial activity can be stimulated by supplying substrates or inhibited by sterilizing the system (Ljungholm et al., 1979). Calorimetry can be used to effectively monitor the effect of adding different reagents, such as fertilizers, biocides, or pollutants, to the system. Valuable information is obtained if microcalorimetric values are correlated with other microbial activities such as ATP, microbial biomass C, and respiration.

### 6.3.5 Enzyme activity

Both intra- and extracellular enzymes can be excreted in the environment. Studies involving enzyme activities can furnish useful information about microbial activities in the environment (Benefield et al., 1977; Trevors, 1984). However, when studying enzyme activity in such samples, it is difficult to determine the proportion of activity due to viable cells as opposed to extracellular activity since some

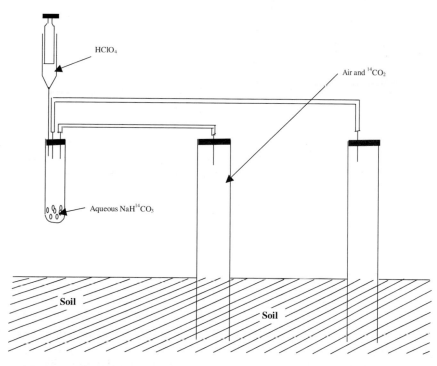

**Fig. 6.3** Simple apparatus for monitoring metabolism by microorganisms *in situ* using $^{14}CO_2$ (adapted from Atlas and Hubbard, 1974 with permission from Springer-Verlag GmbH & Co. KG).

enzymes from dead cells may be excreted and survive for some time in the environment. Most commonly used in enzyme studies is the ability of the environmental sample, e.g., soil, sediments, or water to reduce a tetrazolium salt, which is characteristically a heterocyclic ring containing one C and four nitrogen atoms, as an alternate electron acceptor, converting it to formazan. Formazans are water-insoluble compounds formed when the heterocyclic ring is opened. Common tetrazolium salts include 2,3,5-triphenyltetrazolium chloride (TTC) and 2-(p-iodophenyl)-3-(p-nitrophenyl-5-phenyl) tetrazolium chloride (INT). As expected, enzymatic activity measurements depend on chemical aspects such as $E_h$, oxygen tension, pH, dissolved gases, ion exchange capacity in the environment, and antagonistic as well as synergistic substances that may be present. Furthermore, physical factors, notably temperature, presence or absence of light, surface area, and adsorption also affect enzymatic activity. Thus, their absolute number can be rendered totally meaningless if proper controls to maintain these factors in unison within test samples are not established. Comparative measurements of enzymatic activity are only useful if carried out using identical methods and environmental conditions and their usefulness can be greatly enhanced if other parameters such as respiration, ATP activity, etc, are also determined.

### 6.3.6 Molecular biomarkers

Only a small amount of microbial diversity can be studied through conventional culturing techniques. However, until about a decade-and-a-half ago, most of what is known about the dynamics of microorganisms in the environment was almost exclusively derived from culture-based studies. With recent advances in molecular biology, nonpure culture ecosystems, including nonculturable microorganisms can be studied by amplifying nucleic acids *in situ* using PCR and cloning and sequencing gene fragments encoding 16S rRNA (Fig. 6.4).

As apparent from the summary discussion of other techniques above, this section does not dwell on the numerous step-by-step protocols currently available in molecular biology, but rather gives an overview of the general principles that apply to such methods. Individual protocols are readily obtained from an appropriate laboratory manual such as Sambrook et al. (1989), Akkermans et al. (1995), and Rochelle (2001). Some of the molecular techniques increasingly relied upon for analyzing whole microbial communities and their attributes are summarized in Table 6.6. They can be broadly distinguished into approaches that either require extraction of nucleic acid from environmental matrices, extraction of lipids, or targeting nucleic acids in intact microbial cells.

As alluded to in Chapter 2, the number of known phylogenetic divisions for prokaryotes has continued to increase with the advent of culture-independent techniques. Sequencing without culture has shown that the range of microbes known through conventional cultivation is very small compared to what is out there (Amann et al., 1995; Dojka et al., 2000). The physiological value or role of most of those microorganisms with no known cultured relatives still remains untapped. Except for FISH, the techniques summarized in Table 6.6 initially involve extracting community DNA, amplifying the 16 rRNA genes from the community DNA using group-specific primers, and depending on the technique, separating the products. This approach is limited by differences in the lysis of environmental species, preferential amplification of specific templates due to the primer used, and the possible presence of free DNA in the environmental matrix that may otherwise have no physiological function (Lee et al., 1996; Kuske et al., 1997).

### 6.3.6.1 *Nucleic acid extraction*

To study the structural diversity of microbial communities, nucleic acids are increasingly used because they are universal in all cells. This has enabled analysis of communities with-

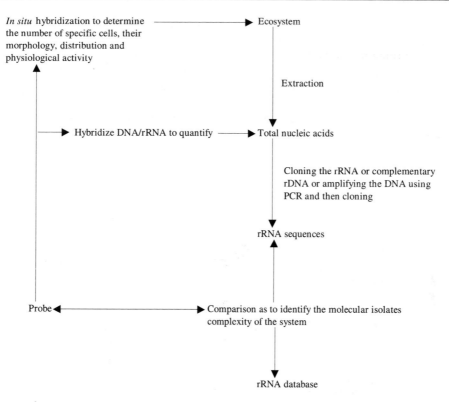

*In situ* hybridization to determine the number of specific cells, their morphology, distribution and physiological activity ⟶ Ecosystem

Extraction

Hybridize DNA/rRNA to quantify ⟶ Total nucleic acids

Cloning the rRNA or complementary rDNA or amplifying the DNA using PCR and then cloning

rRNA sequences

Probe ⟵ Comparison as to identify the molecular isolates complexity of the system

rRNA database

**Fig. 6.4** Diagram of characterization of a mixed biological sample by rRNA sequencing and hybridization without culturing (modified from Amann et al., 1991 with permission from Nature Publishing).

out culturing. These methods require extraction of the genetic material and subsequent sequencing of nucleic acids for genes coding for phylogenetically and taxonomically significant rRNA. The extraction should aim at optimally lysing the cells so as to release the nucleic acids. Unreleased DNA or RNA will not contribute to the final analysis of diversity.

Nucleic acids occur in small amounts in the environment and there is often need to amplify them using the polymerase chain reaction (PCR). Thus, the extracted nucleic acids are typically amplified and then subjected to RFLP, TGGE/DGGE, ARDRA, SSCP, and rep-PCR (Table 6.6). However, it should be emphasized that each of these methods has its strengths and weaknesses as well as appropriate situations under which it is deemed more suitable than the others. That said, the procedures for acquiring such sequence information, for example by using PCR-amplification of 16S rDNA or reverse transcriptase-based synthesis of complementary DNA (cDNA) derived from 16S rRNA followed by cloning to enable the copies of the rDNA molecules to be separated and isolated, are potentially more reliable and less biased than relying on selective growth of particular members of a mixed microbial community. PCR-based approaches may from the onset introduce some biases because:

- primer-template hybrids could be formed with different efficiencies due to differing preferences between primers,
- the efficiency with which PCR products are annealed is affected by the concentration of the template, and
- after denaturation, the templates form secondary structures which may not be readily accessible to the primers.

**Table 6.6** Description of some molecular methods currently used in environmental microbiology

| Common abbreviation | Full name | Description | References |
|---|---|---|---|
| ARDRA | **A**mplified **R**ibosomal **D**NA **R**estriction **A**nalysis | Involves isolating crude DNA from the environment, amplifying it using universal primers, digesting with restriction enzymes that recognize 4 recognition sites, and then analyzing the small subunit rRNA gene sequences (rDNAs) so as to identify the microbial species without culturing. Limitations similar to those of RFLP. Provides a higher resolution than F-RFLP because the sizes of all restriction fragments are used in the analysis for each enzyme. Thus, using this method, a larger portion of the rDNA sequence is sampled. Works well especially when the community of interest is enriched for a few dominant members, e.g., toluene degraders or 2,4-D degraders. | Borreman and Triplett (1997); Kuske et al. (1997); Tiedje et al. (1999) |
| DGGE/TGGE | **D**enatured/**T**hermal **G**radient **G**el **E**lectrophoresis | DNA is amplified using PCR and the product sequences are separated based either on solubility in a denaturant or its melting behavior (thermal) gradient in a gel. The technique is very sensitive and can detect single base differences and circumvents the need for constructing and screening enormous clone libraries, a laborious task. However, it has a low power of resolution and is only qualitative. | Muyzer et al. (1993); Crueng and Kinkle (2001) |
| FISH | **F**luorescent **I**n-**S**itu **H**ybridization | The relatively well-conserved 16S rRNA present in all microorganisms as specific signature sequences is targeted and probed after hybridization. Thus, this approach enables targeting whole cells and identifying organisms up to the species level without isolation or culturing. However, the level of growth and activity of microorganisms prior to fixation and probing correlates with the abundance of ribosomes and thus the content of rRNA is such that in some oligotrophic environments in which most microorganisms exist in an inactive/semidormant state with few ribosomes can be missed as they may not generate a detectable signal. | DeLong et al., 1989; Ward et al., 1992; Amann et al., 1995 |
| RFLP | **R**estriction **F**ragment **L**ength **P**olymorphism | Technique good for detecting changes in diversity but is not quantitative. If the environmental system has many species, the profiles generated are too complex and of little value. Analysis is based on differences in size/length between restriction fragments, which limits resolution of the method in complex communities since sequence information | Lee et al., 1996 |

*Table 6.6 contd.*

*Table 6.6 contd.*

| | | | |
|---|---|---|---|
| | | other than restriction sites cannot be used. Furthermore, a single species can generate 4-6 restriction fragments. | |
| rep-PCR | Repetitive extragenomic palindromic PCR | Chromosome structure is rather variable among strains. This approach provides a fine level of resolution. However, it is culture dependent. | Fulthorpe et al. (1998) |
| SSCP | Single Strand Conformation Polymorphism | DNA is denatured with NaOH (0.1M) and EDTA (20 mM) in the absence of glycerol and the resultant single strands PCR-amplified based on differences in electrophoretic mobility caused by conformational differences in folded single-stranded products. In mixed cultures (typical of environmental samples), each single band corresponds to a single species. However, just like DGGE/TGGE, the approach has low resolution. It is also limited by the fact that single-stranded nucleic acids can have somewhat unstable conformations. | Lee et al., 1996 |
| T-RFLP | Terminal Restriction Fragment Length Polymorphism | The procedure for isolating, amplifying and restriction digestion of DNA is similar to that in ARDRA, the main difference being use of fluorescently tagged primers that can be detected by an ABI-automated sequencer. In this manner, the size of the terminal restriction fragment is quantified. Avoids many of the limitations of RFLP and ARDRA, mentioned above as restriction fragment lengths for the entire ribosome are determined, thus providing a logical phylogenetic starting point. | Liu et al., 1997 |

Furthermore, PCR-amplification based techniques usually generate enormous libraries which are rarely sequenced exhaustively and cannot distinguish between multiple copies of operons from one cell or many from different cells of the same species (Ward, 2002). These limitations have to be taken into consideration when adapting PCR-based methods.

Both thermal gradient gel electrophoresis (TGGE) and denaturing gradient gel electrophoresis (DGGE) are discussed below in greater detail because of their superior potential compared to the other biomarker extraction methods, to provide an insight into the microbial composition of communities. For both techniques, DNA fragments of the same length but differing in sequences are separated according to their melting properties. When electrophoresed through a linearly increasing gradient of denaturants, the fragments remain double stranded until they reach conditions that cause some sections (based on the sequence) to melt. The partially melted domains sharply decrease the mobility of the DNA fragments in the gel. The difference between the two methods is that with DGGE, the double strand is separated in a linearly increasing concentration of urea and formamide at high temperatures whereas with TGGE, separation is effected with a linearly increasing temperature gradient but with the same high concentrations of urea and formamide (Muyzer et al., 1993; Heuer and Smalla, 1997). The polymerase chain reaction amplified fragments of rDNA result in bands demonstrating the different microbial species present in the complex communities. When primers annealing to the conserved regions of the 16S rRNA gene are used, the PCR-amplified fragments subjected to either DGGE or TGGE generate fingerprints for the most dominant constituents of mixed microbial populations.

The technique was recently used by Chueng and Kinkle (2001) to show shifts in microbial populations and composition over time in the presence of organic contaminants, with the rationale that shifts in populations indicate degradation. To be detected by either TGGE or DGGE via PCR, the microbial strain has to comprise at least 1% of the total culturable population (Heuer and Smalla (1997). Thus changes that occur in nonpredominant microbial species will not be detected as their rDNA will not be amplified since their abundance lies below the threshold of the template sequence needed to generate a band. Detection of such species can, however be greatly enhanced if primers specific for those species (or specific for their group) are used. We can use rRNA after reverse transcription as a template for PCR amplification with TGGE or DGGE primers to develop a profile for the dominant metabolically active fraction in a particular environment. This approach was initially used by Teske et al. (1996) and Felske et al. (1996) on water and soil samples respectively, to compare the DGGE patterns derived from reverse transcribed 16S rRNA with those from corresponding genes (16S rDNA). They observed marked differences in the patterns obtained as the former represented the metabolically active fraction but the latter the total population structure.

### 6.3.6.2 Targeting intact cells

There are limitations with methods that involve directly extracting and amplifying nucleic acid from environmental matrices. Notably, extraction procedures may fail to completely lyse microbial cells. In general, spores, tiny cells and cells of gram-positive bacteria are more difficult to lyse. Some cells may also not be easily dislodged from the environmental matrix. For example, different extraction methods differ in efficiency of dislodging bacteria from soil particles (Heuer and Smalla, 1997). Furthermore, some of the retrieved nucleic acids could have originated from naked DNA and PCR-based methods provide no information about the morphology, spatial distribution or number of microorganisms present in a particular environment. To overcome these limitations, whole cells can be targeted and probed directly without altering their morphology or integrity. Initial efforts toward targeting and identifying microorganisms in their natural environment involved the use of immuno-fluorescence using species-specific monoclonal antibodies. However, this approach requires the prior isolation of pure cultures for raising specific antibodies. It is, therefore, not culture independent. FISH targets nucleic acid sequences by a fluorescent-labeled probe that specifically complements sequences within the cell. The technique is based on the premise that the two DNA strands can be separated by heat or alkali denaturation. The process is reversible once denaturing conditions are eliminated which enables the individual strands to reanneal into the original duplex or binding with complementary RNA fragments. Thus, single strands of nucleic acids, mostly DNA, are designed and used as probes to bind specifically to complementary DNA or RNA, enabling us to establish the composition of naturally occurring communities even without culturing them (Amann et al., 1991, 1992, 1995). The nucleic acids within intact microbial cells are probed irrespective of whether the cells are viable or not.

The procedure is shown schematically in Fig. 6.5 and involves fixing the sample, dehydrating it to facilitate entry of the probe into the cells, hybridizing (usually at a high temperature), washing off the excess probe, mounting and then visualizing the intact cells under the microscope. Oftentimes, it is essential to counterstain the cells with a general stain such as DAPI or acridine orange to get a sense of what fraction of the total microbial population is responsive to the probe compared to the total (Jjemba et al., in litt,; Zarda et al., 1997). Figure 6.6A shows the cells that were targeted with a eubacteria probe and Figure 6.6B cells targeted with $\alpha$-proteobacteria probe. It can be seen in Figure 6.6B that only one cell belongs to the $\alpha$-probeobacteria subclass. Typical probes are 15 to 30 nucleotides (Moter and Göbel, 2000) and are readily synthesized commercially. The stability of the probe-target complex depends on the homology (similarity) of the sequence. FISH is also advantageous to the other techniques summarized in Table 6.6 because it is quantitative (Jjemba et al., in litt.). However, the recovery or detection greatly varies with the type of environmental matrix. When actively growing, the number of rRNA molecules can be as many as 71,000 (Ward et al., 1992) and the cellular content of ribosomes indicates the synthesis of proteins and therefore growth (DeLong et al., 1989; Ward et al., 1992; Oda et al., 2000). It is noteworthy that in some environments, particularly soil and sediments, the nutrient status is usually so poor that the exist-

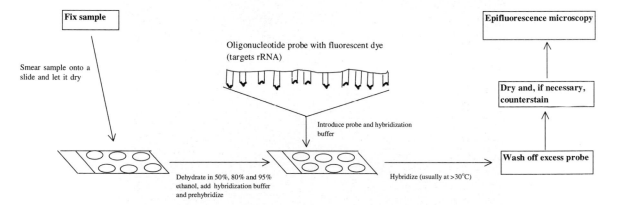

**Fig. 6.5** Fluorescent *in-situ* hybridization (FISH).

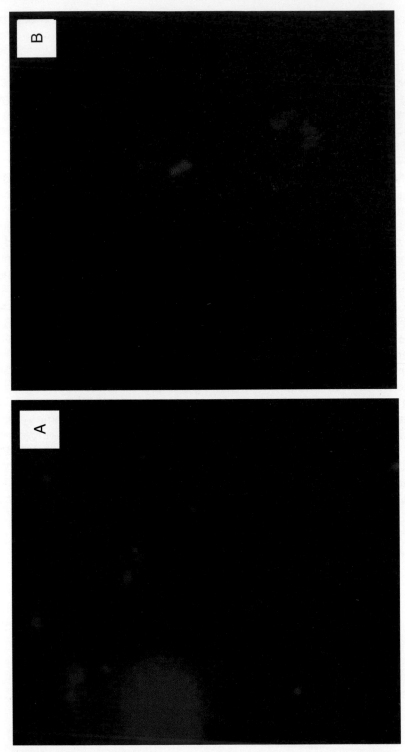

**Fig. 6.6** A mixed soil microbial community that was targeted with (A) a probe that targets all bacteria (EUB338) and (B) a probe that targets only α-proteobacteria.

ing microbes are mostly in an inactive or dormant form, i.e., spores, cysts, etc. (Roszak and Colwell, 1987). The signal intensity during hybridization and probing depends on the content of ribosomes. Inactive or dormant cells (i.e., cells with low metabolic activity) typically contain only about 30-50% of the amount of ribosomes compared to actively metabolizing cells (Quiros et al., 1989; Setlow, 1994). When targeted, these fewer ribosomes may not be readily detected on probing. This limitation, together with autofluorescence from the soil and sediment particles, can result in a weak probe signal or a complete failure of detection of some of the cells (Hahn et al., 1992; Zarda et al., 1997).

Several studies have recently used this reality beneficially by providing specific recalcitrant organic compounds as a carbon source to entire soil (Jjemba et al., in litt.) and groundwater (Bakermans and Madsen, 2000) microbial communities. The communities are also provided antibiotics that inhibit actively growing cells from dividing without affecting RNA and protein synthesis (Kogure et al., 1979; Barcina et al., 1995). The susceptible cells become enlarged or elongated and hence easily distinguishable from their nonmetabolic counterpart. Increasingly, more specific probes are currently being used to identify the major degraders, irrespective of whether they are culturable or not (Jjemba et al., in litt.).

It is important to remember that an oligonucleotide probe is only as accurate as the nucleotide sequence on which it is based. Some organisms, particularly gram-positive bacteria, are not readily permeable to the probe. This fact introduces a difficulty when complex environmental samples containing various bacterial species are analyzed because treatments such as HCl, or treatment with lipase to permeabilize gram-positive bacteria can negatively impact the integrity of some gram-negative bacterial cell wall. Despite these limitations, FISH is widely accepted as a powerful means of obtaining insights into the morphology, number and spatial distribution of microorganisms in their natural environment irrespective of their culturability. Recent modifications such as use of the peptide nucleic acid (PNA) probe instead of rDNA oligonucleotides, better facilitate penetration of the probe into intact cells. PNAs are DNA analogs synthesized with uncharged polyamides instead of sugar phosphates in DNA and can easily penetrate the cell wall of gram-positive bacteria (Moter and Göbel, 2000). However, the cost associated with PNA synthesis compared to rNA probes is still prohibiting their widespread use.

### 6.3.6.3  Signature lipid analysis

Several lipids, including poly-β-hydroxy-alkanoate (PHA), plasmogens, sphingolipids, steroids, quinones, triglycerides, and phospholipid fatty acids (PLFA), have been used as signature biomakers to characterize and even quantify microorganisms in their natural environment (White, 1983; White and Ringelberg, 1997). The lipids are extracted directly from environmental samples using organic solvents and then separated by chromatography. Individual lipids are then identified, providing an insight into the complexity of the microbial community, its nutritional status and biomass. Their identification and nomenclature is based on the total number of carbons followed by the number of double bonds beginning with the position of the double bond closest to the methyl end ($\omega$) of the molecule. The double bond is identified as either a cis (*c*) or trans (*t*). For example, 18:2$\omega$5*t*7*c* is an 18 carbon PLFA with two double bonds located at the 5th and 7th carbon from the methyl end of the molecule. The double bond at the 5th position has a trans and the one at the 7th position has a cis configuration. Different microorganisms synthesize unique types of lipids and can be used to establish phylogenetic relationships comparable to those obtained based on ribosomal RNA (Kohring et al., 1994).

Most widely studied and used as identification signatures are phospholipids because they are essential membrane components of all living cells. Cellular enzymes readily hydrolyze the phosphate group within minutes, leaving diglycerides (Fig 6.7). The rate of hydrolysis is enhanced by the phospholipase enzyme. If calibrated, the ratio of PLFAs to glycerides gives a good estimate of the viable microbial biomass. Storage lipids such as PHAs and some triglycerides are formed during unbalanced growth when sufficient amounts of carbon and electron acceptors but essential nutrients such as phosphate, nitrate, and trace elements are limiting (White et al., 1998). Thus, their presence in relatively large amounts indicates the nutritional status of the cells. Ratios of PHA to

PLFA have also been successfully used to determine the nutritional status of microbial communities in specific environments whereby ratios >1 indicate high levels of PHA and indicate existence under nutrient stress (Balkwill et al., 1988). This approach preserves the microstructure of the microbial consortium. If properly handled, the fatty acid composition remains stable within samples with no dramatic differences detected between samples processed after a week compared to those processed immediately after sampling (Balkwill et al., 1988). If improperly handled, the proportion of trans to cis isomers may increase, as a result of starvation. In healthy growing prokaryotes, a ratio of about 0.01 has been reported (White and Ringleberg, 1997). However, changes in

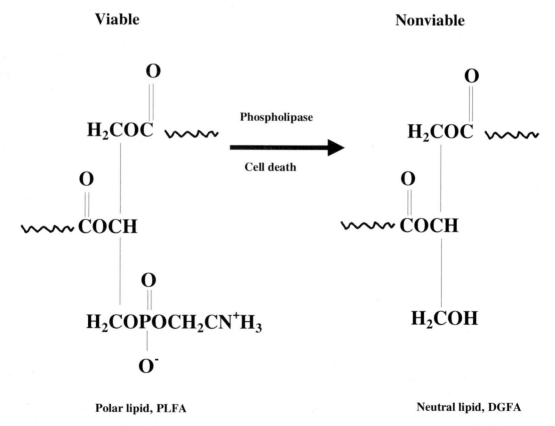

**Fig. 6.7** Conversion of PLFA in viable cells to diglyceride on the death of microbial cells (White and Ringelberg, 1997 with permission from CRC Press).

trans/cis isomers do not appear to be universal as some prokaryotes display no such changes. This observation emphasizes the need to interpret results from environmental samples with caution (see limitations below).

Microbial biomass and activity can similarly be estimated by analyzing the abundance of essential components of cell walls such as muramic acid or diaminopimelic acid, chitin (for fungi), and sterols (for eukaryotes such as algae and protozoa) existing within a particular environment. In prokaryotic cell walls, the molar ratio of muramic acid to glucosamine is 1:1 (Balkwill et al., 1988). The molar ratio of these two cell wall components has been used to estimate prokaryote to eukaryote ratios in the environment without cultivation. It is important to note that this approach gives us insights into the biomass and microbial diversity but because many microbial species contain overlapping fatty acid patterns, it is not possible to identify every species present in a particular environment. However, fatty acid profiles of a number of culturable prokaryotes have been established. Currently, a fatty-acid based Microbial Identification System (MIDI) is widely used, particularly in clinical microbiology. Because the MIDI database was developed based on the isolation and culturing of microorganisms, this system is unable to identify the wide array of uncultured microbes and therefore has limited potential in environmental microbiology since innumerable microorganisms have not yet been cultured. Thus, from a broader perspective, the diversity obtained based on fatty acid analysis is only relative and more useful if the signatures in their entirety are between sites or between different treatments monitored for changes in the microbial composition over time. Furthermore, results obtained by different laboratories may not be easily comparable as the fatty acid extraction protocol, rate of growth of the organisms, incubation temperature, types of growth media, and a variety of other parameters affect the fatty acid patterns (Leclercq et al., 2000). For example,

growth temperature directly affects the fluidity of the fatty acid membranes.

In fungi, a lipid ergosterol widely occurs in membranes of Ascomycotina, Deuteromycotina, and Zygomycotina but is absent or rare in oomycetes and hyphytrichiomycetes (Schnürer and Börjesson, 1996). In pure fungal cultures, its concentration has been found to correlate highly with mycelial dry weight, mycelial protein, hyphal length, chitin, $CO_2$-evolution, mycotoxins, volatile fungal metabolites and sometimes with fungal CFUs. Hence it is used in some instances to quantify fungal biomass and activity in environmental matrices (Miller et al., 1998) although its use in environmental microbiology is not yet as widespread as either PLFA or PHA. Schnürer and Börjesson (1996) found a high correlation between ergosterol concentrations and values of measurements of opaqueness in fungal infested biological materials using near infrared reflectance spectroscopy (NIRS). When adequately researched, this technique will provide an even more rapid quantification of fungal activity in environmental matrices.

## 6.4 CONCLUDING REMARKS

It is apparent that in most instances we need to combine a number of approaches to simultaneously identify and assay microbial activity in the environment. More specifically, the exciting molecular approaches generated cannot be used exhaustively as an alternative to conventional methods, but rather as a complement to them. More innovative methods for culturing some microorganisms are also beginning to emerge. Recently, Kaeberlein et al. (2002) used a diffusion chamber to grow previously uncultured microorganisms from sediments. Some novel approaches have also been successfully employed to grow previously noncultured microorganisms (Connon and Giovannoni, 2002; Zengler et al., 2002), giving some relevance to one school of thought that argues—what are referred to as viable but nonculturable are culturable, but have not received sufficient attention for successful culturing.

# References

Ahmed M., J.M. Oades, and J.N. Ladd (1982). Determination of ATP in soils: Effects of soil treatments. *Soil Biol. Biochem.* **14:** 273-279.

Akkermans A.D.L., J.D. van Elsas, and F.J. de Bruijn (eds)(1995). Molecular Microbial Ecology Manual. Kluwer Acad. Publ., Dordecht, Netherlands.

Amann R.I., W. Ludwig, and K.-H. Schleifer (1995). Phylogenetic identification and *in situ* detection of individual microbial cells without cultivation. *Microb. Rev.* **59:** 143-169.

Amann R.I., B. Zarda, D.A. Stahl, and K.-H. Schleifer (1992). Identification of individual prokaryotic cells by using enzyme-labeled rRNA-targeted oligonucleotide probes. *Appl. Environ. Microbiol.* **58:** 3007-3011.

Amann R., N. Springer, W. Ludwig, H.-D. Görtz, and H.-K. Schleifer (1991). Identification *in situ* and phylogeny of uncultured bacterial endosymbionts. *Nature* **351:** 161-164.

Anderson J.P.E. and K.H. Domsch (1978). A physiological method for the quantitative measurement of microbial biomass in soils. *Soil Biol. Biochem.* **10:** 215-221.

Atlas R.M. and J.S. Hubbard (1974). Applicability of radiotracer methods of measuring $^{14}CO_2$ assimilation for determining microbial activity in soil including a new *in situ* method. *Microb. Ecol.* **1:** 145-163.

Bakermans C. and E.L. Madsen (2000). Use of substrate responsive-direct counts to visualize naphthalene degrading bacteria in a coaltar-contaminated groundwater microbial community. *J. Microbiol. Meth.* **43:** 81-90.

Balkwill D.L., F.R. Leach, J.T. Wilson, J.F. McNabb, and D.C. White (1988). Equivalence of microbial biomass measures based on membrane lipid and cell wall components, adenosine triphosphate, and direct counts in subsurface sediments. *Microb. Ecol.* **16:** 73-84.

Barcina I., I. Arana, P. Santorum, J. Iribern, and L. Egea (1995). Direct viable count of Gram-positive and Gram-negative bacteria using ciprofloxacin as inhibitor of cellular division. *J. Microbiol. Meth.* **22:** 139-150.

Benefield C.B., P.J.A. Howard, and D.M. Howard (1977). The estimation of dehydrogenase activity in soil. *Soil Biol. Biochem.* **9:** 67-70.

Bhupathiraju, V.K., M. Hernandez, P. Krauter, and Alvarez-Cohen. 1999. A new direct microscopy based method for evaluating *in situ* bioremediation. *J. Hazard. Mate.* **B67:** 299-312.

Bloomfield S.F., G.S.A.B. Stewart, C.E.R. Dodd, I.R. Booth, and E.G.M. Power (1998). The viable but non-culturable phenomenon explained? *Microbiology* **144:** 1-3.

Bohlool B.B. and E.L. Schmidt (1980). The immuno-fluorescence approach in microbial ecology. *Adv. Microb. Ecol.* **4:** 203-241.

Bornemann J. and E.W. Triplett (1997). Molecular microbial diversity in soils from eastern Amazonia: evidence of unusual microorganisms and microbial population shifts associated with deforestation. *Appl. Environ. Microbiol.* **63:** 2647-2653.

Chueng, P-Y. and B.K., Kinkle, (2001). *Mycobacterium* diversity and pyrene mineralization in petroleum-contaminated soils. *Appl. Environ. Microbiol.* **67:** 2222-2229.

Connon S.A. and S.J. Giovannoni (2002). High-throughput methods for culturing microorganisms in very-low-nutrient media yield diverse new marine isolates. *Appl. Environ. Microbiol.* **68:** 3878-3885.

Contin M., D.S. Jenkinson, and P.C. Brookes (2002). Measurement of ATP in soil: correcting for incomplete recovery. *Soil Biol. Biochem.* **34:** 1381-1383.

DeLong E.F., G.S. Wickham, and N.R. Pace (1989). Phylogenetic stains: Ribosomal RNA-based probes for the identification of single cells. *Science* **243:** 1360-1363.

Dojka M.A., J.K. Harris, and N.R. Pace (2000). Expanding the known diversity and environmental distribution of an uncultured phylogenetic division of bacteria. *Appl. Environ. Microbiol.* **66:** 1617-1621.

Fairbanks B.C., L.E. Woods, R.J. Bryant, E.T. Elliot, C.V. Cole, and D.C. Coleman (1984). Limitations of ATP estimates of microbial biomass. *Soil Biol. Biochem.* **16:** 549-558.

Felske A., B. Engelen, U. Nübel, and H. Backhaus (1996). Direct ribosome isolation from soil to extract bacterial rRNA for community analysis. *Appl. Environ. Microbiol.* **62:** 4162-4167.

Fischer K., D. Hahn, R.I. Amann, O. Daniel, and J. Zeyer (1995a). In situ analysis of the bacterial community in the gut of the earthworm *Lumbricus terrestris* L. by whole cell hybridization. *Can. J. Microbiol.* **41:** 666-673.

Fischer K., D. Hahn, W. Hönerlager, F. Schönholzer, and J. Zeyer (1995b). *In situ* detection of spores, and vegetative cells of *Bacillus megaterium* in soil by whole-cell hybridization. *Syste. Appl. Microbiol.* **18:** 265-273.

Fulthorpe R.R., A.N. Rhodes and J.M. Tiedje (1998). High levels of endemicity of 3-chlorobenzoate-degrading soil bacteria. *Appl. Environ. Microbiol.* **64:** 1620-1627.

Hahn D., R.I. Amann, W. Ludwig, A.D.L. Akkermans, and K-H. Schleifer (1992). Detection of micro-organisms in soil after *in situ* hybridization with rRNA-targeted, fluorescently labelled oligonuceluotides. *J. Gen. Microbiol.* **138:** 879-887.

Heuer H. and K. Smalla (1997). Application of denaturing gradient gel electrophoresis and temperature gradient gel electrophoresis for studying soil microbial communities. *In:* J.D. van Dirk, J.T. Trevors, and E.M.H. Wellington (eds.). Modern Soil Microbiology. Marcel Dekker, New York, NY, pp. 353-373.

Hugenholtz P., B.M. Goebel, and N.R. Pace (1998). Impact of culture-independent studies on the emerging phylogenetic view of bacterial diversity. *J. Bacteriol.* **180:** 4765-4774.

Hughes J.B., J.J. Hellmann, T.H. Ricketts and B.J.M. Bohannan (2001). Counting the uncountable: statistical approaches to estimating microbial diversity. *Appl. Environ. Microbiol.* **67**: 4399-4406.

Jenkinson D.S. and D.S. Powlson (1976). The effects of biocidal treatments on metabolism in soil, V. *Soil Biol. Biochem.* **8**: 209-213.

Jenkinson D.S. and J.M. Oades (1979). A method for measuring adenosine triphosphate in soil. *Soil Biol. Biochem.* **11**: 193-199.

Jenkinson D.S., S.A. Davidson and D.S. Powlson (1979). Adenosine triphosphate and microbial biomass in soil. *Soil Biol. Biochem.* **11**: 521-527.

Jjemba P.K., B.K. Kinkle and J.R. Shann (in litt.). *In situ* enumeration and identification of pyrene-degrading soil bacteria.

Kaeberloin, T., K. Lewis, and S.S. Epstein. 2002. Isolating "uncultivable" microorganisms in pure culture in a simulated natural environment. *Science* **296**: 1127-1129.

Kamprelyants A.S., J.C. Gottschal, and D.B. Kell (1993). Dormancy in non-sporulating bacteria. *FEMS Microbiol. Rev.* **104**: 271-286.

Kogure, K., U. Simidu, and N. Taga (1979). A tentative direct microscopic method for counting living marine bacteria. *Can. J. Microbiol.* **25**: 415-420.

Kohring L.L., D.B. Ringelberg, R. Devereux, D. Stahl, M.W. Mittelman, and D.C. White (1994). Comparison of phylogenetic relationships based on phospholipid fatty acid profiles and ribosomal RNA sequence similarities among dissimilatory sulfate-reducing bacteria. *FEMS Microbiol. Lett.* **119**: 303-308.

Kuske C.R., S.M. Barns, and J.D. Busch (1997). Diverse uncultivated bacterial groups from soils of the arid southwestern United States that are present in many geographic regions. *Appl. Environ. Microbiol.* **63**: 3614-3621.

Leclercq A., A. Guiyoule, M.E. Lioui, E. Carniel, and J. Decallonne (2000). High homogeneity of the *Yersinia pestis* fatty acid composition. *J. Clin. Microbiol.* **38**: 1545-1551.

Lee C.C., R.F. Harris, D.J.H. Williams, D.E. Armstrong, and J.K. Syers (1971). Adenosine triphosphate in lake sediments. 1. Determination. *Proce. Soil Sci. Soc. Ameri.* **35**: 82-86.

Lee D-H., Y-G. Zo, and S-J. Kim (1996). Nonradioactive method to study genetic profiles of natural bacterial communities by PCR-single-strand-conformation polymorphism. *Appl. Environ. Microbiol.* **62**: 3112-3120.

Liu W-T., T.L. Marsh, H. Cheng, and L.J. Forney (1997). Characterization of microbial diversity by determining terminal restriction fragment length polymorphisms of genes encoding 16S rRNA. *Appl. Environ. Microbiol.* **63**: 4516-4322.

Ljungholm K., B. Norén, R. Sköld, and I. Woldsö (1979). Use of microcalorimetry for the characterization of microbial activity in soil. *Oikos* **33**: 15-23.

Lodge D.J. (1993). Nutrient cycling by fungi in wet tropical forests. *In*: S. Isaac, J.C. Frankland, R. Watling, and A.J.S. Whalley (eds.). Aspects of Tropical Mycology. Cambridge Univ. Press, Cambridge, UK, pp. 37-57.

Madsen E.L. (1995). A critical analysis of methods for determining the composition and biogeochemical activities of soil microbial communities *in situ*. *Soil Biochem.* **9**: 287-370.

Madsen E.L. (1998). Epistemology of environmental microbiology. *Environ. Sci. Tech.* **32**: 429-439.

Miller M., A. Palojärvi, A. Rangger, M. Reeslev, and A. Kjøller (1998). The use of fluorogenic substrates to measure fungal presence of activity in soil. *Appl. Environ. Microbiol.* WN **64**: 613-617.

Moreira D. and P. López-García (2002). The molecular ecology of microbial eukaryotes unveils a hidden world. *Trends Microbiol.* **10**: 31-38.

Moter A. and U.B. Göbel (2000). Fluorescence *in situ* hybridization (FISH) for direct visualization of microorganisms. *J. Microbiol. Meth.* **41**: 85-112.

Muyzer G., E.C. de Waal, and A. Uittelinden (1993). Profiling of complex microbial populations by denaturing gradient gel electrophoresis analysis of polymerase chain reaction-amplified genes coding for 16S rRNA. *Appl. Environ. Microbiol.* **59**: 695-700.

Oda Y., S-J. Slagman, W.G. Meijer, L.J. Forney, and J.C. Gottschal (2000). Influence of growth rate and starvation on flourescent *in situ* hybridization of *Rhodopseudomonas palustris*. *FEMS Microbiol. Ecol.* **32**: 205-213.

Orchard V.A. and F.J. Cook (1983). Relationship between soil respiration and soil moisture. *Soil Biol. Biochem.* **15**: 447-453.

Pažout J. and V. Vančura (1981). Simultaneous measurement of ethylene and $CO_2$ production and consumption in soil and pure microbial cultures. *Plant Soil* **62**: 107-115.

Quiros L.M., F. Parra, C. Hardisson, and J.A. Sales (1989). Structural and functional analysis of ribosomal subunits from vegetative mycelium and spores of *Streptomyces antibioticus*. *J. Gen. Microbiol.* **135**: 1661-1670.

Rochelle P.A. (ed.) (2001). Environmental Molecular Microbiology: Protocols and Applications. Horizon Sci. Press, Wymondham, Norfolk, UK.

Rochelle P.A., B.A. Cragg, J.C. Fry, R.J. Parkes, and A.J. Weightman (1994). Effect of sample handling on estimation of bacterial diversity in marine sediments by 16S rRNA gene sequence analysis. *FEMS Microbiol. Ecol.* **15**: 215-225.

Roszak D.B. and R.R. Colwell (1987). Survival strategies of bacteria in the natural environment. *Microbiol. Rev.* **51**: 365-379.

Roszak, D.B., D.J. Grimes, and R.R. Colwell (1984). Viable but non-recoverable stages of *Salmonella enteridis* in aquatic systems. *Can. J. Microbiol.* **30**: 334-338.

Sambrook J., E.F. Fritsch, and T. Maniatis (1989). Molecular Cloning: A Laboratory Manual. Cold Spring Harbor Laboratory Press, New York, NY. (vols. 1,2, and 3).

Schnürer J. and T. Börjesson (1996). Quantification of fungal growth in the environment. *In*: D.F. Jensen, H-B. Jansson, and A. Tronsmo (eds.). Monitoring Antagonistic Fungi Deliberately Released in the Environment. Kluwer Acad. Publ., Boston, MA, pp. 17-24.

Setlow P. (1994). Mechanisms which contribute to long-term survival of spores of bacillus species. *J. Bacteriol.* **76:** S49-S60.

Sorheim R., V.L. Torsvik, and J. Goksoyr (1989). Phenotypical divergences between populations of soil bacteria isolated on different media. *Microb. Ecol.* **17:** 181-191.

Sparling G.P., T.W. Speir, and K.N. Whale (1986). Changes in microbial biomass C, ATP content, soil phospho-monoesterase and phospho-diesterase activity following air-drying of soils. *Soil Biol. Biochem.* **18:** 363 370.

Stolp H. (1988). Microbial Ecology. Organisms, Habitats, Activities. Cambridge Univ. Press, Cambridge, UK.

Strickland J.D.J., O. Holm-Hansen, R.W. Eppley, and R.T.Linn (1969). The use of a deep tank in plankton ecology. Studies of growth and composition of phytoplankton crops at low nutrient levels. *Limnol. Oceanog.* **14:** 23-34.

Tate K.R. and D.S. Jenkinson (1982). Adenosine triphosphate measurements in soil: An improved method. *Soil Biol. Biochem.* **14:** 331-335.

Teske A., C. Wawer, G. Muyzer, and N.B. Ramsing (1996). Distribution of sulfate-reducing bacteria in a stratified fjord (Mariager Fjord, Denmark) as evaluated by most probable-number counts and denaturing gradient gel electrophoresis of PCR-amplified ribosomal DNA fragments. *Appl. Environ. Microbiol.* **62:** 1405-1415.

Tiedje J.M., S. Asuming-Brempong, K. Nüsslein, T.L. Marsh, and S.J. Flynn (1999). Opening the black box of soil microbial diversity. *Appl. Soil Ecol.* 13: 109-122.

Trevors J.T. (1984). Electron transport system activity in soil, sediment and pure cultures. *CRC Crit. Rev. Microbiol.* **11:** 83-100.

Tuomi T., K. Reijula, T. Johnsson, K. Hemminki, E-L. Hintikka, O. Lindroos, S. Kalso, P. Koukila-Kähkölä, H. Mussalo-Rauhamaa., and T. Haahtela (2000). Mycotoxins in crude building materials from water-damaged buildings. *Appl. Environ. Microbiol.* **66:** 1899-1904.

Wagner M., R. Erhart, W. Manz, R. Amann, H. Lemmer, D. Wedi, and K-H. Schleifer (1994). Development of an rRNA-targeted oligonucleotide probe specific for the genus *Acinobacter* and its application for *in situ* monitoring in activated sludge. *Appl. Environ. Microbiol.* **60:** 792-800.

Waksman S.A. (1927). Principles of Soil Microbiology. Williams and Wilkins. Baltimore, MD.

Ward B.B. (2002). How many species of prokaryotes are there? *Proce. Natl. Acad. Sci. (USA)* **99:** 10234-10236.

Ward D.M., R.Weller, and M.M. Bateson (1990). 16S rRNA sequences reveal numerous uncultured microorganisms in a natural community. *Nature* **345:** 63-65.

Ward D.M., M.M. Bateson, R. Weller, and A.L. Ruff-Roberts (1992). Ribosomal RNA analysis of microorganisms as they occur in nature. *Adv. Microb. Ecol.* **12:** 219-286.

Weiss P., B. Schweitzer, R. Amann, and M. Simon (1996). Identification *in situ* and dynamics of bacteria on limnetic organic aggregates (lake snow). *Appl. Environ. Microbiol.* **62:** 1998-2005.

White D.C. (1983). Analysis of microorganisms in terms of quantity and activity in natural environments. *In:* J.H. Slater, R. Whittenbury, and J.W.T. Wimpenny (eds.). Microbes in Their Natural Environments. *Soc. Gen. Microbiol. Symp.* **34:** 37-66.

White D.C. and D.B. Ringelberg (1997). Utility of the signature lipid biomarker analysis in determining the *in situ* viable biomass, community structure, and nutritional/physiologic status of deep subsurface microbiota. *In:* P.S. Amy and D.L. Haldeman (eds.). The Microbiology of the Terrestrial Deep Subsurface, CRC Lewis Publishers, Boca Raton, FL (USA). pp. 119-136.

White, D.C., C.A. Fleming, K.T. Leung, and S.J. Macnaughton (1998). *In situ* microbial ecology for quantitative assessment, monitoring and risk assessment of pollution remediation in soils, the subsurface, the rhizosphere and in biofilms. *J. Microbiol. Meth.* **32:** 93-105.

Wu J., R.G. Joergensen, B. Pommerining, R. Chaussod, and P.C. Brookes (1990). Measurement of soil microbial biomass C by fumigation-extraction-an automated procedure. *Soil Biol. Biochem.* **22:** 1167-1169.

Zarda B., D. Hahn, A. Chatzinotas, W. Schönhuber, A. Neef, R.I. Amann, and J. Zeyer (1997). Analysis of bacterial community structure in bulk soil by *in situ* hybridization. *Arch. Microbiol.* **168:** 185-192.

Zengler K., G. Toledo, M. Rappé, J. Elkins, E.J. Mathur, J.M. Short, and M. Keller (2002). Cultivating the uncultured. *Proc. Natl. Acad. Sci. (USA)* **99:** 15681-15686.

# 7

# Mechanisms of Adaptation by Microorganisms to Environmental Extremes

On a macroscale, the distribution of organisms is universal. This observation prompted Beijerinck to state that everything is everywhere and the environment selects what organisms prevail. From an oversimplified perspective, if one takes a soil, one is likely to find whatever type of microbe is sought. However, this is an overstatement as the environment tends to select what organisms to favor under different conditions. In natural environments (i.e., soils, sediments, aquatic and aero system), microorganisms may be subjected to various stresses such as nutrient deprivation, osmotic shock, temperature fluctuations, heat and cold shock, desiccation, pH, radiation (UV, X-rays) and hydrostatic pressure, stresses for which, to survive they must adapt to. The selection process can be over an evolutionary period (time) or towards colonization that leads to succession and a subsequent climax community of organisms that work together (i.e., selection over space).

Survival through the selection process by an organism requires the possession of certain fitness traits that endow it with the capability to cope with the physical, biological, and chemical stresses within a specific environment, directly or otherwise affecting their distribution and activity. Fitness traits can be biological (biotic) or nonbiological (abiotic) in nature. Biotic traits

may include a wide range of attributes such as growth, ability of the organism to move from one location to another, access to specific growth factors, ability to store nutrients, and the ability to inflict direct injury to undesirable competitors through the production of toxins and antibiotics, etc. Abiotic factors, on the other hand, are imposed by the environment, e.g. high pressure, low/high temperature, moisture, high salts, etc. Adaptation and activities of microorganisms under some of these extreme conditions are discussed.

## 7.1 PRINCIPLES OF SELECTIVE ENRICHMENT

Under specific conditions prevailing in the natural environment, representative microbial populations can be favored and selected over time. In some habitats, the factors conferring selective advantage may be obvious. For example, in hot springs and thermal vents, temperature would be the factor conferring selective advantage. Enrichment and selection can be demonstrated in the laboratory using a Winogradsky column (Fig. 7.1). The column is assembled from a plastic or glass tube packed with sediments. Within the column, a gradient of oxygen, nutrients, and metabolic products develops. The initially existing organisms are

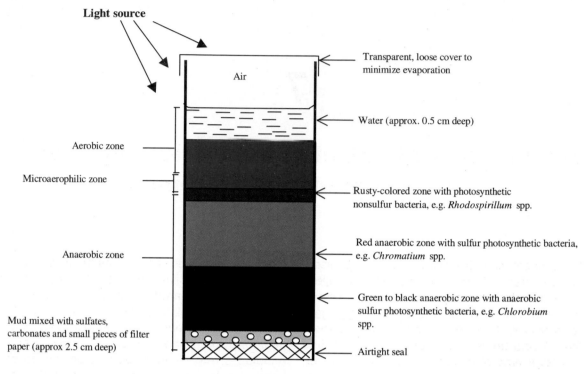

**Fig. 7.1** The Winograsky column (redrawn from Herbert, 1982 with permission from Blackwell Publishers, Oxford, United Kingdom).

enriched within the material used as inoculum, leading to succession and dominance of the best adapted species and their variants (mutants) over time. Since the cover on the column is not airtight, the aerobic zone in the column and the sulfide present in the sediments favor the development of *Thiobacillus* spp. in the oxic zone. The deeper anoxic zone where sulfate levels are high, enhances sulfate-reducing bacteria such as *Desulfovibrio* spp., causing them to predominate. If sulfate is not present, competition from methanogens can outwit the sulfur-reducing bacteria (SRB), generating methane.

Iron-oxidizing bacteria, if present, will cause the formation of a $Fe^{2+}/Fe^{3+}$-layer, with a characteristic rusty appearance, at the oxic-anoxic (i.e., micro-aerophilic) zone. Formation of the rusty appearance at the micro-aerophilic zone

can be induced simply by introducing a nail or a paper clip into the column.

$$Fe^0 \longrightarrow Fe^{2+} \longrightarrow Fe^{3+}(OH)_3$$
$$\text{(rusty appearance)}$$

If a light source is provided, phototrophs such as cyanobacteria will predominate in some layers. Thus, if the properties and requirements of a particular microorganism are known, it is possible to modify the conditions so as to favor its enrichment. For example, *Azotobacter* populations can be enriched by culturing soil under aerobic conditions in a carbon source but devoid of combined nitrogen, providing a chance for *Azotobacter* to fix nitrogen and predominate. Similarly, *Thiobacillus denitrificans* can be enriched under anaerobic conditions with sulfur or thiosulfate as electron donors and nitrate as the electron acceptor.

Other variables, such as a higher temperature can also be introduced as a factor to select for thermotolerant species. This simple enrichment strategy basically represents what occurs naturally all the time in the environment. Some of the scenarios mentioned above also show the variety of metabolic possibilities that microorganisms display to survive and dominate in different environments.

## 7.2 ENERGETICS OF MICROBIAL PROCESSES IN THE ENVIRONMENT

In nature, microbial populations are kept in check by the resistance of the environment. This has been the case since the first life occurred. Resistance of the environment is in the form of food available, predators, climate, etc. An understanding of the energy relationships in the growth of microorganisms in the environment requires us to revisit two of Newton's laws of thermodynamics, i.e. (i) energy is neither created nor destroyed but rather, conserved, and (ii) no process involving an energy transformation occurs unless degradation of energy from a concentrated into a dispersed form has occurred. All chemical reactions are accompanied by changes in energy, the energy being transformed from one form to another as the cells get some work done. All processes that involve energy transformations introduce some level of chaos (entropy). During dispersion some of the energy becomes unavailable. This implies that as reactions take place, the products tend toward a lower energy level (Fig. 7.2). Organisms maintain their highly organized (i.e., low entropy) state by transforming energy from high to low utility states. At fixed energy, only those processes for which there is an increase in entropy will occur spontaneously. Thus, thermodynamically unfavorable reactions can only be conducted at the expense of energy.

Microorganisms perceive thermodynamic gradients through a diversity of biochemical processes, generating ATP, and create linkages between biogeochemical cycles. The ability of an organism to carry out redox reactions depends on the oxidation-reduction state of the environment. Some microorganisms can only be active in oxidizing and others only in reducing environments. The proportion of oxidized to reduced components constitutes the redox potential ($E_h$). A positive $E_h$ value favors oxida-

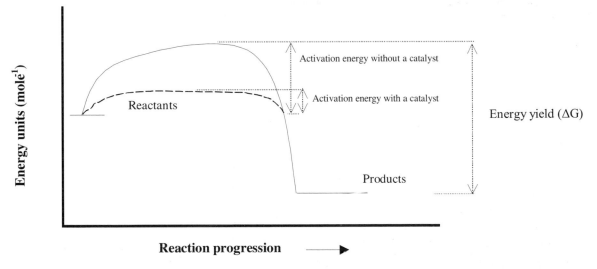

**Fig. 7.2** Changes in free energy as reactants are converted to products. The conversion occurs at a reduced free energy requirement in the presence of a catalyst (enzyme).

tion and a negative $E_h$ value is indicative of a reducing environment. Thus different pairs of substances of metabolic importance in the environment shown in Fig. 7.3 are oxidized or reduced (redox reactions) depending on the electron potential. The hierarchy of biogeochemical regimes is determined by Gibbs free energy ($\Delta G$) (Table 7.1). In mathematical terms:

$$\Delta G = \Delta H - \Delta TS$$

where $\Delta H$ is the change in the total amount of energy released (i.e., enthalpy), T the temperature, and $\Delta S$ the change in entropy. Alternatively, free energy is related to the potential ($E_H$) generated by an electrode immersed in a solution and coupled with a standard hydrogen electrode according to the equation

$$\Delta G = -nFE_H$$

where F is Faraday's constant (96.5 kJ $V^{-1}$ $mol^{-1}$ or 23.1 kcal $V^{-1}$ $mol^{-1}$) and n the number of electrons transferred during the process. We can use thermodynamics to predict direction, $\Delta G$ (+ or −). The $\Delta G$ must be negative for the reaction to be favored. Under constant temperature and pressure, the change in free energy will be negative for a spontaneous process, positive for a nonspontaneous process, and zero for a state of equilibrium. The free energy, at standard conditions (i.e., 1 atm; pH 7; 25°C), is depicted by $\Delta G^0$. The value of $\Delta G^0$ is a function of the number of electrons transferred in a reaction. All of the processes in Table 7.1 are major in heterotrophs. The pE represents the abundance of electrons and at neutrality (pH 7) the electrons become abundant and reduced as the conditions become more and more anaerobic. This governs physiological reactions and also provides predictive power. The $\Delta G$ for oxygenic reactions is very high. Although we can predict the energy relationships of a particular metabolic process

**Table 7.1**   Some chemolithotrophic oxidation reactions and the energy associated from them

| Reaction | Process/typical organisms | $\Delta G$ (kJ $mol^{-1}$ substrate) | Probable mol ATP synthesized/mol substrate |
|---|---|---|---|
| $H_2 + 0.5O_2 = H_2O$ | Aerobic respiration | 237 | 2-3 |
| $5H_2 + 2NO_3^- + 2H^+ = N_2 + 6H_2O$ | Denitrification | 241 | |
| $4H_2 + CO_2 = CH_4 + 2H_2O$ | Methanogenesis by *M. thermoautotrophicum* | 35 | <0.25 |
| $NH_4^+ + 1.5O_2 = NO_2^- + H_2O + 2H^+$ | Deamination/nitrification | 272 | 1-2 |
| $NH_2OH + O_2 = NO_2^- + H_2O + H^+$ | Hydroxylamine oxidation | 288 | 2 |
| $NO_2^- + 0.5O_2 = NO_3^-$ | Nitrification | 73 | 1 |
| $H_2S + 0.5O_2 = S^0 + H_2O$ | Sulfur oxidation | 210 | 1 |
| $S^0 + 1.5O_2 + H_2O = H_2SO_4$ | same | 496 | 3 |
| $HS^- + 2O_2 = SO_4^{2-} + H^+$ | same | 716 | 4 |
| $S_2O_3^{2-} + 2O_2 + H_2O = 2SO_4^{2-} + 2H^+$ | *Thiobacillus* sp. and *Thermothrix* sp. | 936 | 1.5-3.0 |
| $5S_2O_3^{2-} + 8NO_3^- + H_2O = 10SO_4^{2-} + 2H^+ + 4N_2$ | *Thiobacillus denitrificans* | 741 | 4-5 |
| $S_4O_6^{2-} + 3.5O_2 + 3H_2O = 4SO_4^{2-} + 6H^+$ | Fe oxidation/*T. ferroxidans* | 1,654 | 5 |
| $5S_4O_6^{2-} + 14NO_3^- + 8H_2O = 20SO_4^{2-} + 16H^+ + 7N_2$ | same | 1258 | 8-10 |
| $2Fe^{2+} + 2H^+ + 0.5O_2 = 2Fe^{3+} + H_2O$ | *T. ferroxidans* | 47 | 0.5 |
| $4FeS_2 + 15O_2 + 2H_2O = 2Fe_2(SO_4)_3 + 2H_2SO_4$ | same | 1,210 | |
| $Cu_2S + 0.5O_2 + H_2SO_4 = CuS + CuSO_4 + H_2O$ | Oxidation of $Cu^+$ to $Cu^{2+}$ | 120 | 1 |
| $CuSe + 0.5O_2 + H_2SO_4 = CuSO_4 + Se0 + H_2O$ | Oxidation of selenide to selenium | 124 | 1 |

Data modified from Kelly (1992 with permission from Springer-Verlag GmbH & Co. KG)

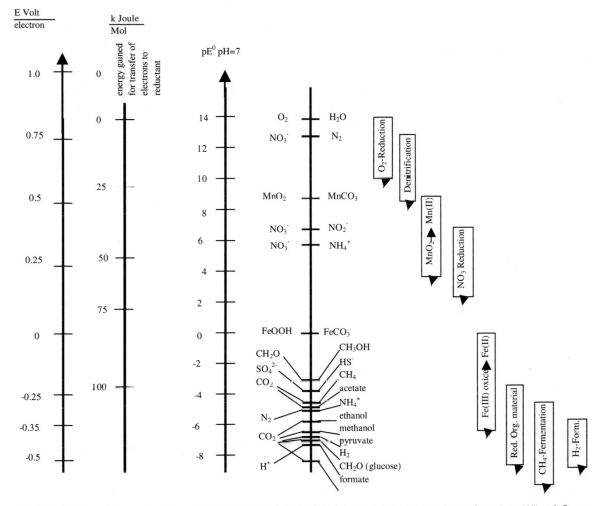

**Fig. 7.3** Electron-free energy diagram (redrawn from Zehnder and Stumm, 1988 with permission from John Wiley & Sons, Inc, New York, NY).

based on free energy, we also need to have some idea about the rate at which the reaction proceeds. Many metabolic reactions, even when chemically favorable, do not occur at appreciable rates. They need some activation from some external energy source (i.e., activation energy). The activation energy can be greatly reduced if appropriate enzymes (catalysts) are present (Fig. 7.2). The success of microorganisms in any environment is influenced by biotic factors, and abiotic factors, or both. We now focus on some of these factors and

how microorganisms adapt to their extremes in the environment.

## 7.3 BIOTIC FACTORS AFFECTING ADAPTATION

Several biologic factors contribute to the adaptation and activity of microorganisms in the environment. These include growth, metabolic plasticity, synthesizing storage structures, as well as antibiosis and allelopathy. Species relationships are appropriately discussed in Chapter 8.

## 7.3.1 Microbial growth

Growth is a characteristic of all living things. For eukaryotes, growth is at the individual level and is manifested as an increase in size and/or complexity. For prokaryotes, however, growth is reflected by an increase in populations as a result of the division of cells by binary fission. Increase in size of the individual prokaryotic cells occurs but is generally of little significance, in terms of growth, compared to the increase in populations. Various techniques are used to measure growth. These include synthesis of nucleic acids (Christensen et al., 1989; Cheung and Kinkle, 2001), synthesis of lipids (Langworthy et al., 1998), respiration, and direct viable counts (Roszak and Colwell, 1987). Growth of bacteria by binary fission has been widely studied compared to that of other prokaryotes and is dealt with here. Binary fission occurs in the presence of a carbon and energy source as well as other essential nutrients. It basically involves one cell dividing into two cells, which in turn divide into 4, and 4 into 8, and so on. Thus, as the populations undergo binary fission, they increase exponentially. Under binary fission, as long as the conditions remain favorable for growth, the population will increase at regular and logarithmic intervals. In mathematical terms, the growth rate (K) can be expressed as

$$N_t = N_0 2^n \tag{i}$$

$$\log N_t = \log N_0 + n \log 2 \tag{ii}$$

$$n = (\log N_t - \log N_0)/\log 2 \tag{iii}$$

$$= (\log N_t - \log N_0)/0.301 \tag{iv}$$

$$= 3.3 (\log N_t - \log N_0) \tag{v}$$

where $N_t$ and $N_0$ are the cell numbers at time (t) and initial time ($t_0$), respectively, and n is the number of generations. Growth rate can also be expressed as the mean doubling time ($t_{gen}$) derived from n/t. In natural environments (as opposed to continuos culture systems), the conditions necessary for exponential growth will not prevail indefinitely. On inoculation, the bacterial cells may undergo a period of inactivity called the lag phase during which no growth occurs (Fig. 7.4). During this phase the cells adjust to the new medium and synthesize the enzymes necessary to metabolize the new substrates encountered. The lag phase could also be due to abiotic alterations in the physical (e.g. aeration, pH, temperature) or chemical (e.g. nutrient) conditions. Under both laboratory culture conditions where the lag phase is only hours, for most microorganisms, it can turn into days or even weeks in the environment. If the enzymes are constitutive, the lag phase is expected to be shorter. The lag phase differences can be seen in the various bacterial strains shown in Figure 7.4. Furthermore, the same bacterial strain can differ in the rate at which it doubles depending on what other microorganisms are present in the system.

After the lag phase, the cells come into the exponential phase during which they divide at a constant rate. Exponential growth continues as long as all the essential nutrients and growth factors required by the organism are present. Growth rate is also affected by the build-up of toxins and waste products that accumulate within the microenvironment. With exhaustion of even one of the nutrients, growth ceases, leading to the stationary phase. The nutrient or factor that triggers the onset of the stationary phase is called the growth limiting nutrient. During the stationary phase the net rate of growth is zero although such residual cellular activity continues utilizing the internal cell reserves and thus the cells remain viable. As viability diminishes in some of the cells, a decline in population is observed, which signals commencement of the death phase.

Since most natural environments contain a variety of microorganisms they do support a mixture of populations. Growth of the individual species that may be present depends on the nature and extent of interactions with the other species present. A broader view of the different

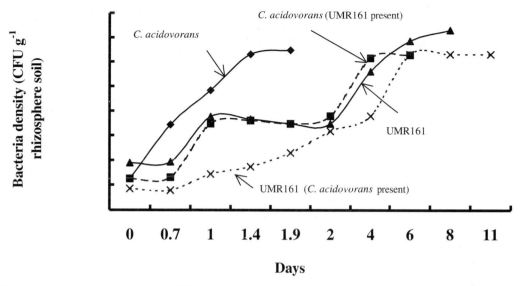

**Fig. 7.4** Typical growth curve for bacteria. Notice the pronounced diauxic growth pattern when *Comomonas acidovorans* is growing in the presence of *Bradyrhizobium japonicum* UMR161 and when UMR161 is growing in a sterile rhizosphere soil (redrawn from Jjemba, 2001 with permission by the Society of Protozoologists).

levels of ecological organization and how differences in the availability of resources affect the interacting populations is thus imperative. It is important to note that microorganisms in the environment are, in most situations, exposed to a variety of sources for any specific nutrient. Thus, substrates occur as a mixture at low concentrations (Egli, 1995). Contrary to the impression portrayed by most degradation studies, the organisms successively exhaust the available substrates, switching from one to the other, a phenomenon called diauxy (Fig. 7.4). The whole mix is further complicated by a variety of microbial species that normally exist at a single microsite. Understanding the growth and behavior of microorganisms in nature mandates a better understanding of their utilization of mixed substrates in the environment. Not many studies have put this important reality into perspective. How different microbial species can interact to exploit the almost always limiting resources is discussed in Chapter 8.

### 7.3.2 Metabolic diversity

Microorganisms are adapted to a wide range of resources. They try to grow, survive and multiply in the environment. Microbial cells comprise C, O, N, H, P, K and S as well as a variety of micronutrients. These nutrients are obtained by the microbes through their normal metabolic processes, transforming the nutrients into cellular constituents. The process by which cells build up from nutrients is called anabolism (biosynthesis). Anabolism requires energy in the form of ATP. On the other side of the spectrum are catabolic reactions through which compounds are broken down, generating energy and waste products. Metabolism is essentially a balance between catabolism and anabolism (Fig. 7.5). In a broad sense, growth and successful metabolism require the organism to extract these nutrients, usually competitively, from the environment for its own requirements. Through this process, the less fit organisms are outwitted by species that can effec-

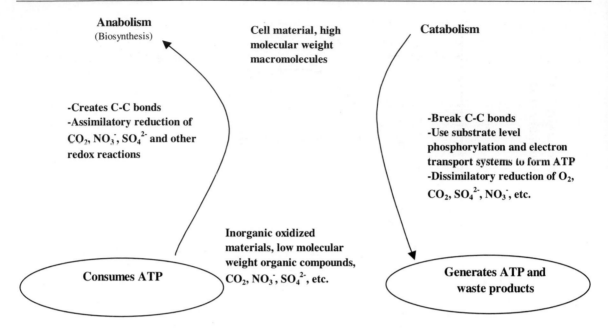

**Fig. 7.5** Schematic representation of metabolism as a net result of anabolism and catabolism.

tively take advantage of the existing nutrients and environmental conditions.

Microorganisms have three different sources of energy from the environment, i.e., light, organic compounds, and inorganic nutrients. Based on these sources, microorganisms are classified as phototrophs, organotrophs, and lithotrophs respectively. Furthermore, all microbial cells require some carbon which is obtained from either an organic source (heterotrophs) or an inorganic source (autotrophs). Metabolic pathways in different environments include fermentation, respiration, phototrophy, and lithotrophy. These are well discussed in biochemistry textbooks and so not repeated here. However, it is important to note that each of these pathways generates the energy required by the organisms in the form of ATP and/or $NAD^+$. In general, generating ATP requires the reduced inorganic substrates (e.g., $H_2$, $H_2S$, $NH_3$, $NH_4^+$) being oxidized and generating inorganic wastes such as $H_2O$, $SO_4^{2-}$, $NO_3^-$, and $CO_2$. The ATP generated by

autotrophs is utilized in turn during anabolism, together with available nutrients, such as C, O, H, N, and S, to generate cellular (i.e., organic) materials. The reduced cellular materials provide substrates for heterotrophs, which in turn generate ATP, $CO_2$, wastes as well as reduced inorganic substrates that are utilized by autotrophs as building blocks for generating biomass. Through this intricate series of metabolic capabilities, microorganisms create linkages between the various biogeochemical cycles. The wastes generated during metabolic processes can, in some cases, also be important as they give a competitive edge to a particular type of organism in that micro-environment.

With the currently high oxygen levels in the atmosphere compared to the Precambrian period, respiration is a very dominant process as it almost always implies availability of oxygen as the terminal electron acceptor. However, if $O_2$ is absent, other electron acceptors, for example, $NO_3^-$, $SO_4^{2-}$, $CO_2$ (in methanogenesis), $FeOOH$, $AsO_4^{2-}$, $NO_2^{2-}$, $MnO_x$, $SeO_4^{2-}$, and $UO_2^{2+}$, and organic acids (e.g., formate in *E. coli*) can be very useful (Fig. 7.3).

### 7.3.3 Synthesis of storage structures

Under nutrient abundance, some well-adapted microorganisms do not utilize the existing substrates to support excessive rates of growth. Rather, they synthesize storage polymers such as poly-hydroxybutyrate (PHB), polyhydroxy alkanoates (PHA), and poly-phosphates (poly-P) to survive environmental extremes such as nutrient stress. Studies by Falvo et al. (2001) indicate that as much as 70% of the readily available substrate can be converted to storage materials, polymers that act as reserves of carbon and energy when starvation sets in. With continued starvation, the storage structures are slowly degraded. Synthesis of these structures does not seem to be universal among all microorganisms and hence can play a pivotal role in determining the most active and better adapted organisms in a particular microenvironment.

### 7.3.4 Antibiotics and other compounds

Microorganisms rarely exist as single organisms in the microenvironment, but rather as complex communities. In the communities, some individual species or strains synthesize complex chemicals that can influence the activity and distribution of their competitors. Most widely documented are antibiotics, a term broadly used in this case to include antibacterial, antifungal, antiprotozoan, and anticyanobacteria agents. Of primary relevance to our case are what environmental conditions elicit the production of antibiotic and how the organisms that produce them protect themselves against the adverse effects of their own biological weapons. Production of these compounds particularly occurs when food and space for further multiplication is limiting (Baba and Schneewind, 1998). Thus, their synthesis at a time of starvation tends to be costly. The mechanisms by which they affect their target organisms range from disrupting the cell wall, changing the membrane conformation, and inhibiting DNA gyrase (thus, preventing cell division).

Recently, a new class of bacterial pheromones known as acyl-homoserine lactones (HSLs), discovered in a wide range of gram-negative bacteria were found to monitor population growth, indicating some sort of quorum sensing (Swift et al., 1994). The full impact of this discovery in environmental microbiology is only beginning to be investigated but preliminary results strongly suggest that bacterial cells can communicate with many of their neighbors that may or may not be of their own species. The discussion of production of chemicals as a means of adapting or surviving in particular environments is greatly intertwined with competition per se. Competition is extensively discussed in Chapter 8.

## 7.4 ABIOTIC (PHYSICOCHEMICAL) FACTORS AFFECTING ADAPTATION

The abundance and activity of microorganisms in an environment are determined not only by biotic factors, but also by physicochemical conditions. For respective organisms to succeed, these conditions have to be within a range they can tolerate. The existence of organisms in extreme environmental conditions requires them to possess specialized adaptive features. We now focus on such adaptations.

### 7.4.1 Acidity and alkalinity (pH)

The concentration of hydrogen ions is measured on a pH scale, a scale that basically expresses the tendency of a solution to donate or accept protons. Thus, for a solution HA which dissociates into $HA \rightleftharpoons H^+ + A^-$, a dissociation constant K is computed, i.e.,

$$K = [H^+][A^-]/[HA]$$

$$pH = pK + \log ([H^+]/[HA])$$

Thus, pH is an indicator of the metabolic activity but organisms can also change it through their excretions.

Various microbial strains have an intrinsic range of pH within which each functions optimally, but most of them grow best in a narrow pH range near neutral (i.e., pH 6.5 - 7.5). Microorganisms are classified into acidophiles, neutraphiles, or alkaliphiles. Under highly acidic or alkaline conditions, the microbial cell components of most microbes are hydrolyzed and the enzymes denatured. However, a few acidophiles grow below pH 4.0 or even lower (Table 7.2). Detailed studies with *Thiobacillus thioxidans* indicate that even when it is growing at such an extremely acidic pH, it functions best at near neutral pH. In contrast, another S-oxidizing bacterium, *Thiobacillus thioparus,* grows optimally near neutral pH. The membrane of prokaryotes is very impermeable to protons. So, to survive under acid conditions, acidophiles have to maintain an internal pH near neutral by actively pumping out protons. Yeasts and molds grow over a wider pH range than prokaryotes but grow optimally at pH 5. They probably use a similar mechanism to survive acidity.

The ability of microorganisms to resist pH extremes has important applications in food preservation and in mining industries. For the latter, which has more relevance to the environment, *T. ferroxidans* naturally occurs in S-containing minerals and obtains its energy by oxidizing inorganic materials such as CuS and FeS (fool's gold), generating sulfuric acid (Brierley and Luinctra, 1993).

$$FeS + 7.5O_2 + 7H_2O \rightarrow 2Fe(OH)_3 + 4H_2SO_4$$

The unfortunate aspect of this oxidation in the mining industry is the piles of such pyrites provide an enrichment for *T. ferroxidans* and the acid generated leaches into the environment (soil, streams, lakes, etc.). This condition is called acid mine drainage. The affected soil becomes quite acidic and if the situation is left uncorrected over time, fails to support any vegetation. Acid leachate into the waterways also adversely affects marine life and can dissolve steel bridges as well as concrete structures once these are exposed. In the past, the state of Ohio suffered great calamities from acid mine drainage. Nowadays, with increased awareness about acid drainage mining problems, the pyrites are promptly covered so as to minimize contact with $O_2$ and water, offsetting oxidation. A critical study of this process led to important industrial applications. For generations, the world depended on high-grade Cu and gold ores that were near the surface. It so happens that these high-grade ores have been naturally oxidized over time by bacteria, sunlight, and water. Over time, these high-grade ores have dwindled and the shift toward low-grade ores that exist at deeper surfaces is evident. Purification of low-grade ores is expensive as it requires heating at high temperatures and pressure to oxidize the sulfides. The reaction by which acid mine drainage can occur has been controlled at an industrial level to facilitate acidophiles, such as *T. ferroxidans*, to extract the pyrites. The copper is trapped by plating on scrap Fe. Before this approach was discovered, the pyrites were regarded as an

**Table 7.2**   pH ranges for several bacteria that grow under extreme conditions and the respective electron donors

| Organism | Lithotrophic electron donor | Range of pH for growth |
|---|---|---|
| *Thiobacillus thioxidans* | $S^0$ | 2-5 |
| *T. ferroxidans* | $S^0$, $S^-$, $Fe^{2+}$ | 1.5-4 |
| *T. intermedian* | $S_2O32^{2-}$ | 3-7 |
| *Sulfolobus* sp. | $H_2S$, $S^0$ | 1-5 |
| *T. denitrificans* | $H_2S$, $S^0$, $S_2O_3^{2-}$ | 6-8 |
| *T. thioparus* | $H_2S$, $S^-$, $S^0$ | 6-8 |

otherwise low-grade ore unworthy of copper extraction. Nowadays biomining of copper through this approach is a multibillion dollar industry.

## 7.4.2 Pressure extremes

The effects of pressure are due either to hydrostatic and atmospheric pressure or to osmotic pressure (i.e., salinity and alkalinity).

### 7.4.2.1 Hydrostatic pressure

On average, the depth of oceans is about 3,800 m (Morita, 1986). Whereas the deep ocean is fairly stable in terms of temperature, organic matter content, salinity and oxygen, it can vary in hydrostatic pressure. At sea level, the pressure is 1 atm but increases by 1 atm for every 10 m increment in depth. Thus, at 3,800 m depth, the pressure translates into approximately 380 atm. In the deep ocean, however, the increment in hydrostatic pressure is more than 1 atm for every 10 m due to increases in compression with depth attributed to the density of sea water. The actual hydrostatic pressure can be precisely determined at a given depth if the density of the water (which depends on the dissolved salts) is known. At elevated hydrostatic pressure the rate at which biochemical reactions occur is reduced accordingly, as expressed in these mathematical terms:

$$(\delta \ln K/\delta P) = -\Delta V/RT \qquad (i)$$

$$(\delta \ln k/\delta P) = -\Delta V^*/RT \qquad (ii)$$

where K is the equilibrium constant, k the reaction rate constant, P the pressure, $\Delta V$ the volume changes of reaction, $\Delta V^*$ the apparent volume change of activation, R the gas constant, and T the temperature (in Kelvin). The effects of pressure and temperature are intertwined through the ideal gas law (PV = nRT).

Low temperature and high pressures of deep-sea environments are highly inhibitory for many microbial processes, resulting in a tremendous slowdown of vital metabolic processes (Jannasch et al., 1971; Jannasch and Wirsen, 1973). More specifically, at an elevated pressure, the biopolymers (i.e., proteins) are denatured due to a disruption in hydrophobic interactions, leading to breakages of electrostatic bonds in proteins (Marquis and Matsumura, 1978). The breakage in bonds in turn exposes the charged groups to the aqueous environment. In protozoa and higher organisms, excessive pressure disintegrates such structures as the microtubules, flagella and "freezes" the mitotic apparatus. Increased pressure also represses the replication of DNA which in turn inhibits cell division, thereby resulting in the formation of filamentous cells (Zobell and Cobet, 1964). Viruses are simpler in structure and organization than cellular microorganisms and are generally less susceptible to the lethal actions of pressure and temperature (Solomon et al., 1966).

The temperature in the deep sea lies in the range of 3-4°C. Depressed temperatures, combined with increased hydrostatic pressure together with a depressed metabolic rate enable microorganisms to survive for long periods in these nutrient-poor environments. At such abyssal depth, the existing microorganisms are classified as either barophilic or baroduric. The former category can only grow at elevated hydrostatic pressure but not at 1 atm. Organisms such as *Hyphomonas jannaschiana* grow more rapidly at 300 atm than at 1 atm (Weiner, 1999). Baroduric organisms, on the other hand, grow at 1 atm but can endure higher pressures as well. To be able to survive under elevated hydrostatic pressure, the organisms have to be able to counterbalance the massive external pressure. The actual mechanisms by which they achieve this are not known. A number of barophilic microorganisms are associated with the gut or surfaces of deep sea animals. Such associations are necessary for growth and multiplication of the microorganisms because the sediments are often low in available nutrients at such depths. The sea animals with which the barophilic organisms are associated can supply some of the limiting nutrients (Deming and Colwell, 1981).

Spores are generally more resistant to pressure than vegetative cells. Spores for some species, e.g. *Bacillus pumilus,* can actually be induced by high pressures to germinate. However, the ecological implications of this induction, in the case of *B. pumilus,* are not very clear because the resultant vegetative cells are barosensitive (Clouston and Wills, 1969).

### 7.4.2.2 Osmotic pressure

Microorganisms obtain most of their nutrients in solution, from the surrounding water. Most cells survive and operate best in a medium whose salinity is 0.16 molar (about 1% by weight in water). Many kinds of cells can still survive the salinity of sea water (0.6 molar), although above 0.8 molar, the membrane that holds the cell contents becomes permeable and disintegrates. The internal osmotic pressure of nonhalophilic bacteria, such as *E. coli*, is in the range of $10^5$ Pa (Henis, 1987). When the solute concentration in the solution surrounding a cell is higher than the concentration in the cell, cellular fluids pass out through the cell membrane to the surrounding solution. Growth of the cell can then be inhibited as the plasma membrane pulls away from the cell wall, causing damage that, if imposed suddenly, is irreversible. The solutes come from the continuous weathering of rocks, flowing into the waterways. In the water system, the weathered compounds dissociate into their respective cation and anion which behave as two quite independent and separate entities. For example, with sodium chloride (NaCl), $Na^+$, like many other cations, has a relatively short residence time in the ocean and is removed by biochemical and chemical processes as well as hydrothermal chemical reactions within the seabed, and deposited as sediments, clays, dolomite and limestone. However, the Cl-anions are quite biochemically inert, creating an ion imbalance that translates into salinity.

Some microorganisms are able to tolerate high salt concentrations. Thus, according to their response to salts, microorganisms are classified as nonhalophiles, moderate halophiles or extreme halophiles. Most represented in extremely saline environments are bacteria such as *Halobacterium* spp. (a gram negative), *Halococcus* spp. (a gram positive), and a few cyanobacteria. *Halobacterium* spp. and *Halococcus* spp. require an absolute concentration for salt (NaCl) in excess of 2 M to sustain their growth. Most of the hypersaline tolerant cyanobacterial isolates are unicellular and can be grouped within the genera *Agmenellum quadriplicatum* (*Synechococcus*) or *Synechocystis*, or filamentous isolates of *Oscillatoria limnetica* and *Phormidium* spp. (Reed, 1986).

More common are facultative halophiles which do not require high salt concentrations but are able to grow at concentrations as high as 2% salt. To be able to withstand these conditions, halophiles adapt at least one of three strategies:

(1) Insulating the interior of the cell from the external saline environment, thereby precluding (or limiting) the entry of the salt. The insulation is in the form of a cell wall or plasma membrane with a lower permeability to toxic $Na^+$ than to $K^+$. The cell walls are composed of glucosaminuronic acid-containing polysaccharides instead of peptidoglycan (Reistad, 1975).

(2) Protecting the intracellular function and metabolic activity from the toxic effects of $Na^+$ since most halotorelant organisms have generally high (i.e., x2 or more) salt levels than their marine or freshwater counterparts. Protection can be in the form of actively excluding $Na^+$ by substituting this ion with other ions such as $K^+$ or by using organic molecules to generate an intracellular osmotic potential (Boyley and Morton, 1978).

(3) Modifying the cellular metabolism to function optimally under high salt environ-

ments. Relevant modifications may specifically include the presence of enzymes that optimally operate at different salt concentrations, or in the presence of substantial amounts of acid proteins, e.g., aspartic acid and glutamic acid in the cell envelope (Kushner, 1978).

Regarding the third strategy, some bacteria actively produce neutral solutes called sulfur and nitrogen betaines which are nontoxic to the cell and substitute for the salts (Imhoff and Rodriguez-Valera, 1984). Betaines are electrically neutral salts which carry a positive charge associated with the sulfate or nitrogen and a negative charge associated with the propionic acid ion on the same molecule. Other organic solutes in salt-stressed microbes include polyols, sucrose, glycerol, proline trehalose, sorbitol, mannitol, cyclohexanetetrol, and galactosylglycerol (Reed, 1986). Polyols are encountered mostly among Archea as this domain does not produce significant amounts of betaines. It is worthy to note that halophiles cannot compete effectively with mainstroam organisms under nonsaline conditions.

Often associated with salinity is alkalinity, a condition that can occur in both aquatic and terrestrial environments as a natural process or due to human activities. Naturally occurring alkaline environments may arise by biological activity such as sulfate reduction, ammonification and oxygenic photosynthesis. In marine environments, alkalinity is characterized by the presence of large amounts of sodium carbonate or its complexes, formed by evaporative concentration. Such conditions are prevalent in most lakes in the East African Rift Valley, the Caspian Sea region (Russia), Wadi Natrum (Egypt), and in other areas in Asia and the Americas (Grant and Tindall, 1986). Where man-made, salinity can occur in terrestrial systems if salty water is used for irrigation, following repeated cycles of evaporation.

Similar to salinity, microorganisms inhabiting alkaline environments are classified as either alkaliphiles or alkali-tolerant. Alkaliphiles have an obligate requirement for alkaline conditions (i.e., pH > 8.0) to grow whereas alkali-tolerant organisms are able to grow at alkaline pH but grow optimally at lower pH. The diversity of microbial species decreases with increasing alkalinity (Vareschi, 1982), and these environments suffer a scarcity of $Mg^{2+}$ and $Ca^{2+}$ due to the presence of high carbonates that tend to precipitate these alkali earth metals as insoluble carbonates. Alkaliphilic organisms tend to require high levels of $Na^+$ and actually use this cation as a symporter in place of the scarce protons, a strategy that enables them to establish a normal proton gradient (i.e., an internal alkaline) at alkaline pH. The internal alkaline pH of a number of alkaliphiles does not exceed pH 9.0-9.5. If the external pH happens to exceed 9.5, the pH gradient is reversed, becoming internal acidic (Guffanti et al., 1978).

### 7.4.3 Moisture content (Water availability)

Moisture content in a sense refers to the availability of water. Most microorganisms are at least 80% water on a weight basis. In the environment, microbes experience periods of water stress and abundance, sometimes over very short cycles. Availability of water for microbial uptake can be indirectly influenced in different environments by the surrounding solid surfaces such as soil and sediments (due to matric forces) and substances such as dissolved salts which can absorb water (due to osmotic forces). Two soils may have the same water content but differ in water availability due to adsorption and solution phenomena. The sum of matric and osmotic forces reflects what is termed the water potential ($\psi$), expressed in negative bars (- bars). One negative bar is equivalent to -100 kPa or 0.1 MPa. Since matric potential results from the adsorption of water molecules to surfaces at interfaces in solid substrates whereas osmotic potential occurs in solution due to interaction between solute and water molecules, water availability requires consideration of the amount of energy which must be expended by an organism to obtain it, referred to as water potential.

The water potential of pure water is zero bars and all other solutions have negative potentials with respect to pure water. Microorganisms require approximately 2 joules (0.5 calories) to assimilate water that is held in soil at -1 bar (Skujinš, 1984). Water availability may also be expressed as water activity ($a_w$) or relative humidity (RH). Water activity is the ratio of the vapor pressure of water interfaced with the solid (p) to the vapor pressure of the saturated air at the same temperature as the solid air ($p_0$). The activity of pure distilled water is 1. Water potential, RH, and $a_w$ are related to each other by the equations:

$$RH = a_w/100 \qquad ...(i)$$

$$\psi = 1000RT/W_A)*\ln(a_w) \qquad ...(ii)$$

where R is the universal gas constant (8.2 MPa cm$^2$/mole K), T the absolute temperature, and $W_A$ the molecular weight of water.

Microorganisms are more sensitive to moisture stress associated with adsorption or surface tension (i.e., matric water stress) than to osmotic water stress (Fig. 7.6). This is possibly due to the likely restriction in the availability of nutrients to the microorganisms from the solid matrix (e.g., soil) that is under moisture stress. Under low osmotic potential, the growth of microorganisms is restricted because the microbial cells will be lacking a positive turgor pressure. For a cell to be turgid, it must possess an internal water potential that is lower than that of its environment or else it will collapse. Moisture stress also increases resistance to diffusion which in turn reduces the flow of substrates to the microbial cell surface (Stark and Firestone, 1995). However, microorganisms greatly differ in their ability to withstand water stress. Based on their ability to tolerate moisture stress, microorganisms are classified as hydrophobic (> −1.5 MPa), xerophilic (−8 to −1.5 MPa), and hyperxerophilic (< −8 MPa) (Table 7.3). Fungi and actinomycetes generally toler-

ate lower water potentials than do prokaryotes. The higher survival of actinomycetes is attributable to the formation of spores as vegetative actinomycetes hyphae are also considerably easily damaged by water stress (Williams et al., 1972). Furthermore, their filamentous nature may allow them to exploit microsites that are isolated by discontinuous or thin water films (Stark and Firestone, 1995). The densities to which both actinomycetes and fungi exist in moisture-stressed environments are lower compared to their densities in terrestrial environments not subject to moisture stress. However, the incidence of many fungal diseases noticeably increases with increasing matric water potential, i.e., greater water stress (Griffin, 1969), possibly due to the reduced resistance of their respective hosts to the disease.

Water activity in soils ranges between 0.9 and 1 (i.e., −15 to −0.003 MPa). The diversity of microbial species, their movement as well as their survival in general are affected by water activity (Zhou et al., 2002). Most microorganisms require a water activity of at least 0.96 for metabolism but some actinomycetes and halotorent prokaryotes can conduct metabolism at activities of 0.75 or below. For most organisms below −82 MPa, the DNA is destroyed, a situation similar to what happens under high pressure conditions (see Section 7.4.2). If osmotically stressed, the cell (or organism) can incorporate some of the solutes and achieve a low internal water potential. In some cases, the low internal water potential is achieved by the synthesis of low molecular weight compounds, e.g., glycerol, betaines, etc. On the other hand, a cell subjected to matric stress in an environment with solute concentration does not have the luxury of assimilating osmotically active substances. Rather, it increases its internal solute molecules, thus counteracting the matric stress by either synthesizing solutes *de novo* or degrading some intracellular macromolecules such as proteins and nucleic acids. The latter option, unless carefully regulated, would be somewhat self-

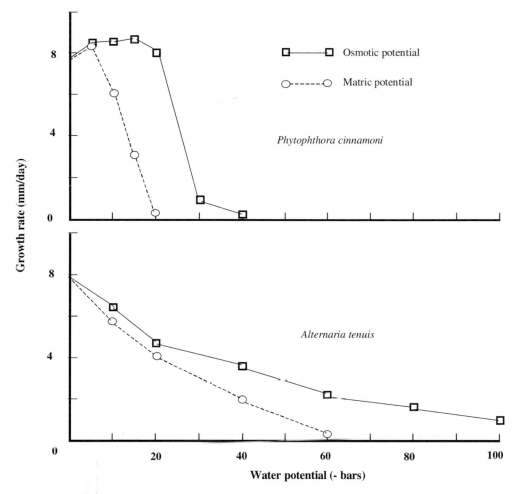

**Fig. 7.6** Relationship between growth rate and water potential for both *Phytopthora cinnamoni* and *Alternaria termis* in a silt loam soil. The relationship was also similar in a sandy clay loam and a sandy soil (Adebayo and Harris, 1971 with permission from the authors).

destructive and may also explain why matric potential is more damaging than osmotic stress.

Microorganisms in desert conditions undergo extreme desiccation due to a persistent lack of rainfall (Skujinš, 1984). The drying in itself may not be lethal but rather the rewetting process. However, even in a dry environment, a layer of water must exist around microorganisms if they are to remain viable. Most notable of desert microbial adaptations and in areas that have endemic moisture stress are lichens

(see Chapter 4). Also prevalent in moisture-stressed environments such as deserts are spore-forming microorganisms. However, nonspore-forming prokaryotes may, under moisture stress, modify their morphology to avoid desiccation. For example, *Arthrobacter* sp. converts its morphology from a rod to a coccus to combat desiccation (Smith, 1978). In the atmosphere the water content is usually quite low, limiting the growth of microorganisms (Gregory, 1973). If high humidity persists, microorganisms may grow on permissible sur-

**Table 7.3** Approximate limiting water potential for microbial growth at 25°C

| Category | Water potential (MPa) | Water activity ($a_w$) | Reference points | Bacteria | Yeasts | Fungi |
|---|---|---|---|---|---|---|
| Hydrophilic | −0.003 | 1.00 | Blood | *Caulobacter* | — | — |
| Xerophilic | −1.5 | 0.989 | Wilting point | *Spirillum* | — | *Sclerotium rolfsii* |
| | −3.0 | 0.978 | Sea water | Chemoautotroph | — | — |
| | −7.1 | 0.95 | — | Most gram-negative rods | Basidiomycetous yeasts | *Rhizopus* sp. |
| Hyperxerophilic | −15 | 0.90 | — | Most cocci; *Bacillus*; *Lactobacillus* | Ascomycetous yeasts | *Fusarium* sp.; *Aspergillus flavus*; Mucorales |
| | −22 | 0.85 | — | *Staphylococcus* | *Saccharomyces rouxii*; *Debaryomyces* | *Aspergillus niger*; *Paecilomyces varioti* |
| | −31 | 0.80 | — | — | — | *Penicillium* sp. |
| | −39 | 0.75 | Salt lake | Halobacteria; Halococci | — | *Sporendonema sebi*; *Wallemia* sp.; *Aspergillus* spp. |
| | −49 | 0.70 | — | — | — | *Aspergillus ruber*; *Chrysosporium* sp. |
| | −59 | 0.65 | — | — | — | *Erotium* sp. |
| | −70 | 0.60 | — | — | — | *Xeromyces bisporus* |
| | −82 | 0.55 | DNA disordered | — | — | — |

*Source:* Skujinš (1984 with permission from Kluwer Academic Publishers, New York, NY).

faces such as rocks, brick/concrete walls and foliage, causing biodeterioration (see Chapter 12). It is important to note that moisture availability also, directly or otherwise, affects or is affected by other parameters such as aeration of the microenvironment and temperature.

### 7.4.4 Nutrient availability

Nutrients are needed by microorganisms for three main purposes: (i) cell structure, (ii) as a source of energy, and (iii) as an electron acceptor. Various nutrients can be measured in soil, water and plant materials. Macronutrients, which include C, O, H, N, P, S, K, Ca, and Mg, make up the biomass whereas micronutrients, for example Fe, Na, Mn, Co, Cu, Mo, Zn, V, Ni, B, Se, Si, Cl and W, as well as substances such as vitamins, are required in minute quantities. Low nutrient environments are prevalent in soil, sea water, and fresh water. In nature, the microbes experience episodes of semi- or total famine, occasionally relieved by brief periods of abundance. Under starvation, bacteria exhibit some morphological and physiological changes which in most cases include an increased scavenging of missing nutrients attained through the reduction of high affinity uptake as well as acquiring the capacity to utilize diverse sources of the nutrient in question. In sea water, organic C is exceedingly low, with the dissolved organic C ranging between 0.35 and 70 $ngL^{-1}$ and the POC from 3 to 10 $ugL^{-1}$ (Morgan and Dow, 1986). Microbial flora in fresh water and its naturally low nutrient status has received less attention except in cases where such fresh water is polluted. Microorganisms mineralize dissolved organic carbon (DOC) and particulate organic carbon (POC), a trait that contributes to the cycling of nutrients. In soil, the organic matter content typically varies from 0.8-2%, of which 50-80% is humus, a recalcitrant molecule with a half-life of 250-5,000 years (Matin et al., 1999). Terrestrial environments are more heterogeneous than aquatic systems. Thus, nutrients in terrestrial environments are obtained by microbial cells through various mechanisms which include diffusion, active transport, and mass flow. In most cases, the availability of nutrients in terrestrial microcosms is also limited.

In soil, the major carbon sources for microorganisms, e.g., cellulose and hemicellulose, are sorbed and not in solution. To take up the nutrients contained therein, the microbes have to normally expend some energy by diffusion (passive or facilitated) or through active transportation. Low nutrient status usually involves a multiple limitation of C, N, P, S and/or trace elements. Under marine environments, stressful levels of 1-15 mg carbon $L^{-1}$ are widespread (Morgan and Dow, 1986). Bacteria that grow at low nutrient concentrations are referred to as oligotrophs. Various bacteria are facultative oligotrophs as they are able to grow at both low and high nutrient levels. Carbon fluxes *in situ* are commonly less than 0.1 mg $L^{-1}$, which by the limits highlighted above is oligotrophic. Furthermore, the existence of truly obligate oligotrophs is questionable since such oligotrophic isolates rapidly "adapt" to high nutrient concentrations on subculturing.

In general, the survival of microorganisms under low nutrients requires the cell to be able to maintain a functional gradient across the membrane and to prevent efflux of the accumulated nutrients. To be able to grow and survive, organisms have specific mechanisms that adapt them for existence under a low nutrient environment. Under nutrient deprivation, three starvation-survival patterns are possible. The viable cell populations may:

- increase initially until a constant number is attained,
- increase initially and then decrease until a constant number, or
- decrease until a constant number.

All three possibilities have been demonstrated in marine organisms (Amy and Morita, 1983). The ability of microorganisms to survive in low nutrient habitats is characterized by small cell size. Miniaturization increases the surface-to-volume ratio, which in turn enables the cells to obtain substrates from a nutrient-poor environment (Morita, 1982). Starving cells also tend to develop the ability to catabolize more than one carbon source simultaneously, and develop low anabolic (metabolic) rates. Through these adaptations, more nutrients are taken up by the cells, enabling them to utilize the nutrients more efficiently. Microbes in low nutrient environments also accumulate excess nutrients under storage as polymers. Such polymers are in the form of polyalkanoates, phosphates (i.e., polyphosphate) and carbon (polyhydroxybutyrate, lipids and carbohydrates) (Weiner, 1999). Furthermore, through such mechanisms, microorganisms are able to channel the nutrients, utilizing them for cellular maintenance, and/or providing for more flexibility as to which nutrients can be used. Other strategies against low nutrients include chemotaxis, phototaxis and aerotaxis from a point/microcosm of low to one of high nutrient content. However, knowledge about these movements and responses in relation to nutrients particularly under soil and sediments is currently limited. The type of survival strategy adapted seems to relate to the cell wall composition (i.e., gram positive or negative). Most notable among the gram positives is the formation of spores. The spore-forming and nonforming phenotypes are discussed separately.

### 7.4.4.1 Formation of nutrient stress-resistance structures

Several microbes have the capacity to respond to stresses, e.g. nutrient limitations by forming a resting stage, which has a greater resistance to environmental stress than the corresponding vegetative cells. Such stress-resistant stages include spores (in bacteria and fungi), cysts (in protozoa and in some bacterial species such as *Azotobacter*), fruiting bodies (in fungi and myxobacteria), oocyts (in protozoa), and alkinates (in some cyanobacterial species, e.g., *Anabaena cylindrica*). Such resting stages tend to be strikingly resistant to other adverse environmental conditions, such as heat, radiation, and desiccation. Most widely studied are endospores of *Bacillus* sp. and other genera of gram-positive bacteria. These studies have shown sporulation to be a very complex process, involving more than 80 genes (Losick et al., 1989). It occurs in several stages (Table 7.4) which, under laboratory conditions at 37ºC, lasts 6-8 hours. It is not certain how long the sporulation process takes in the natural environment but probably varies with prevailing environmental stresses. Growth of the spore back into a vegetative cell can occur readily and possibly involves just two stages, that is, germination and outgrowth once conditions become ideal. Efficient germination of spores almost strictly requires heat-shock and absorption of moisture.

### 7.4.4.2 Nonspore formers

Generally, gram-negative prokaryotes do not form spores. However, some of them display some level of survival even under nutrient stress. Most widely studied is *Vibrio* because of its clinical implications, forming viable but nonculturable pathogens (Colwell and Huq, 1994). *Vibrio* spp. are reported to withstand starvation by fragmenting to produce large numbers of ultramicrobacteria (ultracels) that show a markedly reduced endogenous respiration (Novitsky and Morita, 1976, 1978; Egli, 1995). Work by Amy et al. (1983) showed that the ultracels also have a steady decrease in cellular DNA, RNA, and proteins at the onset of starvation. The ATP also rapidly declines, indicating that the starved cells are still physiologically capable of metabolizing substrates. Similar behavior/strategies have been observed in various other marine isolates under starvation (Amy and Morita, 1983; Weiner, 1999). Ultrabacteria are common in the natural environment in soil and sea water (Morita, 1982) and can increase in size when fed properly (Egli, 1995). Other morphological changes by

**Table 7.4**   Sequence of events in the sporulation of *Bacillus subtilis*

| Stage | Nature of event |
| --- | --- |
| 0 | Vegetative cell |
| I | Chromosome condenses into an axial filament |
| II | An asymmetrically positioned septum is formed which partitions the cell into unequal components. Both components presumably contain identical chromosomes but are divergent in developmental fates. The smaller compartment will result in an endospore whereas the larger component cuts as a mother cell, lysing toward the end of the stage, to liberate the mature endospore. |
| III | The membrane of the mother cell engulfs the smaller compartment (forespore), providing a second membrane layer. The forespore is pinched off the wall and becomes fully internalized in the mother cell. |
| IV | A cell wall-like material (cortex) is deposited between the inner and outer membranes surrounding the internalized forespore. |
| V | A tough protein sheath (coat) encases the outer membrane of the internalized forespore. |
| VI | The encased forespore undergoes maturation during which the characteristic properties of the spore develop. |
| VII | The mother cell lyses, releasing the mature endospore. |

nonspore-forming prokaryotes include shrinkage of the protoplasts and enlargement of the periplasmic space.

## 7.4.5   Radiation

Radiation has various effects on cells, depending on its intensity, wavelength, and duration. Radiation in the environment is classified as either ionizing or nonionizing. Both ionizing and nonionizing rays are hazardous to microorganisms, the latter being more harmful than the former (Tong and Lighthart, 1998).

### 7.4.5.1   Nonionizing radiation

Nonionizing radiation has longer wavelengths that are greater than 1nm. They basically impose their radiation effect by inducing formation of singlet oxygen in the cytoplasm. Damage to microorganisms by nonionizing radiation such as the ultraviolet (UV), visible, and infrared wavelengths, has been documented for many years. In single-celled organisms, the nonionizing radiations give rise to photochemical reactions producing pyrimidine dimers in DNA. The intensity of solar radiation varies with solar angle, altitude, and other environmental factors such as cloud cover, air pollution, atmospheric turbidity, and the ozone concentra-

tion. For a wavelength of 307.5 nm, the annual average radiation at the equator is approximately 56.4 Watt $s^{-1}$ $cm^{-2}$ (Schulze and Gräfe, 1969). Under natural conditions, nonionizing solar radiation has a greater potential to destroy biological systems. For example, solar radiation can potentially destroy *E. coli* and yeast that are deficient in DNA-repair mechanisms (Resnick, 1970). More than 99.9% of the DNA-repair mechanism-deficient *E. coli* strain are killed in 3 minutes on exposure to sunlight. Most harmful wavelengths are below 400 nm (i.e., far and near UV range) but the lethal effect extends up to 700 nm. Most susceptible to damage from radiation are the thymine units of the DNA (see Section 7.4.5.2 below).

The harmful effects of these wavelengths are of ecological significance particularly in aquatic and terrestrial environments. Visible light occupies 400-750 nm (0.39-1 μm) while near UV ranges from 290-400 nm (1-100 μm). All these regions of the visible spectrum initiate light-induced damage to microorganisms. Most of the wavelength that is less than 290 nm (i.e., less than 0.39 μm) does not reach the surface of the earth as the ozone layer effectively absorbs light from this region. However, damage by short wavelength to cells has been demonstrated under laboratory conditions. Thus, under natural conditions, microorganisms are not subjected to these far UV wavelengths.

Damage is largely owing to an inhibition of cell replication due to affected DNA. The lethal effect to near UV seems to be greater in the presence of oxygen than under anoxic conditions, the damage being mediated by a photosensitizer whose reaction involves molecular oxygen. However, inactivation of microorganisms by visible light appears to be largely independent of molecular oxygen except for a few bacterial species, notably *Bacteroides fragilis* (Jones et al., 1980). In freshwater and marine environments, UV has been implicated in the inhibition of photosynthesis and respiration in fresh and marine waters (Abeliovich et al., 1974). More information is available about the damaging effects of UV light on fresh and marine organisms than damage to organisms inhabiting terrestrial environments.

DNA is the critical target to radiation. UV damages DNA of exposed cells by causing bonds to form between adjacent thymines in DNA chains. The dimers inhibit correct replication of the DNA which, after irradiation, forms dimers that cannot form H-bonds with complimentary bases. As a result, abnormal replication of the DNA results in mutations or even death. UV radiation is most lethal at 250-260 nm because at this wavelength, the nucleic acids absorb UV most strongly, forming dimers between adjacent pyrimidines (Häder, 1997; Henis, 1987). Resistance to UV radiation may, in some microorganisms, involve a form of protection to the organism against the induction of damage. Protection can be offered by carotenoids, phycocyanin, or other compounds. Ironically, mutation which can also be caused by radiation, offers a selective advantage. Thus, from an evolutionary perspective, a balance between sensitivity and resistance to radiation is advantageous. Microorganisms are more susceptible to UV damage in the frozen state because DNA damage is less reparable

by the enzymatic system. However, radiation-resistant microorganisms such as *Micrococcus radiodurans* and spores of *Aspergillus nidulans* seem to be capable of this repair even in a frozen state (Ashwood-Smith and Bridges, 1967).

### 7.4.5.2 Ionizing radiation damage and repair of DNA

Ionizing radiations are attributable to long wavelengths from either terrestrial or extraterrestrial sources. Sources of ionizing radiation include gamma ($\gamma$) rays, X-rays, protons, neutrons, fast electrons, $\alpha$-particles, and high-energy electron beams. The extraterrestrial wavelengths primarily originate from cosmic rays in outer space which give rise to secondary cosmic rays to which organisms may be subjected. At sea level, ionizing radiation impacts a dose of approximately 30 mrad year$^{-1}$, doubling every 1,500 m at sea level within the troposphere (Nasim and James, 1978). Terrestrial ionizing radiations are emitted by radioactive nuclides in soil, rocks, and the atmosphere. Some of the ionizing radiation is man-made during nuclear weapon tests, power production from nuclear fission, medical therapy practices, and other sources. Some of these practices, periodically or chronically, expose microorganisms to high levels of ionizing radiations. The long wavelengths basically affect the ionization potential in living matter by randomly ionizing water ($H_2O$), which forms highly reactive free hydroxyl radicals (OH·) and ionized molecules. The generated hydroxyl radicals then react with cellular organic components such as DNA, killing the cell. As in the case of nonionizing (UV) radiation, ionizing radiation also results in damage to the DNA. However, with ionizing radiation, damage to the DNA is not by pyrimidine dimer formation; rather, damage occurs due to breakages in the DNA strands (Henis, 1987).

The sensitivity to ionizing and non-ionizing radiation greatly varies between species (Table 7.5). *Micrococcus radiodurans* can withstand doses as high as 500 krad without considerable inactivation and has been found living in water

**Table 7.5**  Radiation sensitivity of different microorganisms

| Microorganism | Dosage giving approximately 37% survival, $LD_{37}$ | |
| --- | --- | --- |
| | UV light (ergs mm$^{-2}$) | Ionizing radiation (krad) |
| *E. coli* K-12 | 500 | 2 |
| *Micrococcus radiodurans* | 6,000 | 150 |
| *Amoeba* | — | 120 |
| *Paramecium* | — | 350 |
| *Bodo marina* | 50,000 | — |
| *Saccharomyces cerevisiae* | 800 | 3 |
| *Schizosaccharomyces pombe* | 1,350 | 80 |
| *Ustillago maydis* | 1,300 | — |

*Source*: Nasim and James (1978) with permission from Elsevier.

used to cool nuclear reactors. *Deinococcus radiodurans,* on the other hand, can grow in the presence of 6,000 rads h$^{-1}$ under nutrient-rich conditions and its rate of growth, viability, and ability to express cloned genes are unaffected by such high radiation doses (Lange et al., 1998). These radiation-resistant microorganisms can conserve energy to support growth from reduction of uranium and other radioactive metals. In the process, the microorganisms immobilize the radioactive metals, a trait that can be tapped to remediate radioactive materials from the environment (Lloyd and Macaskie, 2000). *Pseudomonas aeruginosa* takes up substantial quantities of uranium which accumulate as intracellular deposits, eventually leading to death of the organism. This bioaccumulation is of biotechnological interest as a metal removal process. As a matter of fact, microbiological treatment has been proposed as one of the viable means of treating some of this waste (Lloyd and Lovely, 2001; Lloyd et al., 1997).

*M. radiodurans* is able to repair breaks in double-strand DNA (Dean et al., 1966). The mechanisms of resistance to radiation have not been fully understood but in general, radiation-resistant organisms often tend to be highly pigmented. Pigmentation acts as an energy sink, preventing the effects of visible light from damaging the DNA. An excessive irradiation of 253.7 nm wavelength over a one-hour duration had no visible negative effect on the dark pigmented desert fungus *Stemphylium ilius*, as opposed to the negative effect incurred by *Aspergillus niger*, damaged within just 2 min exposure (Durrell and Shields, 1960). Other traits that affect cell resistance to radiation include higher levels of catalase, ability of the cells to elongate, and a cell wall structure less permeable to the different radiation wavelengths (Nasim and James, 1978). In most cases, more than one mechanism is involved in imparting resistance to radiation. Under aquatic environments, organisms may adapt circadian rhythms, avoiding the bulk of radiation. Many phytoplanktons move toward the surface early in daylight (at sunrise) and deeper down the water column at times of excessive irradiation (around noon), only to come back toward the surface in the afternoon (Häder, 1997). Their buoyancy is facilitated by gas vacuoles, oil droplets and phototaxis.

### 7.4.6  Temperature

The microbial world displays the greatest ability to exist in a wide range of temperatures compared to other living organisms. Microbial life and growth occur at temperatures below freezing to over 250°C. The upper temperature limit for eukaryotes is approximately 68°C at which only a few species of fungi, algae, and protozoa can grow (Table 7.6). Most peculiar among algae that exist under elevated temperatures is *Cyanidium culdarium* as it can also grow under

**Table 7.6** Uppermost temperature at which growth of various microbial groups has been reported

| Group | Examples | Temperature (°C) | Reference |
|---|---|---|---|
| Viruses | *Sulfolobus shibatae* virus, *S. islandicus* rod virus | 70-92 | Rice et al., 2001 |
| Prokaryotic microorganisms | | | |
|   Cyanobacteria | *Synechococcus* spp. | 70-73 | Brock, 1967 |
|   Archaea | *Thermococcus litoralis* | >250 | Baross and Deming, 1983 |
| | *Thermus aquaticus*, *Pyrococcus* spp. | >90 | Brock, 1967 |
|   Heterotrophic bacteria | *Bacillus* spp. | 40-70 | Brook, 1967 |
| Eukaryotic microorganisms | | | |
|   Protozoa | *Cothuria* spp. | 63 | Fenchel, 1987 |
| | *Chilodon* spp. | 68 | Fenchel, 1987 |
| | *Oxytricha fallax* | 56 | Fenchel, 1987 |
| | *Hyalodiscus* spp. | 50-52 | Fenchel, 1987 |
| | *Cercosulcifer hamathensis* | 56-59 | Fenchel, 1987 |
| | *Vahlkampfie reichi* | 60 | Fenchel, 1987 |
| | *Cyclidium citrullus* | 50-58 | Fenchel, 1987 |
| | *Tetrahymena pyriformis* | 41.2 | Fenchel, 1987 |
| | *Naegleria fowleri* | 45-46 | Fenchel, 1987 |
|   Algae | *Cyanidium culdarium* | 55-60 | Tansey and Brock, 1978 |
|   Fungi | *Dactylaria gullopavo* | 60-62 | Brock, 1967 |
| | *Aspergillus fumigatus*, *Humicola* spp., *Absidia* spp., *Mucor* spp., *Chaetomium thermophile*, *Sporotrichum thermophile*, *Talaromyces* spp., *Thermoascus aurantiacus* | >50 | Jennings and Lysek, 1999 |

extremely acidic conditions. It exists in both aquatic (hot springs) and terrestrial habitats, mostly in association with *Bacillus coagulans* and the fungus *Dactylaria gullopavo*. The bacteria and fungi obtain nutrients from the algae (Belly et al., 1973). Chemolithotrophic and heterotrophic prokaryotes can withstand temperature higher than 90°C, and the temperature limit to prokaryotic growth is still undetermined with certainty. At temperatures as high as 250°C, growth and existence occur under pressure, particularly in deep seas around hydrothermal vents. Viruses have also been reported at temperatures as high as 70-92°C (Rice et al., 2001).

For a particular microorganism, growth increases with temperature until an optimal temperature, above which growth declines. Each species has a minimum, optimum, and maximum temperature that affects growth. The maximum optimal temperature is determined by the sensitivity of the most fastidious enzyme essential for the cells to replicate. Above the optimum temperature, chemical reactions, and growth for that matter, rapidly decline. Based on the temperature range which they can tolerate, microorganisms are classified as psychrophiles (cold-loving), psychrotrophs, mesophiles (moderate-temperature loving), thermophiles (heat-loving), or extreme thermophiles (Fig. 7.7). Changes in temperature cause physiological changes in biological membranes. Functioning membranes tend to have a fluidy liquid-crystalline interior. Chilling

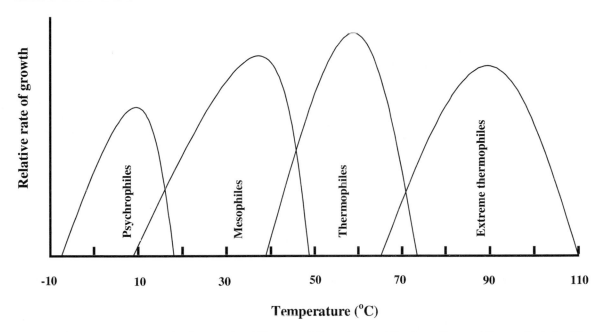

**Fig. 7.7** Organisms are categorized as psychrophilic, mesophilic, or thermophilic. Some thermophiles, particularly of the Archaea domain, grow optimally at extremely high temperatures (i.e., extreme thermophiles).

temperatures change these membranes to a gel phase which impairs function whereas high temperatures tend to disrupt the integrity of the membranes. The temperature at which the membrane gel phase occurs is related to the nature of fatty acids that compose the membrane. Membranes rich in unsaturated or branched-chain fatty acids have a lower transition temperature compared to membranes rich in saturated straight-chain fatty acids. For example, various cold-tolerant microorganisms were observed to possess membranes rich in lipids and polyols (Wynn-Williams, 1990). Thus, microorganisms that tolerate temperature extremes have modified membranes.

Psychrophiles grow best at around 10°C. Mesophiles grow best at about 35°C but can also grow between 15-45°C (Fig. 7.7). Thermophiles thrive best at temperatures greater than 45°C whereas extreme thermophiles typically belonging to the Archaea domain grow optimally at 80-95°C. In each case, the optimum temperature, represented by the peak, is the temperature at which the fastest reproduction occurs, assuming other conditions are suitable. Below and above the optimum temperature range, the reproductive rate is lower. Between the temperature range, the growth rate increases up to a maximum and then drops off rapidly. The increase in rate of growth is due to an increase in rates of reactions. This brings up what is called Shelford's law of tolerance, whereby a minimum and maximum temperature set the "tolerance range" for each organism. Plants and animals have a fairly small temperature tolerance range compared to microorganisms.

Psychrophiles occur mostly in polar regions and in deep ocean waters. Of environmental concern also are psychrotrophs which are mesophiles and grow optimally at 20-25°C but can also grow, albeit slowly, at refrigerator temperatures. They greatly contribute to the spoilage of foodstuffs under refrigeration. Most of the globe has temperatures in the mesophilic range and so it is by no surprise that the bulk of the microbial population belongs to this category. This also happens to be the temperature range favorable to humans, a fact that emphasizes our continued interaction with the majority of microorganisms and its implications to environmental health. Mesophiles have been well studied compared to the other two categories. Thermophiles are mostly important in organic compost piles in which these optimum temperatures are very typical. Extreme thermophiles mostly live near hydrothermal vents and in hot springs. For the former, the high pressure in deep oceans prevents water from boiling at 100°C, enabling temperatures of over 200°C to be attained. The maximum and minimum growth temperatures appear to be limited by liquid water (Brock, 1967). If conditions allow the existence of free water, either through reduction in the freezing point due to salt concentration, or through an elevation of the boiling temperature due to elevated pressure, microorganisms are able to grow. However, despite these modifications, wherein growth can occur at low temperatures, it tends to do so at a comparatively low rate.

Getting biological systems to operate at high temperatures is tantamount to putting into the system more energy, a precedent that tends to increase the rate of reactions (i.e., metabolic activity). Such an increase is also, under aerobic conditions, synonymous to an increase in rate of oxygen consumption. An activity versus temperature relationship ($Q_{10}$ values) has been developed based on the fact that, within the tolerance range, enzymatic activity increases with increase in temperature. The $Q_{10}$ is derived from the formula

$$Q_{10} = \frac{\text{Specific activity at temperature (T°)} + 10°C}{\text{Specific activity at temperature (T°)}}$$

This relationship reflects the changes in enzymatic activity caused by a 10°C increment. $Q_{10}$ values for biological processes range from 1.5 to 3. Generally, most enzymes have a $Q_{10}$ of 2, which implies a doubling in enzymatic activity with a 10°C temperature increment. Above the optimum growth temperature, the necessary enzymes in the cell are inactivated. If $Q_{10}$ is much greater than 3, another environmental parameter besides temperature is controlling the rate of reaction. For example, $Q_{10}$ values of 5 or higher have been reported during the decomposition of petroleum compounds at low temperatures (10 - 20°C) due to the low viscosity and low solubility of petroleum in water (Tate, 2000).

In temperate regions and deserts, temperatures can oscillate over a short period of time (e.g. overnight) between extremes of freezing and thawing. Biederbeck and Campbell (1971) found that one freeze-thaw cycle can result in the death of 92% of a bacterial population, 55% of fungi, and 33% of actinomycetes. The death of the cells is not strictly due to freezing only, but also the thawing of environmental matrices such as water (ice), soil, and sediments. As the matrices thaw, ice crystals form in the microbial cells, disrupting their integrity. In terrestrial environments, for reasons not quite certain, microorganisms residing in disrupted soil crumbs are more susceptible to freeze-thaw cycles compared to those experiencing the same cycles in undisturbed habitats (Cameron, 1971). This has implications in the survival of microorganisms under the till versus no-till farming systems and in ecologically disturbed versus pristine soils in general. On a positive note, the moribund cells resulting from death due to freeze-thaw become a source of nutrients for the surviving cells, which in turn can provide prolific new growth when permissive environmental conditions prevail.

Much of what we know about psychrophiles has been learned from microorganisms based in the polar regions. Needless to say, the diversity in these regions is far less than that in temperate or tropical regions. In general, microbial growth in the polar regions shows a ruderal pattern by growing opportunistically in existing traces of water (Pugh, 1980). Bacteria are rare compared to fungi and algae in polar regions, the latter existing as a mat cemented by organic material (Siegel et al., 1983). In frozen waters of these regions, the existing photosynthetic microbes have been reported to photosynthesize at light intensities as low as 0.5 E m$^{-2}$ s$^{-1}$ and at considerable depths (Vincent, 1981). In reality, tolerance of microorganisms to low temperatures is also governed by several ecological factors, such as high solar radiation, infrequent freeze-thaw cycles, absence of evaporative winds, slow drainage, high humidity, low salinity of the soil and surrounding water, as well as a balanced ionic composition (Wynn-Williams, 1990).

At the other extreme are microorganisms that can withstand high temperatures (i.e., thermophiles). Most thermophiles fall in the Archaea domain. Generally, Archaea have a high guanine plus cytosine (G + C) ratio which provides three H-bonds, as opposed to two H-bonds between adenine plus thymine (A + T) ratio. It takes more energy to break the three versus two bonds. From a biotechnology perspective, considerable interest has arisen in investigating Archaea due to the potential of its enzymatic ability to withstand high temperatures. At an industrial level, higher temperatures translate into enhanced rates of reactions. Most known among these processes is the use of Tac-polymerase, from the methanogen *Thermus aquaticus* (whose enzymes operate at 70°C), in polymerase chain reactions (PCR). As extensively discussed in the last chapter, use of this enzyme opened up new opportunities in environmental microbiology and biotechnology in general. Of even greater potential is the possibility of using the DNA polymerase from *Thermococcus litoralis* and *Pyrococcus* sp., both of which are hydrothermal organisms and can withstand even higher temperatures. The microbiology of hot springs and hydrothermal vents is discussed later in this section.

It is desirable, for various reasons, to be able to control microbial growth and/or invasion. Temperature has long been used to achieve such control, in the form of autoclaving, microwaving, steaming, dry heating, and similar processes. Some pathogens can form spores and thus resist heat. From this perspective, the resistance of microorganisms to high temperatures can be viewed as undesirable. Most control methods utilize temperature as a key component, but better control of microbial invasions is achieved by adapting regimens that use more than one method. For example, in the food canning industry it is often necessary to use high temperature and a low pH or a high temperature combined with a high salt content.

A variety of natural and man-made habitats with high temperatures include hot springs (≥101°C depending on the altitude), volcanoes (1,000°C), compost piles, soil, rocks, domestic hot-water heaters, steam-building heating pipes, and industrial wastewater recharge to damping sites. A couple of these, i.e., composting and hydrothermal vents, have been singled out for further discussion because of their importance in environmental microbiology.

### 7.4.6.1 Composting

Composting has been practiced by mankind for many centuries, so much so that its importance can almost be taken for granted. However, the growing concern in controlling pollution of the environment has elevated interest in this ancient process, with a desire to make it scientific. This enables us not only to understand the process, but also to devise ways of improving it to suit specific situations which range from domestic (backyard composting) to fully fledged industrial settings. Obviously, the level of sophistication of the process varies but a common player in all composting processes is the succession of microorganisms. The process occurs in three stages which hinge on temperature changes, normally mesophilic (20-40°C), thermophilic (>40°C), and stabilization (Fig. 7.8). From the onset, the material to undergo composting is not sterile and contains natural

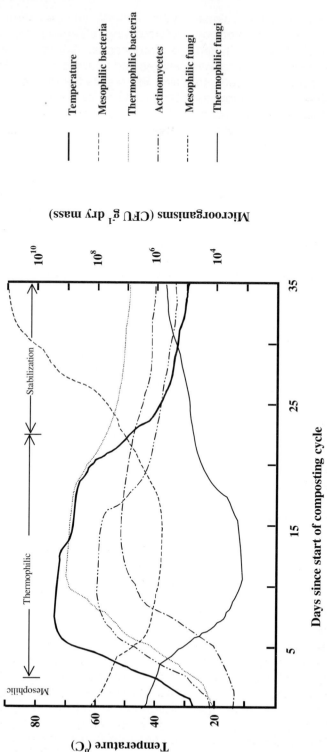

**Fig. 7.8** General sequence of events during the composting process. As the temperature increases, the density of mesophiles (bacteria and fungi) declines but rebound later during the stabilization phase.

populations of bacteria, actinomycetes, protozoa, fungi, and, in some cases, viruses as well as algae.

While the role of protozoa, viruses, and algae in the composting process, if any, has not been extensively studied, the successive role of the other microorganisms is quite well established. The initial decomposition is facilitated by mesophilic bacteria which increase in density as they rapidly break down and utilize the simple carbohydrates. The rate and duration of this initial decomposition depends on the nature of the material being decomposed. During this initial process, energy is dissipated as heat but because of the excellent insulation in the compost pile, most of it gets trapped within the pile, resulting in significant temperature increases. This scenario sets the stage for enhancing the activity of thermotolerant and thermophilic fungi (e.g. *Aspergillus* spp., *Mucor* spp., *Absidia* spp., *Chaetomium* spp., *Sporotrichum* spp., *Humicola* spp., *Talamyces* spp., and *Thermoascus* spp.) and actinomycetes (e.g.,*Streptomyces* spp., *Thermostreptomyces* spp.) which can utilize the more complex substances such as cellulose, hemicellulose, and lignin as substrates (Jennings and Lysek, 1999). During the thermophilic stage temperatures can reach 80°C as a direct result of microbial activity. A few bacterial species such as *Bacillus* can, through sporulation, survive these temperatures and would still be detectable in large numbers (Chang,1967; Chang and Hudson, 1967).

Other factors which influence succession include oxygen supply and pH changes. The increase in heat eventually becomes self-limiting, the temperature gradually stabilizing close to that of the surroundings. During the stabilization process the pile naturally becomes re-inoculated with microorganisms from the surrounding vegetation, burrowing vermin, dust, air, etc.

### 7.4.6.2  Hydrothermal vents

At the ocean floor where no light penetrates, photosynthesis is not possible. However, ocean beds are beaming with life as was discovered by Corliss et al. (1979). Corliss' group set out to validate the plate tectonics theory initially proposed by a German meteorologist Alfred Wagner at the beginning of the 20th century. The theory stipulates that the continents are not fixed in position, but constantly experience crustal movements. Thus, the solid plates of the Earth's lithosphere slide over the asthenosphere at rates of a few centimeters per year. This slight movement causes the flow of hot magma at the convergent plate boundaries on the ocean floor. Through these ridges and thermal springs on the ocean floor called hydrothermal vents, the oozing hot magma supplies new crustal material. These sites also harbor a whole range of peculiar animals, many of which are just being discovered.

The hot vents have flow rates of up to 2 ms$^{-1}$ and their plume rises several hundred meters above the sea floor. At these hydrothermal vents temperatures as high as 340°C have been reported. In the emitted vent water, microbial populations in the range of $10^5$ - $10^8$ bacterial cells ml$^{-1}$ have been recorded, which are about four times greater than in water further from the vents. Even higher cell densities occur in cooler, shallower vents where slow emissions of fluids at the rate of 1 - 2 cm s$^{-1}$ make slowly moving plumes. The ATP in warm vents is 2-3 times that in surface water, indicating that hydrothermal vents are a very productive area.

Around these vents, the trophic web develops in the total absence of sunlight. The chemosynthetic autotrophic microorganisms in suspension obtain ATP from inorganic oxidations and use it to fix $CO_2$, generating biomass. The zooplanktons feed on the microorganisms whereas the suspension feeding animals eat plankton. Thus, in this unique environment, it is the chemoautotrophic bacteria, and not the photosynthetic organisms, that constitute the first trophic level. The chemoautotrophs in this habitat use inorganic compounds such as $H_2S$ as the energy source and $CO_2$ as the C-source, thus creating an environment that supports higher forms of life. Both the $H_2S$ and $CO_2$ are generated from the hydrothermal vents as the superheated water from within the Earth's crust rises through the vents, reacting with the surrounding rock, dissolved metal ions, and sulfides (Fig. 7.9).

$$H_2S + 2O_2 \rightarrow SO_4^{2-} + 2H^+ \qquad ...7.1$$

$$S^0 + H_2O + 1.5O_2 \rightarrow SO_4^{2-} + 2H^+ \qquad ...7.2$$

$$S_2O_3^{2-} + H_2O + 2O_2 \rightarrow 2SO_4^{2-} + 2H^+ \quad ...7.3$$

The $SO_4^{2-}$ generated is chemically reduced to sulfide ($H_2S$), sustaining the cycle. As a matter of fact, the abundance of life around the vents corresponds to the concentration of $H_2S$ to the extent that when the vent stops jetting the magma, the nonmotile animals around the vent die rather quickly due to the absence of $H_2S$. Most notable among the sulfur oxidizers in this environment is *Thiomicrospira* sp. Methane is also generated and some existing methanotrophs use methane as a C and energy source.

$$2CH_4 + 2O_2 \rightarrow 2[CH_2O] + 2H_2O \qquad ...7.4$$

These may cooxidize carbon monoxide as a fortuitous reaction. The hydrogen oxidizers

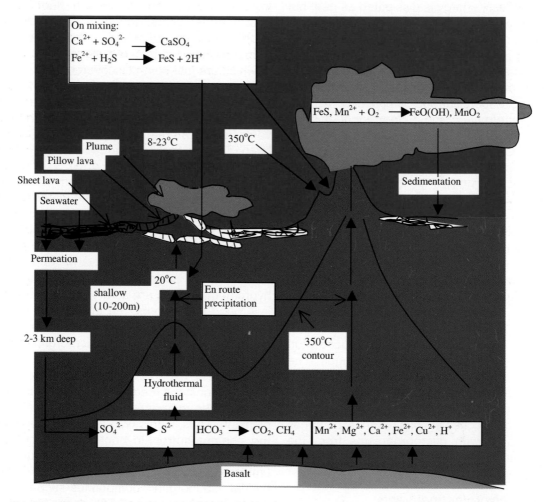

**Fig. 7.9** Geochemical reactions at two types of vents in the Earth's crust at the ocean floor during the hydrothermal cycling of seawater (Redrawn from Jannasch, 1989 with permission from John Wiley & Sons, Inc, New York, NY).

(hydrogenomonas) also generate energy, as shown in eqn 7.5:

$$6H_2 + 2O_2 + CO_2 \rightarrow CH_2O + 5H_2O \quad ...7.5$$

Nitrate ($NO_3^-$) can substitute for oxygen but there is so little $NO_3^-$ in this water that this possibility may not be very important. Oxygen can also be substituted with $S^0$ and $SO_4^{2-}$, but these alternatives are not yet well studied in the vent communities. The oxygen at such deep ocean levels is basically from recycling from the top layers. The potential for the ATP-generating nitrifying bacterial reactions is present but neither $NH_4^+$ nor $NO_2^-$ are abundant in the hydrothermal fluids.

$$NH_4^+ + 1.5O_2 \xrightarrow{\text{Nitrobacter}} NO_2^- + 2H^+ + H_2O \quad ...7.6$$

$$NO_2^- + 1.5O_2 \xrightarrow{\text{Nitrosomonas}} NO_3^- \quad ...7.7$$

Deposits of Fe and Mn cover most of the surfaces near the vents. $Fe^{2+}$ oxidizes to $Fe^{3+}$ spontaneously at the neutral pH of sea water. Therefore, iron chemolithotrophy is unlikely. Oxidation of $Mn^{2+}$ to $Mn^{4+}$ has been shown to occur at seawater pH as a bacterially mediated process that generates ATP. Therefore, $Mn^{2+}$ oxidation may contribute to vent metabolism. The hydrothermal fluids leach minerals from rocks (sea water is $Na^+$, $Mg^{2+}$, $Cl^-$ and $SO_4^{2-}$). At temperatures of 300°C or lower, flow rates of approximately 1 cm s$^{-1}$ prevail, causing a high degree of mixing and generating what are referred to as "white smokers". At 350°C or higher, a low degree of mixing occurs at flow rates of approximately 1 m s$^{-1}$ causing "black smokers" (Jannasch, 1989). The black smokers are due to calcium sulfate (anhydrite) and metal sulfides causing black particles and forming chimneys of brown-black rock.

To date, more than 290 species, most of which were previously unknown, have been found in hydrothermal vents, 97% of which dwell specifically in this habitat. These cover a range of invertebrate animals such as *Riftia pachpktila*, clams, mussels, mollusks, and a variety of polychaete worms and crabs. The best studied of these are large clams (0.3-0.4 m long) and giant tubeworms more than 2 m long. The vents also have extensively attached microbial mats with chemosynthetic autotrophs on which grazing animals (gastropods) such as snails, feed.

The symbiosis between vent invertebrates and chemolithotrophs account for approximately 90% of the organic C production at the vents. Most widely studied is the giant worm *Riftia pachyptila* which contains a chemolithotrophic bacterial symbiont in its tissue (Cavanaugh et al., 1981). *R. pachyptila* lost all vestiges of ingestive ancestry and does not have a mouth, gut, or anus. It has a modified gastrointestinal tract (trophosome) which comprises 50% of the total body weight. The gastrointestinal tract is packed with S granules and harbors a high density of bacteria (i.e., about $10^9$ bacterial cells/g tissue). The bacteria include S-oxidizers, nitrifiers, methylotrophs, methanogens, H-oxidizers, as well as Fe- and Mn-oxidizers. They fix $CO_2$ and hence have RUBISCO similar to chloroplasts. The worm also contains high levels of bacterial enzyme for $CO_2$ fixation, indicating the presence of $CO_2$ fixing bacteria inside the trophosome. The bacteria found herein have rarely been successfully cultured from worms pulled from the deep oceans and their discovery has opened up a totally new area of research in symbiosis. Based on electron microscopy, the associated bacteria resemble *Thiovulum*, a marine sulfur oxidizing bacterium. Sequencing 16S rRNA indicates that they belong to the γ-proteobacteria (Distel et al., 1988).

The blood of *R. pachyptila* contains a sulfide-binding protein that appears to concentrate sulfide from the environment, making it available to the microendosymbiont (Powell and Somero, 1983). The worm has a striking red color from special hemoglobin which transports $O_2$ and $H_2S$, the two gases that symbiotic bacteria need to fix $CO_2$, through the bloodstream to the bacteria in the trophosome. Unlike other types of hemoglobin that are known, this type is not poisoned by $H_2S$. The nature of this symbiosis is that the worm receives nourishment and the bacterium escapes competition in a comfortable habitat.

Another hydrothermal vent-based symbiont, the large clam (*Calyptogena magnifica*) also has a striking red color caused by the presence of the $O_2$- and $H_2S$-transporting hemoglobin. Its digestive system is intact but the gills contain large numbers of $H_2S$-oxidizing bacteria. Symbiosis is also such that the endosymbiont provides carbon and energy whereas the host provides raw materials and housing. Thus, in both *R. pachyptila* and *C. magnifica*, there is a strong coevolutionary linkage between the prokaryotes and the respective animal host. It is important to note that the entire ecosystem at this depth evolved in the absence of light, and at the expense of chemically unstable inorganic compounds, mostly sulfides. At the anaerobic-aerobic interface, microorganisms exploit these inorganic resources and are at the base of the food chain. They generate ATP and fix $CO_2$. These ecosystems are certainly important "oases" of the deep sea, supporting a biomass hundreds of times higher than the surrounding deep sea areas. Other hydrothermal vent-based animals with chemosynthetic symbionts which have not been widely studied include *Bathymodiolus* sp. and *Lucinoma* sp., both of which are mollusks.

The bacteria isolated from the vents grow at 110°C and the enzymes isolated function at 140-150°C. There is potential for using DNA polymerase from hydrothermal vent microorganisms such as *Therococcus litoralis* (Vent$_R$) and *Pyrococcus* sp. (Deep Vent$_R$) for PCR. The polymerase from these organisms is not denatured at 98°C, unlike *Thermus aquaticus* for which the PCR-cycle has to be reheated each time, i.e., taken through heating and cooling cycles and adding a fresh enzyme.

# References

Abeliovich A., D. Kellenberg, and M. Shilo (1974). Effect of photooxidative conditions on levels of superoxide dismutase in *Anacystis nidulans*. *Photochem. Photobiol.* **19:** 379-382.

Adebayo A.A. and R.F. Harris (1971). Fungal growth responses to osmotic as compared to matric water potential. *Soil Sci. Soc. Amer. Proc.* **35:** 465-469.

Amy P.S. and R.Y. Morita (1983). Starvation-survival patterns of sixteen freshly isolated open-ocean bacteria. *Appl. Environ. Microbiol.* **45:** 1109-1115.

Amy P.S., C. Pauling, and R.Y. Morita (1983). Starvation-survival processes of a marine vibrio. *Appl. Environ. Microbiol.* **45:** 1041-1048.

Ashwood-Smith M. and B.A. Bridges (1967). On the sensitivity of frozen microorganisms to ultraviolet radiation. *Proc. Roy. Soc. London Series B* **168:** 194-202.

Baba T. and D. Schneewind (1998). Instruments of microbial warfare: bacteriocin synthesis, toxicity and immunity. *Trends Microbiol.* **6:** 66-71.

Baross J.A. and J.M. Deming (1983). Growth of "black smoker" bacteria at temperatures of at least 250°C. *Nature (London)* **303:** 423-426.

Belly R.T., M.R. Tansey, and T.D. Brock (1973). Algal excretion of $^{14}$C-labelled compounds and microbial interactions in *Cynadium caldarium* mats. *J. Phycol.* **9:** 123-127.

Biederbeck V.O. and C.A. Campbell (1971). Influence of simulated fall and spring conditions on the soil system. I. Effect on soil microflora. *Soil Sci. Soc. Amer. Proc.* **35:** 474-479.

Boyley S.T. and R.A. Morton (1978). Recent developments in the molecular biology of extremely halophilic bacteria. *Crit. Rev. Microbiol.* **6:** 151-205.

Brierley J. and L. Luinctra (1993). Heap concept for pretreatment of Refractory gold ore. *In:* A.E. Torma, J.C. Wey and V.I. Lakshmanan (eds.). Biohydrometallargical Techniques, vol 1. Minerals, Metals, and Materials Society, Warrendale, PA.

Brock T.D. (1967). Life at high temperatures. *Science* **158:** 1012-1019.

Cameron R.E. (1971). Antarctic soil microbial investigations. *In:* L.Q. Quan and H.D. Porter (eds) Research in the Antarctica. Amer. Assoc. Advancement of Science, Washington DC, pp. 137-189.

Cavanaugh G.M., S.L. Gardner, M.L. Jones, H.W. Jannasch, and J.B. Waterbury (1981). Prokaryotic cells in the hydrothermal vent tube worm *Riftia pachyptila* Jones: possible chemoautotrophic symbionts. *Science* **213:** 340-342.

Chang Y. (1967). The fungi of wheat straw compost, II. Biochemical and physiological studies. *Trans. Brit. Mycol. Soc.* **50:** 667-677.

Chang Y. and H.J. Hudson (1967). The fungi of wheat straw compost, I. Ecological studies. *Trans. Brit. Mycol. Soc.* **50:** 649-666.

Cheung P-Y. and B.K. Kinkle (2001). *Mycobacterium* diversity and pyrene mineralization in petroleum-contaminated soils. *Appl. Environ. Microbiol.* **67**: 2222-2229.

Christensen H., D. Funck-Jensen, and A. Kjoller (1989). Growth rate of rhizosphere bacteria measured directly by the triated thymidine incorporation technique. *Soil Biol. Biochem.* **21**: 113-118.

Clouston J.G. and P.A. Wills (1969). Initiation of germination and inactivation of *Bacillus pumilus* spores by hydrostatic pressure. *J. Bacteriol.* **97**: 684-690.

Cohen Y. et al. (1989). Interaction of sulphur and carbon cycles in microbial mats. *In*: P. Brimblecombe and A.Y. Lein (eds.). Evolution of the Global Biogeochemical Sulphur Cycle. John Wiley & Sons, New York, NY, pp.191-238.

Colwell R.R. and A. Huq (1994). Vibrios in the environment: viable but nonculturable *Vibrio cholerae*. *In*: I.K. Wachsmuth, P.A. Blake, and Ø. Olsvik (eds.). *Vibrio cholerae* and *Cholera*: Molec. Global Perspect. Washington DC, pp. 117-133.

Corliss J.B., J. Dymond, L.I. Gordon, R.P. van Herzen, R.D. Ballard, K. Green, D. Williams, A. Bainbridge, K. Crane, and T.H. van Andel (1979). Submarine thermal springs on the Galapagos Rift. *Science* **203**: 1073-1083.

Dean C.J., P. Feldschreiber, and J.T. Lett (1966). Repair of x-ray damage to the deoxyribonucleic acid in *Micrococcus radiodurans*. *Nature (London)* **209**: 49-52.

Deming J.W. and R.R. Colwell (1981). Barophilic bacteria associated with deepsea animals. *Bioscience* **13**: 507-511.

Distel D.L., D.I. Lane, G.J. Olcon, S.J. Giovannoni, B. Pace, N.R. Pace, D.A. Stahl, and H. Felbeck (1988). Sulfur-oxidizing bacterial endosymbionts: analysis of phylogeny and specificity by 16S rRNA sequences. *J. Bacteriol.* **170**: 2506-2510.

Durrell L.W. and L.M. Shields (1960). Fungi isolated in culture from soils of the Nevada test site. *Mycologia* **52**: 636-641.

Egli T. (1995). The ecological and physiological significance of the growth of heterotrophic microorganisms with mixtures of substrates. *Adv. Microb. Ecol.* **14**: 305-386.

Falvo A., C. Levantes, S. Rossetti, R.J. Seviour, and V. Tandoi (2001). Synthesis of intracellular storage polymers by *Amaricoccus kaplicensis*, a tetrad forming bacterium present in activated sludge. *J. Appl. Microbiol.* **91**: 299-305.

Fenchel T. (1987). Ecology of Protzoa. The Biology of Free-Living Phagotrophic Protists. Science Tech Publ., Madison, WI.

Grant W.D. and J.B. Tindall (1986). The alkaline saline environment. *In*: R.H. Herbert and G.A. Codd (eds.). Microbes in Extreme Environments. Acad. Press, New York, NY, pp. 25-54.

Gregory P.H. (1973). The Microbiology of the Atmosphere. John Wiley & Sons, New York, NY.

Griffin D.M. (1969). Soil water in the ecology of fungi. *Ann. Rev. Phytopathol.* **7**: 289-310.

Guffanti A.A., P. Susman, R. Blanco, and T.A. Krulwich (1978). The protonmotive force and aminoisobutyric acid transport in an obligately alkalophilic bacterium. *J. Biol. Chem.* **253**: 708-715.

Häder D.P. (1997). Effects of UV radiation on phytoplankton. *Adv. Microb. Ecol.* **15**: 1-26.

Henis Y. (1987). Survival and dormancy of bacteria. *In*: Y. Henis (ed.). Survival and Dormancy of Microorganisms. Wiley and Sons, New York, NY, pp. 1-108.

Herbert R.A. (1982). Procedures for the isolation, cultivation and identification of bacteria. *In*: R.G. Burns and J.H. Slater (eds.). Experimental Microbial Ecology. Blackwell Sci. Publ., Boston, MA, pp. 3-21.

Imhoff J.F. and F. Rodriguez-Valera (1984). Betaine is the main compatible solute of halophilic eubacteria. *J. Bacteriol.* **160**: 478-479.

Jannasch H.W. (1989). Sulphur emission and transformations at deep sea hydrothermal vents. *In*: P. Brimblecombe and A.Y. Lein (eds.). Evolution of the Global Biogeochemical Sulphur Cycle. John Wiley & Sons, New York, NY, pp. 181-190.

Jannasch H.W. and C.O. Wirsen (1973). Deep-sea microorganisms; *in situ* response to nutrient enrichment. *Science* **180**: 641-643.

Jannasch H.W., K. Eimhjellen, C.O. Wirsen and A. Farmanfarmaian (1971). Microbial degradation of organic matter in the deep sea. *Science* **171**: 672-675.

Jennings D.H. and G. Lysek (1999). Fungal Biology: Understanding the Fungal Lifestyle. Springer-Verlag, New York, NY.

Jjemba P. K. (2001). The interaction of protozoa with their potential prey bacteria in the rhizosphere. *J. Eukary. Microbiol.* **48**: 320-324.

Jones D.F., F.T. Robb, and D.R. Woods (1980). Effects of oxygen on the survival of *Bacteroides fragilis* after far-UV irradiation. *J. Bacteriol.* **144**: 1178-1181.

Kelly D.P. (1992). The Chemolithotrophic Prokaryotes. *In*: A. Balows et al. (eds.). The Prokaryotes: A Handbook on the Biology of Bacteria: Ecophysiology, Isolation, Identification, Applications, vol. I. Springer Verlag, New York, NY.

Kushner D.J. (1978). Life in high salt and solute concentrations: Halophilic bacteria. *In*: D.J. Kushner (ed.). Microbial Life in Extreme Environments. Acad. Press, New York, NY, pp. 317-361.

Lange C.C., L.P. Wackett, K.W. Minton, and M.J. Daly (1998). Engineering a recombinant *Deinococcus radiodurans* for organopollutant degradation in radioactive mixed waste environments. *Nature Biotech.* **16**: 929-933.

Langworthy D.E., R.O. Stapleton, G.S. Sayer, and R.H. Findlay (1998). Genotypic and phenotypic responses of a riverine microbial community to polycyclic aromatic hydrocarbon contamination. *Appl. Environ. Microbiol.* **64**: 3422-3428.

Lloyd J.R. and L.E. Macaskie (2000). Bioremediation of radioactive metals. *In*: D.R. Lovely (ed.). Environmental Microbe-Metal Interactions. ASM Press, Washington DC, pp. 277-327.

Lloyd J.R. and D.R. Lovely (2001) Microbial detoxification of metals and radionuclides. *Curr. Opin. Biotech.* **12:** 248-253.

Lloyd J. R., J.A. Cole, and L.E. Macaskie (1997). Reduction and removal of heptavalent technetium from solution by *Escherichia coli. J. Bacteriol.* **179:** 2014-2021.

Losick R., L. Kroos, J. Errington and P. Youngman (1989). Pathways of developmentally regulated gene expression in *Bacillus subtilis. In:* D.A. Hopwood and K.F. Chatel (eds.). Genetics of Bacterial Diversity Acad. Press, New York, NY, pp. 221-242.

Marquis R.E. and P. Matsumura (1978). Microbial life under pressure. *In:* D.J. Kushner (ed.). Microbial Life in Extreme Environments. Acad. Press, New York, NY, pp.105-158.

Matin A., M. Baetens, S. Pandïa, C.H. Park, and S. Waggoner (1999). Survival strategies in the stationery phase. *In:* E. Rosenberg (ed.). Microbial Ecology and Infectious Disease. ASM Press. Washington DC, pp. 30-48.

Morgan P. and C.S. Dow (1986). Bacterial adaptations for growth in low nutrient environments. *In:* R.H. Herbert and G.A. Codd, (eds.). Microbes in Extreme Environments. Acad. Press, New York, NY, pp. 187-214.

Morita R.Y. (1982). Starvation-survival of heterotrophs in the marine environment. *Adv. Microb. Ecol.* **6:** 171-198.

Morita R.Y. (1986). Pressure as an extreme environment. *In:* R.H. Herbert and G.A. Codd (eds.). Microbes in Extreme Environments. Acad. Press, New York, NY, pp. 171-185.

Nasim A. and A.P. James (1978). Life under conditions of high irradiation. *In:* D.J. Kushner (ed.). Microbial Life in Extreme Environments. Acad. Press, New York, NY, pp. 409-439.

Novitsky J.A. and R.Y. Morita (1976). Morphological characterization of small cells resulting from nutrient starvation of a psychrophilic marine vibrio. *Appl. Environ. Microbiol.* **32:** 617-622.

Novitsky J.A. and R.Y. Morita (1978). Possible strategy for the survival of marine bacteria under starvation conditions. *Mar. Biol.* **48:** 289-295.

Powell M.A. and G.N. Somero (1983). Blood components prevent sulfide poisoning of respiration of the hydrothermal vent tube worm *Riftia pachyptila. Science* **219:** 297-299.

Pugh (1980). Strategies in fungal ecology. *Trans. Brit. Mycol. Soc.* **75:** 1-14.

Reed R.H. (1986). Halotolerant and halophilic microbes. *In:* R.H. Herbert and G.A. Codd (eds.). Microbes in Extreme Environments. Acad. Press, New York, NY, pp. 55-81.

Reistad R. (1975). Amino sugar and amino acid constituents of the cell walls of extremely halophilic cocci. *Arch. Microbiol.* **102:** 71-73.

Resnick M.A. (1970). Sunlight-induced killing in *Saccharomyces cerevisiae. Nature (London)* **226:** 377-378.

Rice G., K. Stedman, J. Snyder, B. Wiedenheft, D. Willits, S. Brumfield, T. McDermott, and M.J. Young (2001). Viruses from extreme thermal environments. *Proc. Natl. Acad. Sci. (USA)* **98:** 13341-13345.

Roszak D.B. and R.R. Colwell (1987). Metabolic activity of bacterial cells enumerated by direct viable count. *Appl. Environ. Microbiol.* **53:** 2889-2893.

Schulze R. and K. Gräfe (1969). Consideration of Sky Ultraviolet Radiation in the Measurement of Solar Ultraviolet Radiation. In: F. Urbach (ed). The Biologic Effects of Ultraviolet Irradiation with Emphasis on the Skin. pp. 359-373. Pergamon, New York.

Siegel B.Z., S.M. Siegel, J. Chen, and P. La Rock (1983). Extraterrestrial habitat on earth: The algal mat of Don Juan Pond. *Adv. Space Res.* **3:** 39-42.

Skujinš J. (1984). Microbial ecology of desert soils. *Adv. Microb. Ecol.* **7:** 49-91.

Smith D.W. (1978). Water relations of microorganisms in nature. *In:* D.J. Kushner (ed.). Microbial Life in Extreme Environments. Acad. Press, New York, NY, pp. 369-380.

Solomon L., P. Zeegen, and F.A. Eiserbing (1966). The effects of high hydrostatic pressure on coliphage T-4. *Biochem. Biophys. Acta* **112:** 102-109.

Stark J.M. and M.K. Firestone (1995). Mechanisms for soil moisture effects on activity of nitrifying bacteria. *App. Environ. Microbiol.* **61:** 218-221.

Swift S., N.J. Bainton, and M.K. Winson (1994). Gram negative communication by N-acyl homoserine lactones: a universal language? *Trends Microbiol.* **2:** 193-198.

Tansey M.R. and T.D. Brock (1978). Life at high temperatures: Ecological aspects. *In:* D.J. Kushner (ed.). Microbial Life in Extreme Environments Acad. Press, New York, NY, pp. 159-216.

Tate R.L. (2000). Soil Microbiology. John Wiley & Sons, New York, NY ( 2nd ed.).

Tong Y. and B. Lighthart (1998). Effect of simulated solar radiation on mixed outdoor atmospheric bacterial populations. *FEMS Microbiol. Ecol.* **26:** 311-316.

Vareschi E. (1982). The ecology of Lake Nakuru (Kenya), III: Abiotic factors and primary production. *Oecologia* **32:** 81-101.

Vincent W.F. (1981). Production strategies in Antarctic inland water: Phytoplankton eco-physiology in a permanently ice-covered lake. *Ecology* **62:** 1215-1224.

Weiner R. (1999). The plasticity of marine bacteria. Adaptations to high- and low-nutrient habitats. *In:* E. Rosenberg (ed.). Microbial Ecology and Infectious Disease. ASM Press, Washington DC, pp. 17-29.

Williams S.T., M. Shameemullah, E.T. Watson, and C.I. Mayfield (1972). Studies on the ecology of actinomycetes in soil. VI. The influence of moisture tension on growth and survival. *Soil Biol. Biochem.* **4:** 215-225.

Wynn-Williams D.D. (1990). Ecological aspects of Antarctic microbiology. *Adv. Microb. Ecol.* **11:** 71-146.

Zehnder A.J.B. and W. Stumm (1988). Geochemistry and biogeochemistry of anaerobic habitats. *In:* A.J.B. Zehnder (ed.). Biology of Anaerobic Microorganisms. John Wiley and Sons, New York, NY, pp. 1-38.

Zhou J., B. Xia, D.S. Treves, L.-Y. Wu, T.L. Marsh, R.V. O'Neill, A.V. Palumbo, and J.M. Tiedje (2002). Spatial and resource factors influencing high microbial diversity in soil. *Appl. Environ. Microbiol.* **68:** 326-334.

Zobell C.E. and A.B. Cobet (1964). Filament formation by *Escherichia coli* at increased hydrostatic pressures. *J. Bacteriol.* **87:** 710-719.

# 8

# Ecological Relationships in Exploiting Resources

## 8.1  LEVELS OF ECOLOGICAL ORGANIZATION

In nature, populations of various species are kept in check by the resistance of the environment in the form of predation, climatic limitations, and the availability of resources such as organic and inorganic nutrients. However, nature is not entirely cruel because allies and enemies are fostered in some form of association. Whereas some associations are purely accidental, a number of them are obligate or facultative, requiring the participants to share the resources at their disposal. Associations may occur between individuals of the same (intraspecific) or differing (interspecific) species and may be mutually beneficial or one-sided, forming populations. With increasing complexity, the populations form communities, which in turn comprise ecosystems (Fig. 8.1). A definition of these levels of complexity is in order.

An ecosystem is a group of self-sustaining communities as they interact with their physical environment and with each other as energy flows and cycles within the ecosystem (Odum, 1983). Thus the members within an ecosystem feed themselves and remove the wastes they generate. No export or import of matter is needed to sustain the members. A **community** (guild) is an assemblage of populations of all the organisms in an area, with individuals of different species living in a designated place at the same time and exploiting the same resources. In other words, the individuals of different species in a community occupy the same niche and compete for the same resources, for example S as the electron donor. A community, therefore, consists of species that are, in a sense, functionally similar. The community and its surroundings comprise an ecosystem. **Population** refers to a group of individuals of the *same* species living together in a specific location at the same time. If many populations occur in a given location, they may be genetically related organisms or unrelated organisms performing a similar function. For example, anthracene degraders in a particular system may comprise populations of various species (fungi and bacteria). Populations within the same community can interact with each other, a phenomenon called **synecology**, but not with populations in other communities. **Autecology**, the opposite of synecology, emphasizes the behavior and life history of individual populations adapting to the physical or abiotic environment, for example, temperature, pH, radiation. Populations may be autochthonous (indigenous) or allochthonous (alien), or zymogenous (transformed due to internal changes). Individual refers to a single organism of a particular species.

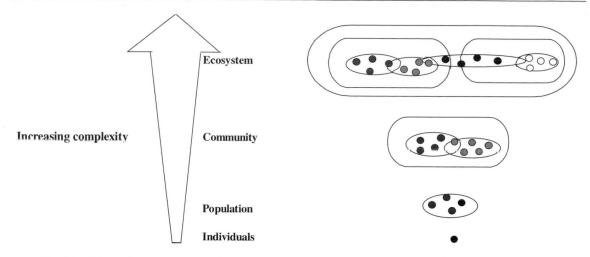

**Fig. 8.1** Schematic representation of the different levels of ecological organization with increasing complexity.

Further clarification of terminology appears appropriate here. **Niche** is often misused. The niche of an organism refers to its way of life, and specifically its method of acquiring resources. For example, the niche of sulfur-reducing bacteria (SRB) in the Winograsky column is to reduce sulfur versus a **habitat** which signifies where an organism lives, e.g. SRB habitat is the anaerobic zone. Thus, niche and habitat signify physiological properties and locale respectively. **Succession** refers to the directional change in the composition, relative abundance, and spatial pattern of species comprising a community. Changes are both quantitative and qualitative with time. Succession involves stages of modifying the environment until a climax community (i.e; stable state in which there is a dynamic balance between the members of the community) develops. However, it is not clear whether the early successional stages are necessary to condition the environments for those that follow. The apparent succession may be due to simply the rapid growth characteristics and reproductive abilities of the pioneering species.

## 8.2  RESOURCES

A resource is any substance or factor consumed by an organism that leads to an increase in its rate of growth paralleling the availability of the resource in the environment (Keddy, 1989). Resources include a wide array of nutrients and abiotic factors such as C, N, S, and prey species. Electron acceptors such as $O_2$, $NO_3^-$, $SO_4^{2-}$ and $Fe^{2-}$ as well as space pH, temperature, $E_h$, $O_2$ availability, salinity, etc. also often limit organisms. They are scarce and therefore limiting, but play a significant role in the organization and coexistence of communities. Tilman (1982) defined six different categories of resources (Table 8.1), the possible extremes being either cardinal or perfectly substitutable. Essential resources, contrarily, cannot be substituted. They are neither synthesized nor destroyed by the cell and include inorganic elements such as N, P, K, Mg, S, Fe, and important organic substances such as vitamins that are mandatory for growth. Resources within a particular habitat are repeatedly consumed and renewed over time. If essential resources are limiting, microbial growth will follow Liebig's law of the minimum in the sense that only one cardinal resource will be limiting at any one time and the growth rate of the organisms will be determined by the least available resource. On the other hand, substitutable resources are substrates which occur in different chemical forms that are closely replaceable with another and will not, therefore, affect growth rates if a

**Table 8.1**  Types of resources

| Resource type | Description | Schematic representation[a] |
|---|---|---|
| Perfectly substitutable | Two resources can substitute for each other with equal effect at all abundances. Thus, consuming an amount of resource 1 ($R_1$) is equivalent to consuming $R_2$ for all values of $R_1$. | |
| Complementary | Two resources, each containing different proportions of two nutritionally essential elements, leading to a higher rate of growth when they are consumed together than would be the case if each is consumed alone. For example, $R_1$ may be rich in lysine and $R_2$ rich in sulfur, with one proportion of each leading to maximum complementarity to rate of growth. | |
| Antagonistic | Two resources whereby an organism consumes more of one resource to attain a given reproductive rate if both resources are consumed together due to toxic effects of one of the resources. Thus, if consumed together, both resources may have a toxic effect. | |
| Essential | Resources that are absolutely necessary for the organism to grow and maintain metabolic functions. Such resources include N, P, K, C, Ca, H, S, etc. For a pair of cardinal resources, the growth rate of an organism is determined by either one of them. | |
| Hemi-essential | One resource may be nutritionally complete while the other lacks some nutritional element(s) available in relative excess in the first resource. | |
| Switching | The growth rate is determined by the availability of one resource out of two, which leads to the largest reproductive rate. Thus, in the areas labeled 2 on the panel, $R_2$ will be consumed because it gives higher reproductive/growth rates Likewise, in the areas marked 1, $R_1$ will be used for the same reasons. | |

[a]The vectors represent the amount of resources ($R_1$ and $R_2$) consumed per individual per unit time in the habitat with the available resource at that point on the growth isocline. The slope of each vector is the ratio of the amount of $R_2$ consumed to the amount of $R_1$ consumed

substitute is available. Examples of substitutable resources include glucose substituting for maltose, nitrates for ammonium, acetate for thiosulfates, etc. If the two substrates yield an identical growth rate, they are considered to be perfectly substitutable. All other categories of resources fall within these two extremes.

Equally relevant in environmental microbiology are interactive essential resources, that is, resources to which the transformation from limitation by one resource to a second resource is not abrupt. They are called interactive essential because the population growth is determined by the availability of both resources rather than the availability of just one of the two growth-limiting resources at a time. Exploitation of resources is a selective force for evolution and is always a relentless challenge to microbial cells in sediments, waters, within host organisms and in soil. The abundance of resources can increase, decrease, pulsate with time, or remain fairly stable as a result of constant replenishment. Thus, microbial activity and dominance change along resource gradients in natural environments. The Monod (1950) model of resource-limited bacterial growth relates the biomass-specific microbial growth rate ($\mu$, $hr^{-1}$) to the concentration of a growth-limiting resource (R, mM) and the maxi-

mum biomass-specific growth rate ($\mu_{max}$) as well as the half-saturation constant ($K_s$) for growth on the limiting resource (Fig. 8.2). Mathematically,

$$\mu = \mu_{max}[R/(K+R)]$$

On the whole, limitation of resources to the growth of microorganisms, and other organisms for that matter, is widespread in the biosphere. Since organisms forage on more than one resource, they respond to changes in ratios of such resources, a fact that puts more logic into looking at resource utilization in terms of a resource competition theory (Tilman, 1982). This theory will be fully explored when we discuss competition.

The influence from the concentration of nutrients can be assessed by considering the growth rate of two species, A and B, in media containing increasing concentrations of a nutrient. In this hypothetical situation, we assume that both species are competing for a single resource and that species A has a high growth rate compared to that of species B at a low nutrient concentration. Species A ends up prevailing better under these conditions of nutrient stress (Fig. 8.3). With more abundant nutrients, however, the situation is reversed and favors species B which despite a lower growth rate

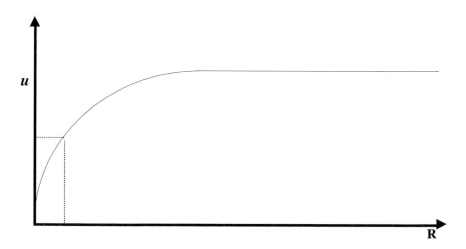

**Fig. 8.2**  Effect of resource concentration on specific growth rates.

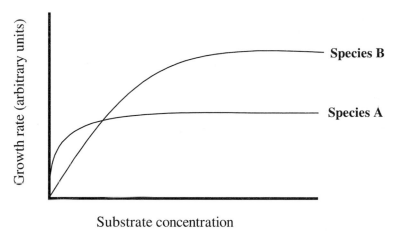

Substrate concentration

**Fig 8.3** Growth rate of two microbial species at different nutrient concentrations. The substrate concentration on the x-axis represents the availability of resources (R).

compared to A, becomes dominant. As the concentration of the limiting resource increases, the specific growth rate also increases.

## 8.3 TYPES OF INTERACTIONS

Interactions in naturally occurring ecosystems are usually highly specific and quite dependent on both biotic and abiotic conditions. Interactions can be at an interpopulation, intrapopulation or an interspecies level. Within a single population (i.e., intrapopulation) they can be either positive or negative. Change in the rate of growth with increasing population under a positive or negative interaction, is shown in Fig. 8.4. A negative interaction occurs when the cells are genetically identical or if the ecological niche is shared. Intensive intrapopulation interactions provide a basis for genetic changes and niche differentiation at the level of individual cells. However, across species and/or populations (interpopulations) interactions are generally classified as negative, positive, or neutral (Fig. 8.5). The principles of interpopulation competition are similar to those of intrapopulation competition except that competing interpopulations comprise more genetically diverse individuals. It can be seen in Figure 8.5 that the majority of interactions have some negative influence. In some cases, the

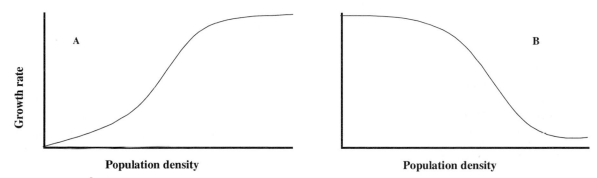

**Fig. 8.4** Changes in growth rate with increasing populations in cases of (A) positive and (B) negative interaction.

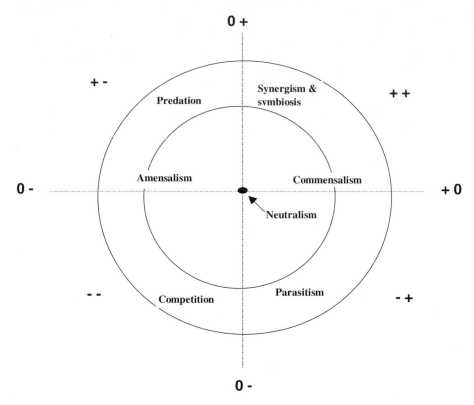

**Fig. 8.5** Coordinate model of a two-species interaction showing the possible relationship (modified from Odum, 1989 with permission from Sinauer Associates, Inc., Sunderland, MA).

interacting populations could be responding to toxins produced by one of the competing individuals or another individual in the ecosystem, a toxin which affects some but not all individuals. For example, in anaerobic environments $H_2S$ may be produced. The $H_2S$ could be toxic to some individuals in that environment. However, the boundary between some particular types of interactions may not always be clear cut. For example, mycorrhizae are known to enhance the uptake of nutrients, particularly P. However, as discussed later, mycorrhizae may end up drawing nutrients away from a less photosynthetic plant and literally passing the photosynthates to a more photosynthetic plant (see Section 8.4.2). Individual types of interactions are discussed below.

### 8.3.1 Negative interactions

Negative interactions include amensalism, competition, parasitism, and predation. All of the negative interactions reflect some form of competition, but to different degrees. They tend to display self-regulation mechanisms, preventing overpopulation, destruction of habitats, and ultimate extinction.

#### 8.3.1.1 Amensalism

Amensalism is the ecological term for toxins in soil. It involves a form of unilateral antagonism, with an individual or population producing an inhibitory substance or substances that are antagonistic to other populations. The population that produces the substance may be modified by the inhibitory substance or given a competitive edge over its antagonists. Coevolution between the two or more amensals is either weak or totally absent. Production of inhibitory substances under amensalism explains why some microorganisms fail to establish in some environments. It is the basis for finding antibiotics which have played a major role in the advancement of modern medicine. Antibiotics are organic compounds produced by one organism that in low concentrations eliminates/prevents the growth of another organism. During the exploration of antibiotics, soil is inoculated onto a sterile medium. The antibiotic prevents the existing organism from growing, thus resulting in a clearing zone. Besides antibiotics, other substances that can cause antagonism include nitrogenous substances ($NO_2^-$ and $NH_3$) which are otherwise toxic, and acids, e.g. acetic, propionic, and sulfuric acid. Sulfuric acid is produced by *Thiobacilllus* sp. in sediments and acid mine drainage sites. As the acid depresses the pH to levels as low as pH 1, the other organisms are eliminated, giving a competitive edge to the acid-tolerant *Thiobacillus* sp. Besides the antibiotics and mining industries, amensalism has had considerable impact on human life in various other processes, such as cheese making, wine manufacturing, and prevention of colonization of the skin by noxious microbes (Table 8.2).

### 8.3.1.2  Competition

Competition is the struggle between two or more organisms over a limiting resource. Competition for resources holds a central place in environmental microbiology and the evolution theory and plays an important role in the organization and function of microbial communities in a variety of natural environments. It represents a mutually negative interaction among the two or more species (or populations) involved, causing a reduction in growth rate in each of them. The magnitude of the effect need not have the same impact on the species (or populations) involved. The respective competitors negatively affect each other but one may be less affected than the other(s). Within a population, competition is due to both a scarcity of resources and an overlap in the niche. This relationship is expressed by the Lotka-Volterra model (Keddy, 1989; Grover, 1997) or resource competition (also known as the resource-ratio) model (Tilman, 1982).

**Table 8.2**  Examples of amensalism in various environments

| Environment | Process | Responsible microorganisms |
|---|---|---|
| Crop residues | A plant pathogen, produces antifungal substances and competes well between cropping periods in the field | *Cephalosporium graminea* |
| Human oral cavity | Bacteria produce hydrogen peroxide in the saliva. Catalase detoxifies the $H_2O_2$. Catalase and the organisms survive, that is $H_2O_2 + H_2O_2 \rightarrow 2H_2O + O_2$ | *Actinomyces* spp.; *Lactobacillus* spp. |
| Skin | Volatile fatty acids secreted by the skin microflora inhibit new colonization of the skin by foreign organisms | *Micrococcus luteus, Staphylococcus epidermidis, Corynebacterium* spp. *Brevibacterium* spp., *Propoinibacterium* spp. |
| Soil | Fungicidal and fungistatic substances that inhibit germination of fungal spores | *Trichothecium roseum, Gliocladium fimbriatum, Penicillium frequentans, Mucor ramannianus* |
| Vinegar | Acetic acid prevents spoilage | *Acetobacter* spp.; *Gluconobacter* spp. |
| Wine | Ethanol produced by yeast (must remain anaerobic or else the ethanol is oxidized to $CH_3COOH$ (ethanoic acid)) | *Acetobacter* spp. |
| Aquatic/sediments | In mining sites, acid mine drainage due to greatly lowered pH during oxidation of sulfur | *Thiobacillus thioxidans* |

The diversity that exists among inter-populations increases the intensity of competitions and the likelihood of further increase in differentiation of the niche. Intercompetition also increases the range of possible outcomes for the various competitors. A possible outcome is that one or more of the populations may become extinct from the ecosystem (competitive exclusion). Alternatively, the respective populations may persist and coexist. In the latter situation, one population may become more abundant (dominant) while the other becomes rarer or even, in the long run, extinct. To demonstrate extinction or coexistence, the Lotka-Volterra model is based on the logistic equation

$$\delta N/\delta t = rN(1-N/K)$$

where N is the population density, t the time, r the intrinsic rate of increase in the population, and K the carrying capacity of the respective ecosystem. Since competition only applies to two or more individuals, species or populations, this equation is applied by assigning each population an intrinsic rate of increase ($r_i$) and a carrying capacity ($K_i$). If we consider populations 1 and 2, the growth rate of each of the competitors will be negatively influenced by the other as is reflected in the coupled equations

$$\delta N_1/\delta t = r_1 N_1 [1-(N_1/K_1)-(\alpha_{12} N_2/K_1)]$$

$$\delta N_2/\delta t = r_2 N_2 [1-(N_2/K_2)-(\alpha_{21} N_1/K_2)]$$

where $\alpha_{12}$ and $\alpha_{21}$ are the competition coefficients that describe the effect of population 2 on 1 and 1 on 2 respectively, arising from competition. If populations 1 and 2 influence each other to the same extent, the coefficients will be equal, indicating that the two populations are indistinguishable. This occurs during intraspecies competition but is rare if competition is across species.

Based on this model, a two-species competitive interaction can be depicted using isoclines which represent all the possible pairs of population densities for one population if the growth

rate of its competitor is zero. For population 1, this situation will be depicted by the equation

$$\delta N_1/\delta t = r_1 N_1 [(K_1-\alpha_{11} N_1-\alpha_{12} N_2)/K_1] = 0$$

The above equation is satisfied if $r_1 = 0$, $N_1 = 0$ or if $K_1-\alpha_{11}N_1-\alpha_{12}N_2 = 0$. The possibility of $r_1$ or $N_1$ being zero is not important but if it is zero, setting the density of population 2 to zero is the same as

$$K_1-\alpha_{11}N_1-\alpha_{12}N_2 = 0$$

Thus
$$K_1 = \alpha_{11}N_1$$

Finally
$$N_1 = K_1/\alpha_{11}$$

So, to cause extinction of population 2 in the system, population 1 has to be at its carrying capacity per unit of its own competition coefficient. A similar relationship can be derived for the second population, i.e., $N_2 = K_2/\alpha_{22}$, by setting $N_2$ equal to zero.

From these equations, an isocline can be developed (Fig 8.6) which depicts the carrying capacity. The two axes represent the two competing species or populations. Below the isocline, the population of 2 is below the carrying capacity and increases with time (solid arrows). Above the isocline, population 2 has exceeded its carrying capacity and declines with time (dotted arrows). Furthermore, as $N_2$ increases,

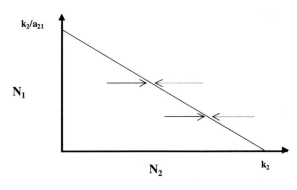

**Fig. 8.6** Isocline depicting the carrying capacity of the ecosystem for a population under the Lotka-Volterra theory. Above the isocline, the population of species 2 ($N_2$) has exceeded its carrying capacity and declines with time (dotted arrows). Below the isocline, its population is below the carrying capacity and increases with time (solid arrows).

$N_1$ decreases and vice versa. Within the ecosystem, different populations have different isoclines. As indicated earlier, only two outcomes are possible, that is, either one of the populations excludes the other (competitive exclusion) or the two populations coexist. These possibilities are represented in Fig. 8.7 whereby population 1 excludes population 2 (Fig. 8.7A) or population 2 excludes population 1 (Fig. 8.7B). An interesting situation is depicted in Figure 8.7C because either of populations 1 or 2 can exclude the other, the outcome being determined by the respective initial population densities in the ecosystem. If the initial populations of 1 are comparatively large, 1 excludes 2 and vice versa. Since more similar individuals are expected to need more identical resources, competition is expected to be more intense within (intra-) than between (inter-) species. This assumption may not always hold true. For example, one of the competitors may exert its superior competitive ability through the production of compounds deleterious to its competitor (allelopathy). Where it holds true, however, it has ramifications in the establishment of coexisting populations, the coexistence occurring at equilibrium (Fig. 8.8). The ability of these two populations to coexist is more a result of the intense competition within each species than between the two populations (or species, i.e., interspecific competition). Unfortunately, the Lotka-Volterra equations have no predictive power because the coefficients cannot be predicted but are rather estimated after competition has occurred. Thus, the Lotka-Voltera model is more exploratory compared to the resource-ratio theory.

The resource ratio theory, developed by Tilman (1982), is based on these facts:
(1) organisms require resources for their growth and reproduction;
(2) different species or phenotypic strains of the same species often differ in their efficiency to take up and utilize the resources that potentially limit growth; and

(3) the physiological difference in uptake and utilization of resources results in differences in competitive ability among organisms that share the same resources.

For two populations ($P_1$ and $P_2$) utilizing a pair of nonsubstitutable resources ($R_1$ and $R_2$), we can construct an isocline (Fig. 8.9). Axes represent the abundance of the respective resources. In this case, the critical level is reached as a result of the intensity of intrapopulation competition. Above the critical concentrations of the two resources ($R_1^c$ and $R_2^c$), represented by the shaded area, the populations thrive (if they meet certain conditions, see further). Owing to intraspecific competition, the resources are consumed to some critical levels below which the population cannot grow because mortality exceeds the rate at which the resources are replaced (unshaded area). Along the isocline, the net growth in population is zero (i.e., N/dt = 0) since the population growth will be balanced by population losses due to mortality (Smith, 1993).

Similar to the Lotka-Volterra theory, under the resource-ratio competition theory, each population has a different isocline for a particular habitat. In a competitive situation with two populations ($P_1$ and $P_2$) competing for the same resource, coexistence or exclusion occurs. For $P_2$, if one of the resources ($R_2$ or $R_1$) is consumed to below the critical level $R_2^{c2}$ or $R_1^{c1}$ respectively, the mortality rate for $P_2$ will exceed the per capita rate at which $P_2$ can grow, leading to its being outwitted by $P_1$ which can still tap the existing pair of resources (Fig. 8.10A). Similarly, if either of the resource levels falls below $R_1^{c1}$ or $R_2^{c1}$, $P_2$ excludes $P_1$ from the habitat. Above resource levels $R_1^{c1}$, $R_1^{c2}$ and $R_2^{c1}$, $R_2^{c2}$, both populations can exist competitively. In other words, $P_1$ is a superior competitor for resource 2 and an inferior competitor for resource 1. On the other hand, $P_2$ is a superior competitor for resource 1 ($R_1$) but an inferior competitor for resource 2 ($R_2$).

The situation depicted in Figure 8.10C deserves to be described further as it displays

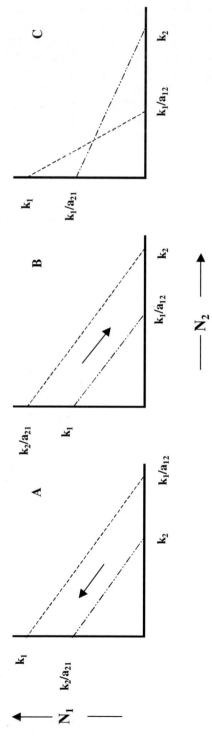

**Fig. 8.7** Isoclines depicting possible scenarios resulting in the exclusion of one population by another. In A, population 1 exceeds population 2 and in B, population 2 excludes population 1. In C, either population 1 or 2 can exclude the other, depending on the initial population density in the ecosystem.

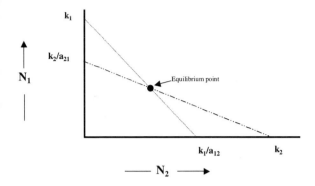

**Fig. 8.8** Isoclines depicting the coexistence of two populations at equilibrium (dot) as a result of the competitors requiring more identical resources.

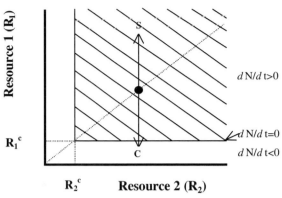

**Fig. 8.9** Under the resource-ratio theory, the isocline showing the region above the critical resource levels ($R_1^c$ and $R_2^c$) below which no growth will occur. The resource consumption vector (diagonal dotted line) passes through the intercept for the two resources and the intercept for the isoclines. Considering a supply point (solid dot) resources are either supplied (S) or consumed (C).

two possible scenarios, i.e., exclusion or coexistence depending on the environmental conditions. At any one time, the supply of resources will fall in one of the regions A-F. If the concentration of resource supply falls anywhere in region A, the habitat is too impoverished and the resources will collectively be insufficient. Thus both populations will be unable to survive since for both of them the per capita rate of loss due to mortality will be greater for each of the populations than their per capita rate of growth. If the concentration of resources supplied is in region B, population 1 will exclude population 2 due to insufficient nutrients for population 2. Alternatively, if the concentration of resources supplied falls in region F, population 2 will exclude population 1 for the same reasons (Fig. 8.10C). If either of the two populations is by itself and the concentration of resources supplied falls in regions C, D or E, that population will survive because there is no competition and the resources are sufficient. However, if the two populations are present together, $P_1$ wins if the supply level is in region C, and $P_2$ wins if the supply level falls in region E because in either case, the winning population has a lower critical resource level threshold for the resources that exist. If the supply level falls in region D, population 1 may exclude population 2 or vice versa depending on the initial conditions, e.g., the ini-

tial populations of the respective species (Fig. 8.10C), a situation similar to that encountered under Figure 8.7C by the Lotka-Volterra theory. This last situation, that is, Fig. 8.10C, is important for environments inoculated for a particular purpose. Alternatively, both populations may simply coexist under equilibrium conditions imposed by a greater intra- than interspecific competition, as is the case under the Lotka-Volterra theory (Fig. 8.8).

For interactive essential resources and perfectly (or linearly) substitutable resources, the isocline(s) will not be L-shaped but rather range from semicircular to hyperbolic and negatively diagonal respectively (Fig. 8.11). For interactive-essential resource isoclines, the point nearest to the intercept of the two resource concentrations represents equal limitation by the resources in that a unit increase in either resource gives the same increment to population growth. Further away from this point along the isocline, population growth responds greatly to an increase in one of the resources compared to the other. Thus, it is possible to characterize the population as linked more or less by a particular resource or equally limited by

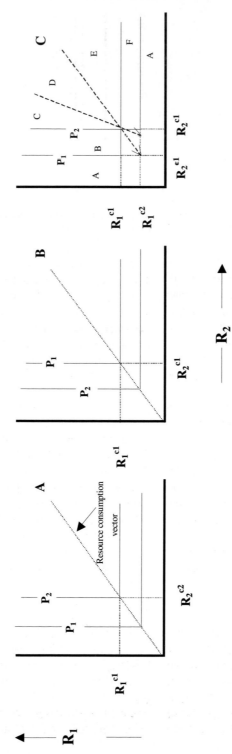

**Fig. 8.10** Isoclines depicting either competitive exclusion of (A) $P_2$ by $P_1$ and (B) $P_1$ by $P_2$ or (C) the exclusion/coexistence based on the availability of resources. $R_1^{c1}$, $R_2^{c1}$, $R_1^{c2}$, and $R_2^{c2}$ are the critical resources 1 and 2 for populations 1 and 2 respectively.

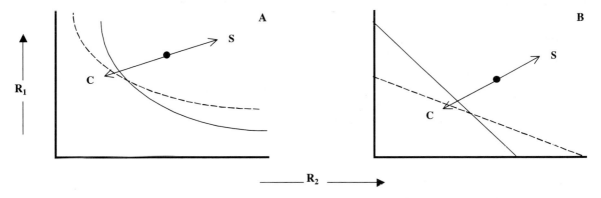

**Fig. 8.11** Isoclines of a population (species) that depends on (A) two interactive-essential resources $R_1$ and $R_2$ or (B) two perfectly substitutable resources $R_1$ and $R_2$.

both. Successful coexistence will in this case require that each population consume proportionally more of the resource that greatly limits its growth than the resource that greatly limits its competitor. Depending on the extent of curvature, the isoclines for interactive-essential resources may intercept each other more than once (Taylor and Williams, 1975), complicating predictions of the outcome further.

Thus both the Lotka-Volterra and the resource-ratio theories, despite fundamental differences in approach, display similar conclusions about coexistence and exclusion of populations or species. However, the latter theory has some predictive potential for the outcome of competitive interactions if both the requirements for these resources and the motility rates from the lack of competitive ability are known. The resource competition model can shape the community structure of fresh water, marine, terrestrial plants, and natural microbial communities (Smith, 1993). In practice, application of either theory is more complex because organisms compete for a multiplicity of resources. Owing to phenotypic tradeoffs, a species or population that is competitively superior for one resource will typically not be competitively superior for all possible resources (Tilman, 1982). In nature, gradients for resources can occur over time from associated macromolecules such as root exudates, shed plant tissue, pulses of ingested food into the gastrointestinal tract, etc. In such situations, the rate of growth indeed becomes a key selective trait during the various episodes of resource availability (feast). On the other extreme, in situations of scarcity (famine), dormancy would be a key selective trait.

### 8.3.1.3 Parasitism

Parasitism occurs if one organism (the parasite) withdraws body substances from another (the host). The parasite therefore typically depends on its host, benefiting from the association at the host's expense since it uses the host as both a habitat and energy source. A closer distinction between parasitism and commensalism is that the former thrives on the notion that one organism—the parasite, debilitates another—the host, over a prolonged period of time. Commensalism, on the other hand, is one-sided, benefiting one of the organisms but not harming the other (see Section 8.3.2.3). The organism does not live very long if it kills or damages its host. Thus a well-adapted parasite tends not to destroy its host but rather develops feedbacks that benefit it and its host so that both can thrive. If it destroys its host, it runs the danger of destroying itself. In the words of Croll (1966), parasites must live on interest and not on capital. Individual parasites would not cause disease to the host. Rather, disease is usually a result of hyperinfestation which suggests a density-dependence to cause disease. If parasitism leads to host mortality, death usually occurs if the parasite completes its development or when other circumstances add to the injury caused by the parasite. Although not always the case, the parasite is usually smaller than its host.

In the association, the host normally provides nutrition to the parasite but may also supply $O_2$, vitamins, water, metabolic control, or transportation. The relationship may be highly specific and involve a high level of coevolution between the involved parties. In an evolutionary sense, specificity results from the loss of adaptability of the parasite to other potential hosts oftentimes owing to a single profound structural or physiological feature of the parasite or the host. Indeed, in a number of cases, parasites are inherited phylogenetically, that is, passed on from one ancestral host to its offsprings. However, some parasites are ecologically separated from their potential host initially getting picked up only by accident.

Parasites are either (i) facultative or obligate, (ii) endoparasitic (i.e., live within the host) or ectoparasitic (attached externally), and (iii) balanced or destructive. Facultative parasites can survive and grow in the absence of a host and this creates problems of definition for it is unjustified to refer to such organisms when free-living as parasites. Obligate parasites, on the other hand, need the host for survival. Ectoparasites feed on the outside of their host whereas endoparasites live inside their hosts. Endoparasites can be further classified as superficial if they live within the wall and in the gut or as somatic endoparasites if they live in deeper tissue (Croll, 1973). In an evolutionary sense, parasites with a balanced relationship are more long-lasting and the majority of parasites live in harmony with their hosts. The association only becomes fatal (destructive) at times of environmental stress, a situation that may provide for parasites taking more than their metabolic quota from the host.

**Types of hosts:** Successful transmission of the parasite requires contact between the infective stage and the host. Thus the ecological adaptations between the host and the parasite cannot be disregarded in the overall ecology of parasitism. Most parasites have an independent, free-living stage in their life cycle, which enables them to disperse and make contact with the next host. Parasites, depending on the type, may have definitive hosts, intermediate hosts, ecological hosts, reservoir hosts or vectors (Table 8.3). Definitive hosts are those in which a parasite reaches maturity. Contrary to what the name suggests, however, a definitive host is not necessarily the most important host in the life cycle nor more essential than the intermediate hosts. Intermediate hosts are those other than "definitive" hosts, in which a particular phase to the parasite occurs. Furthermore, a definite structural or metabolic change must occur in the intermediate host. A parasite may have more than one intermediate host. For example, *Plasmodium vivax* matures in humans, the Anopheles mosquito only acting as an intermediate host (see Chapter 14). Such a strategy of the parasite isolating itself from its host into a vector to facilitate some portions of its life cycle ensures successful transmission from one host to another and avoids hyperinfestation of the host. From the perspective of parasite control, it creates avenues for distribution and thus complicates control measures. For those parasites to which the host is essential, prevalence is naturally limited to regions in which the hosts occur. If they require more than one host in their life cycle, they are restricted to regions where both hosts occur. Similarly, regions with a diverse number of hosts will also contain large numbers and varieties of parasites.

Ecological hosts, also called carriers, are unessential. No parasitic development occurs in ecological hosts but their presence provides useful ecological links between the parasite and the host. Ecological hosts may over time evolve into intermediate hosts. Vector hosts facilitate the transmission of parasites from one individual to another but here again the parasite undergoes no development within the carrier (Table 8.3). Reservoir hosts are somewhat similar to vector hosts but exhibit some apparent differences. Reservoir hosts harbor the parasite without endangering themselves. If attempts to eliminate the parasite from a particular target species (or populations) are made without tackling the same parasite in other related species/populations (i.e., reservoirs), the target species may repeatedly be reinfected by the parasites that harbor in the reservoir. A well-studied example of this situation is a parasitic protozoan *Trypanasoma brucei* which causes nagana in cattle and its close relative *T. gambiense* which causes sleeping sickness in humans. Transmitted by the tsetse fly (*Glossina palpalis*) as it sucks blood from cattle and wild ungulates, *T. brucei* causes no disease in wild ungulates since they are naturally resistant to it. However, the resistant ungulates can act as carriers of the pathogen, thus providing a continual source of infection. In a somewhat similar mode, tsetse flies, and many domestic animals such as pigs, goats, and dogs are infected by *T. gambiense* but do not contract sleeping sickness, unlike man who has a high rate of infection. Tsetse flies, pigs, goats, and dogs thus act as reservoirs for *T. gambiense* and its associated disease.

Unlike ecological hosts, vectors are absolutely essential for the parasite to be transmitted from one host to the next. Vectors are of particular relevance in environmental microbiology because various microbes, in particular protozoa, are mechanically transmitted by vectors, such as tabanid flies for *Trypanasoma evansi* to humans and a variety of domestic animals (cattle, camel, horses) causing various diseases.

Once established on or inside a host, the parasite can be carried passively by it, facilitating distribution of the parasite's eggs and infective stages. From an ecological perspective, a concentration of hosts means an increased likelihood of successful transmission. Such a mechanism of spread can thus affect not only local movements, but also intercontinental migration of parasites. In the case of parasites that infect warm-blooded animals, the parasite, once established on its host, is not only assured of a source of nutrition, but also a stable temperature environment that is important to the production of eggs. The host also offers protection to the parasite, particularly endoparasites, from potential predators and physical factors such as moisture stress.

### 8.3.1.4 Predation

Predation was first formally studied by a Russian ecologist, G.F. Gause, in the early half of the 20th century (Gause, 1934). It basically involves a large predator eating a small prey. Feeding is usually by abruptly engulfing the prey and involves some level of selectivity. Microbiological predators include protozoa (Ciliates, Flagellates and Amoeba), cellular Slime molds (Eukaryotic amoebae), Myxobacteria (gliding bacteria that form aggre-

**Table 8.3** Relationship between hosts and parasites

| Host | Stage of parasite in host | Necessity of host in life cycle | Role in life cycle |
|------|---------------------------|----------------------------------|---------------------|
| Definitive | Mature | Essential | Maturation |
| Intermediate | Juvenile | Essential | Development and transmission |
| Ecological | Juvenile | Not essential | Ecological link |
| Vector | Any | Essential | Mechanical transmission |
| Reservoir | Any | Not essential but its presence enhances continuity | Alternative host |

*Source*: Modified from Croll (1973).

gate fruiting bodies), some prokaryotes and Nematodes (trapping fungi). Predation involves three principles, that is, it:

(i) is density dependent,

(ii) requires the coexistence of the predator and prey, and

(iii) impacts both the predator and prey.

**Density dependence of predation:** The rate of ingestion of the prey by the predator is proportional to the size and density of the prey. Thus, the more the prey, the more intensively it will be grazed. Furthermore, the higher the density of the prey in the medium, the more diminished the distance between the prey cells (Table 8.4). These observations are critical in aquatic but also have some implications in terrestrial. They represent a cost-benefit analysis for predation in form of the distance which has to be traveled before the predator comes into contact with its prey. The growth rate and abundance of bacteria are consistently reduced if the soil or the rhizosphere contains abundant protozoan predators (Jjemba and Alexander, 1999; Jjemba, 2001). Therefore, prey situated far from its potential predators is likely to escape predation as the predators may deem the energy expended in reaching distant prey not worth the energy benefit expected from feeding on them (Habte and Alexander, 1975). Filter-feeding on surfaces is only profitable at a high prey density.

Studies indicate that the larger the prey, the more likely it will be readily caught by the predator. Thus, the biggest and most active prey are the ones mostly consumed though if extremely large (>20μm) they may not be readily ingested (Pussard et al., 1994). Extremely large prey is likely to put up a greater struggle

**Table 8.4** Relationship between prey density and proximity between bacterial cells

| Prey density (cells ml⁻¹) | Distance between cells (μm) |
|:---:|:---:|
| $10^9$ | 11 |
| $10^7$ | 32 |
| $10^5$ | 240 |
| $10^3$ | 1,100 |

before it succumbs to its predator. Each protozoan division requires an ingestion of $10^3$-$10^4$ bacteria. Hence, a high density of predators suggests a rapid generation of prey. The inedibility of some bacteria seems to be related to some extent to pigment production. Bacteria producing red, violet, blue, green or fluorescent pigments were observed to be inedible by two unidentified amoebae species (Singh, 1942, 1945) and to *Leptomyxa reticulata* (Singh, 1948). The surface properties of bacteria might also be important in determining what is edible. Gurijala and Alexander (1990) observed that of several bacterial strains tested, those that survived predation by protozoa mostly had highly hydrophobic cell surfaces.

Most of the predator-prey research has involved protozoa but nematodes, bacteriophage, and some bacteria, notably *Bdellovibrio*, also prey on other microorganisms in the environment. *Bdellovibrio* and bacteriophages have been found in a variety of soils, natural waters, and sewage systems and it might be expected that they would have a significant impact on the organisms they attack. *Bdellovibrio* attaches to the host, penetrates the prey's periplasm from which it grows and, within a matter of hours lyses into other *Bdellovibrio* progeny, rupturing and destroying its host (Keya and Alexander, 1975; Martin, 2002). In a sense, predation of bacteria by *Bdellovibrio* can be regarded as some form of parasitism but the invading microbe does not appear to use any carbohydrates from its host, using instead its host's nucleic acids, proteins, and lipids. Just like protozoa-bacteria prey systems, predation by *Bdellovibrio*-bacteria prey is also density dependent, the predator leaving some bacterial prey cells unattacked. Such a relationship where in the prey is not totally destroyed by the predator is quite expected or else *Bdellovibrio* would also be indirectly destroying itself. However, neither *Bdellovibrio* nor bacteriophages were found to be native during the decline of large populations of *X. campestris* in soil (Habte and Alexander, 1975). Other predator-prey relationships that involve prokaryotic predator and prey, e.g. *Ensifer* spp. preying on *Micrococcus* as well as *Chromatiaceae* preyed on by both *Vamprirococcus* and *Daptobacter* in marine environments have not been studied to the same extent as *Bdellovibrio* but offer exciting opportunities in environmental microbiology (Martin, 2002). In Chapter 14, we draw on predatory prokaryotes to shed some light on some of the intricate mechanisms used by intracellular pathogens to live in eukaryotic cells, causing diseases.

**Coexistence of predator and prey:** A characteristic of predators is not to eliminate its prey, lest it eliminate itself in the process. There are questions as to how the surviving prey is able to do so. Studies in solution culture done by concentrating the bacterial prey survivors and then reintroducing the predator showed that the predator fed on the survivors and reduced their numbers to the original population that prevailed before their concentration (Danso and Alexander, 1975). Thus, the surviving prey concentrated by centrifugation and provided to protozoa were still susceptible to grazing, indicating that they are not intrinsically resistant to predation by protozoa. After re-attaining these low prey population densities, one of these possibilities occurs:

(i) the predator spends more energy catching the prey cells left, thus sustaining a high prey population in the ecosystem;

(ii) the few prey cells left remain physically inaccessible in the soil micropores because the protozoa are too large for some of the much smaller soil pores into which the remaining bacterial prey have access; or

(iii) the protozoan predator may form cysts, a process of hibernation during which metabolism is at a minimal level.

One, or all three of these possibilities may prevail concomitantly and explain how the remaining prey population escape total elimination. The possibility of remaining inaccessible has received more scrutiny and during periods of stress, some of the prey have been demonstrated to take refuge in heterogeneous habitats, remaining inaccessible to the predators (Keya and Alexander, 1975; Heynen et al., 1988; Postma et al., 1990). Postma et al. (1990) enumerated the surviving *Rhizobium leguminosarum* bv. *trifolii* cells in different soil fractions with and without adding the flagellate *Bodo sultans*. Using this approach, both free organisms and organisms associated with particles or aggregates were separated. The percentage of particle-associated bacteria was higher in the presence of the flagellate predator, indicating that the surviving bacteria were present in small pores in the aggregates or associated with the surface of soil particles. De Weger et al. (1995) used *lux*-marked bacteria to show that cells present in the small soil pores were better protected from protozoan predation than those in larger pores.

Only some small amoebae and small flagellates are able to enter smaller pores of 3-6 μm diameter (Hattori, 1994). A study by Vargas and Hattori (1986) showed that *Colpoda* spp. (32 × 17 × 10 μm) were only able to devour *Klebsiella aerogenes* cells in the outer microhabitat (OM) and could not feed on those in the inner microhabitat (IM) as *Colpoda* could not penetrate the smaller pores (3-6 μm) in the IM. The OM in this instance refers to aggregates with larger pores in which microbial migration and interchange between sites and the surrounding environments is easy. In the IM, on the other hand, microbial migration and interchange is more restricted. Predation within the IM can, however, be increased as the soil moisture content tends toward maximum water-retention capacity (Hattori, 1994), possibly due to an increase in size of the water film, which in turn enhances contact between separate aggregates. Field studies have also shown a heavy dependency of predation on soil moisture content (Anderson, 2000).

A high predator population is indicative of *in-situ* growth of prey. Some predators go into dormancy when the prey population becomes low. This is typical of protozoa which, in most cases, form cysts. As a matter of fact, cysts are the most predominant form in which protozoa occur in soil. Ciliates, which are filter feeders, have a high capacity for ingesting bacteria in water (Danso and Alexander, 1975), but in soil their large size restricts their active feeding to periods with a high water content (i.e., high soil moisture content). This probably diminishes their overall importance under actual field conditions.

However, the observation that protozoa do not eliminate their prey is mainly related to grazing of single species populations. Mallory et al. (1983) demonstrated that *Salmonella typhimurium* could be reduced to very low levels in sewage when large numbers of *Enterobacter agglomerans* cells were added. In that instance, it is possible that *E. agglomerans* served as alternative prey and when present in the sewage the protozoa could continue feeding even if the numbers of *S. typhimurium* were low. In mixed culture, therefore, continuous grazing may ultimately result in elimination of bacteria present in low numbers. Ramirez and Alexander (1980) also studied the survival of *Rhizobium* spp. inoculated on legume roots and found that the addition of an alternative prey enhanced reduction in number of test rhizobia.

In situations of stress, some prey can adapt methods of avoiding predation, such as assuming a filamentous or heterogeneous (rather than homogeneous) shape. Alternatively, some predators can adopt alternative prey. When two or more bacteria prey species are present in the environment and at least one of these is present in densities above the threshold value for predation while the other grows at a rate less than the predation rate, it is possible that the latter will be eliminated by predation (Mallory et al., 1983). Sinclair and Alexander (1989) likewise found that bacteria with fast growth rates survived protozoan predation in higher numbers than those with slow rates. Grazing has also been seen to favor bacteria which are

inedible to protozoa because of their shape or size. Gude (1979) found that flagellates grazing on bacteria in activated sludge resulted in increased abundance of "grazing-resistant" spiral-shaped and filamentous bacteria. Bianchi (1989) observed that grazing by flagellates and ciliates allowed growth of starlike and filamentous bacteria in sea water. The aforesaid experiments of Sinclair and Alexander (1989) versus those of both Gude (1979) and Bianchi (1989) indicated that protozoan predation acts as a selection factor whereby bacteria can either avoid elimination by growing very fast, thereby replacing cells lost through predation (r-selection) or they can grow slowly and avoid decimation by attaining large sizes instead (K-selection).

Under marine and freshwater conditions, chemoreception is a powerful mechanism by which predatory protozoa identify possible prey (Seravin and Orlovskaja, 1977). By using chemical food models, these workers demonstrated that chemoreception operates in ciliates, for example *Didinium nasutum*, *Dileptus anser*, *Lacrymaria olor*, *Coleps hirtus* and in amoeba, such as *Peranema trichophorum* and *Amoeba proteus*. Prey mobility may also be important in attracting predators to their food. In water, bacteria-feeding ciliates were reported to respond positively to oscillating particles simulating motile bacteria. However, reports by Nisbet (1984) indicated that this taxis does not normally result in a complete feeding response unless followed by the appropriate chemical stimulation.

Under predator-prey conditions, a Monod function was suggested by Bazin et al. (1990) for predator specific growth rate:

$$\gamma = (\gamma_m X)/(K_2 P + X)$$

where X is prey population density, $\gamma_m$ the max specific growth rate for the predator $K_2$ the saturation constant, and P the predator density. Ingham et al. (1985) investigated predation of fungi and bacteria by nematodes in microcosms, with and without plants present. They found that bacteria, fungi, and nematodes were more abundant in the rhizosphere than in root-free soil, with bacteria density higher in the presence of nematodes, suggesting that nematodes impact fungi more than they impact bacteria in the environment. Bacterial feeding nematodes at forest sites have been calculated to consume just about 2% of the bacterial population (Clarholm and Rosswall, 1980).

**Impact of predation:** Predation accelerates the rate at which nutrients are cycled because it keeps the bacteria in a log phase of growth. The rapidly growing prey cells quickly turnover nutrients, such as N, P, and C. Clarholm (1985a, b) found that net mineralization to produce inorganic N from SOM was strongly enhanced in the presence of protozoa and living roots (soluble sugar condition). Apparently the root exudates being excreted stimulate bacterial growth, which in turn increases the SOM degradation rate and subsequent assimilation of nitrogen into the bacterial biomass. Eventually, protozoan populations increase due to increased bacterial numbers and as the bacteria are consumed, their cell N is remineralized by the protozoa and taken up by the plant roots.

Generally, gram-negative bacteria are more delectable to protozoa than gram-positive, mycobacteria, and actinomycetes. Protozoa also prefer zymogenous bacteria (*Pseudomonas* spp., *Enterobacteriaceae* spp.) over autochthonous bacteria (*Arthrobacter* spp., *Micrococcus* spp., *Corynebacterium* spp., *Bacillus* spp.), although there are exceptions to these generalizations (Pussard et al., 1994).

Slow-growing bacteria are likely to disappear more rapidly in the presence than in the absence of faster growing bacteria that are able to support growth of indigenous protozoa (Mallory et al., 1983). For example, the establishment of a *R. phaseoli* population in soil was shown to be more difficult in the presence of large numbers of faster growing bacteria which could support large protozoan populations (Ramirez and Alexander, 1980). However, recent studies using a mixture containing *Comamonas acidovorans*, *Bradyrhizobium japonicum* UMR161, and *Acanthamoeba castellani* showed that even the slow-growing UMR161 attains population densities in the rhizosphere comparable to those attained if no grazers are present (Fig. 7.4). Under such rich conditions, the rhizosphere provides a rich milieu offering sufficient carbon to replenish the prey density at a rate faster than they are being consumed by the predator.

### 8.3.2 Positive interactions

Positive interactions reflect some form of cooperation between species or populations. They tend to prevail in stable environments, foster coevolution and create new niches ("new" organisms), as exemplified in the legume-*Rhizobium* symbiosis. Under this symbiosis, neither the *Rhizobium* spp. nor the plant can fix nitrogen alone; they need each other to do so. Positive interactions between two or more species or populations take three forms, namely symbiosis commensalism, and protoco-operation.

#### 8.3.2.1 Symbiosis

Symbiosis is that situation wherein two or more populations inhabit the same living space and interact in a mutually beneficial manner. It involves a highly specific bilateral exchange of "materials" between organisms to overcome physiological limitations. It comprises either endo- or ectosymbionts (as already mentioned, the former live inside the host cell, the latter on its surface). Like most biological relationships, the classification is not always clear cut. It often involves two species that differ taxonomically. It is important to observe that in nature, no organism exists in total isolation. Symbiosis enables one of the organisms involved in the relationship to acquire novel metabolic capabilities from its partner (Table 8.5). Usually, the recipient of the novel metabolic capacity is the

**Table 8.5**   Symbiosis as a source of novel metabolic capabilities

| Capability acquired | Donor of capability | Recipient of capability |
|---|---|---|
| Photosynthesis | Chloroplasts | Algae and higher plants |
| | Algae and cyanobacteria | Various protists, vertebrates, and lichens |
| Nutrients, e.g. amino acids and vitamins | Mycorrhizal fungi and various bacteria, e.g. *Buchnera* sp. | Terrestrial plants (for mycorrhiza); insects and some animals |
| Chemosynthesis | Bacteria | Various invertebrates, e.g. deep-ocean invertebrates |
| Respiration | Mitochondria | Eukaryotes |
| Nitrogen fixation | *Rhizobia, Frankia,* and cyanobacteria | Plants |
| Methanogenesis | Methanogens | Anaerobic protists in rumen, swamps, etc. |
| Cellulose degradation | Bacteria, e.g. *Ruminococcus* Hypermastigote protists | Vertebrates (especially herbivores) Lower termites |
| Luminescence | *Vibrio* and *Photobacterium* | Marine cephalopods and teleost fish |

Modified from Douglas (1994) with permission from Oxford University Press, Oxford, United Kingdom.

host. Symbiosis is somewhat similar to commensalism but symbionts are usually not able to survive independently, nor if separated, capable of performing the processes achieved in a symbiotic relationship.

Symbiosis, as shown above, traditionally has been restricted to the coexistence of organisms that are mutually beneficial to each other. That view is increasingly being abandoned because instances of associations nonbeneficial to at least one of the parties have been observed (Douglas, 1994). For example, two or more plants can be interconnected by the same mycorrhizal fungus network. The plant with the highest photosynthetic capacity may end up drawing some photosynthates through the mycorrhiza from the other associated plants to meet its photosynthate requirements. The plant that is losing photosynthates through the mycorrhiza to the highly photosynthetic plant is doing so by virtue of its participation in a symbiotic relationship from which it is not benefiting. Some symbiotic relationships have exerted a dramatic impact on the morphology of the parties involved. Like competition, the establishment and maintenance of a symbiotic relationship is also dependent on the resource consumption ratio theory, that is on the availability

of resources that are commonly utilized by the respective symbionts. For example, the *Hydra-Chlorella* symbiosis competes for nitrogen and dissolved organic carbon (Smith, 1993). Enrichment of the environment with nitrogen results in an increase in the density of the algae within the *Hydra* host cells. A decrease of nitrogen in the presence of light (and thus with increasing availability of dissolved organic C to the *Hydra* due to continued photosynthesis by the algae) will gradually eliminate the algae from the *Hydra*, causing it to become nonsymbiotic-carrying (Fig. 8.12). The *Chlorella* isocline is parallel to DOC because *Chlorella* is autotrophic and the exogenous supply of DOC would not be necessary. The consumption of N ($C_N$) in the presence of light continues to support photosynthesis which in turn increases the DOC but is detrimental to *Chlorella* which is wiped out. The supply of N ($S_N$) favors *Chlorella* which in turn increases photosynthesis. The rate of growth of the symbionts has to be harmonious with that of its host, to facilitate a stable coexistence. If the symbiont grows too rapidly, it can overgrow its host, eventually smothering it. On the other hand, if it grows too slowly, it may be diluted out, failing to keep pace with the rate of growth

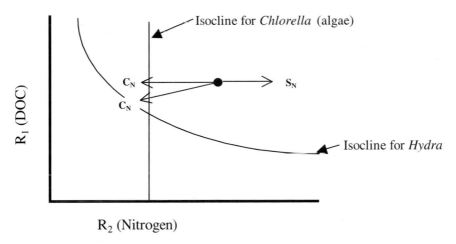

**Fig. 8.12** Hypothetical isoclines for DOC and nitrogen in a symbiotic relationship between *Hydra* and *Chlorella*.

of the hosts. This observation gives possible credibility to the contention that symbiosis evolved from parasitism.

A classic example of symbiosis is also displayed by lichens which are basically a combination of fungus and algae (or cyanobacteria). The fungus provides housing, growth factors, protection from the sun, minerals and resistance to desiccation for the system. The algae or cyanobacteria, on the other hand, fix nitrogen and carbon which are utilized by the associated fungus. Under ideal conditions, both components of the symbiosis can exist individually but their existence is more robust when they associate into a lichen (Croll, 1973). Lichens were discussed in detail in Chapter 4.

Protozoa endosymbionts also display a unique type of symbiosis in anoxic sediments, landfills, and the rumen of various herbivores. Microbial-mediated cellulose and hemicellulose degradation is fairly widespread in herbivores. The rumen contains approximately $10^{10}$-$10^{11}$ bacteria $ml^{-1}$ such as *Ruminococcus albus*, *R. flavefaciens*, and *Fibrobacter succinogenes*, $10^5$-$10^6$ rare ciliates $ml^{-1}$ and some chytrid protists (Ellis et al., 1989). The various microorganisms present in the microphilic portion of the gut degrade both cellulose and hemicellulose into short-chain fatty acids such as acetic acid. The short-chain fatty acids are then utilized by the aerobic tissue of the host animal as an energy source. If the microorganisms were to degrade these polymers aerobically to $CO_2$ and water, the host animal would derive no nourishment from the process.

The host protozoa, *Paramecium*, can associate with the alga *Chlorella*, providing the algae with physical protection, growth factors, and motility. The intracellular *Chlorella*, on the other hand, provides organic C and $O_2$ in anaerobic habitats, enabling the *Paramecium* to survive under these conditions. Other examples include Plagiophyla and Metapus from landfills containing intracellular *Methanobacterium formicicum* and the amoeba *Pelomyxa palustris* containing various methanogens (van Bruggen et al., 1988). With the advent of techniques that detect nonculturable microorganisms such as FISH, intriguing protozoa-prokaryote relationships have been found in unexpected instances. Some examples of these symbiotic relationships include *Paramecium caudatum* with *Holospora*, *P. caudatum* with *Caedibacter caryophila*, *Acanthamoeba* spp. with *Rickettsia*-like bacteria, and *A. castellani* as well as *Tetrahymena pyriformis* with *Legionella pneumophila* (Moter and Göbel, 2000). It is noticeable that most of these prokaryotes and some of the protozoa are important pathogens. Thus, these relationships have significant implications in the survival and control of these pathogens and are discussed in Chapters 11 and 14.

The symbiosis between microbes of the genus *Rhizobium* and leguminous plants has been extensively studied. Nitrogen fixation is widespread in bacteria but absent in eukaryotes. The N-fixation process is of prime importance in agriculture and in the biogeochemical cycling of N. In the legume-rhizobium symbiosis, virtually all the nitrogen fixed by the rhizobia is released to the host plant, with none being directly channeled into the bacterium for its proliferation. The process involves the development of nodules on leguminous roots, powerhouses inside which, through a series of intricate physiological processes and chemical reactions, atmospheric $N_2$ is converted into ammonia and subsequently into amino acids (see Section 9.3.1). The process starts with a chemostatic attraction of the rhizobia to the root hair. The bacterial cell and root hair cell are then attached through plant-produced lectins that are quite specific. The root hairs also excrete an amino acid, tryptophan, which is converted by the rhizobium to a plant-growth hormone, indoleacetic acid (IAA) that changes the root hair morphology, causing it to curl. During the curling process, the root hair cell wall is softened by plant or bacterial coenzymes. The bacterium enters the root cell and from thereon the root cell nucleus directs the growth of an infection thread into the root cortex. The rhizobium cells proliferate and are delivered to tetraploid root cells, subsequently becoming nodules. The rhizobia are released from the infection thread, lose their rod shape and become irregularly formed bacteroids inside which N-fixation occurs. Upon the death of the root or the entire plant, dormant rhizobia that remain in the root tissue are released. These live to begin symbiosis all over again, once host plant root hairs become available.

In the nodule, four-carbon organic acids are delivered to the bacteriod. The ATP generated therefrom is used to supply the nitrogenase enzyme so as to meet the energy requirements for fixing nitrogen. The enzyme is sensitive to oxygen but leghemaglobin regulates $O_2$ in the nodule in such a manner that the ratio of bound to free oxygen is 10,000:1. The heme is from the rhizobium whereas the globin is from the plant, a fact that displays extreme cooperation between the two parties. The net result of all this is an incredible display of evolutionary dialogue between the two symbiotic partners, the plant providing refuge and C to the bacterium while the bacterium fixes and provides nitrogen to its host.

Other symbioses include bacteria such as *Enterobacter agglomerans* and *Citrobacter freundii* in the hindgut of termites, *Teredinibacter* in shipworms, cyanobacteria in lichens and between planktonic diatoms of genera *Rhizosolenie* and *Hemiaulus*. A wide range of insects also symbiotically depend on various bacteria to derive vitamins and essential amino acids (Douglas, 1994). In most cases, the bacterial symbionts in these insects are located in specialized cells called mycetocytes or in the insect gut. They are further discussed in Section 8.4.3. As seen in Table 8.5, chloroplasts and mitochondria are included as symbionts, making all terrestrial-based primary production systems symbiotic. As alluded to in Chapter 1, evidence from 16S rRNA shows that these organelles were once bacteria that lived independently but later coevolved to comprise the present-day eukaryotes. The importance of symbiosis cannot be overemphasized.

### 8.3.2.2 *Protocooperation*

Protocooperation is also referred to as synergism or mutualism. It is a two-way positive exchange of materials, between the two species involved which benefit each other.

Thus both species grow more prolifically in the presence of each other, than if each was growing by itself. However, the two synergists are not essential to one another and each is capable of surviving by itself. Their relationship, therefore, although highly specific, is not obligate.

Protocooperation is displayed in cases where a photosynthetic sulphide oxidizer, such as *Chlorobium*, supplies sulfate and organic C to the sulfate-reducing *Desulfovibrio*. The *Desulfovibrio* in turn generates $CO_2$ and $H_2O$ during its S-reduction process, products required by *Chlorobium* (Fig. 8.13). Nitrogen-fixing heterocysts of cyanobacteria (Anabeana) leak ammonia to the surface-colonizing heterotrophs (Pseudomonads). The heterotrophs also consume $O_2$, thus reducing the ability of the $O_2$ to reach the heterocysts where it would otherwise inhibit the fixation of nitrogen. Synergism is also displayed by complementary enzyme systems. For example, fungi together with bacteria degrade lignin into simpler products which are then readily degraded further by a wider array of bacterial species (Fig. 8.14).

### 8.3.2.3 Commensalism

Commensalism is a low specificity relationship which displays a one-way positive exchange of materials. Thus, under commensalism, one species benefits and the other wholly unaffected. Commensalism is from the Greek *co*, together and *mesa*, table and literally means "together at the table". Thus, the relationship between the two commensals is based on feeding habits, although it can sometimes include other intimacies. In a commensal relationship, the nonbenefiting organism is likely to lose something from the association. It is common to find the benefiting organism existing without its commensal associate. Its associate, on the other hand, almost exclusively exists in the presence of the commensal partners, suggesting that the association is necessary for one but not for the other species. The commensal thrives on body pieces or wastes of another organism, such as organic carbon, macronutrients, growth factors, genetic material (DNA) or inhibitors (Table 8.6). In doing so, the commensal causes no significant harm. Commensalism therefore portrays the loosest and least obligate interspecific interaction as both partners can often survive independent of their association but perform better in their association compared to individuals of the same species not in association.

The supply of organic carbon (OC) and energy to bacteria by phytoplanktons through extracellular organic carbon excretions (EOC)

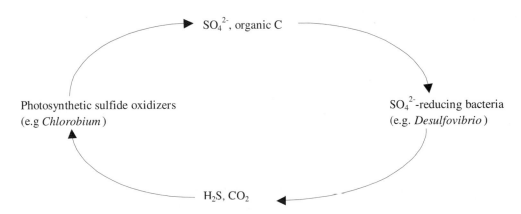

**Fig. 8.13** Synergistic cycling of sulfur by the photosynthetic S-oxidizer, *Chlorobium* spp., and the S-reducing bacteria, *Desulfovibrio* spp.

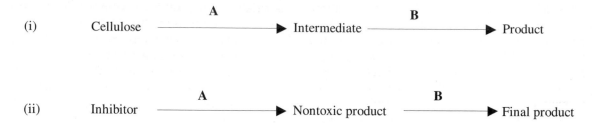

**Fig. 8.14** Complementary enzyme systems of cellulose-degrading bacteria.

**Table 8.6** Examples of different types of materials exchanged between commensal organisms

| Type of exchange | Examples |
|---|---|
| Macronutrients | Photosynthetic algae excrete organic C and provide a habitat to aquatic heterotrophic bacteria |
| | Mineralization of organic N to ammonia (ammonification) by *Nitrosomonas* spp. which in turn support nitrifiers as they obtain a raw material |
| | Methanogens generating $CH_4$ for use by methanotrophs |
| Growth factors | Soil bacterial populations excreting vitamins B, cysteine, siderophores (to chelate Fe), or acid phosphatase to solubilize $PO_4^{2-}$ |
| Genetic material | Natural transfer of plasmids that code for resistance to different antibiotics or for the ability to degrade specific pollutants |
| Inhibitors | Heterotrophs consuming $O_2$ which in turn favors anaerobes at the expense of aerobes |

under low nutrient concentrations has been well studied (Bratback and Thingstad, 1985). Besides C, some algae also provide $O_2$ to epiphytic bacterial commensal partners. The bacteria are expected to increase in density as long as dissolved organic C is available. However, under commensalism, the bacteria cannot outcompete the algae as doing so would reduce their source of C supply.

### 8.3.3 Neutral interactions

Neutralism represents a minimal interaction between the species or populations involved and hence fairly independent ecological niches. Under neutral interaction, the growth of two species in a mixed culture is exactly the same as growth of each individual species alone. Usually, it involves no overlapping niches, low nutrient concentrations, low growth rates, and some physical separation between the interacting populations. An example of a neutral inter-

action is the destruction by soil pathogens of the rooting system of a tree, thereby causing decay. In turn, the decay helps other soil organisms existing in dormant stages some distance from the microenvironment.

## 8.4 TYPICAL MICROBIAL INTERACTIONS

Understanding the nature of microbe-microbe and microbe-host plant or animal interactions as well as the associated ecological relationships is important in controlling microorganisms and their associated deleterious activities.

### 8.4.1 Microbe-microbe interactions

Competition for nutrients occurs between microorganisms but the relevance of interspecific competition can also play a pivotal role in determining the outcome of disease processes (Smith, 1993). During the development of disease, competition occurs not only between microorganisms, but also between microorganisms and the host plant or animal cells. In terrestrial and aquatic environments, sudden changes in environmental conditions, for example moisture or temperature, can easily affect supply vectors and lead to the competitive exclusion of some microorganisms. Because vertebrates have a relatively constant temperature, they provide some environmental stability. Such stability could enable the invading pathogen to thrive, providing a milieu for disease to develop (Jones, 1980). The assumption here is that the establishment and persistence of a pathogen within the host can only be precluded if both the host and the pathogen are limited by the same resource. In cases where the host is a superior competitor, the disease will be suppressed and vice versa.

### 8.4.2  Microbe-plant interactions

Microbial activity can be strongly influenced by the host plant and its resource supply. Given its importance, the plant-mycorrhizal interaction is discussed in some detail. In this interaction, the fungus obtains photosynthates (C-source), vitamins, and a safe haven that enable it to escape competition. The plant, on the other hand, vastly increases surface contact with the soil for nutrient and water uptake, defense against other fungal pathogens and, where applicable, resistance to low soil pH and other stresses. Mycorrhizae are of four types, viz., ectomycorrhiza, vesicular-arbuscular mycorrhiza (VAM), ericoid, and orchid. The last three are sometimes lumped together as endomycorrhizae. Ectomycorrhizae comprise Basidiomycotina and Ascomycotina. There are more than 2,100 ectomycorrhizae species, many of which have been successfully grown in pure culture. They thrive on gymnosperms (conifers) and angiosperms (flowering plants) but not on monocotyledonous plants. They are particularly abundant in temperate and boreal forests, especially those dominated by beech (Fagace-

ae) and pine (Pinaceae). Plants infected by ectomycorrhizae often have very few or no root hairs and the roots tend to be short and thick. The ectomycorrhizae are external root sheaths and have a limited ability to penetrate intercellularly. If they penetrate the root, they form a complex branching called a Hartig net which mediates the transfer of nutrients between the fungus and the plant.

Vascular-arbuscular mycorrhizae (VAM) are Zygomycotina. They thrive on a wider range of gymnosperms and angiosperms than all the other types of mycorrhizae, including soybean, corn, wheat, herbs, coffee, rubber, conifers, and shrubs. At least 75% of all vascular plant species bear VAM (Trappe, 1987). VAM are especially abundant in tropical forests, savannah and temperate grasslands. VAM lack regular septa, are obligate, and very few have been grown in pure culture. Morphologically, they form a loose network on the root surface that penetrates intracellularly, forming either tree-like brown spheres (arbuscules) or terminal spheres (vesicles). The arbuscules act as the principal sites of nutrient exchange. Many VAM produce large chlamydospores in soil which enable them to survive adverse environmental conditions.

Ericoid mycorrhizal fungi belong to Ascomycotina and infect only Ericales, such as *Azalea* sp., *Rhododendron* sp., and *Calluna* sp.—plants particularly associated with acidic and low nutrient environments. Some ericoid mycorrhizae such as *Hymenoscyphus ericae* have been isolated in culture. In ericoid infections, the hyphae particularly infect lateral roots around which they form a dense mycelium (Read, 1983). Ericoid mycorrhizae occasionally penetrate cells but no vesicles or arbuscules are formed. They are able to use chitin as the sole source of nitrogen (Leake and Read, 1990), a trait that could give them a competitive advantage over other types of fungi under low nutrient environments.

Orchid mycorrhizae belong to the Basidiomycotina and their infection is restricted to the family Orchidaceae. Several genera, including *Armillaria* and *Ceratobasidium*, have been identified. The orchid mycorrhiza is absolutely required for germination of orchid seeds and development of the seedling. This is due to the fact that the orchid plant derives carbon from the fungus, unlike other mycorrhizae in which the nutrient flux is bidirectional, with the photosynthates flowing from the plant to the fungus and the nutrients from the fungus to the plant. If the developing orchid seedling remains nonphotosynthetic, it continues to draw its carbon from the fungus throughout its entire life.

Under the plant-mycorrhizae interaction, there seems to be a long varied dialogue between the plants and fungi. Some root systems of single plants may support several fungal endosymbionts. The role of mycorrhizae diminishes if the nutrients are abundant (Table 8.7). Hyphae of mycorrhizae can also translocate other nutrients, including heavy metals (Gadd, 1986). This has implications for the growth of plants in polluted environments since toxicity from these metals to the host plants can result. If a specific plant species can withstand the metals, however, it could be used to remediate sites contaminated by the heavy metals. This approach is already in use for cleaning up sites contaminated by organic pollutants (Aitchinson et al., 2000)

### 8.4.3  Microbe-animal interactions

A wide range of ruminants, including deer, goat, sheep, cattle, antelope, caribou, giraffe, and moose eat vegetation that has high levels of cellulose and hemicellulose. However, these animals lack the necessary enzymes to break down the complex polysaccharides and are therefore unable to directly digest what they consume. Their digestive system comprises a rumen subdivided into a reticulum, omasum, and abomasum, and the small intestine. Food travels from the esophagus to the rumen and is then regurgitated, rechewed, and swallowed, traveling to the reticulum, omasum, abomasum and intestines in that order. Rechewing the partially digested food (clumps and cuds) formed in the rumen increases the surface area of the food already exposed to microorganisms and enzymes in the rumen. The addition of saliva during mastication also helps to neutralize the cuds, which are then passed on to the omasum and abomasum. In the abomasum, the routine digestive stomach processes begin, the fully digested food being readily absorbed in the intestine.

The rumen basically operates like an anaerobic fermentor. It harbors unique protozoa, mainly ciliates and chytrid protists, which break down the cellulose and hemicellulose into simpler compounds that the host ruminant can further degrade or assimilate directly. The ciliates and chytrids can only survive in an anaerobic environment provided by the rumen and reticulum. They have $\beta$-1,4-endoglucanase, a cellulase which attacks cellulose and hemicellulose. The chytrid protists colonize the fibrous plant fragments in the rumen, and weaken the structure of the plant material, providing a new surface for colonization by cellulytic bacteria. The ciliates attack the larger and structurally more complex

**Table 8.7**  Effect of mycorrhizae on the growth of *Festuca ovina*[a]

| Nutrient status | Relative growth rate mg g$^{-1}$ d$^{-1}$ | | Level of significance |
|---|---|---|---|
| No additional nutrients | Without mycorrhiza | With mycorrhiza | |
| No additional nutrients | 0.76 | 1.24 | $p < 0.01$ |
| 7 ppm N + 4 ppm P | 1.16 | 2.27 | $p < 0.01$ |
| 14 ppm N + 8 ppm P | 2.56 | 2.73 | $p > 0.05$ |
| 28 ppm N + 16 ppm P | 2.87 | 2.94 | $p > 0.05$ |

[a]After mycorrhizal infection in a sterile rendzina soil, the seedlings were transplanted into a 1:1 mixture of horticultural moss-peat and acid-washed sand.
*Source*: Koucheki and Read (1976) with permission from Blackwell Publishers, Oxford, United Kingdom.

cellulytic material by either attaching their pseudopodia on the material (as in the case of *Epidinium* sp.) and releasing cellulases or engulfing the intact plant fibers (as in *Polyplastron* sp.). The microbial population in the rumen anaerobically breaks down these complex polysaccharides to acetic, butyric, and propionic acids. Some $H_2$ and $CO_2$ are also generated in the abdominal chambers. The organic acids are absorbed through the rumen wall into the host's blood stream. The food chain within the rumen therefore runs from cellulose to bacteria to protozoa to the ruminant animal. When the microbial symbionts enter the omasum and beyond, they are digested and this supplies the host animal with vitamins and amino acids. The uptake of organic acids also eliminates toxins and shifts the equilibria toward organic acid formation. Despite the sink for organic acids, some are converted to methane by methanogens in the rumen. Within the rumen, methane is also formed from $H_2$ and $CO_2$. Almost 30% of the total global methane emissions to the atmosphere are derived from ruminants, a fact that has implications on the global biogeochemical cycling of C (see Chapter 9). The net result is that every ruminant has an entire ecosystem in its gut, with a blatant coevolution. The ruminant gains C, N, and vitamins whereas the endosymbionts gain a habitat, stability, and food, displaying a perfect symbiosis.

Cellulose-degrading microorganisms also exist in phytophages (insects that feed on plant materials) such as grasshoppers and locusts, and in xylophagous (wood-eating insects) such as cockroaches and termites. The adult *Trichonympha campanulla* digests the wood, making the products of degradation available to the termite. In *T. campanulla*, degradation is possibly aided by symbiotic bacteria. Termites and roaches are incapable of synthesizing the cellulases required to break down cellulose. The flagellates are able to digest such celluloses, excreting metabolites that provide nourishment for the host. Some xylophagous

termites die if their symbiotic protists are eliminated (Yamin and Trager, 1979) indicating an obligate existence. Not many studies have been done to elucidate the mechanisms involved in these symbioses but the biochemistry of the process is probably generally similar to that in ruminants.

Many organisms generate light by oxidation of a substrate (luciferin) with molecular oxygen, in a reaction catalyzed by the enzyme luciferase. Generation of light (luminescence) is widespread among protists, fungi, and invertebrates, but among bacteria is known to be expressed among members of genera *Alteromonas*, *Photobacterium*, *Vibrio* and *Xenorhabdus*. A symbiotic dialogue goes on between a bioluminiscent bacterium (*Vibrio fisheri*) and the squid (*Euprymna scolopes*). Newborne squids lack the endosymbiont. The light organism is only infected by certain *V. fisheri* strains but the mechanisms of specificity are still not known. Light organs include the cavities for bacteria, reflector muscle tissue, and lenslike sacs. The high surface area of the cavity fosters regulation of the nutrients exchanged. The cells grow to $10^8 ml^{-1}$ in pure culture at the expense of the host's nutrients. If not infected, the light organs develop differently and in that case *V. fisheri* loses polar flagella. The physiological functions are coordinated in the population. The net outcome of this relationship is that the squid gets protection as it is able to avoid predators by controlling the intensity of ventrally projected light. In this way, the squid matches moonlight to prevent shadows. On the other hand, the bacterium gets food and housing. The *lux* gene, central in this process, is currently being used as a molecular marker in a variety of biotechnological processes (see Chapter 15). A whole range of other instances of microbe-animal interactions that are of economic importance in environmental microbiology because of their parasitic nature, is listed in Table 8.8. Those of medical consequence are discussed in Chapter 14.

**Table 8.8**　Other examples of microbe-animal interactions

| Microbe | Animal host | Remarks |
| --- | --- | --- |
| *Trypanosoma equiperdum* | Cattle, Horses | Sexually transmitted parasite |
| *Trichomonas vaginalis* | Cattle | Protozoon that causes stillbirth in cattle. Sexually transmitted. |
| *Babesia bigemina* | Tick (*Biphilus* sp.) | Protozoon that causes "Red water fever" or hemoglobineric fever in cattle following bites by infected ticks. |
| *Herpetomonas* sp. | Insects | Parasitic flagellate in the gut of flies and other insects. It is thought to be an evolutionary intermediate between *Phytomonas* (which lives in some plants, e.g. coconuts) and *Trypanosoma* sp. |
| *Iodamoeba butschlii* | Swine | Transmits trichinosis |
| *Eimaria* spp. | Chickens | Protozoa that cause coccidiosis in poultry |

# References

Aitchinson E.W., S.L. Kelley, P.J.J. Alvarez, and J.L. Schoor (2000). Phytoremediation of 1,4-dioxine by hybrid polar trees. *Water Environ. Res.* **72:** 313-321.

Anderson O.R. (2000). Abundance of terrestrial gymnoamoebae at a northeastern US site: A four-year study, including the El Niño winter of 1997-1998. *J. Eukary. Microbiol.* **47:** 148-155.

Bazin M.J., P. Markham, E.M. Scott, and J.M. Lynch (1990). Population dynamics and microbial interactions. *In*: J.M. Lynch (ed.). The Rhizosphere. John Wiley and Sons, New York, NY, pp 99-127.

Bianchi M. (1989). Unusual bloom of star-like prothecate bacteria and filaments as a consequence of grazing pressure. *Microb. Ecol.* **17:** 137-141.

Bratbak G. and T.F. Thingstad (1985). Phytoplankton-bacteria interactions: an apparent paradox? Analysis of a model system with both competition and commensalism. *Mar. Ecol. Progr. Series* **25:** 23-30

Clarholm M. (1985a). Possible roles for roots, bacteria, protozoa and fungi in supplying nitrogen to plants. *In* A.F. Fitter, (Ed.). Ecological Interactions in Soil. Blackwell, Oxford, UK pp. 355-365.

Clarholm M. (1985b). Interactions of bacteria, protozoa and plants leading to mineralization of soil nitrogen. *Soil Biol. Biochem.* **17:** 181-187.

Clarholm M. and T. Rosswall (1980). Biomass and turnover of bacteria in a forest soil and tundra peat. *Soil Biol. Biochem.* **12:** 49-51.

Croll N.A. (1966). Ecology of Parasites. Harvard Univ. Press. Cambridge, MA.

Croll N.A. (1973). Parasitism and Other Associations. Pitman Publ. Corp., New York, NY.

Danso S.K.A. and M. Alexander (1975). Regulation of predation by prey density: the protozoa-rhizobium relationship. *Appl. Microbiol.* **29:** 515-521.

de Weger L.A., A.J. van der Bij, L.C. Dekker, M. Simons., C.A. Wijffelman, and B.J.J. Lugtenberg (1995). Colonization of the rhizosphere of crop plants by plant-beneficial pseudomonads. *FEMS Microbiol. Ecol.* **17:**221-228.

Douglas A.E. (1994). Symbiotic Interactions. Oxford Univ. Press, New York, NY.

Ellis J.E., D. Lloyd, and A.G. Williams (1989). Protozoal contribution to ruminal oxygen utilization. *In*: D. Lloyd, G. Coombs and T.A. Paget (eds.). Biochemistry and Molecular Biology of "Anaerobic" Protozoa. Harwood Acad. Publ., New York, NY, pp. 32-41.

Gadd G. M. (1986). Fungal responses towards heavy metals. *In*: R.A. Herbert and G.A. Codd, (eds.). Microbes in Extreme Environments. Acad. Press, New York, NY, pp. 83-110.

Gause G.F. (1934). The Struggle for Existence. Williams and Wilkins Co., Baltimore, MD.

Grover J.P. (1997). Resource Competition. Chapman and Hall, New York, NY.

Gude H. (1979). Grazing by protozoa as selection factor for activated sludge bacteria. *Microb. Ecol.* **5:** 225-237.

Gurijala K.R. and M. Alexander (1990). Effect of growth rate and hydrophobicity on bacteria surviving protozoan grazing. *Appl. Environ. Microbiol.* **56:** 1631-1635.

Habte M. and M. Alexander (1975). Protozoa as agents responsible for the decline of Xanthomanas campestris in soil. Appl. Microbiol. **29:**159-164.

Hattori T. (1994). Soil micronutrients. *In*: J.F. Darbyshire, (Ed.). *Soil Protozoa*. CAB International, Wallingford, UK, pp. 43-64.

Heynen C.E., J.D. Vanelsas, P.J. Kuikman, and J.A. van Veen (1988). Dynamics of *Rhizobium leguminosarum* biovar *trifolii* introduced into soil: The effect of bentonite clay on predation by protozoa. *Soil Biol. Biochem.* **20:** 483-488.

Ingham R.E., J.A. Trofymow, E.R. Ingham, and D.C. Coleman (1985). Interaction of bacteria, fungi, and their nematode grazers effects on nutrient cycling and plant growth. *Ecol. Monog.* **55:** 119-140.

Jjemba P.K. (2001). The interaction of protozoa with their potential prey bacteria in the rhizosphere. *J. Eukary. Microbiol.* **48:** 320-324.

Jjemba P.K. and M. Alexander (1999). Possible determinants of rhizosphere competence of bacteria. *Soil Biol. Biochem.* **31:** 623-632.

Jones G.W. (1980). Some aspects of the interaction of microbes with the human body. *In*: D.C. Ellwood, N.J. Latham, J.N. Hedger, J.M. Lynch, and J.H. Slater (eds.). Contemporary Microbial Ecology. Acad. Press, New York, NY, pp. 253-282.

Keddy P.A. (1989). Competition. Chapman and Hall, New York, NY.

Keya S.O. and Alexander M. (1975). Regulation of parasitism by host density: the *Bdellovibrio-Rhizobium* interrelationship. *Soil Biol. Biochem.* **7:** 231-237.

Koucheki H.K. and D.J. Read (1976). Vesicular-arbuscular mycorrhizas in natural vegetation systems, III. The relationship between infection and growth in *Festuca ovina* L. *New Phytol.* **77:** 655-666.

Leake J.R. and D.J. Read (1990). Chitin as a nitrogen source for mycorrhizal fungi. *Mycolog. Res.* **94:** 993-995.

Mallory L.M., C.S. Yuk, L.N. Liang, and M. Alexander (1983). Alternative prey: A mechanism for elimination of bacterial species by protozoa. *Appl. Environ. Microbiol.* **46:** 1073-1079.

Martin O.M. (2002). Predatory prokaryotes: an emerging research opportunity. *J. Molec. Microbiol. Biotech.* **4:** 467-477.

Monod J. (1950). La technique de culture continue; théorie et applications. *Ann. Inst. Pasteur* **79:** 390-410.

Moter A. and U.B. Göbel (2000). Fluorescence *in situ* hybridization (FISH) for direct visualization of microorganisms. *J. Microbiol. Methods* **41:** 85-112.

Nisbet B. (1984). Nutrition and Feeding Strategies in Protozoa. Croom Helm Ltd., London, UK.

Odum E.P. (1983). Basic Ecology. Saunders College Publ., Philadelphia, PA.

Odum E.P. (1989). Ecology and Our Endangered Life-Support Systems. Sinauer Assoc. Publ., Sunderland, MA.

Postma J., C.H. Hok-a-hin, and J.A. Van Veen (1990). Role of microniches in protecting introduced *Rhizobium leguminosarum* biovar *trifolii* against competition and predation in soil. *Appl. Environ. Microbiol.* **56:** 495-502.

Pussard M., C. Alabouvette, and P. Levrat (1994). Protozoan interactions with the soil microflora and possibilities for biocontrol of plant pathogens. *In*: J.F. Darbyshire, (ed.).

Soil Protozoa. CAB International, Wallingford, UK, pp. 123-146.

Ramirez C. and M. Alexander (1980). Evidence suggesting protozoan predation on *Rhizobium* associated with germinating seeds and in the rhizosphere of beans (*Phaseolus vulgaris* L.). *Appl. Environ. Microbiol.* **40:** 492-499.

Read D.J. (1983). The biology of mycorrhiza in the Ericales. *Can. J. Bot.* **61:** 985-1004.

Seravin L.N. and E.E. Orlovskaja (1977). Feeding behavior of unicellular animals, 1. The main role of chemoreception in the food choice of carnivorous protozoa. *Acta Protozo.* **16:** 309-332.

Sinclair J.L. and M. Alexander (1989). Effect of protozoan predation on relative abundance of fast- and slow-growing bacteria. *Can. J. Microbiol.* **35:** 578-582.

Singh B.N. (1942). Toxic effects of certain bacterial metabolic products on soil protozoa. *Nature* **149:** 168.

Singh B.N. (1945). The selection of bacterial food by soil amoebae, and the toxic effects of bacterial pigments and other products on soil protozoa. *Brit. J. Experim. Pathol.* **26:** 316-325.

Singh B.N. (1948). Studies on giant amoeboid organisms, 1. The distribution of *Leptomyxa reticulata* Gooday in soils of Great Britain and the effect of bacterial food on growth and cyst formation. *J. Gen. Microbiol.* **2:** 8-14.

Smith V.H. (1993). Implications of resource-ratio theory for microbial ecology. *Adv. Microb. Ecol.* **13:** 1-37.

Taylor P.A. and P.J. Williams (1975). Theoretical studies on the coexistence of competing species under continuous-flow conditions. *Can. J. Microbiol.* **21:** 90-98.

Tilman D. (1982). Resource Competition and Community Structure. Princeton Univ. Press, Princeton, NJ.

Trappe J.M. (1987). Phytogenetic and ecological aspects of mycotrophy in the angiosperms from an evolutionary standpoint. *In*: C.R. Safir, (ed.). Ecophysiology of VA-Mycorrhizal Plants. CRC Press, Boca Raton, FL, pp. 1-25.

Van Bruggen J.J.A., G.L.M. van Rens, E.J.M. Geertman, K.B. Zwart, C.K. Stumm and G.D. Vogels (1988). Isolation of a methanogenic endosymbiont of the saprophytic amoeba *Pelomyxa palustris* Greef. *J. Protozool.* **35:** 20-23.

Vargas R. and T. Hattori (1986). Protozoa predation of bacterial cells in soil aggregates. *FEMS Microbiol. Ecol.* **38:** 233-242.

Yamin M.A. and W. Trager (1979). Cellulolytic activity of an axenically cultivated termite flagellate, *Trichomitopsis termopsidis*. *J. Gen. Microbiol.* **113:** 417-420.

# 9

# Microorganisms in Biogeochemical Cycling

Biogeochemistry is the study of the exchange of materials between the living and nonliving components of the biosphere. The biogeochemical cycling of nutrients involves the physical transportation of the nutrient in question, or its chemical and biochemical transformation. In most instances, these transformations of nutrients are cyclic in nature. Thus, biogeochemical cycling describes the regional and global cycling of life elements between the living organisms and the respective reservoirs—including the whole or part of the atmosphere, sediments, and oceans. Physical transportation is facilitated by the hydrologic cycle, including evaporation and precipitation, the flow of streams, groundwater, and ice as well as the effects of wind, ocean currents, and geologic movements, especially at the boundaries of tectonic plates. Of relevance to environmental microbiology are the activities and movement of organisms which feed on other organisms, subsequently cycling the nutrients.

The cycling of nutrients is determined by the nature and distribution of the sources and sinks, specifically oceans, soil, and the atmosphere. Most recognized in the cycling of nutrients in these systems is photosynthesis by prokaryotes, algae, and especially plants. There are fundamental differences in the cycling of nutrients between aquatic and terrestrial systems. Whereas in aquatic systems, the distribution of water is continuous, its distribution in terrestrial ecosystems is discontinuous both in time and space. Thus, under terrestrial ecosystems, life cycles of organisms that facilitate the chemical cycles tend to have intermittent periods of activity and inactivity. For bacteria these periods involve the formation of spores and other survival structures whereas fungi tend to form cysts and excyst once conditions become more favorable. In this chapter, we focus on the role of microbes in cycling of nutrients.

Microorganisms have a higher intrinsic rate of growth despite a lower metabolic rate compared to macroorganisms. This, coupled with a shorter generation time, enables microorganisms to turn over nutrients rapidly. In comparison, macroorganisms require a greater proportion of their assimilated energy for maintenance, a requirement that makes them less efficient in conserving the nutrients they cycle. The distribution of microorganisms and their ability to recycle inorganic elements exerts a tremendous effect on the local constitution of various environments (i.e., aquatic systems, rocks, soils, and sediments). Thus, managing the activity of microbes plays a significant role in the recycling of elements that are vital to life, sustaining the biosphere and its diversity.

Biogeochemical cycles fall into two groups, (i) gaseous types with a large reservoir and (ii) sedimentary types with a reservoir in the soils and sediments of the Earth's crust. The chemistry of the atmosphere and the Earth's crust are therefore mainly a result of the Earth's biosphere and the microbial processes that take place. Most of the mineral cycles involve both physicochemical (chemical and photochemical) and biological reactions, releasing gases into the atmosphere and/or dissolved elements deposited in soil, water or sediments. Hence biogeochemical cycling involves the study of the transport and transformation of substances in the natural environment, at a global scale. Transformation can be microbial in origin, making microorganisms an important component of biogeochemical transformations and cycling. It is difficult to explicitly discuss the role of microorganisms in the cycling of nutrients, however, without highlighting the importance or detrimental effects of excess concentrations of some of these nutrients and their effect on the global climate.

## 9.1 SIGNIFICANCE OF BIOGEOCHEMICAL CYCLING OF NUTRIENT ELEMENTS

Biogeochemical cycles are important in understanding such ongoing activities such as global warming, depletion of the ozone layer, and climatic change. These topics are discussed in detail in Chapter 11. In soil and aquatic environments, three subcompartments of material exist, i.e., living organic material, dead organic material, and inorganic material, each of which can easily be further subdivided for convenience to address the roles played by plants, animals, and microorganisms. Common to all biogeochemical cycles are the sources of nutrients, reservoirs, sinks, fluxes, and turnover times. A **reservoir** is an amount of material that is defined by certain biological, chemical or physical characteristics which are considered reasonably homogeneous, e.g., $CO_2$ in the atmosphere, S in sedimentary rocks, etc. **Flux** is the amount of material transferred from one reservoir to another per unit time. It generally denotes an element of mass per unit time. **Source** and **sink** refer to the flux of materials out of and into a reservoir respectively, whereas turnover time is the ratio of the reservoir to the sum of its sink or sources. **Turnover time** reflects the duration it will take to empty the reservoir in the absence of sources if the sink remains constant. These parameters can be used as the basis for quantitative modeling and identifying gaps in our knowledge about the cycling of nutrients. For simplicity, we shall focus on one element or class of compounds at a time. However, it is important to keep in mind that the respective elements or compounds do not exist and cycle individually but rather always interact and overlap with other geochemical cycles as the processes involve one element or set of compounds. Furthermore, the Earth's natural environment can be divided into the atmosphere, hydrosphere and pedosphere (soil), all three of which are involved in the cycling of nutrients. We shall primarily concern ourselves with the cycling of major nutrients (macronutrients), in particular C, O, N, P, S, and one important micronutrient (Fe).

## 9.2 CARBON CYCLE

Carbon is very central to nutrient cycling because of its relation to energy and also it comprises a substantial bulk of protoplasm. Furthermore, its decomposition affects the fate of all the other major nutrients such as nitrogen, phosphorus, and sulfur. In fact, without carbon, there would be no N, P, or S cycles. The carbon cycle determines the amount of $CO_2$ in the atmosphere as well as the rate of turnover of organic matter in soils and sediments. Carbon cycling is fairly complex since it includes all life on earth as well as inorganic C reservoirs and the links between them. The basic question then becomes what the carbon pools are, where they are, and how they change. Microorganisms play an important role in regulating the pools.

Carbon atoms have oxidation states ranging from +4 to −4, the most common state being +4 in $CO_2$ and carbonates whereas methane is the most reduced form of carbon (Fig. 9.1). Thus, the cycle has an 8-electron difference. Carbonates exist either in the oceans as dissolved carbon ($H_2CO_{3(aq)}$, $HCO_{3\ (aq)}^-$, $CO_3^{2-}{}_{(aq)}$) or in the lithosphere as carbonate minerals such as calcium carbonate ($CaCO_3$) and dolomite ($CaMg(CO)_3$) and serve as an important reservoir of this element. The reservoir is occasionally tapped through abiotic processes to introduce $CO_2$ in the environment.

The initial part of the cycle involves the oxidation or reduction of detritus that initially has an oxidation/reduction potential of zero (Fig. 9.1) and contains simple sugars as well as complex polymers such as cellulose, lignin, waxes, and phenols. The large amounts of C found in terrestrial detritus can be rapidly exchanged into the atmosphere, predominantly as $CO_2$. However, a substantial amount of detrital residue that is complex and resistant to microbial degradation builds up in soil. This material is collectively referred to as humin and can be characterized as the alkali-insoluble fraction. It may consist of both humic and fulvic acids that are bound to minerals, the binding rendering them insoluble during alkali extraction (Fig. 9.2). Humic substances consist of aromatic rings originating from lignin, phenols, and quinones that are synthesized by microorganisms and later polymerized with nitrogenous compounds. These substances are recovered by alkali extraction and comprise three fractions, i.e., humic acids, fulvic acids, and humin.

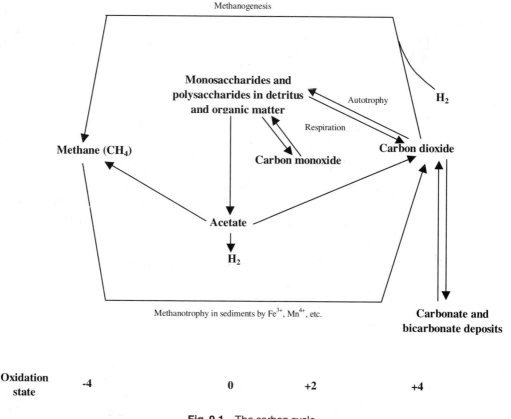

**Fig. 9.1**   The carbon cycle.

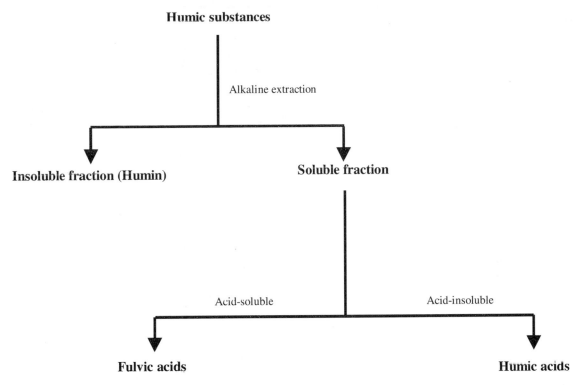

**Fig. 9.2** Schematic representation of the composition of humin.

Humic acids are the acid-insoluble fraction of humic substances whereas the acid-soluble fraction comprises the fulvic acids. The former have a very long turnover time in the environment (>1,000 years). This turnover duration can be even longer under anaerobic conditions such as swamps, sediments, and other water-logged environments. Over geological time, some of this material can be abiotically transformed into coal, petroleum, and natural gas whose degradation is discussed in the next chapter.

Primary production through photosynthesis and significant processes, notably respiration, and the decay of detritus affect the levels of $CO_2$ in the atmosphere through photosynthesis by plants, algae, and cyanobacteria, maintaining the C flux from the atmosphere to the biota as expressed by the simplified equation.

$$CO_2 + H_2O \rightarrow (CH_2O)_n + O_2 \qquad ...(1)$$

Both decomposition and respiration are almost balanced with photosynthesis. The terrestrial biomass is estimated at 560 Pg C, 90% of which is possibly contained in forest vegetation and associated litter fall. For the forest-based C, almost half is in wood that is basically composed of recalcitrant fractions such as lignin, waxes, and phenols. Release of these C fractions by microbial activity can take years, depending on their complexity and ambient temperatures (Fig. 9.3). The most abundant polymer, cellulose, consists of glucose with $\beta$-(1-4) linkages which are unbranched. Cellulose is often intimately associated with lignin, making the cellulose even less accessible by cellulases. The $\beta$-(1-4) linkage distinguishes cellulose from the relatively easily mineralized polymers such as starch, pectin, and laminarin

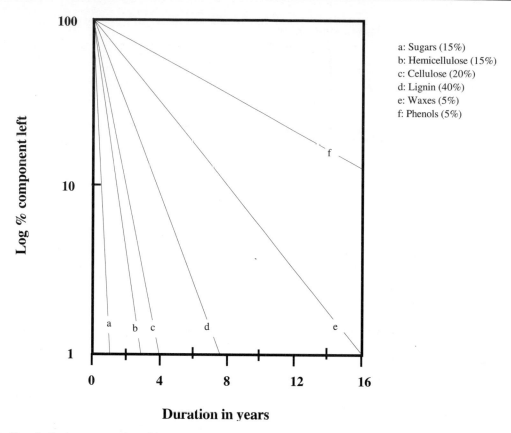

a: Sugars (15%)
b: Hemicellulose (15%)
c: Cellulose (20%)
d: Lignin (40%)
e: Waxes (5%)
f: Phenols (5%)

**Fig. 9.3**  Hypothetical representation of the breakdown of several organic polymers in soil. The percentages in brackets are the typical amounts of the component in the litter (Source: Stout et al., 1976 with permission from Blackwell Publishers, Oxford, United Kingdom).

which are nonrigid and without a $\beta$-1,4 linkage. This type of linkage also gives cellulose a higher molecular weight ($10^6$) with 1,200 monomers.

The transformation of cellulose is schematically depicted in Figure 9.4. Its degradation is an extracellular process that typically requires at least three types of enzymes, i.e., endo-1,4,-glucanase, exo-1,4-glucanase (cellobiohydrase), and cellubiose ($\beta$-glucosidase). The cellobiohydrase modifies "native" cellulose, although it is not clear what the modification is but it is still a long chain that is subsequently broken down into short glucose chains by endo-$\beta$-1,4-gluconase. Endo-$\beta$-1,4-glucanases

do not act on dimers and trimers but act on other polymers. The short glucose chains (cellubioses) are eventually broken down by $\beta$-glucosidase into glucose units. Some organisms can transform the modified intermediate straight into glucose units. Sometimes the organisms make proteins called isozymes (or isoenzymes) which perform exactly the same degradation by snipping off short dimers from the cellulose chain at one time, thus the exoenzyme (Fig. 9.4). Since there are only two ends at any instant per chain of 10,000 or more C molecules, this reaction is quite slow. If the isozymes attack the interior carbon and not the ones at the end, they are called endozymes.

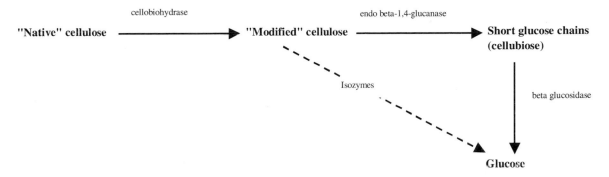

**Fig. 9.4** Schematic representation of the degradation of cellulose by microorganisms.

Cellulose degradation has been well studied in *Clostridium thermocellum*, an important cellulose degrader in the environment (Felix and Ljungdahl, 1993). *Clostridium* can grow at even high temperatures (i.e., above 45°C). Of interest also are *Cytophaga* and *Sporocytophaga,* which can degrade cellulose at high pH. An *Alteromonas* sp. (Strain 2-40) isolated by Whitehead et al. (2001) can degrade cellulose and several other insoluble complex polysaccharides such as chitin, pectin, xylan, alginate, and glucan. Typically, cellulose does not occur alone but together with hemicellulose and lignin, two polymers that influence its decomposition.

*Clostridium thermocellum* degrades cellulose by synthesizing substrates (cellusomes) that arrange degradative enzymes in a surface array together with binding proteins (Béguin and Aubert, 1994). The synthesis of cellusomes is highly regulated but their formation provides for direct contact with the substrates, subsequently solubilizing it.

Hemicellulose contains polysacchrides with two types of glycans, i.e., homoglycans and heteroglycans (with 2-5 sugars), both of which have pentose sugars, hexoses, and uronic acids. Its decomposition is relatively rapid but Ca, Fe, and Al slow it down, possibly by forming complexes. Many species of fungi and bacteria decompose hemicellulose, characteristically giving products similar to those of cellulose.

Lignin, a high molecular weight complex insoluble aromatic polymer present in all vascular plants (both terrestrial and aquatic angiosperms), has a major impact on the cycling of C in the environment. Lignin protects and strengthens vascular plants. It does not have a regular structure with an essential unit formed from condensed phenylpropane units. This irregularity makes it difficult for microorganisms to evolve hydrolytic enzymes to target its degradation. Its persistence in the environment is also attributed to its insolubility, toxicity, and its large molecular size (Colwell and Sayler, 1978). It is extremely resistant to biodegradation, its degradation being largely attributed to fungi (basidiomycetes and ascomycetes) and actinomycetes (Table 9.1) using nonspecific enzymes (ligninase, laccase, and peroxidaes). However, its degradation is enhanced in fungal-bacterial consortia compared to fungi alone (Fenchel et al., 1998). Fungi convert the lignin polymers to simple sugars, carboxylic acids, amino acids, purines, pyrimidines, etc., which bacterial cells can readily absorb. Most widely studied are two lignolytic organisms, i.e., the actinomycete bacterium *Streptomyces viridosporus* and the white-rot fungus *Phaenerochaete chrysosporium*. In *P. chrysosporium*, three enzymes—ligninase (lignin peroxidases), laccase acid, and manganese peroxidases—are important in the degradation of the lignin molecule.

**Table 9.1** Organisms that modify and/or degrade lignin

| Organism | Degradation | |
| --- | --- | --- |
| | Partial | Complete |
| White rot fungi, e.g. *Pleuritus, Phanerochaeta, Coriolus* | + | + |
| Brown rot fungi, e.g. *Lentinus, Lenzites,Poria* | + | − |
| Soft rot fungi, e.g. *Preussia, Thielavia* | + | ND' |
| Actinomycetes (Norcardia, Streptomycetes) | + | ND |
| Others, e.g. *Pseudomonas, Bacillus* | + | ND |

'ND: Not determined

They can be induced by nitrogen starvation due to increased production of $H_2O_2$ under low nitrogen (Reid, 1995; Moore, 1998). The degradation of this polymer greatly depends on the presence of oxygen. Thus, under anaerobic conditions such as soils with excessive moisture or sediments, degradation of lignin is quite limited and slow (Benner et al., 1984). Other equally complex and recalcitrant organic compounds in plants include phenols and waxes (Fig. 9.3) although these are not very abundant and hence have a minimal impact on the C-cycle.

Because of its abundance, oxygen is typically used as the oxidant. However, in environments where it is limiting, other oxidants such as $NO_3^-$, $Mn^{4+}$, $Fe^{3+}$, and $SO_4^{2-}$ are used in successive order, generating $CO_2$. The consumption of these alternative oxidants affects other cycles as is discussed below. Microorganisms play a central role in this respiration process. The $CO_2$ generated can, especially under marine conditions, dissolve, i.e., dissolved inorganic carbon (DIC) forms carbonic acid ($H_2CO_3$). Losses of $CO_2$ from marine environments depend on the concentration of $CO_2$ in the atmosphere in relation to that in the water. The carbonic acid formed can dissociate to bicarbonates ($HCO_3^-$) and carbonates ($CO_3^{2-}$) which can subsequently become part of the long-term C-deposits in the lithosphere (shales, igneous and metamorphic rocks) where the turnover time of C is many orders of magnitude longer than in any of the other reservoirs. Thus, the natural fluxes between this reservoir and the atmosphere, hydrosphere or biosphere are

small although mankind has, since the advent of the Industrial revolution, tapped into this reservoir to obtain fossil fuels, thus slightly altering the fluxes between this and the other three reservoirs.

Under anaerobic conditions, $CO_2$ can undergo methanogenesis becoming reduced to methane (Fig. 9.1), a process attributed entirely to microorganisms-specifically, methanogens which exist in great diversity in anaerobic environments (eqn. 2).

$$CO_{2(g)} + 4H_{2(g)} \rightarrow CH_{4(g)} + 2H_2O_{(l)} \quad ...(2)$$

The above reaction is energetically favorable but requires enzymatic activation to break the H-H and C-O bonds. Anoxic soils, especially in wetlands and rice paddies, provide an excellent habitat for methanogens and are the most important sources of atmospheric methane, contributing about 60% (Conrad, 1995). Previously, Archaea, to which methanogens belong, were believed to be prevalent only in extreme environments such as hot springs and thermal vents. However, recent evidence using rRNA and rDNA indicates that they also occur in non-extreme environments, including garden soils (Rondon et al., 2000). Most of the methane ends up in the atmosphere, but in some instances, when it reaches the aerobic zone, it is oxidized to carbon dioxide and water by methanotrophs, a process from which the methanotrophs generate ATP. Two types of methanotrophic bacteria oxidize methane in oxic soils. The most common methanotrophs utilize only single-C compounds such as methane, methanol, and methylamines, and are predominant in oxic soils. They have a low affinity for methane and metabolize this gas only at elevated concentrations. They are reportedly able to form cysts or exospores and they may just be present in a resting stage, only becoming active when exposed to sufficiently high methane concentrations, a situation which may arise when the well-aerated soils suddenly become anoxic, for example during flooding or soil compaction. The second type of methanotrophs is less known but has a high affinity for methane and is more able to metabolize the gas even at atmospheric concentrations of 1.8 ppm. Ammonium oxidizers (nitrifying bacteria) also fortuitously oxidize methane (Conrad, 1995). Some of the methane from sediments may also be attributed to the endosymbiotic methanogens in the existing protozoa such as *Pelomyxa* sp. and other "methanogenic protozoa" (see chapter 8, endosymbiotic protozoa). *Pelomyxa* sp. can contain as many as $10^5$-$10^9$ methanogenic bacteria per cell (Finley, 1990). These methane transformations reduce the net amounts of methane reaching the atmosphere, although other major sources of methane in the environment such as ruminants, landfills, and termite mounds contribute to the ultimate quantities generated.

## 9.3  NITROGEN CYCLE

Nitrogen is the most abundant gas in the atmosphere and the second most important element after carbon. It is coupled with other elements of living matter, notably C, S, and P. Some of the common nitrogen compounds include ammonium, nitrite, nitrate, HCN, and SCN. The important biological processes of nitrogen transformation, all of which are catalyzed by microorganisms, are summarized in Fig. 9.5, the products ranging in oxidation states from $-3$ to +5. Notice that similar to the carbon cycle, the N-cycle also has an 8-electron difference between the most reduced ($NH_4^+$) and the most oxidized ($NO_3^-$) inorganic molecules. Changes

in oxidation yield energy.

The cycle can be considered to start from the N-fixation process, when an inert gas with an oxidation state of zero is reduced to an ammonium ion (or ammonia) which is more utilizable by organisms. The nitrogen-fixation process requires much energy in terms of ATP and is discussed in greater detail later; in general, on a global scale, any increase in the rate of conversion of N-compounds along one path within the cycle is quickly compensated by adjustments along other paths. Locally, however, nitrogen often becomes limiting to biological systems either because regeneration from its less available to a readily available state is too slow. The ammonium can be assimilated into organic nitrogen (which basically translates into the synthesis of amino acids) or undergo nitrification, the latter process occurring in two steps: the first step is conducted by organisms such as *Nitrosomonas* and *Nitrocystis* (an autotrophic organism) in an aerobic environment and the second step by *Nitrobacter, Nitrospina,* and *Nitrococcus* (Table 9.2), ultimately forming nitrates. Once fixed into $NH_3$/ $NH_4^+$, nitrogen is assimilated into the microbial biomass (ammonia assimilation) or oxidized to $NO_2^-$/$NO_3^-$ for uptake by plants. This gives microorganisms that can assimilate ammonia a competitive advantage. The ammonia is incorporated into organic compounds of the cell with glutamine synthetase and glutamate synthase involved in its incorporation. The fixation process is suppressed at high ammonium concentrations as the production of both nitrogenase and glutamine synthetase are repressed. The organic nitrogen can be decomposed by heterotrophic bacteria back into ammonium, a process called ammonification (Fig. 9.5). Some of the ammonium-N formed in this form is volatilized, but the majority remains in the biosphere. The mineralization of N (as well as a number of other nutrients) by microorganisms is, in most instances, coupled with the mineralization of C. Thus, a C:N ratio of >30 decreases mineralization and increases assimilation. In

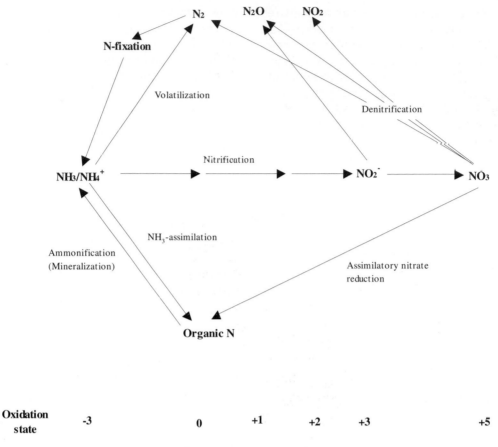

**Fig. 9.5** The nitrogen cycle.

contrast, a low C:N ratio increases mineralization of N to inorganic forms, nitrification, and subsequent denitrification.

Nitrate is the most prevalent form of nitrogen in aerobic soils and aquatic environments and thus several organisms (the nitrifiers) are able to utilize nitrification as an energy source (Table 9.2). However, these organisms are very poor competitors against fungi and other bacteria in soil. In contrast, free $NH_4^+$ does not exist for long in aerobic soils before nitrification occurs. Furthermore, the phylogeny and distribution of ammonia-oxidizing bacteria are more limited although anything that increases ammonia or ammonium increases their activity (Bédard and Knowles, 1989). To oxidize $NH_4^+$, the nitrifiers utilize $CO_2$ as the carbon source although some

heterotrophs utilize organic compounds rather than $CO_2$ to perform nitrification albeit less effectively/efficiently than autotrophs. In soils, the rate of nitrification decreases exponentially with decreasing water pressure (Stark and Firestone, 1995), possibly due to a decrease in microbial activity under water stress. Nitrification in wetlands is only possible in the shallow oxic zones at the soil-root-water interface. This is the same site where ammonium is oxidized to nitrate which can then diffuse into the adjacent anoxic zone to be reduced and allow the production of $N_2O$ as discussed later here and in Chapter 11. Thus, it is important that paddy fields be fertilized with urea or ammonium rather than nitrate fertilizers.

**Table 9.2** Various chemical conversions of nitrogen attributed to microorganisms

| Step | Reaction | Organism |
|------|----------|----------|
| Ammonia assimilation | $2NH_3 + 2H_2O + 4CO_2 \rightarrow 2CH_2NH_2COOH + 3O_2$ | Many bacteria |
| Fixation | $2N_2 + 6H_2O \rightarrow 4NH_3 + 3O_2$ | *Azotobacter, Rhizobium*, etc. |
| Nitrification (1st step) | $2NH_4^+ + 3O_2 \rightarrow 2NO_2 + 4H^+ + 2H_2O$ | *Nitrosomonas, Nitrocystis, Nitrosospira, Nitrolobus, Nitrosovibrio* |
| Nitrification (2nd step) | $2NO_2- + O_2 \rightarrow 2NO_2^-$ | *Nitrobacter, Nitrospina, Nitrococcus* |
| Denitrification | $2NO_3^- + 10H^+ \rightarrow N_2 + 9H_2O + 2OH^-$ | *Pseudomonas* |
| Denitrification | $5S + 6KNO_3 + 2CaCO_3 \rightarrow 3K_2SO4 + 2CO_2 + 3N_2$ | *Thiobacillus denitrificans* |
| Denitrification | $C_6H_{12}O_6 + 6NO_3^- \rightarrow 6CO_2 + 3H_2O + 6OH^- + 3N_2O$ | Many types of bacteria |
| Dissimilatory nitrate reduction | $NO_2^- + 9H^+ \rightarrow NH_3 + 3H_2O$ | *Clostridium* and many other bacteria |

In the soil or aquatic systems, nitrate has two major pathways. It can be assimilated into the organism's biomass (assimilatory nitrate reduction), a process likely to dominate under aerobic conditions or if nitrogen is in low supply. Alternatively, under anaerobic conditions, it is lost to the atmosphere (denitrification). Generally, $N_2$ is the end product of denitrification but other gases, in particular $N_2O$ and NO are also common (see Fig. 10.5).

Denitrification occurs under available carbon and anaerobic conditions and, contrary to what the name suggests, is not a reverse process of nitrification. It is an energy-requiring reaction and is performed by bacteria, especially those of genera *Pseudomonas, Acinetobacter, Alcaligenes, Azospirillum, Bacillus, Cytophaga, Flavobacterium, Gluconbacter, Hyphomicrobium, Micrococcus, Paracoccus, Rhizobium, Rhodopseudomonas,* and *Spirillum.* Thus, denitrifiers are very versatile, belonging to diverse phylogenetic groups and are widely distributed in soils and switch to anaerobic respiration [when conditions become anaerobic] due to the presence of nitrites (Fenchel et al., 1998). The factors affecting rates of denitrification in soil include poor drainage, temperatures, pH and the carbon:nitrogen (C:N) ratio. Denitrification is maximal around 25°C and neutral pH. It is totally inhibited below 2°C and under acidic conditions (pH 4.5). In the presence of vegetation, the denitrification process is enhanced by root exudates and dead root

tissue that provide energy for the growth of the denitrifiers. Furthermore, the respiration of existing roots generates a demand for oxygen, particularly under waterlogged conditions. However, some aquatic plants such as rice can channel oxygen directly around the rhizosphere, thus reversing the otherwise anoxic conditions in the rhizosphere. Such a reversal in turn controls (or at least reduces) denitirification within the rhizosphere in the presence of such aquatic plants.

As depicted in Figure 9.5, both nitrite and nitrate can also be lost to the atmosphere after they are converted to NOX, more specifically nitrous oxide ($N_2O$) and nitric oxide (NO) by nitrifiers, denitrifiers, and a variety of other prokaryotes such as *Azotobacter vinelandii, Erwinia carotovora, Serratia marcescens, Klebsiella pneumoniae, E. coli, Enterobacter aerogenes,* and fungi such as *Fusarium oxysporium* as well as some yeasts (*Rhodotorula* sp.) (Smith and Zimmerman, 1981; Bleakley and Tiedje, 1982, Mosier et al., 1998). However, it is important to point out the fact that most of the NOX gases in the atmosphere are from abiotic sources, in particular the anthropogenic combustion of both fossil fuels and biomass. Significant amounts of nitric oxide may be produced by the chemical decomposition of nitrite, especially under acidic conditions due to the drying and wetting of soils which may cause enrichment of $H^+$ and nitrite ions at microsites (Davidson, 1991). Whether of biotic or abiotic origin, these oxides contribute to the destruction of the stratospheric ozone layer, as discussed in Chapter 11. For the purposes of this Chapter, it should be noted that the abundance of $N_2O$ is currently increasing at about 0.2-0.3% per annum (Robertson, 1993) and most of the increases in NOX gases in general are attributable to intensive agriculture, which is almost always associated with an increase in amounts of N-fertilizer inputs in agricultural soils and subsequent microbial activity (Table 9.3). The production of $N_2O$ during denitrification increases at lower pH values and at higher $O_2$ concentrations. In agricultural systems, 10-75% of N-fertilizers may be lost through denitrification and various crop management practices such as splitting the applications over time, use of nitrification inhibitors to slow down the formation of nitrates from ammonium, timing applications, etc. have been widely researched.

The winter period has traditionally been assumed to be of minor importance in the emission of NOX gases but recent work by Röver et al. (1998) showed that 67-78% of the $N_2O$ emissions from arable land occur during winter in temperate regions. Furthermore, the rates of $N_2O$ emissions are also higher during winter than during the regular cropping season. The increased $N_2O$ in winter is possibly due to the temporary increase in the carbon available from the detritus and microorganisms killed by freezing (Christensen and Tiedje, 1990; Christensen and Christensen, 1991). In vegetated wetlands, such as paddy fields, $N_2O$ is emitted into the atmosphere through the plants' gas vascular system.

Looking at the sinks, some trivial amounts of $N_2O$ are consumed by denitrifiers as an intermediate in the metabolism of denitrifiers but is also reduced by anaerobes, such as *Wollinella succinogenes* which have the same affinity ($K_m = 7.5 \ \mu M$) for $N_2O$ as denitrifiers (Conrad, 1995) thus making these anaerobes equally competitive for reducing $N_2O$ under anaerobic condtions. Even though soils are a major source of NOX gases, they can also be a temporary sink although the mechanisms of consumption of these gases are still not clear. Potential candidates for NOX consumption in soils include *Nitrobacter*, methanotrophic bacteria, heterotrophic *Pseudomonas* sp. and nitrite-oxidizing autotrophic nitrifiers, for example *Nitrobacter winogradskii*.

The above discussion may leave the impression that most of the nitrates are lost through denitrification. However, unlike ammonium ions which are easily bound to negatively charged organic and inorganic particles (soil, clay, and colloids) and therefore not rapidly lost, nitrates, due to their negative charge, are easily leached into groundwater and surface water causing eutrophication. Nitrates can also be assimilated into organic nitrogen, which can in turn undergo ammonification (Fig. 9.5). It is important to realize that ammonification requires oxic conditions whereas denitrification occurs under anoxic conditions. This observation has implications in soils and sediments and shows that these two processes interact with

**Table 9.3** Nitrous oxide emission from fertilized and unfertilized plots during the cultivation of winter wheat (March, 1995-February, 1996; n=5).

| Period (days) | Sampling days | Fertilized plot[1] | | Unfertilized plot[1] | |
|---|---|---|---|---|---|
| | | Mean rate | Total loss | Mean rate | Total loss |
| Mar-Nov (275) | 39 | 17.5 a | 1155 (33) | 6.3 b | 413 (22) |
| Dec-Feb (90) | 27 | 108.3 a | 2340 (67) | 66.3 a | 1431(78) |

[1]Mean rate in $\mu g$ $N_2O$-N $m^{-2}$ $h^{-1}$; Total loss in g $N_2O$-N $ha^{-1}$; numbers accompanied by the same letter within the row are not significantly different ($p \leq 0.05$); numbers in parentheses are percentage.
*Source*: Röver et al. (1998) with permission from Elsevier.

the moisture content. Thus, with more pore spaces in the soil and sediment matrix, both ammonification and nitrification predominate, whereas at low pore space, denitrification becomes predominant (Fig. 9.6).

## 9.3.1 Nitrogen fixation

With a high abundance of 79% of atmospheric gases, nitrogen gas is unavailable for direct use by the primary producers, and has to be converted to ammonium, a form which is more readily utilizable. Fixation is through lightning, an industrial process (Haber process), or a biological process. Approximately 200 million tons of N are fixed per annum, of which only 2.4% is through lightning and 29% through the Haber process. The rest is fixed by bacteria. Nitrogen fixation is the only biological process by which the abundant atmospheric N source is availed. Prokaryotes have the ability to fix nitrogen into ammonia through either symbiotic or free-living systems. Fixation occurs in a number of bacteria and algae. Both aerobes (e.g. *Azotobacter, Cyanobacteria, Azospirillum, Klebsiella*) and

anaerobes (e.g. *Clostridium*) as well as bacteria of all major physiological groups, including methanogens, methyltrophs, and sulfur-reducing bacteria have been implicated (Table 9.4). The bulk of fixation is attributed to rhizobia in association with legumes, which fix as much as 140 kg N ha$^{-1}$ y$^{-1}$ but substantial amounts are also fixed in rice paddies by the cyanobacteria *Anabaena azollae* which lives in symbiosis with the water fern (*Azolla*). Cyanobacteria also play a role in N-fixation in freshwater aquatic systems where they are fairly widespread. The free-living in grasslands, meadows, and forests contribute 25 kg/ha/yr (Table 9.5). N-fixation also occurs in oceans although the mechanisms of its fixation in this environment have not been widely studied.

Thermodynamically, the N-fixation process requires energy for its activation (Fig. 9.7; eqn.3).

$$N_2 + 3H_2 + 2H^+ \rightarrow 2NH_4^+ - \Delta G = 18 \text{ kCal} ...(3)$$

At 25°C, $\Delta G$ for the production of ammonium ($NH_4^+$) from $N_2$ and $H_2$ is negative and the N-fixation reaction is thermodynamically feasible

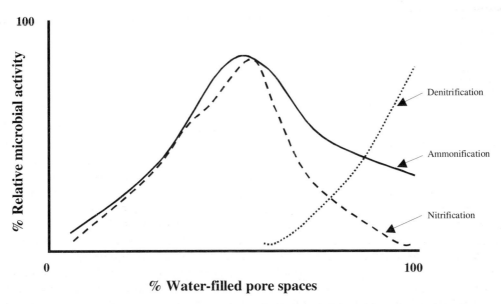

**Fig. 9.6**  Relative microbial activity with ammonification, denitrification, and nitrification as a function of soil water content (Reproduced from Linn and Doran, 1984 with permission from the authors).

**Table 9.4** Nitrogen-fixing organisms

| System | Type of mechanism | Examples |
|---|---|---|
| Free-living anaerobes | Chemoorganotrophs | *Clostridium* spp., *Desulfotomaculum* spp., *Desulfovibrio* spp. |
| | Chemolithotrophs | *Methanococcus* and *Methanosarcina* (both are Archea). |
| | Phototrophs | *Chlorobium* spp., *Chromatium* spp., *Heliobacillus* spp., *Heliobacterium* spp., *Rhodobacter* spp., *Rhodomicrobium* spp., *Rhodopseudomonas* spp. |
| Free-living aerobes | Chemoorganotrophs | *Azospirillum lipoferum*, *Azotobacter* spp., *Bacillus polymyxa*, *Citrobacter freundii*, *Klebsiella* spp., various methylotrophs |
| | Chemolithotrophs | *Alcaligenes* spp., *Thiobacillus* spp. |
| | Phototrophs | Cyanobacteria |
| Symbiotic | Leguminous plants | Alfalfa, common beans, clover, peas, soybeans, etc. in association with *Rhizobium* spp., *Bradyrhizobium* spp., and *Azorhizobium* spp. |
| | Nonleguminous plants | *Alnus* spp., *Casuarina* spp., *Ceanothus* spp., *Comptonia* spp., *Myrica* spp., etc. in association with *Frankia* (an Actinomycetes) |
| | Other nonleguminous associations | Cyanobacteria of the family Nostocaceae in *Azolla-Anabaena*, some lichenized fungi |

**Table 9.5** Dinitrogen fixation in various land-use types

| Land use | Ha (x $10^6$) | Nitrogen Fixed (kg ha$^{-1}$ y$^{-1}$) | Nitrogen Tg[a] |
|---|---|---|---|
| Legumes | 250 | 140 | 35 |
| Paddy (*Azolla-Anabaena* relationships) | 135 | 30 | 4 |
| Other cultivated crops | 1,015 | 5 | 5 |
| Permanent meadows, grasslands | 3,000 | 15 | 45 |
| Forest and woodland | 4,100 | 10 | 40 |
| Unused | 4,900 | 2 | 10 |
| Ice covered | 1,500 | 0 | 0 |
| Total land | 14,900 | ? | 139 |
| Sea | 36,000 | 1 | 36 |

*Source*: Schlesinger (1997) with permission from Elsevier.
Note : [a]1 Tg = $10^{12}$ g

but the triple bond in nitrogen (N ≡ N) is very stable. Activation of N ≡ N, therefore, requires a lot of energy. For comparison, activation of this triple bond under industrial conditions occurs at extremely high temperatures (400-450°C) and pressure (250-300 atm), with iron oxide as a catalyst (Kotz and Purcell, 1987; Jaffe, 1992). In contrast, the symbiotic and nonsymbiotic microbial systems can conduct this process at atmospheric pressure (1 atm) and normal atmospheric temperatures of approximately 15 - 35°C with nitrogenase as the catalyzing enzyme. The nitrogenase enzyme is highly conserved and is composed of two metalloproteins, the larger Fe-Mo protein (MW 220,000-240,000) and the smaller Fe-protein (MW 60,000-65,000), which act in concert. Thus molybdenum (Mo) is very specific for nitrogen fixation and is required for this process, as shown by studies in which Mo-requirements in N-fixing symbiotic plants were compared to those growing on $NO_3^-$. Recent reports indicate that the role of molybdenum can be played by vanadium as an alternative, forming an iron-vanadium-cobolt (FeVCo) complex (Stacey et al., 1992).

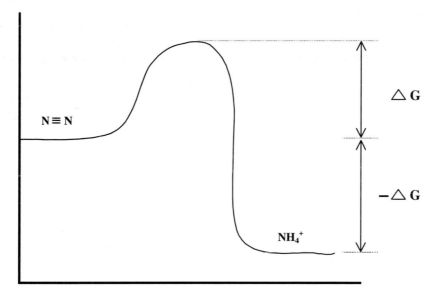

**Fig. 9.7**  Activation of nitrogen.

The biochemistry of the enzyme has been widely studied and is described in appropriate books (e.g. Stacey et al., 1992) but briefly, the Fe-Mo protein binds the $N_2$ to the enzyme, probably at the metal site, whereas the Fe-protein supplies electrons for the reduction. The reduction of nitrogen requires ATP and reducing power (ferredoxin, flavodoxin) forming ammonia as the first stable product. The electrons are derived from ferredoxin or its analog flavodoxin which utilizes ATP. A total of 16 ATPs are required per molecule of $N_2$ fixed, making the fixation process very expensive in terms of energy. It is no surprise, therefore, that it is tightly controlled and only expressed when the organism has no choice but to fix atmospheric nitrogen. If mineral nitrogen (ammonium or nitrate) is available to a fixing organism, the *nif* genes are repressed, causing the fixation process to be switched off instantly and the organism utilizes the readily available N instead.

The reductive process is also highly sensitive to $O_2$. Thus, many of the fixing organisms live in an anaerobic environment or somehow have to provide an anaerobic environment so as to protect the enzyme and continue to fix N (Table 9.6). As an example of a mechanism, *Azotobacter,* through respiratory protection has an enormous respiratory demand, such that at the center of the cell where nitrogenase is found, oxygen cannot reach. In contrast, blue-green algae restrict oxygen photosynthesis such that where photosynthesis occurs nitrogen fixation does not. Its site of N-fixation (heterocysts) is impermeable to oxygen.

## 9.4  PHOSPHORUS CYCLE

The abundance of phosphorus in the Earth's crust is almost equal to that of sulfur although the former is known to be more limiting in natural systems. The C:N:S:P ratio is 100:10:1:1 but P differs from all the other three in the sense that it does not occur in a gaseous form in significant quantities in natural environments. The limited atmospheric reservoirs of P are contained in dust particles. It has a very short residence time. Phosphorus also occurs in the atmosphere as phosphine, which is generated by

**Table 9.6**   Strategies adopted by various N-fixing organisms to protect their nitrogenase from oxygen

| Strategy | Explanation of strategy | Example of organism |
|---|---|---|
| Anaerobes | Fix only under anaerobic conditions. Obligate anaerobes do this naturally but facultative anaerobes fix only under anaerobic conditions. | *Klebsiella pneumonae* |
| Microaerophilly | Limited ability to protect their nitrogenase by fixing only under extremely low levels of $O_2$. | *Thiobacillus* spp. |
| Respiratory protection | Can fix aerobically but have an enormously high respiratory rate. The high rate of $O_2$ consumption ensures that no $O_2$ ever gets inside the cell where the nitrogenase is located. | *Azotobacter* sp. is the only known example. It actually only fixes when provided with excessive nutrient that can support the high respiration rate. |
| Compartmentalization | Form specialized compartments with a thick membrane impermeable to $O_2$. | Cynobacteria form heterocysts which are the specific N-fixation sites. Heterocysts are terminal and cannot grow unless supplied with $NH_4^+$. |
| Temporal separation | Separate in time the evolution of $O_2$ from $N_2$ fixation. Can only fix in darkness when there is no photosynthesis and thus no $O_2$ production. | Some cynobacteria, e.g. *Gloeotheca* sp. |
| Protective proteins | Proteins such as hemoglobin and myoglobin help to so bind the $O_2$ in the cell that no free $O_2$ can get inside. | *Rhizobium* sp. |

obligate anaerobic bacteria reducing organic phosphorus compounds (Jenkins et al., 2000; eqns. 4 and 5).

$$H_2PO^{4-} + H^+ + 4H_2 \rightarrow PH_3 + 4H_2O$$
$$\Delta G° = 244 \text{ kJ} \quad ...(4)$$

$$H_2PO_4{}^{2-} + H_2 \rightarrow HPO_3{}^{2-} + H_2O + H^+$$
$$\Delta G° = 46 \text{ kJ} \quad ...(5)$$

Phosphine is very toxic and concentrations of 20 ppm are highly toxic. A very highly ignitable gas, diphosphine ($PH_2PH_2$), can also be formed but information about its formation and impact in the environment is extremely scant. Some of the bacterial strains implicated in phosphine production include *Clostridium butyricum*, *C. cochliarium*, and *C. acetobutyricum*. However, evidence of the production of this gas in the environment is still controversial (Roels and Verstraete, 2001). Part of the controversy has, it appears, been attributed to the lack of sensitive methods to detect the gas. When detected, it is usally in small quantities (in ng $m^{-3}$ or ng $kg^{-1}$matrix).

Thus, the atmosphere plays a minor role in the global P cycle and most of the discussion about cycling this nutrient focuses on its cycling in aquatic and terrestrial systems. Furthermore, P is almost exclusively present in the V oxidation state as $PO_4{}^{3-}$, indicating that oxidation-reduction reactions play a minor role in controlling the distribution and activity of P in the natural environment.

Phosphorus plays a central role in the transmission and control of chemical energy via the hydrolysis of the terminal phosphate ester bond of the ATP, a high-energy molecule with either a diester or triester bond which is stable (Fig. 9.8). Each of these bonds carries some amount of free energy that can be passed onto other systems to drive reactions. Phosphorus is also a component in nucleic acids (DNA and RNA) and in phospholipids. Nucleic acids are important in the transfer of genetic information across generations whereas phospholipids are an important component of the cell membranes that allows selective permeability across mem-

Fig. 9.8   Ester and triester of ATP.

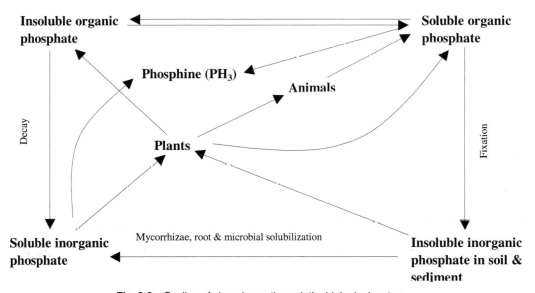

**Fig. 9.9**   Cycling of phosphorus through the biological system.

branes. As apartite [Ca10(PO$_4$)6X$_2$], P also forms structural body parts such as scales, bones and teeth. Considering its interconnection with essential life processes, its cycling is closely linked with biological processes, the organisms directly controlling its distribution in the environment.

The cycling of phosphorus through biological systems is shown in Figure 9.9. Soluble organic phosphorus compounds such as inositol phosphates, phospholipids, and phytins as well as nucleic acids are routinely mineralized to insoluble organic phosphates by a wide range of microorganisms such as *Arthrobacter, Bacillus subtilis, Aspergillus, Penicillium,* and *Streptomyces.* During mineralization the enzyme phosphatase is quite essential. P is assimilated into several macromolecules in the cell, in particular nucleic acids, and constitutes about 1-3% of the dry weight. However, several microorganisms, e.g. *Acinetobacter, Aerobacter, Beggiatoa, E. coli, Moraxella,* and *Mycobacterium* can accumulate P in excess of their normal cell requirements. The excess P is stored in polyphosphate granules such as volutin granules which can be observed under the bright field of the microscope after staining with Neisser's stain (Meganch and Faup, 1988). The P stored in these granules serves as an energy and phosphorus source during stress.

Soluble organic phosphate can alternatively be fixed in soils and sediments, forming insoluble inorganic fractions, or readily be taken up by plants (Fig. 9.9). Thus, both the insoluble and the soluble organic phosphate are big plant, animal, and microbial pools. As a matter of fact, the insoluble inorganic phosphate in soil is the biggest pool that may not be readily available to organisms, making this nutrient almost always limiting to primary production. In a number of instances where it is present but not readily available, inorganic P can be solubilized by some microorganisms, root exudates, or mycorrhizae. Thus, both *Pseudomonas cepacia* E-37 and *Erwinia herbicola,* through a direct oxidative pathway, convert glucose into gluconic acid and 2-keto-gluconic acid, the gluconic acid subsequently solubilizing the phosphates (Goldstein et al., 1993). Phosphorus is also solubilized by organic root exudates which in turn stimulates microbial metabolism and growth. Protozoa also contribute to the P cycle as they remineralize the phosphorus in ingested bacterial cells. Abiotic factors, especially pH, also play an important role in changing the dynamics of phosphorus bioavailability and subsequent cycling. In acid soil conditions, P is adsorbed to aluminum forming insoluble aluminum phosphate complexes whereas in alkaline soils, apatite, which is equally insoluble, limits the availability of P. Therefore, its availability controls the cycling of other bioactive elements as well. Organisms have devised mechanisms for sequestering and concentrating this nutrient to levels that are usually many times higher than the surrounding environment (e.g. soil, water, sediments, etc.).

In marine environments, the bioavailability of P is dependent on surface water temperatures. In summer, the warming of the surface water layers produces strong stratification with the cooler, denser water. During photosynthesis by marine phytoplanktons, the dissolved P in the photic zone is incorporated into phytoplanktons and eventually transported below the thermocline in sinking particles (see Fig. 4.2). Since the deep ocean is devoid of light, the P in this zone is not significantly incorporated into ocean biota but simply stored until it is eventually transported back into the surface ocean layers via upwelling or diffusive mixing, some of the P eventually settling into the sediments. In the long run, however, this P in the sediments is slowly transported back to the surface waters by physical processes. Under situations in which vast amounts of P-rich wastes, such as sewage and sludge, are flushed into the water system, the distribution described for marine environments above, does not apply. If unchecked, this can, with continued primary production in the surface water, generate large amounts of organic materials that sink to the bottom and, during decomposition, deplete the $O_2$, subsequently harming marine life.

The surface ocean reservoir extends to about 300 m deep and roughly corresponds to the top of the main thermocline which restricts exchange between surface and deep waters, thereby representing a natural boundary. To this layer, the dissolved $PO_4^{3-}$ added from rivers, lakes, and erosion material is introduced. Although organisms reside at all depths in oceans, the majority reside within the surface ocean zone where phytoplanktons are predominant. Fish-eating birds also return many tons of P in the P-rich guano they deposit on coastal nesting grounds.

## 9.5   SULFUR CYCLE

Sulfur compounds occur as gaseous, aerosol, aqueous, mineral, and biological forms in several oxidation states, which vary between −2 and +6 (Fig. 9.10). In this sense, the sulfur cycle shows some similarity to both C and N cycles, also spanning a range of 8 electrons, with elemental S having an oxidation/reduction

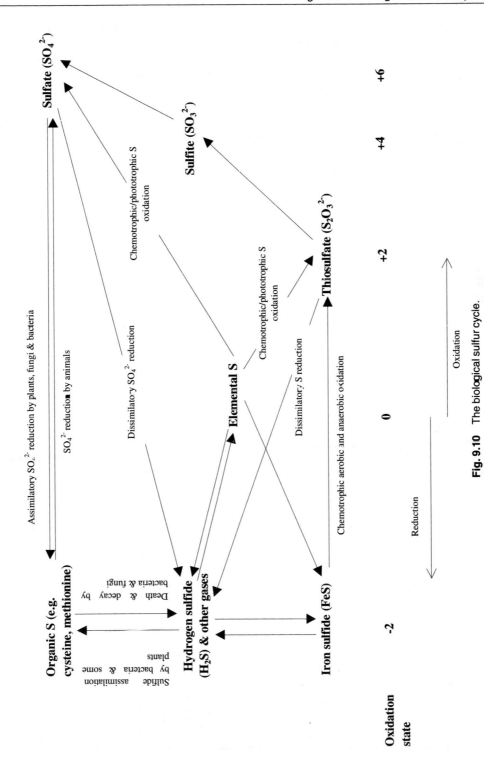

**Fig. 9.10**   The biological sulfur cycle.

state of zero. The biological forms and gaseous forms in the atmosphere are generally of the lowest oxidation state of $-2$. Most prevalent in both aquatic and terrestrial systems is the $+6$ oxidation state of sulfate ion. The major biological forms of sulfur include amino acids (i.e., methionine, cysteine, cystine, and glutathione) and coenzyme A as well as other molecules of biological importance, notably thiamine, biotin, and lipoic acid. Highly oxidized forms of organic S also exist in folic acid and sulfolipids.

In the environment, sulfur pools include oceans, swamps and marshes, soil, plants, and volcanoes. Unlike N and P, the abundance of S is often large enough for biota and does not typically limit growth. Thus, in soil and fresh water, it is present at a concentration of approximately 1 mM whereas concentration in sea water can be as high as 28 mM. The sulfur cycle involves solid (sulfur and metal sulfides), volatile sulfur compounds ($H_2S$, $SO_2$, methylsulfide), and water-soluble sulfur species such as sulfate ($SO_4^{2-}$). Although the greater amount of S occurs in the lithosphere as a component of sedimentary and other types of rocks, a variety of sulfur gases also occur in the environment from both biotic and abiotic (anthropogenic and natural) sources. Biological processes result in the production of a variety of reduced S-containing gases, notably $H_2S$, carbon disulfide ($CS_2$), carbonyl sulfide (OCS), methyl metacarptan ($CH_3SH$), dimethyl sulfide

($CH_3SCH_3$ or DMS), and dimethyl disulfide ($CH_3SSCH_3$ or DMDS) released into the atmosphere. The estimated emissions of volatile sulfur compounds on a global scale are presented in Table 9.7. Even though S fluxes in the atmosphere are large, the amount of S at any given time is small due to the fact that most S compounds in the air have a short life span of a matter of days. An important exception to this is carbonyl sulfide which has a comparatively long life span of 1-5 years (Schlesinger, 1997).

Human activities such as the burning of coal and smelting of sulfide ore cause major perturbations in the S cycle, elevating the total anthropogenic flux to about 80 Tg S $y^{-1}$, making this cycle one of the most perturbed by human activity. Different groups of microorganisms are involved in the biological cycling of S (Fig. 9.10). These can be classified into five metabolic types: (i) dissimilatory sulfate reduction, (ii) dissimilatory elemental S reduction, (iii) phototrophic anaerobic oxidation of S compounds, (iv) chemotrophic aerobic S oxidation, and (v) chemotrophic anaerobic S oxidation (Fauque, 1995). Dissimilatory sulfate reduction is quite important in anaerobic environments and at oxic-anoxic aquatic and soil interfaces where large sulfate reservoirs are located, yielding HS$^-$ and $CO_2$ (Jørgensen and Bak, 1991). The process has been well studied with increasingly better anaerobic techniques and include genera such as *Desulfomonas*,

**Table 9.7** Estimated emissions of volatile sulfur compounds to the atmosphere from natural biotic and abiotic sources

| Source | Sulfur compounds emitted (Mt $y^{-1}$) | | | | | | |
|---|---|---|---|---|---|---|---|
| | SO$_2$ | H$_2$S | DMS | DMDS & others | CS$_2$ | OCS | Total |
| Oceans | | 0-15 | 38-40 | 0-1 | 0.3 | 0.4 | 38.7-56.7 |
| Marshes | | 0.8-0.9 | 0.58 | 0.13 | 0.07 | 0.12 | 1.7-1.8 |
| Swamps | | 11.7 | 0.84 | 0.2 | 2.8 | 1.85 | 17.4 |
| Soil and plants | | 3-41 | 0.2-4.0 | 1 | 0.6-1.5 | 0.2-1.0 | 5.0-48.5 |
| Burning of biomass | 7 | 0-1 | | 0-1 | | 0.11 | 7.1-9.1 |
| Volcanoes | 8 | 1 | | 0-0.02 | 0.01 | 0.01 | 9.0 |
| Total | 15 | 16.5-70.6 | 39.8-45.4 | 1-3.4 | 3.8-4.7 | 2.7-3.5 | 78.9-142.6 |

*Source*: Kelly and Smith (1990) with permission from Kluwer Academic/Plenum Publishers, New York, NY. DMS=dimethyl sulfide, DMDS=dimethyl disulfide, CS$_2$=carbon disulfide, OCS=carbonyl sulfide (also abbreviated as COS in some literature).

*Desulfobulbus, Thermodesulfobacterium, Desulfobacter, Desulfococcus, Desulfuromonas Campylobacter* and at least one member of Archaea, i.e., *Archaeoglobus* (Widdel and Bak, 1992; Fauque, 1995). Thiosulfate and sulfite can also undergo dissimilatory reduction by some of these microorganisms, generating hydrogen sulfide and other gaseous forms of S. Sulfate-reducing bacteria are fairly widely distributed in anaerobic terrestrial and aquatic environments. Their presence is easily revealed by the signatory rotten egg smell of hydrogen sulfide in swamps, ponds, paddy fields, etc. The process can be influenced by temperature, with lower rates of microbial activity associated with low temperatures (Westrich and Berner, 1988). However, a number of sulfate-reducing microorganisms have been found in geothermal environments, clearly indicating that the process can occur under a wide temperature range.

Similarly, elemental S undergoes dissimilatory reduction by several Bacteria (e.g. *Wolinella succinogenes, Thermotoga, Desulforomonas, Sulfurospirillum*) and some members of Archaea in a respiratory type of metabolism, generating hydrogen sulfide. Elemental S can also undergo chemotrophic and phototrophic oxidation, generating sulfate (Fig. 9.10). Phototrophic oxidation is attributed to cyanobacteria and some phototrophic bacteria (i.e., purple bacteria and green bacteria) which use elemental S (as well as sulfides and thiosulfate) as electron donors for photosynthesis under anaerobic conditions. Reduced inorganic sulfur compounds such as iron sulfide can also undergo chemotrophic anaerobic oxidation or chemotrophic aerobic oxidation, subsequently generating sulfates (Fig. 9.10). Anaerobic oxidation is effected by a few bacteria, notably *Thiobacillus denitrificans* and *Thiomicrospira denitrificans* and aerobic oxidation by a diverse range of microorganisms, including *Thiobacillus* spp., *Sulfolobus, Pseudomonas, Thiomicrospira, Beggiatoa, Thiothrix, Thiovulum, Achromatium,*

*Macromonas,* and *Paracoccus denitrificans* (Fauque, 1995).

In biological systems S is basically needed as $H_2S$ and not as the common $SO_4^{2-}$ anion. Therefore, its utilization requires activation of the sulfate by ATP in the presence of the enzyme ATP sulfurlylase, forming adenosine 5'-phosphosulfate (APS) (Fig. 9.11). The reaction is energetically favorable through a reversible conversion. Under anaerobic conditions, APS is catalyzed by APS reductase to sulfite and subsequently, by sulfite reductase, with $H_2$ as the electron donor, to hydrogen sulfide. This process has been well studied in Archaea, in particular *Archeaoglobus fulgidus* (Thauer and Kunow, 1995) and sulfur-reducing Bacteria. Under aerobic conditions, APS is converted to phosphoadenosine-5'-phosphosulfate (PAPS) with APS kinase as the enzyme. PAPS is subsequently converted to sulfite and eventually reduced to hydrogen sulfide. This assimilatory reduction is very common in bacteria and algae.

As pointed out earlier, several other trace gases, besides $H_2S$, are also generated from the metabolism of sulfur, with microorganisms playing a central role in their impact on the environment. Two of these gases are discussed below because of their more significant impact on the environment.

### 9.5.1 Carbonyl sulfide (OCS)

Carbonyl sulfide is the most stable trace sulfur gas in the troposphere (Kelly and Smith, 1990; Conrad, 1995). It is about 2.5 times the concentration of DMS and $CS_2$, exhibiting a lifetime of 1-5 years, an attribute that enables it to enter the stratosphere. However, its natural abundance is very low compared to that of the inorganic S gases, $H_2S$, and $SO_2$, suggesting that it plays a minor role in the global sulfur cycle and turnover. Both the oxic and anoxic soils serve as carbonyl sulfide sources from thiocyanate due to the action of microbial thiocynate hydrolases which hydrolyzes SCN- to $NH_3$ and carbonyl sulfide (Kelly et al., 1993). The en-

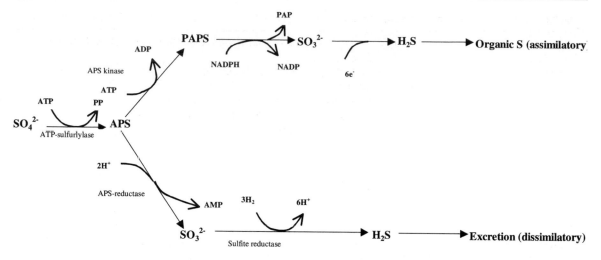

**Fig. 9.11** Schemes of assimilatory and dissimilatory sulfate reduction. (APS = adenosine-5'-phosphosulfate; PAPS = phosphoadenosine-5'-phosphosulfate)

zyme is present in *Thiobacilli thioparus*, an obligate aerobic chemolithoautotroph. Oxic flooded soils and coastal marshes tend to be stronger sources of carbonyl sulfide than upland soils, whereas wetland paddy fields are only small sources. Plants are a major sink for this gas, potentially consuming 2-5 million tons per annum (Schlesinger, 1997) and fortuitously metabolizing it to $H_2S$ and $CO_2$ in enzymatic reactions involved in the fixation of $CO_2$.

### 9.5.2   Dimethyl sulfide (DMS)

Dimethyl sulfide is ubiquitous in the biosphere and is actually the most abundant form of sulfur emitted. Emissions are in the range of 39.6 to 45.4 Mt $y^{-1}$ (Table 9.7). It is generated from many living systems, especially in marine environments, swamps, and marshes, by some algae and marsh plants, notably *Spartina alteriflora* and *Zostera marina*. In these instances, the gas is generated by the enzymatic cleavage of dimethyl sulfonio propionate (dimethyl–$\beta$–propionate) an osmoregulator. Even higher levels of DMS are produced due to the grazing of algae by zooplanktons (Table 9.8) and the physical as well as bacterial breakdown of senescent algal

cells which contain methyl sulfoniopropionate as a betaine. The sulfur betaine decomposes easily to yield the acrytic acid ion and dimethyl sulfide. DMS is also generated from the direct conversion of dimethyl sulfoniopropionate (DMSP) by both bacteria (e.g. *Clostridium* sp.) and phytoplanktons (e.g. *Gyrodinium* sp.)(Kadota and Ishida, 1972) and from the reduction of dimethyl sulfoxide (DMSO) by some bacteria such as *Hyphomicrobium* sp. under low-oxygen or anaerobic conditions that are typical of marine environments. To this effect, therefore, DMS occurs in relatively high concentrations in some aquatic systems than the atmosphere (Zinder and Brock, 1978).

DMS production is constantly occuring in soils (Table 9.9) under aerobic conditions and has also been reported in ruminants and in the breath of human beings due to microbial activity (Cooper, 1983; Lorenzen et al., 1994). At much higher concentrations in the atmosphere, it can undergo photochemical oxidation and leads to $SO_2$, dimethyl sulfoxide, sulfur dioxide, methane sulfonic acid and sulfuric acid (Ferek et al., 1986; Hatakeyama et al., 1982). The two acids can rise to high elevations in the atmos-

**Table 9.8** Algae and cyanobacteria that contain DMSP or produce DMS

| | |
|---|---|
| Cyanobacteria | *Anacystis, Microcoleus, Oscillatoria, Pharmidium, Plectonema, Synechococcus* |
| Bacillariophyta | *Skeletonema, Thalassiosira* |
| Chlorophyta | *Carteria, Chlorococcum, Cladophora, Codium, Enteromorpha, Micromonas, Monostroma, Ulva, Ulothrix* |
| Dinophyta | *Amphidium, Gymnodinium, Gyrodinium, Prorocentrum* |
| Haptophyta | *Emiliania, Hymenomonas, Phyaeocystis, Syrachosphaera* |
| Phaeophyta | *Dictyopteris, Egregia, Endatachne, Halidrys, Laminaria, Macrocystis, Pelvetia* |
| Prasinophyta | *Platymonas* |
| Rhodophyta | *Ceramium, Corallina, Gelidium, Gigartina, Gracilaria, Poicamium, Polysiphonia, Soliera* |

*Source*: Kelly and Smith (1990) with permission from Kluwer Academic/Plenum Publishers, New York, NY.

**Table 9.9** DMS emission in water or in soils from different locations

| Soil/water sample | Number of measurements | Emission rate ($10^{-12}$ g g$^{-1}$ h$^{-1}$)[a] |
|---|:---:|:---:|
| Washington DC | 6 | 47 (26) |
| Walnut Creek, CA | 11 | 54 (31) |
| Seattle, WA | 5 | 32 (20) |
| Baton Rouge, LA | 8 | 66 (42) |
| Lewiston, ID | 15 | 84 (51) |
| Pullman, WA | 7 | 21 (14) |
| Atlantic Ocean | 23 | 1.2 (2.7)[b] |

[a]Numbers in parentheses are the standard deviation.
[b]Emissions in water are in $10^{-11}$ g ml$^{-1}$.
*Source*: Lovelock et al. (1972) with permission from Nature Publishing.

phere where, because of supersaturation with water vapor, they act as cloud droplet nuclei forming acid rain.

DMS is degraded by both aerobic and anaerobic bacteria. Under anaerobic conditions, it can be degraded by both methanogens and sulfur-reducing bacteria to generate $CH_4$, $CO_2$, and $H_2S$ (Kiene et al., 1986) whereas degradation under aerobic conditions yields $CO_2$ and sulfate.

## 9.6  IRON CYCLE

Iron is the fourth most abundant element in the Earth's crust. It exists as insoluble ferric sulfides (e.g. ferric sulfide [FeS]; marcasite [FeS$_2$]; silicates (e.g. olivine [(Mg,Fe)$_2$SiO$_4$]); phosphates (e.g. vivianite [Fe$_2$(PO$_4$)$_2$.H$_2$O]); oxides (e.g. Fe$_2$O$_3$); hematite [Fe$_2$O$_3$]; illite [2K$_2$O.3(Mg,Fe)0.8(Al,Fe)2O$_3$.12H$_2$O];

ilmenite [FeTi(O$_3$)]); oxyhydroxides (e.g. geothite [FeOOH]), and carbonates (e.g. siderite [FeCO$_3$]). A broad range of microorganisms can adapt strategies to obtain nutrients from these insoluble compounds. It does not occur in significant amounts in the Earth's surface in its metallic form (i.e., oxidation state zero), but rather as Fe$^{2+}$ and Fe$^{3+}$, the two oxidation states between which it is cycled (Fig. 9.12). To form highly reduced metallic iron, energy has to be expended to transfer two or three electrons to Fe$^{2+}$ or Fe$^{3+}$ respectively. The ability of iron to cycle between the ferric (Fe$^{3+}$) and ferrous (Fe$^{2+}$) states appears to have been intimately connected to the origin of life (see Chapter 1). In general, large amounts of iron exist in soils and sediments.

All microorganisms, except lactobacilli which lack heme, require iron for their growth and therefore, at least to some extent, partici-

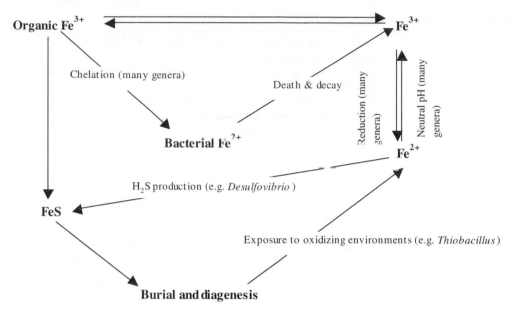

**Fig. 9.12** The microbial iron cycle.

pate in Fe-cycling. However, the majority require iron only in minute quantities ranging between $10^{-7}$ and $10^{-5}$ M (Guerinot and Yi, 1994) for their respiration as an electron acceptor in cytochromes and in enzymes such as ribonucleotide reductase and nitrogenase. Where present in low concentrations insufficient to meet the requirements of bacteria in the environment, siderophores are deliberately synthesized and secreted by the bacteria to solubilize the iron, making it more available even to higher trophic levels. Siderophores are small (usually less than 1,000 Da) nonproteinaceous compounds with a very high affinity for iron. They are produced by a wide range of microorganisms.

Iron acts both as an electron donor and acceptor for many microbial reactions. It also has a high propensity to form complexes with ligands that contain N, S, and O, exerting influence on the cycling of these nutrients. When serving as an electron donor, ferric compounds undergo dissimilatory reduction to Fe(II) (eqn.6). The process is conducted by a variety of microorganisms, including *Pseudomonas*

sp., *Desulfovibrio* sp., *Desulfuromonas acetoxidans*, *Clostridium pasteuranium*, *Shawanella putrefaciens*, and *Geobacter metallireducan* (GS-15), all of which have been reported to reduce $Fe^{3+}$.

$$Fe^{2+} + \tfrac{1}{4}O_2 + H^+ \rightarrow Fe^{3+} + \tfrac{1}{2}H_2O \ \ ...(6)$$

Both *S. putrefaciens* and *G. metallireducan* (GS-15) are fairly widespread but the former is more frequently recovered in sedimentary environments where it accounts for approximately 2% of the microbial population whereas the latter is most predominant in soil (Lovley, 2000). *Geobacter metallireducan* also occurs in freshwater and marine habitats, reducing $Fe^{3+}$ to $Fe^{2+}$ forming a mixed mineral, magnetite $(Fe^{2+}Fe_2^{3+}O_4)$, which is extracellular and is used by these organisms around their cells for orientation. Both organisms resemble sulfur-reducing bacteria in that they use metabolic products, i.e., amino acids, formate, acetate, butyrate, propionate, long-chain fatty acids, etc. to oxidize them to $CO_2$ and $Fe^{2+}$. Many sulfate-reducing microorganisms such as *Desulfovibrio* also have the ability to reduce

Fe(III) to Fe(II) and might be major players in Fe(III) reduction under anaerobic conditions (Coleman et al., 1993; Lovley et al., 1993).

A majority of the microorganisms that reduce iron are also capable of using Mn(IV) as an electron acceptor, reducing it to Mn(II) (Lovley, 1995). The reduction process can occur both under aerobic and anaerobic environments but is of greater significance under anaerobic (anoxic) conditions. Under oxic conditions, the reduced Fe(II) generated is quickly converted back to Fe(III), resulting in no net reduction of Fe(III). Under anaerobic conditions, the reduction of Fe(III) is curtailed if nitrates are present due to the preferential reduction of $NO_3^-$ by iron oxidizing bacteria (Straub et al., 1996). The reduction of Fe(III) by microorganisms has also been linked to the oxidation of organic matter (Canfield et al., 1993) and the removal of organic pollutants from soil and sediments (Lovley, 2000). The role of iron-reducing bacteria in the degradation of organic compounds is discussed further in Chapter 10.

Microorganisms implicated in the oxidation process include *Leptothrix ochracea*, *Thiobacillus denitrificans*, *Ferroglobus placidus*, and *Pseudomonas sturtzeri* (Hafenbrandt et al., 1996; Emerson, 2000). *Leptothrix ochracea* forms visible brown fluffy filamentous layers of $Fe^{2+}$ oxides. They are the most common visible inhabitants of many freshwater iron seeps. However, no pure cultures of *Leptothrix* species have been obtained. They also oxidize $Mn^{2+}$ to manganese dioxide as a deposit. Other iron-oxidizing microorganisms include *Acholeplasma, Actinomyces, Arthrobacter, Ferrobacillus, Hypomicrobium, Leptospirillum, Metallogium, Pedomicrobium, Tonothrix, Gallionella,* and *Thiobacillus* spp. These, together with manganese oxidizing/depositing microorganisms are unique in the sense that they directly transform metals and are involved in microbial corrosion. *Gallionella* are very unusual organisms which excrete a gellike substance with high levels of iron hydroxides that grow into a stalk mainly composed of iron hydroxides. The cells are bean-shaped. They are thought to catalyze this substance to obtain energy from it but no one has demonstrated this to date. *Gallionera* also grow on other compounds such as $H_2$. They are called metal depositing because like *Pedomicrobium* they deposit both Fe and Mn, but unlike the latter genus do not deposit gold as well. *Gallionella* sp. is widely distributed and readily isolated from environments where iron hydroxide and ferromanganese oxides are found. Isolation is accomplished following a gradient enrichment technique in tubes containing FeS generating opposing $O_2$ and Fe(II) gradients (Hallbeck and Pederson, 1990). Under the electron microscope, *Pedomicrobium* cells have a shiny (goldlike) appearance, possibly from the ability of their body polymers to attract metals.

## References

Bédard C. and R. Knowles (1989). Physiology, biochemistry, and specific inhibitors of $CH_4$, $NH_4^+$, and CO oxidation by methanotrophs and nitrifiers. *Microbiol. Rev.* **53**: 68-84.

Béguin P. and J.-P. Aubert (1994). The biological degradation of cellulose. *FEMS Microbiol. Rev.* **13**: 25-58.

Benner R., A.E. Maccubbin, and R.E. Hodson (1984). Anaerobic biodegradation of the lignin and polysaccharide components of lignocellulose and synthetic lignin by sediment microflora. *Appl. Environ. Microbiol.* **47**: 998-1004.

Bleakley B.H. and J.M. Tiedje (1982). Nitrous oxide production by organisms other than nitrifiers or denitrifier. *Appl. Environ. Microbiol.* **44**: 1342-1348.

Canfield D.E., B.B. Jorgensen, H. Fossing, R. Glud, J. Gundersen, N.B. Ramsing, B. Thamdrup, J.W. Hansen, L.P. Nielsen, and P.O.J. Hall (1993). Pathways of organic carbon oxidation in three continental margin sediments. *Mar. Geol.* **113**: 27-40.

Christensen S. and J.M. Tiedje (1990). Brief and vigorous $N_2O$ production by soil at spring thaw. *J. Soil Sci.* **41**: 1-4.

Christensen S. and T. Christensen (1991). Organic matter available for denitrification in different soil fractions: effects of freeze/thaw cycles and straw disposal. *J. Soil Sci.* **42**: 637-647.

Coleman M.L., D.B. Hedrick, D.R. Lovley, D.C. White, and K. Pye (1993). Reduction of Fe(III) in sediments by sulfate-reducing bacteria. *Nature* **361**: 436-438.

Colwell R.R. and G.S. Sayler (1978). Bacterial degradation of industrial chemicals in aquatic environments. *In:* R. Mitchell (ed.). Water Pollution Microbiology. Wiley Intersci., New York, NY, pp. 111-134.

Conrad R. (1995). Soil microbial processes involved in production and consumption of atmospheric trace gases. *Adv. Microb. Ecol.* **14**: 207-250.

Cooper A.J.L. (1983). Biochemistry of sulfur-containing amino acids. *Ann. Rev. Biochem.* **52:** 187-222.

Davidson E.A. (1991). Fluxes of nitrous oxide and nitric oxide from terrestrial ecosystems. *In:* J.E. Rogers and W.B. Whitman (eds.). Microbial Production and Consumption of Greenhouse Gases: Methane, Nitrogen Oxides and Halomethanes. Amer. Soc. Microbiol., Washington DC, pp. 219-235.

Emerson D. (2000). Microbial oxidation of Fe(II) and Mn(II) at circumneutral pH. *In:* D.R. Lovley (ed.). Environmental Microbe-Metal Interactions. ASM Press, Washington DC, pp. 31-52.

Fauque G.D. (1995). Ecology of sulfate-reducing bacteria. *In:* L.L. Barton (ed) Sulfate-Reducing Bacteria. Plenum Press, New York, NY, pp. 217-241.

Felix C.R. and L.G. Ljungdahl (1993). The cellulosome: the extracellular organelle of Clostridium. *Ann. Rev. Microbiol.* **47:** 791-819.

Fenchel T., G.M. King, and T.H. Blackburn (1998). Bacterial Biogeochemistry: The Ecophysiology of Mineral Cycling. Acad. Press, New York, NY.

Ferek R.J., R.B. Chatfield, and M.O. Andreae (1986). Vertical distribution of dimethylsulphide in the marine atmosphere. *Nature* **320:** 514-516.

Finley B.J. (1990). Physiological ecology of free-living protozoa. *Adv. Microb. Ecol.* **11:** 1-35.

Goldstein AH., R.D. Rogers, and G. Mead (1993). Mining by microbes. *Bio/Tech.* **11:** 1250-1254.

Guerinot M.L. and Y. Yi (1994). Iron: nutritious, noxious and not readily available. *Plant Physiol.* **104:**815-820.

Hafenbrandt D., M. Keller, R. Dirmeier, R. Rachel, P. Roßnagel, S. Burggraf, H. Huber, and K.O. Stetter (1996). *Ferroglobus placidus* gen. nov., sp. nov., a novel hyperthermophilic archaeum that oxidizes $Fe^{2+}$ at neutral pH under anoxic conditions. *Arch. Microbiol.* **16:** 308-314.

Hallbeck L. and K. Pederson (1990). Culture parameters regulating stalk formation and growth rate of *Gallionella ferruginea. J. Gen. Microbiol.* **136:** 1675-1680.

Hatakeyama S., M. Okuda and H. Akimoto (1982). Formation of sulfur dioxide and methanesulfonic acid in the photooxidation of dimethyl sulfate in the air. *Geophys. Res. Lett.* **9:** 583-586.

Jaffe D.A. (1992). The nitrogen cycle. *In:* S.S. Butcher, R.J. Charlson, G.H. Orians, and G.V. Wolfe (eds). Global Biogeochemical Cycles. Acad. Press, New York, NY, pp. 262-284.

Jenkins R.O., T.-A. Morris, P.J. Craig, A.W. Ritchie, and N. Ostah (2000). Phosphine generation by mixed- and monoseptic-cultures of anaerobic bacteria. *Sci. Total Environ.* **250:** 73-81.

Jørgensen B.B. and F. Bak (1991). Pathways and microbiology of thiosulfate transformation and sulfate reduction in a marine sediment (Kattegat, Denmark). *Appl. Environ. Microbiol.* **57:** 847-856.

Kadota H. and Y. Ishida (1972). Production of volatile sulphur compounds by microorganisms. *Ann. Rev. Microbiol.* **26:**127-138.

Kelly D.P. and N.A. Smith (1990). Organic sulfur compounds in the environment: Biogeochemistry, microbiology, and ecological aspects. *Adv. Microb. Ecol.* **11:** 345-385.

Kelly D.P. G. Makin, and A.P. Wood (1993). Microbial transformations and biogeochemical cycling of one-carbon substrates containing sulfur, nitrogen or halogens. *In:* J.C. Murrell and D.P. Kelly (eds.). Microbial Growth on C1 Compounds. Intercept, Andover, MA pp. 47-63.

Kiene R.P., R.S. Oremland, A. Catena, L.G. Miller, and D.G. Capone (1986). Metabolism of reduced methylated sulfur compounds by anaerobic sediments and a pure culture of an estuarine methanogen. *Appl. Environ. Microbiol.* **52:** 1037-1045.

Kotz J.C. and K.F. Purcell (1987). Chemistry and Chemical Reactivity. Sanders College Publ., New York, NY.

Linn D.M. and J.W. Doran (1984). Effect of water-filled pore space on carbon dioxide and nitrous oxide production in tilled and nontilled soils. *Soil Sci. Soc. Amer. J.* **48:** 1267-1272.

Lorenzen J., S. Steinwachs, and G. Unden (1994). DMSO respiration by the anaerobic rumen bacterium *Wolinella succinogenes. Arch. Microbiol.* **162:** 277-281.

Lovelock J.E., R.J. Maggs, and R.A. Rasmussen (1972). Atmospheric dimethyl sulfide and the natural sulfur cycle. *Nature* **237:** 452-453.

Lovley D.R. (1995). Microbial reduction of iron, manganese, and other metals. *Adv. Agron.* **54:** 175-231.

Lovley D.R. (2000). Fe(III) and Mn(IV) reduction. *In:* D.R. Lovley (ed.). Environmental Microbe-Metal Interactions. ASM Press, Washington DC, pp. 3-30.

Lovley D.R., E.E. Roden, E.J.P. Phillips, and J.C. Woodward (1993). Enzymatic iron and uranium reduction by sulfate-reducing bacteria. *Marine Geology* **113:** 41-53.

Meganch M.T.J. and G.M. Faup (1988). Enhanced biological phosphorus removal from waste water. *In:* D.L. Wise (ed.). Biotreatment System. CRC Press, Boca Raton, FL, vol.3, pp. 111-203.

Moore D. (1998). Fungal Morphogenesis. Cambridge Univ. Press, Cambridge, UK.

Mosier A.R., J.M. Duxbury, J.R. Freney, O. Heinemeyer, and K. Minami (1998). Assessing and mitigating $N_2O$ emissions from agricultural soils. *Clim. Change* **40:** 7-38.

Reid I.D. (1995). Biodegradation of lignin. *Can. J. Bot.* **73:** S1011-S1018.

Robertson G.P. (1993). Fluxes of nitrous oxide and other nitrogen trace gases from intensively managed landscapes: A global perspective. *In:* L.A. Harper (ed.). Agricultural Ecosystems Effects on Trace Gases and Global Climatic Change. Amer. Soci. of Agron., Madison, WI, pp. 95-108.

Roels J. and W. Verstraete (2001). Biological formation of volatile phosphorus compounds. *Bioresource Tech.* **79:** 243-250.

Rondon M.R, P.R. August, A.D. Bettermann, S.F. Brady, T.H. Grossman, M.R. Liles, K.A. Loiacono, B.A. Lynch, I.A. MacNeil, C. Minor, C.L. Tiong, M. Gilman, M.S. Osburne, J. Clardy, J. Handelsman, and R.M. Goodman (2000). Cloning the soil metagenome: A strategy for accessing the genetic and functional diversity of uncultured microorganisms. *Appl. Environ. Microbiol.* **66:** 2541-2547.

Röver M., O. Heinemeyer, and E-A. Kaiser (1998). Microbial induced nitrous oxide emissions from an arable soil during winter. *Soil Biol. Biochem.* **30:** 1859-1865.

Schlesinger W.H. (1997). Biogechemistry: An Analysis of Global Change. Acad. Press, New York, NY.

Smith M.S. and K. Zimmerman (1981). Nitrous oxide production by nondenitrifying soil nitrate reducers. *Soil Sci. Soc. Amer. J.* **45:** 1545-1547.

Stacey G., R. Burris, and H.J. Evans (1992). Biological Nitrogen Fixation. Chapman and Hall, New York, NY.

Stark J.M. and M.K. Firestone (1995). Mechanisms for soil moisture effects on activity of nitrifying bacteria. *Appl. Environ. Microbiol.* **61:** 218-221.

Stout J.D., K.R. Tate, and L.F. Molloy (1976). Decomposition processes in New Zealand soils with particular respect to rates and pathways of plant degradation. *In:* J.M. Anderson and A. MacFadyen (eds.). The Role of Terrestrial and Aquatic Organisms in Decomposition Processes. Blackwell Sci. Publ., Oxford, UK, pp. 97-114.

Straub K.L., M. Benz, B. Schink, and F. Widdel (1996). Anaerobic, nitrate-dependent microbial oxidation of ferrous iron. *Appl. Environ. Microbiol.* **62:** 1458-1460.

Thauer R.K. and J. Kunow (1995). Sulfate-reducing Archaea. *In:* L.L. Burton (ed.). Sulfate-Reducing Bacteria. Plenum Press, New York, UK, pp. 33-48.

Tiedje J.M. (1988). Ecology of denitrification and dissimilatory nitrate reduction to ammonium. *In:* A.J.B. Zehnder (ed.). Biology of Anaerobic Microorganisms. John Wiley and Sons, New York, NY, pp. 179-244.

Westrich J.T. and R.A. Berner (1988). The effect of temperature on rates of sulfate reduction in marine sediments. *Geomicrobiol. J.* **6:** 99-117.

Whitehead L., S.K. Stosz, and R.M. Weiner (2001). Characterization of the agarase system of a multiple carbohydrate degrading marine bacterium. *Cytobios.* **106:** 99-117.

Widdel F. and F. Bak (1992). Gram-negative mesophilic sulfate-reducing bacteria. *In:* A. Balows, H.G. Trüpper, M. Dworkin, W. Harder, and K.-H. Schleifer (eds.). The Prokaryotes: A Handbook on the Biology of Bacteria: Ecophysiology, Isolation, Identification, Applications. Springer-Verlag, New York, vol. IV, pp. 3352-3378 (2nd ed.).

Zinder S.H. and T.D. Brock (1978). Dimethyl sulfoxide reduction by microorganisms. *J. Gen. Microbiol.* **105:** 335-342.

# 10

# Microbial Interaction with Organic Pollutants

With the advent of the Industrial revolution and modern civilization, new polluting compounds emerged, some of which have threatened to push the Earth out of equilibrium. For example, it is estimated that only a minor fraction (less than 0.1%) of applied pesticides reaches the target pest species, leaving the excess pesticide to move through the environment and potentially contaminate soil, water, and nontarget biota. Contaminants can be either organic or inorganic in origin. Organic pollutants include fungicides, pesticides, solvents, pharmaceutical compounds, dyes, petroleum-based fuels, halogenated aromatics and aliphatics, nitroaromatics from munitions, and feedstuff. Inorganic pollutants include chromium (Cr), copper (Cu), nickel (Ni), cadmium (Cd), lead (Pb), mercury (Hg), selenium (Se), and cyanides, to name but a few. Microbial interaction with inorganic pollutants is extensively discussed in Chapter 12. Here we shall focus on how microorganisms interact with various organic contaminants.

Various organic compounds persist in the environment and their fate is of concern. Persistence is often attributed to:

- environmental conditions (temperature, pH, moisture, nutrients, etc.) being unfavorable to their degradation;
- the compounds being completely novel structures; and/or

- lack of a point of attack for some polymers such as polystyrene, polyethylene (i.e., $[-CH_2]_n$), polypropylene, polyvinylchloride (PVC, i.e. $[-CH_2-CHCl-]_n$), polyurethane, and teflon (i.e. $[-CF_2-CF_2-]_n$) due to the presence of long chains.

However, it is widely accepted that most of these pollutants can be transformed by microbial activity, given sufficient time and appropriate environmental conditions. Transformation based on biological (microbial) activity is referred to as biodegradation.

From a microbiological perspective, the cell's primary tasks are to survive, maintain itself, and to grow. Biodegradation is a manifestation of these tasks and refers to the physiological reactions resulting from the complex action of living organisms. It involves reducing the compound (substrate) of interest to an environmentally acceptable extent, eliminating the undesirable properties. It is important to realize that disappearance of these pollutants can be attributed to both biological (biotic) and nonbiological (abiotic) processes. Abiotic processes include volatilization, photooxidation, sorption, and leaching. Even though degradation due to abiotic processes occurs to a considerable extent, it usually does not result in the complete destruction of pollutants and, in some cases, activates organic compounds, forming more toxic products. Thus, degradation due to biotic processes is more desirable than abiotic processes and is likely to occur when the following conditions prevail:

(1) the chemical is accessible (bioavailability may be reduced by sorption, insolubility, micropores, partitioning into nonaqueous phase liquids (NAPLs), or humification);
(2) microorganisms capable of degrading the compound must be present or incapable organisms must develop the capability to degrade the compound in question;
(3) the degrading organism must have the enzymes requisite for the biodegration pathway;
(4) if degradation is growth-linked, the chemical must enter the cell.

These conditions constitute de facto a subset of biogeochemical cycles except that the focus is on compounds of public health and regulatory concern. It is a subset of transformation. Transformation in this case refers to alteration of molecular structure or a change in functional groups and may result in an increase or decrease in complexity. Transformations that result in degradation of organic compounds can be broadly categorized as **mineralization** or **cometabolism**. Mineralization releases inorganic components ($CO_2$, $N_2$, halide and/or methane) from the organic molecule, providing energy and becoming incorporated into the degrading organism's cellular material. Thus, mineralization of organic compounds serves two roles in the global ecosystem. First, it releases the energy stored in the organic molecules but also returns essential nutrients such as $CO_2$, halides, methane, and nitrogen into the normal biogeochemical nutrient pool (carbon and nitrogen cycles).

Microorganisms play a major role in the transport and transformation of organic compounds in the environment. To be degraded by microorganisms, the compound must be in a form suitable and accessible to the degraders. Thus, insoluble, sorbed, and complexed compounds may owe their persistence to being relatively unavailable for microbial attack. Most compounds almost certainly undergo some changes during the long period of transit, sometimes generating compounds that are even more toxic than the parent compound. Thus, concern about organic compounds stems not only from their persistence, but also their ability to form more toxic metabolites after transformation.

Microorganisms are constantly evolving in the environment to handle the various organic pollutants to which they are exposed. Thus, in general pollution of an environmental matrix decreases the number of species present in that particular ecosystem (i.e., species diversity) but concomitantly increases the number of individuals within taxa that can degrade the pollutant at hand. The diversity can therefore be assessed over time both at polluted and nearby nonpolluted sites and diversity indices computed as an indicator of the extent of pollution. Most commonly relied on in environmental microbiology for this kind of study is the Shannon-Weaver (Weiner) diversity index (Shannon and Weaver, 1963; eqn. 1).

Shannon's diversity index

$$(H) = (n \log n - \sum_{i=1}^{k} f_i \log f_i)/n \quad ...(1)$$

where H is the diversity index, n the sample size, k the number of categories, and $f_i$ the number of observations in category i. To be more meaningful, indices between two (or more) habitats, or before and after a particular (pollution) event, have to be compared. The index varies from zero for habitats with only a single species to a high value for habitats with many species. Thus the closer the diversity index to zero, the more severely the environment in question is seemingly polluted. Organic compounds are mineralized and recycled more efficiently when a diverse microbial population is present (Torsvik and Øvreås, 2002). However, the relationship between microbial diversity and biodegradation is not linear as many metabolic processes are carried out by a consortium rather than individual microbial species.

Other indices, such as the Simpson index, and the log series index, have also been used to assess pollution (Zhou et al., 2002). Irrespective of the index chosen, however, caution should be exercised when applied to microbial systems because of difficulties in speciating the large number of microorganisms which may be present. Oftentimes microbial diversity using these indices has depended on populations established using culturing techniques. As detailed in Chapter 6, this approach reflects only a small fraction of the microbial diversity. Combined with molecular gene probes, a better, more precise assessment of the influence of pollution due to microbial diversity is assured.

## 10.1 GROWTH-LINKED BIODEGRADATION

A bacterial cell generally comprises carbon (50%), oxygen (20%), nitrogen (14%), hydrogen (8%), phosphorus (3%), potassium (2%), sulfur (1%), and other elements (2%) (Neidhart et al., 1990). Other microorganisms are also generally containing the same elements but in slightly different proportions. The carbon is contained in all cellular components namely, proteins, nucleic acids, lipids, lipopolysaccharides, cell walls, etc. whereas nitrogen is a component of proteins, nucleic acids, and peptidoglycans. During the biodegradation process, the compounds serve as a source of carbon, nitrogen, and phosphorus and as final electron acceptor. Thus, under nontoxic conditions, biodegradation is linked to the growth of the degrading organisms, their cells accommodating some of the organic pollutant in biopolymers. Concomitantly the initial concentration of the organic pollutant will not be affected until the cell biomass is in equilibrium with the chemical biomass. After a substantial increase in the biomass of the degrader(s), the chemical rapidly disappears (Fig. 10.1). A typical aerobic conversion of carbon substrates generates about 5% cell carbon and about the same amount of $CO_2$ with just a trace of organic by-products but the exact proportion varies with the substrate, cells, and environmental conditions. The rate at which organic compounds are degraded also depends on the metabolic abilities of the existing microbial diversity. Thus, maximum biodegradation is expected to occur when species that are most effective in metabolizing the existing hydrocarbon metabolism are dominant.

Growth-linked biodegradation is concentration dependent. Thus, at low concentrations of the compound (substrate), the rate of growth will be low due to the limited substrate that may be quite dilute and unable to induce enzymes. With increasing concentrations of the substrate, however, the rate of growth increases accordingly but ultimately becomes limited at high concentrations of the compound due to other nutrients, or toxicity that results in the inability to trigger genes for degradation. Growth-linked biodegradation has been associated with the loss of efficacy of pesticides and enhanced mineralization of the pesticides (2,4-D and glyphosphate) after repeated applications (Robertson and Alexander, 1994). In some instances, some acclimation phase is required before any reasonable degradation occurs. Under growth-linked biodegradation, the organic compound serves as a C-source for the organisms. Thus, if the microbial population is initially low, it increases as the compound is degraded (Robertson and Alexander, 1994). That increase may further accelerate biodegradation. An increase in the concentration of the organic compound, for example through an additional spill, second application of a herbicide/pesticide, or releasing some additional discharge, increases the rate of microbial growth but only to a certain point at which further additions will not increase the growth rate further. This process is explained by Monod kinetics (Monod, 1949) which in mathematical terms states that:

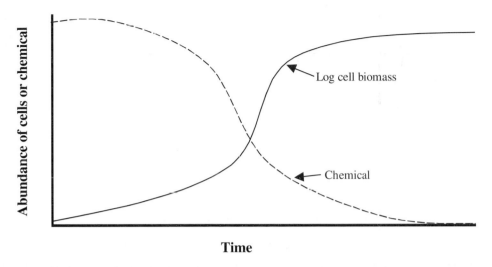

**Fig. 10.1** Disappearance of organic compound with a corresponding increase in biomass, signifying typical growth-linked biodegradation.

$$\mu = \frac{\mu_{max}S}{K_S + S} \qquad ...(2)$$

where $\mu$ is the specific growth rate of the organisms, $\mu_{max}$ the maximum specific growth rate, S the concentration of the compound (substrate), and $K_S$ a constant that represents the concentration of the substrate at which the rate of growth is half the maximum rate. The maximum growth rate ($\mu_{max}$) occurs at the highest range of concentration of the compound (substrate) at which growth still occurs whereas $K_S$ represents the affinity of the organism for the growth-linked compound (Alexander, 1994). The lower the $K_S$ value, the higher the affinity of the organism for the compound. This kinetic assumes a hyperbolic curve (Fig. 10.2).

The role of other microorganisms, other than bacteria, is rarely realized in the biodegradation process. Fungi metabolize organic compounds, possibly more extensively than bacteria in many soils although their contribution remains largely unexplored (Alexander, 1994). The grazing activity of protozoa correlates positively with biodegradation activities of microorganisms (Madsen et al., 1991). Enhanced degradation in the presence of protozoa may be attributed to the rapid turnover of bacterial cells as a result of predation by protozoa. Recent studies also highlight the importance of viruses in the cycling of nutrients due to their ability to lyse bacteria, causing rapid turnover in bacteria, particularly under marine conditions (Fuhrman, 1999). The impact of viruses in biodegradation is virtually an uninvestigated area. These oversights are further evidenced by the fact that the growth-linked kinetics of biodegradation are derived from Monod-based binary fission concepts, typical of bacterial growth. As pointed out in Chapter 4, fungal growth is exponential (see Section 4.1.1) and may not adequately fit a binary fission model.

Enrichment cultures are widely used to search for organisms that degrade compounds. The substrate of interest is supplied as the sole growth-limiting nutrient in a culture medium enabling the organisms with the ability to degrade the limiting nutrient to grow in culture medium and eventually become dominant. Selective enrichments prove physiological capabilities of microorganisms. This observation has led to a growing interest in reliance

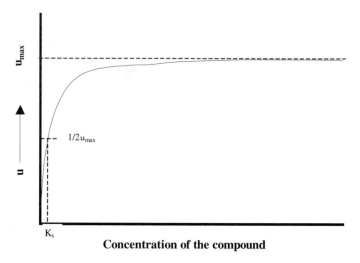

**Fig. 10.2** Hyperbolic relationship between growth rate of bacteria and concentration of the substrate (Monod kinetics model).

upon natural populations and subjecting them to selection pressure to degrade pollutants in the environment. However, this approach greatly relies on microorganisms evolving degradation abilities. Such evolution can be slow, especially when the number of genetic events involved in the degradation pathway are multiple and/or the selection pressure low. To "accelerate" the evolution of some degradation processes, various laboratories have investigated engineering microorganisms with either new degradation pathways or modifying existing ones (Ramos et al., 1987; Lau and de Lorenzo, 1999; Sayler and Ripp, 2000). However, the success of most engineered organisms under field conditions has been limited so far (see Chapter 15).

## 10.2 COMETABOLISM

While mineralization converts the organic molecule to its inorganic components, cometabolism does not result in growth or energy generation. Thus, cometabolism (or cooxidation) is the fortuitous modification of one molecule by an enzyme that routinely acts on another, its breakdown resulting in no assimilable carbon or energy for growth to the cometabolizing organism (Alexander, 1967; Hovarth, 1971). Thus, under cometabolism, although the organism is unable to use the compound as the sole carbon source, it changes the compound, for example through oxidation. During cometabolism another substrate is needed to meet the growth and energy requirements of the cometabolizing organism. If we take an example of a compound X being broken down under a standard metabolism, it could produce an intermediate Y and then a final product Z, generating $CO_2$, biomass and ATP (Fig. 10.3a). The process requires the presence of both enzymes *x* and *y*. Under cometabolism this picture changes completely since the organism lacks the enzyme *y* that facilitates the conversion of the intermediate Y' to the end products. The pool of Y' builds up and could be hazardous, thus worsening the situation. Its build-up would then inhibit the further conversion of X' to Y' by feedback mechanism resulting in an unnoticeable change in the abundance of X' (Fig. 10.3b). Alternatively, Y' could be a resource for other organisms. Thus X' is modified but with no benefit to the cell. These changes are effected because some of the enzymes involved in degrading the organism's main carbon source are not totally

**a) Standard metabolism**

X ———— $\xrightarrow{\quad x \quad}$ Y ———— $\xrightarrow{\quad y \quad}$ Z ———→ ———→ CO$_2$, ATP, biomass

**b) Cometabolism**

X' ———— $\xrightarrow{\quad x \quad}$ Y' ————//————→ Y' pool builds up or is a resource to other organisms

**Fig. 10.3**  Schematic representation of (a) standard metabolism versus (b) cometabolism (Modified from Alexander, 1994 reprinted with permission from Elsevier).

specific for their substrates but can act very slowly on other compounds. Cometabolic processes usually occur via nonspecific enzymatically mediated transformations (Stoner, 1994). In some however, the enzymes for cometabolism may be inducible rather than constitutive, such that the inducer has to be present for cometabolism to occur.

In the environment, cometabolism causes problems as it may fail to significantly reduce the concentration of the chemical due to the inherently low level of the microbial populations undergoing the process. Thus, it results in no significant increase in the biomass of the cometabolizing organisms (Fig. 10.4). Since the compound is not completely mineralized under cometabolism, more toxic metabolites may accumulate. For example, in the cometabolism of 2,3,6-trichlorobenzoate to 3,5-dichlorocatechol by the *Bravibacterium* sp., the accumulating 3,5-dichlorocatechol becomes toxic to *Bravibacterium* sp., arresting the cometabolizing process. An example of cometabolism is the breakdown of DDT to DDE, the latter building up in the food chain. In another instance, both carbofuran and simazine were cometabolized in soil (Robertson and Alexander, 1994). All these instances demonstrate that cometabolism is

fairly widespread and an important mechanism by which a number of recalcitrant organic compounds are degraded. Recalcitrance in this case refers to resistance to biological degradation rather than photooxidative or chemical processes.

## 10.3 INFLUENCE OF PHYSICOCHEMICAL PROPERTIES ON ORGANIC COMPOUND BIODEGRADATION

The disappearance of the compound does not necessarily indicate biodegradation as the disappearance of the compound can also be, at least in part, a result of volatilization, sorption on soil colloids, photolytic decomposition or chemical degradation. Thus, when assessing the degradability of organic (and inorganic) pollutants and laying out a remediation strategy, their physicochemical properties should be considered. Such physicochemical properties include pH, redox potential ($E_h$), electron acceptors, nutrients (especially N and P), solubility, volatility, sorption, temperature, structure, vapor pressure, and the mobility of the compound in question. Furthermore, the oxidation/reduction reactions, solubility, partitioning reactions, concentration, and the toxicity of

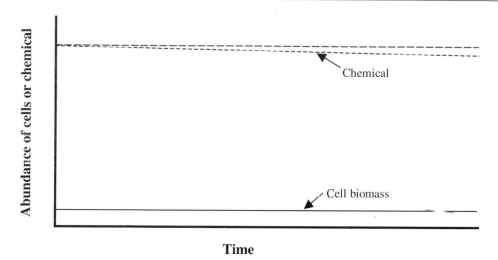

**Fig. 10.4** Abundance of chemical and microbial cells under cometabolism.

associated co-contaminants also affect degradability. Some of these properties are discussed below.

## 10.3.1 Solubility

The solubility of a compound is the amount that dissolves in a given amount of liquid at a given temperature. Most organic pollutants of concern have a low solubility in the universal solvent (water) whereas others, such as gasoline, are completely insoluble. Thus, a variety of organic compounds are immiscible with water and end up forming a separate liquid phase in an aqueous system called a nonaqueous phase liquid (NAPL), impeding mineralization of the organic phase. In general, the solubility and biodegradability of organic compounds declines as the number of carbons in the molecule and/or the length of the chain increases. Insoluble and low-solubility compounds are more resistant to biodegradation for two main reasons:

- insolubility renders them less accessible to reaction sites within the microbial cells, and
- they are more liable to sorb to inert materials which reduces their bioavailability.

From basic chemistry, we know that like solutes dissolve like solvents. Thus, polar compounds dissolve more readily in water (a polar solvent). This has implications on the sorption bioavailability, and subsequently, biodegradability of such compounds as microorganisms require water to metabolize substrates, including organic compounds (Fig. 10.5). For biodegradation to occur, the potential degrader has to come into contact with the organic substrate. Furthermore, the organic substrate typically has to be dissolved in order to be taken up through cell membranes into the cytoplasm. Thus, although some components of petroleum, for example, are readily biodegradable, their degradability in the environment can be greatly reduced because of low solubility and subsequent bioavailability to potential microbial degraders (Radehaus, 1998).

The availability of organic compounds depends on their partition coefficient or, more specifically, the octanol-water partition constant ($K_{ow}$). $K_{ow}$ reflects the partitioning of the organic compound between the natural organic phases (arbitrarily chosen to be *n*-octanol) and water. A high $K_{ow}$ is typical of hydrophobic (i.e., water-hating) compounds and, therefore, more soluble in octanol than water whereas a low $K_{ow}$ signifies a compound highly soluble in water (compared to octanol). Some large hydrophobic

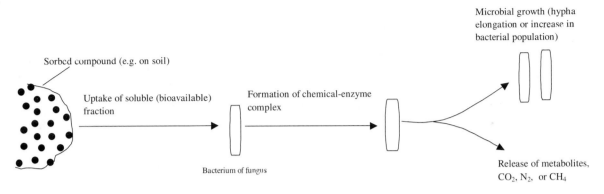

**Fig. 10.5**  Sequence of events in the bioavailability and subsequent biodegradation of sorbed compounds.

organic compounds such as the highly chlorinated biphenyls and polyaromatic hydrocarbons (PAHs) have large $K_{ow}$ values. The solubilities of some pollutants of interest are listed in Table 10.1. Slightly different values are available from different sources, possibly because the quality of water used to determine the values may differ (e.g., deionized water versus distilled water).

Herbicides, pesticides, and pharmaceutical compounds are usually quite soluble, with a comparatively low $K_{ow}$. However, they are also designed to be biologically very active and toxic to microorganisms (Jjemba and Robertson, 2003). Thus, their low $K_{ow}$ will facilitate the transfer of the polar compound into microbial cells leading to their bioaccumulation.

It should be noted that the biodegradation processes also involve enzymatic activity. The enzymes may be either inducible or constitutive. Inducible enzymes are formed in substantial amounts only when the substrate or structurally related compound is present whereas constitutive enzymes are expressed regardless of whether the substrate (or related compound) is present or not. Furthermore, some microorganisms produce surfactants (emulsifying agents) such as lipoproteins that improve on the interface (and thus reduce the $K_{ow}$ on a microscale) between the organic compound and the microbial cells, lowering the interfacial

energy and subsequent uptake by the cell (Phale et al., 1995). $K_{ow}$ also affects sorption of the compound (see below).

### 10.3.2  Sorption
Sorption collectively refers to both adsorption on surfaces and absorption of a liquid into solids. A wide range of organic compounds sorb to matrices in the environment (soil, sediments, waste water, etc.). Organic compounds have either a polar or nonpolar structure and the former are not significantly adsorbed on soil colloids and sediments, a trait that makes them more labile in the environment. Nonpolar compounds on the other hand tend to be more liable to sorption onto sediments. Low $K_{ow}$ facilitates the transfer of polar compounds into biological cells, as the compounds are less sorbed, leading to their bioaccumulation.

In soils, sewage, sediments, and other particles in the environment, sorption greatly depends on the organic matter content as expressed from the following mathematical relationship

$$K_{oc} = K_d \times 100/\%OC \qquad ...(3)$$

where $K_{oc}$ is the distribution coefficient normalized to the percent organic carbon (%OC) present in the sorbent and $K_d$ is the ratio between the concentration of the compound to the sorbent ($C_s$) to its concentration in solution ($C_{aq}$) (Jjemba, 2002a). Put differently,

**Table 10.1**   Physicochemical characteristics and persistence of herbicides insecticides, and PAHs in soils

| Compound | Molecular formula | Solubility (mg L$^{-1}$)[a] | log K$_{ow}$ | Vapor pressure (atm) |
|---|---|---|---|---|
| **Chlorinated insecticides** | | | | |
| Aldrin (Octalene) | $C_{12}H_8Cl_6$ | 0.2 | 6.5 | $1.6 \times 10^{-7}$ |
| DDT (2,2-bis(*p*-chlorophenyl)-1,1,1-trichloroethane) | $C_{14}H_9Cl_5$ | 0.0015 | 6.36 | $2.1 \times 10^{-10}$ |
| Chlordane | $C_{10}H_6Cl_8$ | 0.05 | 6.22 | $1.3 \times 10^{-8}$ |
| Heptachlor | $C_{10}H_5Cl_7$ | 0.18 | 3.9–4.5 | $4.0 \times 10^{-7}$ |
| Lindane (hexachlorocyclohexane) | $C_6H_6Cl_6$ | 6.5 | 3.3 | $1.2 \times 10^{-8}$ |
| **Organophosphate insecticides** | | | | |
| Dieldrin | $C_{12}H_8Cl_6O$ | 0.186 (at 20°C) | 4.56 | $2.4 \times 10^{-10}$ (at 20°C) |
| Malathion (Sumitox) | $C_{10}H_{19}O_6PS_2$ | 145.3 | 2.7 | $4.0 \times 10^{-9}$ |
| Parathion | $C_{10}H_{14}NO_5PS$ | 14.59 | 3.81 | $8.0 \times 10^{-9}$ |
| **Chlorinated herbicides** | | | | |
| 2,4-D (2,4-dichlorophenoxyacetic acid) | $C_8H_6Cl_2O_3$ | 500 | 2.81 | $5.3 \times 10^{-4}$ (at 160°C) |
| 2,4,5-T (2,4,5-trichlorophenoxyacetic acid) | $C_8H_5Cl_3O_3$ | 240 | 6.8 | $9.86 \times 10^{-7}$ |
| Dalapon (2,2-dichloropropanoic acid) | $C_3H_4Cl_2O_2$ | 800,000 | 1.68 | $8.93 \times 10^{-4}$ |
| Atrazine | $C_8H_{14}Cl_6N_5$ | 33 | 2.65 | $4 \times 10^{-10}$ (at 20°C) |
| Simazine (2-chloro-4,6-bis-ethylamino-1,3,5-triazine) | $C_7H_{12}ClN_5$ | 6.2 | 2.18 | $2.9 \times 10^{-11}$ |
| **Aliphatic hydrocarbons** | | | | |
| Trichloroethylene (TCE) | $C_2HCl_3$ | 1,100 | | $0.026^b$ |
| Tetrachloroethene (PCE) | $C_2Cl_4$ | 150 | | $0.018^b$ |
| **Alicyclic and unsaturated hydrocarbon** | | | | |
| Benzene | $C_6H_6$ | 1746.2 | 2.17 | $0.10^b$ |
| Toluene (Methylbenzene) | $C_7H_8$ | 547 | 2.69 | $0.029^b$ |
| Styrene (Vinylbenzene) | $C_8H_8$ | 306.8 | 2.95 | $0.0066^b$ |
| Ethylbenzene | $C_8H_{10}$ | 168 | 3.20 | $0.0092^b$ |
| **Polyaromatic hydrocarbons (PAHs)** | | | | |
| Naphthalene | $C_{10}H_8$ | 32 | 3.33 | $0.00014^b$ |
| Fluorene | $C_{13}H_{10}$ | 0.19 | 4.32 | $0.001$ to $0.01^c$ |
| Phenanthrene | $C_{14}H_{10}$ | 1.12 | 4.57 | $0.00094^c$ |
| Anthracene | $C_{14}H_{10}$ | 0.045 | 4.68 | $0.00024^c$ |
| Pyrene | $C_{16}H_{10}$ | 0.14 | 5.13 | $0.0000025^c$ |
| Fluoranthene | $C_{16}H_{10}$ | 0.22 | 1150 | $0.000005^c$ |
| Benzo(a)pyrene | $C_{20}H_{12}$ | 0.002 | 6.13 | $5.5 \times 10^{-9\ c}$ |
| **Chlorobenzenes** | | | | |
| Chlorobenzene | $C_6H_5Cl$ | 461 | 2.78 | |
| 1,2,3-Trichlorobenzene | $C_6H_3Cl$ | 20.82 | 4.14 | |
| 1,2,3,4-Tetrachlorobenzene | $C_6H_2Cl_4$ | 4.41 | 4.64 | |
| Pentachlorobenzene | $C_6HCl_5$ | 0.65 | 5.18 | |
| Hexachlorobenzene | $C_6Cl_6$ | 0.0062 | 5.80 | |

*Table 10.1 contd.*

*Table 10.1 contd.*

| | | | | |
|---|---|---|---|---|
| **Selected polychlorinated biphenyls (PCB) congeners** | | | | |
| Biphenyl | $C_{12}H_{10}$ | 7.04 | 4.01 | |
| 2-Chlorobiphenyl (PCB1) | $C_{12}H_9Cl$ | 5.43 | | 0.4 |
| 4,4'-Chlorobiphenyl (PCB 15) | $C_{12}H_8Cl_2$ | 0.06 | | 3.5 |
| 2,4,4'-Chlorobiphenyl (PCB28) | $C_{12}H_7C_3$ | 0.16 | | 0.02 |
| 2,2',5,5'-Chlorobiphenyl (PCB52) | $C_{12}H_6Cl_4$ | 0.03 | | |
| 2,2',4,4',5,5'-Chlorobiphenyl (PCB153) | $C_{12}H_4Cl_6$ | 0.001 | | |
| **Pharmaceutical compounds** | | | | |
| Ciprofloxacin | $C_{17}H_{18}FN_3O_3 \cdot HCl$ | 10,000 | 1.32 | |
| Tylosin (Tilmicosin) | $C_{46}H_{90}N_2O_{13}$ | $\geq$1,500 | | |
| Metronidazole (Flagyl) | $C_6H_9N_3O_3$ | 100,000 (at 20°C) | 0.02 | |
| Carbamazepine | $C_{15}H_{12}N_2O$ | 10 | 2.93 | |
| Sulfomethazine | $C_6H_8N_2$ | 1500 (at 37°C) | | |
| Avermectin (Abamectin) | $C_{48}H_{74}O_{14}$, 'R=$C_2H_3$) | 0.000078 | | 3.19 |
| Flumequine | $C_{14}H_{12}NO_5F$ | | | 1.7 |
| Oxytetracycline (Terramycin) | $C_{22}H_{24}N_2O_3$ | 100,000 (at 21°C in $H_2O$; but <1,000 at 17.5°C in organic solvents e.g. alcohol) | | -1.22 |
| Tetracycline | $C_{22}H_{24}N_2O_8$ | 50,000 (at 22°C) | -1.19 | |
| **Other solvents** | | | | |
| Carbon tetrachloride | $CCl_4$ | 8,200 | | |
| Phenol | $C_6H_6O$ | 92.92 | | 0.529[b] |
| Metyl-t-butyl ether (MTBE at 298 K) | $C_5H_{12}O$ | 4,800 | 1.2 | 250[c] |

[a]Most of the solubilities were calculated from values provided by Ruelle and Kesselring (1997).

[b]Vapor pressure at 20°C and [c]at 25°C

*Other sources*: http://www.usace.army.mil/inet/usace-docs/eng-manuals/em1110-1-400'/a-b.pdf and http://chemfinder.cambridgesoft.com

$$K_d = C_s/C_{aq} \qquad ...(4)$$

The normalized distribution coefficient ($K_{oc}$) has a quantitative relationship with the 1-octanol-water partition coefficient ($K_{ow}$), a more readily available parameter. A low $K_{ow}$ results in a relatively low affinity to sorb on organic matter and minerals. Based on $K_{ow}$ values, $K_d$ is estimated and used to compute $K_{oc}$ from the relationship (Stuer-Lauridsen et al., 2000)

$$\log K_d = \%OC \times 0.41 \times \log K_{ow} \qquad ...(5)$$

The more hydrophobic a compound, the more liable it is to sorb passively to colloids. Sorption directly affects bioavailability of any compound prior to degradation by microorganisms. Note that sorption does not completely inhibit biodegradation but rather reduces its rate and extent. For example, our studies with pyrene indicate that the compound is rapidly sorbed onto soil colloids but a small and fairly constant amount remains in solution and thus bioavailable to the degraders (Jjemba et al., in litt). The sorbed and solution fractions therefore exist in equilibrium (eqn. 6).

$$\text{Sorbed chemical} \underset{k_2}{\overset{k_1}{\rightleftharpoons}} \text{chemical in solution}$$
$$\text{(bioavailable)}$$
$$\downarrow k_3$$
$$\text{degradation product} \qquad ...(6)$$

The rate of degradation will depend on the rate at which the compound diffuses into the solution fraction, i.e., the desorption rate ($k_1$). The adsorption rate is represented by $k_2$ whereas $k_3$ is the biodegradation rate. These three constants ($k_1$, $k_2$, and $k_3$) have been used to model the kinetics of biodegradation for various environments with the assumption that desorption is the rate-limiting step. However, modeling can be complicated by the possibility of microorganisms excreting metabolites (i.e., surfactants or emulsifying agents). As pointed out above, some metabolites lower the interfacial energy and increase solubility. It is also plausible that some of the sorbed compounds

may come into direct contact with some organisms that adhere to the same surface, and penetrate the cell without coming into the surrounding liquid first. This scenario is not built into most models either.

### 10.3.3 Volatility

Volatility is the tendency of a chemical to evaporate from its liquid or solid phase. It is expressed as the density of pure vapors to the density of the air. The vapor pressure values of some common organic hazardous compounds are listed in Table 10.1. A high vapor pressure signifies high volatility. Several organic compounds are readily volatile under normal atmospheric conditions but the vapor pressure depends on the structure and temperature of the environment. Thus, increasing temperatures increase the vapor pressure, subsequently increasing the volatility. This in turn may reduce the availability of the compounds to microorganisms for degradation. For example, chlorobenzenes and alkyl halides are highly volatile, a trait that plays a major role in their disappearance from their point of introduction, particularly where temperatures are favorable (>20°C). For biodegradation to proceed, temperatures have to be higher than 10°C.

### 10.3.4 Transport

The extent to which organic compounds can be spatially distributed greatly depends on the environmental matrix in which the compounds occur. For example, under marine conditions distribution will depend on the flow of the water and the wave currents, creating some turbulent mixing. Such mixing and movement also offers an aerobic or semiaerobic environment which could enhance biodegradation. In terrestrial environments, on the other hand, the dispersal of organic compounds can be quite limited due to sorption, limiting bioavailability and degradation. That said, however, organic chemical plumes can, in some instances, drift through soil fractures, channels, and crevices into the groundwater. In this manner, NAPLs can be formed,. NAPLs are also formed in sediments, soils, and other water columns such as estuaries. The solubility of organic compounds that comprise a NAPL is often quite low, such that the NAPL becomes a long-term source of contamination as small amounts of the organic compound solubilize, becoming transported to other sites where it may or may not undergo any significant biodegradation.

## 10.3.5  Acidity and alkalinity (pH)

Organic compounds have functional groups that have either proton-donor or acceptor properties. A wide range of functional groups (e.g. the $SO_2R$, $NH_3^+$, $NR_2$, CN, F, Cl, Br, $NO_2$, COOR, COR, OH, OR, I, SR, and phenyl) have the ability to withdraw electrons and a few (i.e., $O^-$, $NH^-$ and alkyl) have the ability to donate electrons (Schwarzenbach et al., 2002). Since organic matter is naturally negatively charged, it readily interacts with positively charged functional groups, facilitating sorption. Furthermore, pH directly influences the growth rate of microorganisms, with most bacteria optimally growing around neutral pH whereas fungi optimally grow under slightly acidic conditions (pH 5-6). Thus, biodegradation is typically rapid between pH 6-8 (Radehaus, 1998).

## 10.3.6  Redox potential ($E_h$)

The ability of an organism to carry out redox reactions depends on the oxidation-reduction state of the environment (see Chapter 7, Section 7.2). In organic compounds, most redox reactions involve three molecules, notably C, N, and S. These reactions are biologically mediated and redox potential is very important in considering the biodegradation of organic compounds in the environment. Oxygen is one of the strongest electron acceptors (oxidants) and thus, biodegradation is typically more pronounced in the presence of oxygen. This observation has important ramifications in biodegradation and in designing conditions to enhance this process. Most redox reactions in the environment are biologically mediated as electron transfer to or from the organic compound may not be thermodynamically feasible under abiotic conditions. Nevertheless, thermodynamic considerations are important as a first step in considering redox conditions under which a compound might undergo oxidation or reduction. What is apparent is that most biodegradation processes require cooperation between several species. Furthermore, biodegradation occurs in steps. For example, petroleum, which is comprised of straight chain alkanes, cycloalkanes and aromatic components in various fractions (Table 10.2), is degraded by mineralization or cometabolism. The metabolism of alkanes occurs in three distinct steps (Fig. 10.6a). The alkanes are initially converted to an alcohol, which is then converted back to an aldehyde and subsequently to a carboxylic acid. The resultant acid then undergoes β-oxidation (Fig. 10.6b), generating acetate units that "feed" the TCA cycle, releasing $CO_2$. It is therefore obvious that each of these respective components will be metabolized separately. The acetyl CoA generated gets into the tricarboxylic acid (TCA) cycle. As discussed below, a variety of other compounds, including aromatic hydrocarbons and chlorinated herbicides undergo β-oxidation during biodegradation.

It should be noted that even though this sequence of events is shown at the terminal $-CH_3$ group for simplicity, oxidation can also occur on the subterminal methyl groups or on the branches. However, branching increases the resistance of alkanes to microbial attack. The more complex cyclic alkanes are initially

**Table 10.2**  Components of petroleum

| Component | Percentage of | |
| --- | --- | --- |
| | Crude oil | Gasoline |
| Straight chain alkane | 30 | 17 |
| Cycloalkane | 50 | 54 |
| Aromatic | 20 | 10 |

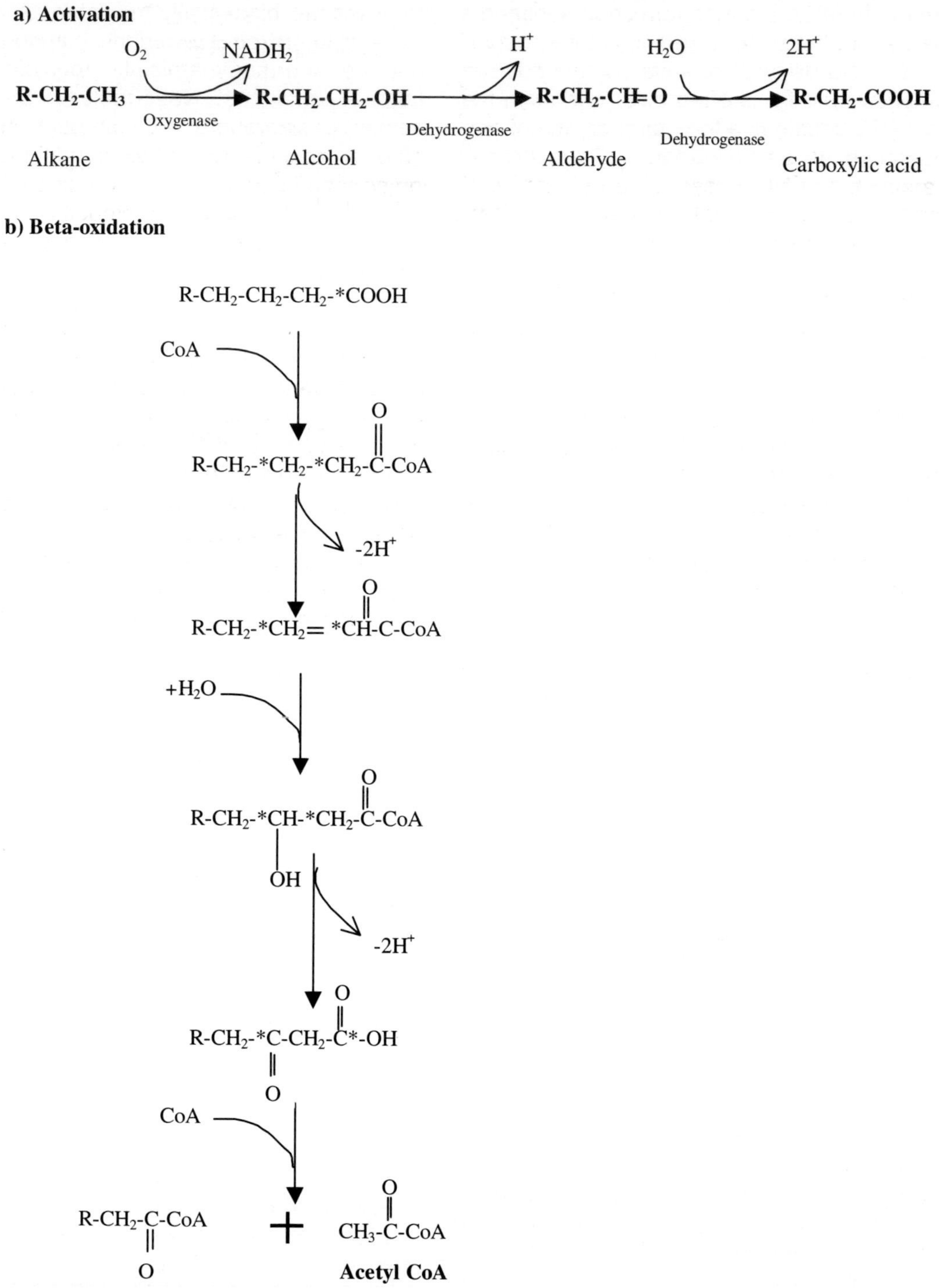

**a) Activation**

R-CH$_2$-CH$_3$ $\xrightarrow[\text{Oxygenase}]{\text{O}_2 \quad \text{NADH}_2}$ R-CH$_2$-CH$_2$-OH $\xrightarrow[\text{Dehydrogenase}]{\text{H}^+}$ R-CH$_2$-CH=O $\xrightarrow[\text{Dehydrogenase}]{\text{H}_2\text{O} \quad 2\text{H}^+}$ R-CH$_2$-COOH

Alkane           Alcohol           Aldehyde           Carboxylic acid

**b) Beta-oxidation**

R-CH$_2$-CH$_2$-CH$_2$-*COOH

CoA

R-CH$_2$-*CH$_2$-*CH$_2$-C-CoA (C=O)

$-2\text{H}^+$

R-CH$_2$-*CH$_2$= *CH-C-CoA (C=O)

$+\text{H}_2\text{O}$

R-CH$_2$-*CH-*CH$_2$-C-CoA (C=O)
|
OH

$-2\text{H}^+$

R-CH$_2$-*C-CH$_2$-C*-OH (C=O)
||
O

CoA

R-CH$_2$-C-CoA (C=O)    **+**    CH$_3$-C-CoA (C=O)

                               **Acetyl CoA**

**Fig. 10.6** Mineralization of alkanes in petroleum by bacteria and fungi (a). The carboxylic acids generated undergo β-oxidation (b).

degraded through cometabolic processes, followed by mineralization of the cometabolites. This observation emphasizes the importance of looking at the degradation process as a holistic event involving a mixed microbial community (consortium) in the environment. Many biodegradation studies have used a single taxon and/or compound of interest. In contrast, most compounds occur as complex mixtures in the environment.

Under anaerobic or microaerophilic conditions, other electron acceptors such as nitrate, ferric, sulfate, etc. (see Fig. 7.3) come into play, with the more oxidized compounds (i.e., compounds with a higher $E_h$) being used first. The composition of the microbial community in a particular environment will, therefore, change depending on the predominant electron acceptor. Heavily contaminated sites tend to become anaerobic because the ongoing respiration subsequently depletes the available oxygen rapidly (Radehaus, 1998).

## 10.4 MECHANISMS OF METABOLIZING DIFFERENT CLASSES OF ORGANIC POLLUTANTS

Microbial attack is crucial in the degradation of organic compounds in the environment. The compounds in question are then oxidized or reduced. The first step in metabolism often involves hydrolysis, dehalogenation, hydroxylation, demethylation, methylation, acetylation, ether cleavage, nitro reduction, or the conversion of nitrile to an amide (Table 10.3). The degradation of specific classes of organic compounds is discussed below.

### 10.4.1 Chlorinated aliphatic compounds

Chlorinated hydrocarbons are of particular interest because of their high toxicity and also high persistence in the environment. The C-Cl bond is rare in nature but not absent. For example, algae make chlorinated lipids and fungi

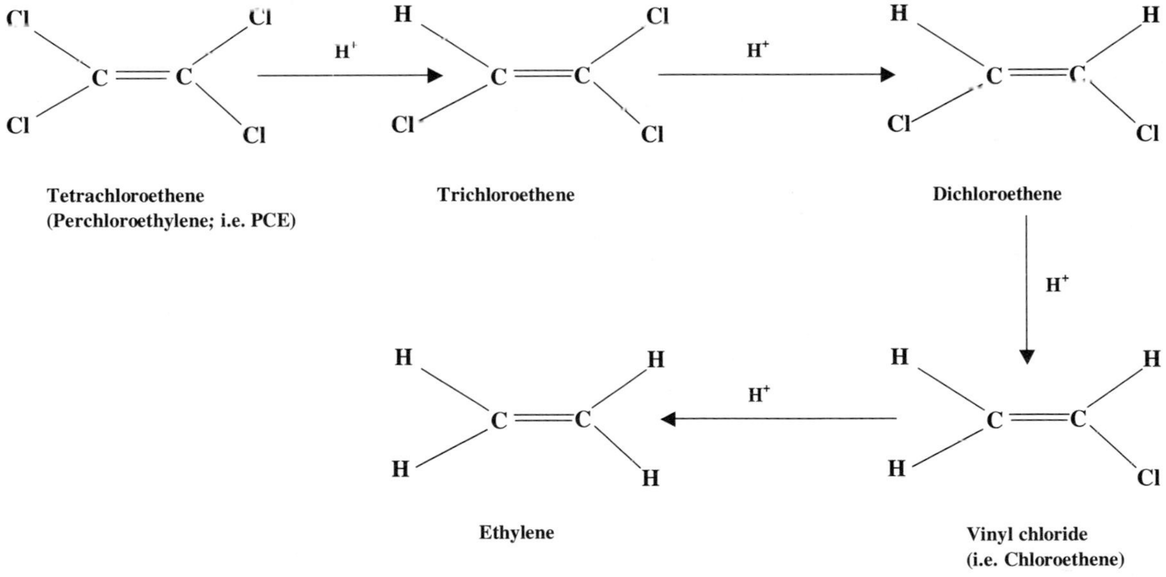

**Fig. 10.7** General pathway for biodegradation of tetrachloroethene (TCE) in the environment.

**Table 10.3** Typical microbially mediated mechanisms of biodegradation

| Mechanism | Description | Remarks/References |
|---|---|---|
| Dehalogenation | Involves replacement of a halogen (i.e. F, Cl, Br, I, or At) by hydrogen (reductive dehalogenation) or a hydroxyl group (hydrolytic dehalogenation). $R\text{-}Cl \rightarrow RH$ or $R\text{-}Cl \rightarrow ROH$ | Reductive dehalogenation is very important under anaerobic conditions and leads to the degradation of a variety of alkyl halides such as trichloroethene and tetrachloroethene (see Fig. 10.7), chlorinated phenols (e.g. pentachlorophenol). Hydrolytic dehalogenation of phenolics subsequently leads to a catechol. Reductive dehalogenation has also been reported for pentachlorophenols under anaerobic conditions (Suflita et al., 1982), DDT and DDD (Guenzi and Beard, 1967). |
| Demethylation (or dealkylation) | Involves the removal of a methyl ($-CH_3$) or alkyl ($C_nH_{2n+1}$). Alkyl groups are the most ubiquitous hydrocarbon substituents in environmental organic chemicals and include ethyl ($CH_3CH_2\text{-}$), propyl ($CH_3CH_2CH_2\text{-}$), and butyl ($CH_3CH_2CH_2CH_2\text{-}$) groups and are straight chains or branched. | Most pesticides and herbicides contain a methyl group that is connected to N or O atoms. Some microorganisms can demethylate or dealkylate, leading to loss of toxicity and subsequent degradation. For example, atrazine is dealkylated resulting in the loss of its N-ethyl or N-isopropyl groups by various microorganisms including *Rhodococcus, Ralstonia, Agrobacterium radiobacter, Norcadioides, Pseudoaminobacter,* and *Pseudomonas* (Falebitso et al., 2002), indicating that the genes for this mechanism are fairly widespread. |
| Ether cleavage | Cleavage of the C-O-C bond. Despite the fact that the C-O bond is a high energy one, some micro-organisms can cleave the bond. | Ether bonds are common in some herbicides such as phenoxy herbicides 2,4-dichlorophenoxyacetic acid (2,4-D) and 2,4,5-trichlorophenoxyacetic acid (2,4,5-T) Biodegradation of these herbicides is discussed in Section 10.5.2. |
| Hydroxylation | Addition of an OH group to the aliphatic or aromatic molecule ($RH \rightarrow ROH$). | Usually the first step in the biodegradation of alkanes and aromatic compounds by bacteria (see Section 10.5.3 and Fig. 10.5b). It is carried out by a membrane-bound system and subsequently leads to the formation of a catechol or diol. Many pesticides are degraded or detoxified by hydroxylation. |
| Hydrolysis | Cleavage of a bond by the addition of water. $RCOOR' + H_2O \rightarrow RCOOH + HOR'$ | Affects various functional groups such as esters, amides and anhydrides. For example, methyl chlorpyrifos is hydrolyzed forming diesters, monoesters, and inorganic phosphothioates that are highly soluble (Stone, 1998). |
| Methylation (and acetylation) | In contrast to demethylation, some microorganisms methylate organic compounds generating less toxic products which may be even more recalcitrant, however. $R\text{-}OH \rightarrow R\text{-}CCH_3$ | Chlorophenols are mostly initially methylated. Methylation is by *Athrobacter* sp. and various fungi such as *Aspergillus, Paecilomyces, Penicillium* and *Scopulariopsis* spp. (Curtis et al., 1974; Neilson et al., 1983). |
| Nitrile conversion | Structurally diverse nitriles (organic cyanides) are converted to amides. $R\text{-}C \equiv N \rightarrow R\text{-}CO\text{-}NH_2$ | Most organic cyanides are very poisonous. Thus, their conversion to amines generates less toxic products which can subsequently be biodegraded. This detoxication can be caused by microorganisms in soil (Alexander, 1994) |
| Nitro reduction | Nitro compounds are reduced to amino groups. $R\text{-}NO_2 \rightarrow RNH_2$ | Nitroaromatic compounds are widely used in dyes, munitions, herbicides, pesticides and solvents. They are resistant to biodegradation but some microorganisms degrade them by initially reducing the nitro group to an amino group (Robertson and Jjemba, in litt.). |

excrete a considerable amount of methyl chlorides. Organochloride compounds constitute a large and important group of environmental pollutants as halogenated aliphatic compounds widely used as pesticides, herbicides, pharmaceuticals, semiconductors and dry-cleaning agents. Thus, chlorinated compounds such as trichloroethane, tetrachloroethane, dichloroethane, vinyl chloride and tetrachloroethene (i.e., perchloroethylene, PCE) are detected in soil, surface water, sediments, groundwater, foods, and in the atmosphere, and are major pollutants, particularly in industrialized countries. PCE is a major pollutant in the US. Its carbon has an oxidation state of +2 and plays a physiological role as a final electron acceptor.

The C-Cl bond can be attacked through three methods, i.e., reductive dechlorination, hydrolytic dechlorination and elimination dechlorination. Hydrolytic dechlorination is also reductive but uses $H_2O$ instead of protons whereas in elimination dechlorination oxygen is introduced, thus destabilizing the C-Cl bond. Reductive dechlorination (also known as reductive dehalogenation) has been more widely studied compared to the other two types of chlorination and is discussed here in some detail with specific reference to several common pollutants. It usually occurs under anaerobic conditions but can also occur under aerobic conditions. For example, it is generally agreed that under both aerobic and anaerobic conditions, PCE and a variety of other chlorinated aliphatic compounds can undergo reductive dechlorination. The process involves several microbial species acting synergistically to transform the compound but it is not clear whether microorganisms catalyze the dehalogenation enzymatically or only generate the reductant that reacts nonenzymatically to dehalogenate the chlorinated molecule (Alexander, 1994). In some instances, tetrachloroethene can be transformed into chloroethene (vinyl chloride, i.e., $CHCl=CH_2$), which is a proven carcinogen and is far more toxic than its parent compound (Fig. 10.7). However, if given sufficient duration, vinyl chloride can be metabolized under anaerobic conditions to ethylene ($C_2H_4$), a nontoxic analogue that is readily metabolized under aerobic conditions. Ethylene has carbon with an oxidation state of $-2$ and is a plant hormone. This pathway is seemingly more prevalent in the presence of zero-valent iron (Fe(0)), zero-valent zinc (Zn(0)), iron sulfide (FeS) or a variety of other reductants (Schwarzenbach et al., 2002). As of now, it is not clear whether this reaction can be facilitated in nature, for example through plant-microbe remediation systems (phytoremedation).

Like tetrachloroethylene, trichloroethylene used in degreasing, dry cleaning, and semiconductors, is one of the most common pollutants in the United States and several other industrialized countries. It is a highly potent carcinogen which is not readily biodegradable. In the presence of microorganisms that have methane monooxygenases, trichloroethylene can undergo cometabolism under aerobic conditions to form biodegradable low molecular weight alcohols (Fig. 10.8). The chemically unstable intermediate that is generated is short lived as it has a half-life of only 12 seconds.

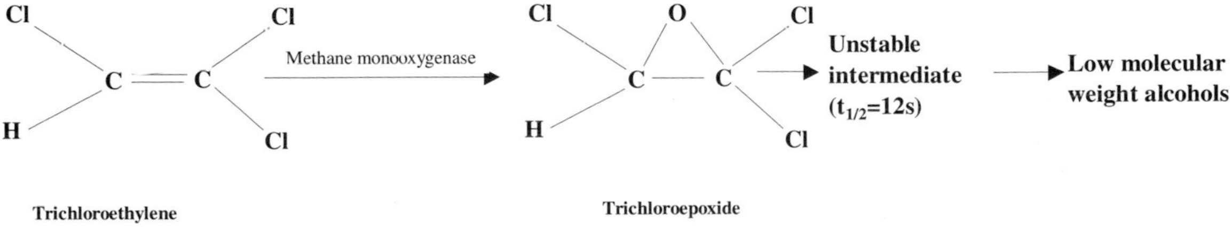

**Fig. 10.8** General pathway for degradation of trichloroethlyene to low molecular weight products.

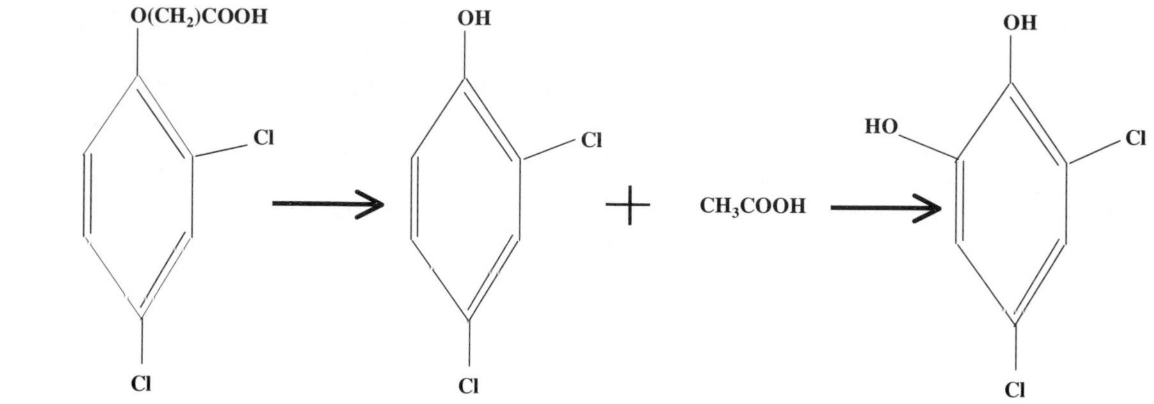

2,4-dichlorophenoxyacteic acid (2,4-D)       2,4-dichlorophenol                                              3,5-dichlorocatechol

**Fig. 10.9**   General pathway for degradation of 2,4-D.

Under anaerobic conditions, methanogens and methanotrophs such as *Methylosinus* sp. degrade TCE. However, the methanogens must have either methanol or acetate to grow and degrade this compound whereas methanotrophs require low levels of methane. Thus, a delicate balance for methane has to exist for the latter to degrade TCE.

Similarly, degradation of chlorinated herbicides such as chlorophenoxy herbicide 2,4-dichlorophenoxyacetic acid (2,4-D) has been quite extensively studied. Most widely implicated in the degradation of this compound in soils and sediments are *Arthrobacter, Pseudomonas*, and *Achromobacter* sp. as well as fungi, in particular *Aspergillus niger* and *Phytophthora megasperma*. These microorganisms are also often associated with the degradation of other chlorophenoxyl herbicides such as chloro-2-methylphenoxyacetic acid (MCPA) and 2,4,5-trichlorophenoxyacetic acid (2,4,5-T). The general mechanism of 2,4-D degradation involves the initial cleavage of the molecule at the ether linkage between the side chain and the aromatic ring forming 2,4-dichlorophenol and destroying the phytotoxicity of the molecule. The dichlorophenol is subsequently converted to a catechol which is further degraded (Fig. 10.9).

2,4,5-T is more persistent in the environment compared to 2,4-D. Its rate of disappearance increases with increasing temperature and moisture, possibly due to an increased rate in biological activity of degraders. It can be converted by *Pseudomonas cepacia* AC1100 and *P. fluorescens* into 2,4,5-trichlorophenol which is then readily metabolized, liberating free chlorine (Rosenberg and Alexander, 1980a, b). Generally, chlorophenoxy herbicides are degraded through β-oxidation.

## 10.4.2 Metabolism of aromatic compounds

The simplest types of aromatic pollutants are commonly referred to as BTEX compounds, namely benzene, toluene, ethylbenzene and xylenes (Fig. 10.10a). Xylene has three isomers, namely ortho-, meta- and para-dimethylbenzene. BTEX compounds are quite common mobile contaminants in groundwater due to leaking underground storage tanks (LUST). Similar to alkanes in petroleum, BTEX metabolism also occurs in three steps. As a model, the metabolism of the simplest 6-carbon aromatic organic compound (benzene) is shown in Fig. 10.10b. Its six aromatic electrons are delocalized and this confers the stability

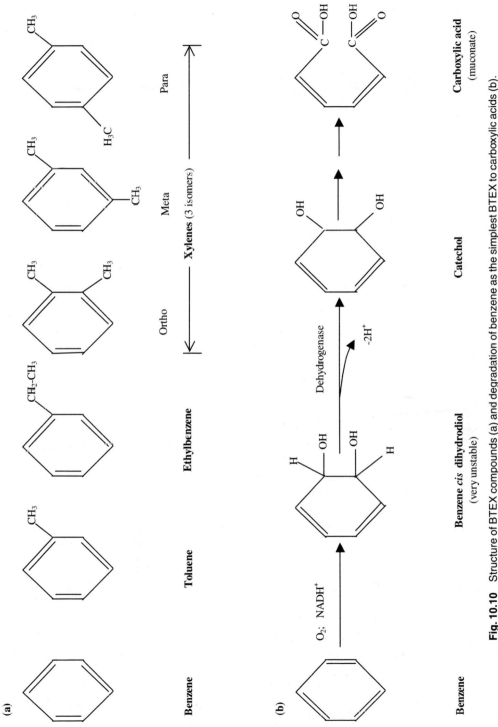

**Fig. 10.10** Structure of BTEX compounds (a) and degradation of benzene as the simplest BTEX to carboxylic acids (b).

inherent in all aromatic structures. Thus, benzene, just like other aromatics, is very difficult to attack and biodegrade until the molecule is activated. Initially, microorganisms, in particular bacteria, hydroxylate the ring under aerobic conditions, forming a very unstable dihydrodiol intermediate (Fig. 10.10b). In contrast, fungi utilize a monooxygenase enzyme to degrade aromatic compounds, the monooxygonase incorporating one atom of molecular oxygen into the benzene ring while converting the other to water. The benzene dihydrodiol intermediate is eventually converted into a catechol. Catechols are very important compounds in biodegradation as they are more easily cleaved and oxidized to carboxylic acids which are then readily metabolized by β-oxidation (Fig. 10.6b) and incorporated into the TCA cycle with the uptake of some ATP.

### 10.4.2.1 Aromatic pesticides, fungicides, and herbicides

Farmers often apply pesticides, fungicides or herbicides on a regular basis to control pests, fungal infections, and weeds respectively. It is desirable that the specific compound degrade after it has served its intended purpose. This usually translates into a desire that the compound degrades within the cropping season in which it is applied. However, some compounds have been observed to degrade rather quickly, even before they impact their intended target, due to accelerated degradation. Accelerated pesticide, fungicide, and herbicide degradation has been the subject of intensive research and has been linked to the possibility that with repeated application, the abundance of microorganisms that can use the compound as a substrate increases. On the other extreme are compounds that will not readily degrade months or even years after they have served their purpose. Most notable is DDT [i.e. 1,1,1-trichloro-2,2-bis (*p*-chlorophenyl)ethane], one of the most persistent pesticides in the environment. It is lipophilic and undergoes biomagnification in higher organisms in the food chain. It is reductively dechlorinated under anaerobic conditions to 1,1-dichloro-2,2-bis(*p*-chlorophenyl)ethane (DDD) in the nonaromatic portion of the molecule or to 1,1-dichloro-2,2-bis(*p*-chlorophenyl)ethylene (DDE) under aerobic conditions (Fig. 10.11). However, its degradation is also dependent on the redox potential of the environment. In soils, the rate of degradation was highest under a low redox potential ($E_h = -90$ to $-25mV$) and lowest at high redox potential ($E_h = +350mV$) (Glass, 1972; Parr and Smith, 1974). Its degradation is effected by various actinomycetes such as *Streptomyces aureofaciens*, *S. cinnamoneus*, *S. viridochromogenes*, *S. albus*, *S. antibioticus*, and *Norcadia crythropolis* also reported to degrade DDT to DDD under aerobic conditions, especially in the presence of high organic matter (Rochkind-Dubinsky et al., 1987). Various plant pathogens and saprophytes can also degrade DDT to DDD under anaerobic conditions (Johnson et al., 1967). It is noteworthy that the resultant DDD and DDE are also highly persistent in the environment.

### 10.4.2.2 Microbial metabolism of polycyclic aromatic hydrocarbons (PAHs)

PAHs are hydrophobic organic compounds with two or more fused benzene rings arranged in a linear, angular or clustered manner (Fig. 10.12). They are generated from petroleum, incomplete combustion of any organic material (e.g., soot in chimneys), coal tar, etc., are insoluble, and have strong binding (sorption) properties. Thus, in soils and sediments, they tend to associate with organic matter. Most of them are persistent, toxic, carcinogenic, or mutagenic (Arif et al., 1999; Tang et al., 2002) and bind to nucleophilic sites in proteins and nucleic acids, thereby precluding replication.

The three simplest polyaromatic compounds, i.e., anthracene, naphthalene and phenanthrene (Fig. 10.12) are more easily metabolized by bacteria. The degradation of PAHs is generally similar to that of monoaromatic hy-

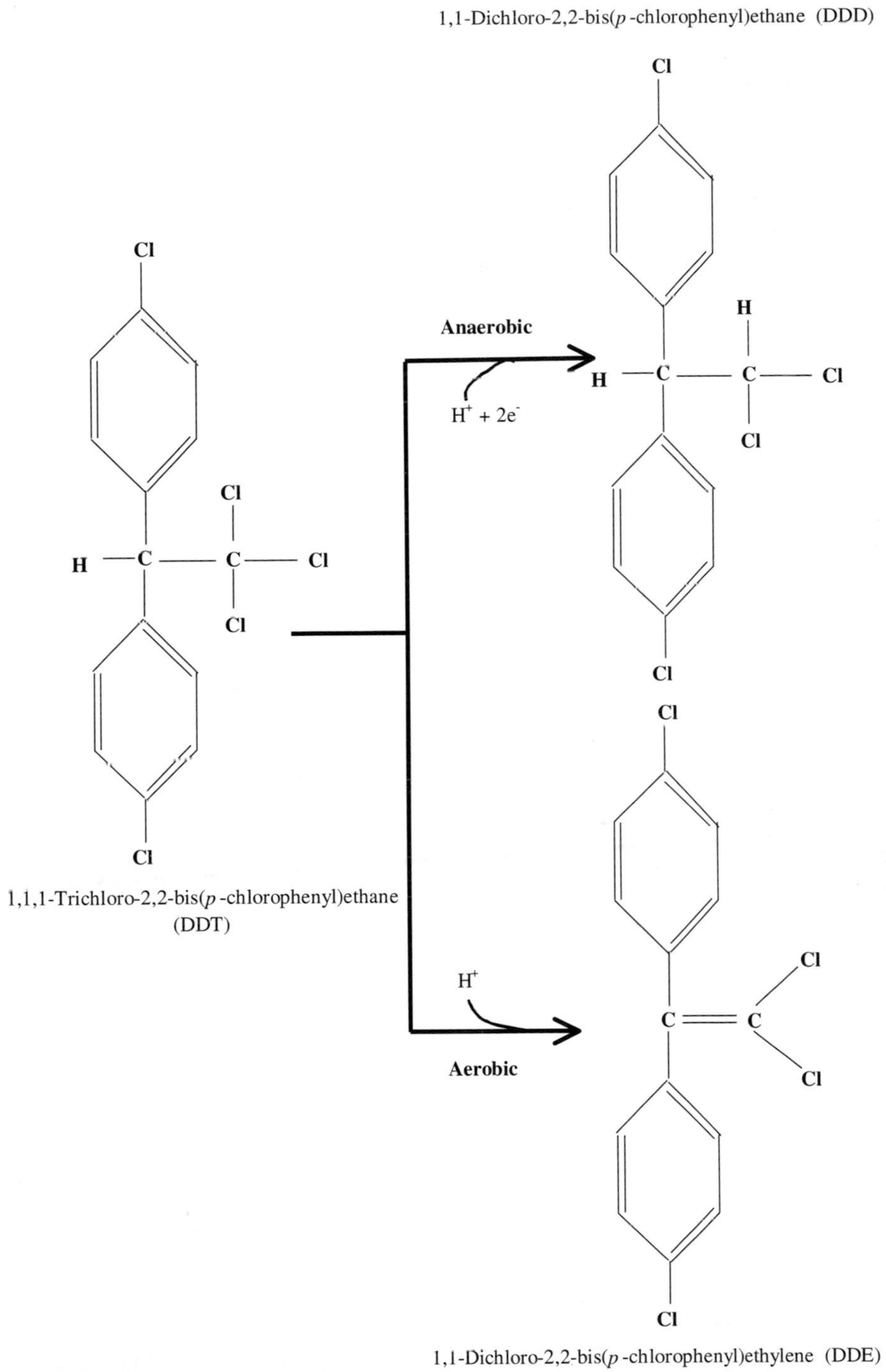
1,1-Dichloro-2,2-bis(p-chlorophenyl)ethane (DDD)
Cl
H
H
C
C
Cl
Cl
Cl
Anaerobic
H+ + 2e-
Cl
Cl
H
C
C
Cl
Cl
Cl
1,1,1-Trichloro-2,2-bis(p-chlorophenyl)ethane
(DDT)
H+
Aerobic
Cl
Cl
C
C
Cl
Cl
1,1-Dichloro-2,2-bis(p-chlorophenyl)ethylene (DDE)

**Fig. 10.11** Dehalogenation of DDT to DDE or DDD. The fate greatly depends on the oxygen status in the contaminated environment as DDE is formed under aerobic conditions whereas DDD is formed under anaerobic.

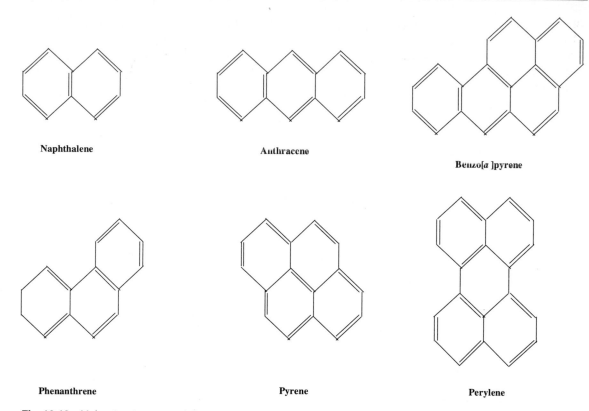

Naphthalene          Anthracene          Benzo[a]pyrene

Phenanthrene          Pyrene          Perylene

**Fig. 10.12**  Molecular structures of some polycyclic aromatic hydrocarbons (PAHs) of significance in the environment.

drocarbons in the sense that the molecules are initially oxidized by the microorganisms, incorporating two oxygen atoms by a dioxegenase enzyme, forming a dihydrodiol. Further oxidation leads to the formation of catechols that undergo fission, cleaving to form linear metabolites that enter the central metabolic pathway. As the number of rings increases, solubility decreases, sorptive reactions increase, and accessibility to enzyme attack decreases, thus decreasing biodegradability. The net result with increasing complexity (number of rings) is that they become persistent carcinogens (Table 10.4). Thus, the 3-ring PAHs were degraded to some extent (11-41%) when the nutrient status in the soil was enhanced with artificial rhizosphere exudates over a 30-day period whereas the 4-, 5-, and 6-ring PAHs were only minimally degraded under the same condi-

**Table 10.4**  Percent reduction in increasingly complex PAHs after 30 days due to both biodegradation and sorption in a column provided with artificial root exudates[1]

| Group | Compound | % decrease |
|-------|----------|-----------|
| 3 rings | Fluorene | 39.4 |
|  | Phenanthrene | 41 |
|  | Anthracene | 11 |
| 4 rings | Fluoranthene | 3 |
|  | Pyrene | 0 |
|  | Benzo[a]anthracene | 0 |
|  | Chrysene | 0 |
| 5 rings | Benzo[a,k]fluoranthene | 0 |
|  | Benzo[a]pyrene | 3.5 |
| 6 rings | Benzo[g,h,i]perylene | 0 |

[1]Based on data from Joner et al. (2002) with permission from Elsevier.

tions. The limited biodegradation of the 4 or more ring compounds is typically due to cometabolism rather than direct metabolism as not many microorganisms have been clearly shown to use these compounds as a source of C and energy. Recent studies by Jjemba et al. (in litt.) using fluorescent *in-situ* hybridization combined with substrate-responsive direct viable counts (SR-DVC) in one soil with a long history of PAH contamination show that pyrene is mostly degraded by α-Proteobacteria and β-Proteobacteria. Work to identify the degraders at the generic level is continuing. The only consolation is that increased complexity is also accompanied by a decrease in bioavailability (Hatzinger and Alexander, 1995). Bioavailability Is affected by the presence of a variety of components in the environment such as proteins, collagen, cellulose, chitin, lignin, fulvic acids, humic acids, humin, and kerogen, which sorb organic pollutants (Schwarzenbach et al., 2002). Thus, studies have clearly shown that progressive reduction in bioavailability corresponds to the respective increase in soil organic matter (Fig. 10.13). High organic matter content is associated with large amounts of

humic, humin, and fulvic acids (Tate, 1987).

Some PAHs, especially the 2- and 3-ringed PAHs, can also be degraded under anaerobic conditions, especially by sulfur-reducing bacteria, methanogens, and nitrate-reducing bacteria (Chang et al., 2002; Hayes and Lovley, 2002). However, their degradation pathways have not been clearly elucidated. Furthermore, their biodegradation under anaerobic conditions occurs more slowly compared to aerobic conditions.

### 10.4.2.3 Microbial metabolism of polychlorinated biphenyls (PCBs)

Polychlorinated biphenyls are very inert, have a high thermal stability, have good electrical insulation properties and are very resistant to acids and bases. These traits make them suitable for use in large transformers as a coolant. Because of their high reflective index, they were also

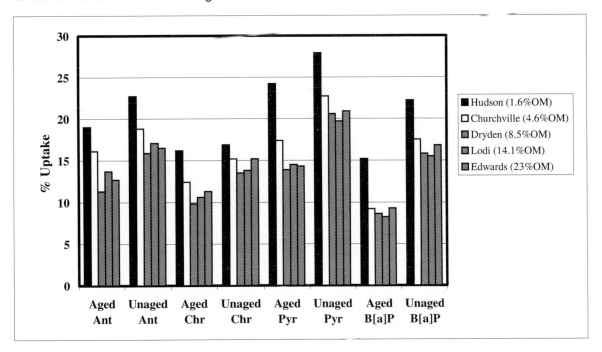

**Fig. 10.13** Bioavailability of pyrene to earthworms (*Eisenia fetida*) in five soils that differed in organic matter content. Note that Ant = Anthracene, Chr = Chrysene, Pyr = Pyrene and B[a]P = Benzo[a]phyrene. (Based on data from Tang et al., 2002; reprinted with permission from Elsevier).

used prior to 1972 in microscopy as immersion oil. Given these "ideal" properties, they were widely employed internationally and released into the environment worldwide. It is estimated that more than 100 million pounds were released into soil and water before the environmental effects were widely considered. Biphenyls are lipophilic and sorb strongly to animal fat tissue where they can undergo biomagnification. The properties that made them popular in industry are also responsible for the persistence of these compounds in the environment where they bind to soils and sediments, resisting biodegradation. Notice that there are 10 sites (Fig. 10.14) on which chlorine can attach, giving more than 200 possible PCB congeners containing zero to 10 chlorine atoms per biphenyl molecule. They can exist as complex mixtures (aroclors) and are quite toxic. The most common isomer of commercial importance is aroclor 1242. Most toxic are those chlorinated at the 3$^{rd}$ and 4$^{th}$ positions as they adversely affect cells.

The US-based General Electric Company used to make large amounts of these compounds which in the past anyway, were routinely spilled in waterways. A widely known casualty from this catastrophic practice is the Hudson River in New York State. They have a very low rate of biological degradation and thus accumulated in the environment when they were still in use. Where it occurs, biodegradation is extremely slow and tapping the potential of microorganisms to degrade PCBs in the environment has great potential (Unterman et al., 1988). Unterman and colleagues tested *Pseudomonas putida* LB400, one of a dozen bacterial strains they isolated that grow on biphenyls and transform PCBs albeit field trials to enhance the degradation of these compounds proved largely unsuccessful (Unterman et al., 1988). Available evidence shows that where degradation occurs, the nonchlorinated ring is attacked (hydroxylated) and cleaved, forming a chlorobenzioc acid that is degraded further. In cases where chlorine occurs on both

rings of the biphenyl, the rings may be metabolized into two separate chlorobenzoates. For example, 2,4'-dichlorobiphenyl can be metabolized by a pseudomonad to turn both into 2-chlorobenzoate and 4-chlorobenzoate (Rochkind-Dubinsky et al., 1987). In such cases, hydroxylation and subsequent cleavage tend to occur more readily on the ring with the fewer number of chlorine atoms. Under anoxic conditions such as sediments and subsurface soils, reductive dechlorination of PCBs may occur, possibly from the spontaneous elimination of a halide ion from an unstable molecule with a monooxygenase present. Once only two chlorides are left after the dechlorination process, the PCBs are then metabolized under aerobic conditions just like other aromatic compounds described earlier, their carbon becoming an energy source for the degraders. Considering the wide range of PCB congeners, it is likely that the remediation strategy which works for one congener will not necessarily work on another.

Also implicated in the biodegradation of PCBs are mycorrhizae, especially those associated with plants (Donnelly and Fletcher, 1995). Studies showed that congeners with less chlorination degrade more readily, the possibility of metabolizing them decreasing with increased chlorination. In the Donnelly and Fletcher studies, no congener was degraded by all of the mycorrhizal species. Mycorrhizae typically exist in association with plant roots and to degrade these compounds underscores the importance of phytoremediation approaches to degrade the recalcitrant compound. In this instance, the mycorrhizae channel nutrients into the plant and receive carbon substrates from it in return. Their presence also enhances the region under the direct influence of the rhizosphere, which largely results in a higher density and diversity of microorganisms and their enhanced activity (Yoshitomi and Shann, 2001; Schipper et al., 2001). Enhanced activity in the rhizosphere region is attributable to the induced changes in pH, $O_2$, and $CO_2$

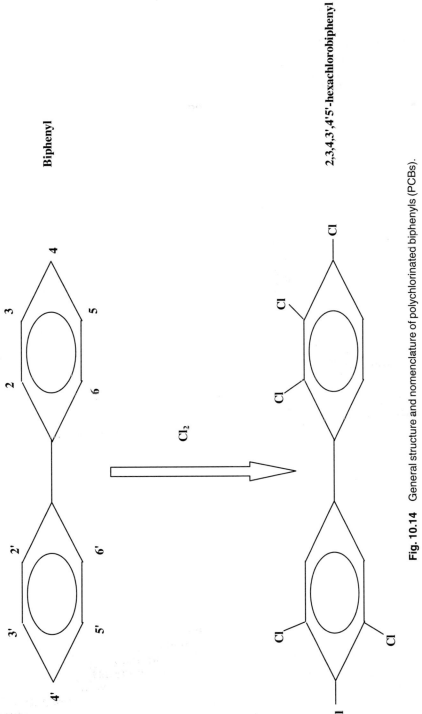

**Fig. 10.14** General structure and nomenclature of polychlorinated biphenyls (PCBs).

partial pressures, input of oxidative enzymes that directly help in the degradation process, as well as the large quantities of exudates and/or leakage of readily degradable organic substrates by the plants and microorganisms.

### 10.4.3 Metabolism of pharmaceutical compounds

Pharmaceutical compounds are used at three different levels—therapeutic, growth promotion, and preventive (prophylactic)—by both humans and in agriculture. For example, a recent report estimated that 17.8 - 24.6 million pounds of antibiotics are used for agricultural purposes in the United States annually (Isaacson and Torrence, 2002). The major pathways through which pharmaceutical compounds are introduced into the environment involve excretion by both humans and livestock (Fig. 10.15). However, their role as environmental pollutants has only surfaced recently compared to other organic pollutants (Richardson and Bowron, 1985; Halling-Sørensen et al., 1998; Daughton and Ternes, 1999; Jørgensen and Halling-Sørensen, 2000). Most of the excreted pharmaceutical compounds and/or their metabolites are persistent in the environment (Jjemba, 2002a).

Pharmaceutical compounds occur as aromatic or aliphatic moieties and have either a polar or nonpolar structure. Their characteristics in the environment are expected to be similar to those of pesticides, fungicides, herbicides, and other organic pollutants in general. However, they are biologically active substances specifically designed to harm microorganisms such as bacteria, fungi, and protozoa (parasites) in infectious disease settings even at low dosage. Since a substantial fraction of the parent compound is typically excreted unchanged (Jjemba, 2002a), this designed trait is expected to persist by the time they reach the environment, e.g., through excretion. Polarity affects their sorption potential to matrices in the environment; nonpolar compounds tending to be less persistent but are

more liable to sorption onto soil colloids and sediment, and also accumulate in the biota (Huberer, 2002). Polar pharmaceutical compounds, on the other hand, are not significantly adsorbed to matrices such as soil colloids, sludge, and sediments, a trait that makes them more labile in the environment. A low $K_{ow}$ will also facilitate the transfer of polar compounds into biological cells, leading to their bioaccumulation (Jjemba and Robertson, 2003). Sorption is based on the organic matter content of the sorbent as expressed by the mathematical relationship described in eqn (5) above to calculate $K_d$. $K_{ow}$ values of a few common pharmaceutical compounds are given in Table 10.1 and reflect the hydrophobicity of pharmaceutical compounds. The more hydrophobic a compound is, the more liable it is to sorb passively onto colloids. Based on these $K_{ow}$ values, some pharmaceutical compounds such as ciprofloxacin, metronidazole, and tetracycline are not very hydrophobic and hence may not extensively sorb to soil colloids and sediments. It is important to note that $K_d$ and organic carbon relationships may greatly underestimate the sorption of pharmaceutical compounds in some environmental matrices, particularly in sewage and sludge (Stuer-Lauridsen et al., 2000) and soils with a high organic matter content (Tolls, 2001) because a number of other mechanisms such as pH, mineral concentration, clay composition and soil temperature interact to affect sorption and thereby bioavailability (Gavalchin and Katz, 1994; Holten Lützhøft et al., 2000; Marengo et al., 1997; Thiele, 2000). For example, the presence of phenols, lignin monomers and dimers, as well as lipids and alkylaromatics as SOM components in a Chernozem soil, enabled sulfapyridine to adsorb more extensively to soil than when these components were low or totally absent (Thiele, 2000).

It is evident from the above discussion that we still lack reliable methods for predicting the bioavailability of various pharmaceutical compounds under a wide range of environmental

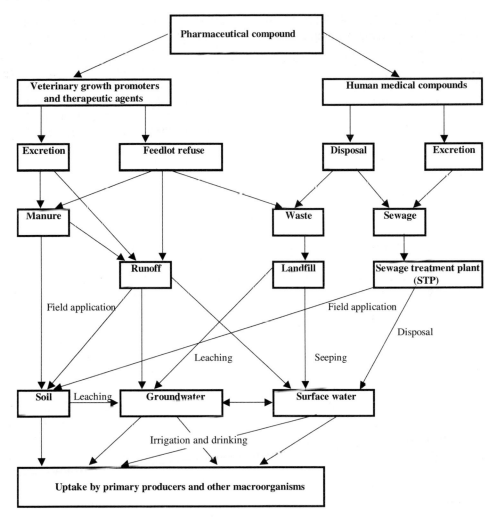

**Fig. 10.15** Possible pathways for therapeutic agents in the environment (Modified from Jjemba, 2002a; reprinted with permission from Elsevier).

conditions and, further, predicting their subsequent biodegradation.

Information about the degradability of pharmaceutical compounds in the environment is only beginning to accumulate. The limited information has mostly come from sewage, water, and wastewater treatment processes, using simple experimental set ups such as the closed bottle test (CBT) (Marengo et al., 1997) or benchtop bioreactors (Ingerslev and Halling-Sørensen, 2000). However, in most instances, the degradation pathways and/or microorganisms responsible for degradation have not been elucidated. In preliminary studies to determine the effects of chloroquine, metronidazole and quinacrine on both soil microbiota (bacteria and protozoa), the presence of only 0.5 mg metronidazole g$^{-1}$ soil reduced the density of protozoa by a tenfold whereas 4 mg quinacrine dihydrochloride g$^{-1}$ soil significantly increased the density of protozoa and their bacterial prey in the rhizosphere (Jjemba, 2002b).

## 10.5 LINKING KNOWLEDGE OF MICROBIAL METABOLISM TO THE FATE OF POLLUTANTS IN FIELD SITES

More often than not, organic contaminants are composed of a mixture of compounds rather than pure/single compound types, a reality that possibly challenges the potential degradors even more. This observation probably also explains, at least in part, why microorganisms confirmed as excellent degraders of specific compounds under laboratory conditions often fail to satisfactorily display such traits under field conditions where they have to work on the chemical as it appears in nature. From the above discussion, it is apparent that multiple compounds may be handled through different degradation pathways.

Bioremediation is the use of living organisms to eliminate or reduce environmental contaminants. Bioremediation of environments polluted by organic compounds should aim at directing microbial activities toward the pollutant and the subsequent destruction of such pollutants either through *in-situ* or *ex-situ* approaches. This may involve supplying nutrients or taking care of other biotic or abiotic limitations. *Ex situ* involves moving the contaminants away from the site, e.g., to a bioreactor while *in situ* technically involves converting the landscape into a microbial culture flask. Whatever strategy is adapted, it is important to note that a bacterium never functions in nature as an individual but rather as a team of several kinds of bacteria, living together (as biofilms in some instances) and aiding each other with complementary enzymes.

Remediation should ideally follow a systematic approach as outlined in Fig. 10.16. Many microorganisms require a period of acclimation before biodegradation occurs. Biodegradation is often limited by the lack of nutrients, particularly nitrogen and phosphorus. Thus, the microbial degradation of organic compounds can be enhanced by the application of macronutrients (Atlas and Bartha, 1973; Joner et al., 2002). Whatever strategy is adapted, some key issues have to be resolved, viz.

- fairly complete understanding of the local geology; an appreciation of the heterogeneity of the site (pollutant intractability);
- complete control of the local hydrology;
- insolubility and inaccessibility of some pollutants; and
- the fact that some pollutants are only released slowly from soils, sediments, and aquifers.

Implicit in the description of bioremediation is the presence of a regulatory end point above which the contaminants are hazardous to organisms that are or may come into contact with the polluted environment. Once a site is polluted, there would almost always remain the question of how well cleaned it has been. This has to be guided by existing regulations and technology. Efforts to clean a particular environment have to be realistic.

Bioremediation requires sampling of the site/fields and the chemical analysis of field samples as well as application of the discipline of environmental engineering to design a remediation strategy (Fig. 10.16). Both high-pressure liquid chromatography (HPLC) and $^{14}C$ analyses provide an excellent material balance for biodegradation of a parent substrate and the accumulated metabolic products. For example, the abundance of PCBs (and other organic compounds) can be monitored using gas chromatography to separate the various gases of the biphenyls. The lightly chlorinated compounds come off first (Fig. 10.17). Utilization of a substrate by microorganisms usually involves utilization of $O_2$ and certainly the production of $CO_2$. Mineralization assays, as a measure of the total biodegradation must include an absolute mass balance for the system under study (Bailey and McGill, 2002). Measurement of $CO_2$ production is the most common measure of the rate and extent of min-

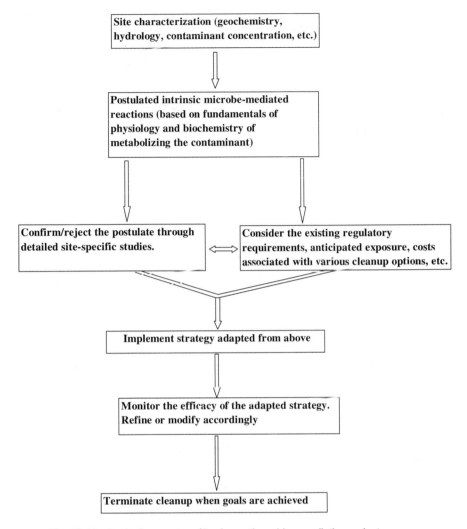

**Fig. 10.16** Logical sequence of implementing a bioremediation project.

eralization of organic pollutants. More commonly used under laboratory conditions are [14]C-labeled analogs. The radiolabeled substrate is added to the reaction system containing the microbiological population of interest and, over intervals, the balance parent compound and [14]C are determined conveniently by liquid scintillation analysis of the radioactive decay of [14]C. Mineralized products such as [14]$CO_2$ and [14]$CH_4$ can also be collected and analyzed by liquid scintillation or titration. Mass spectroscopy (MS) or GC/MS can also be used to identify and confirm the biodegradation products. Designing remediation strategies may include digging up the contaminated soil, providing nutrients, creating a bioreactor, or predicting the flow of fluids (Alexander, 1994). To link biodegradation and remediation adequately, we need to distinguish biotic from abiotic processes in the field.

Finally, most studies on the interaction between microorganisms and organic compounds have been carried out in temperate climates, in particular North America and West-

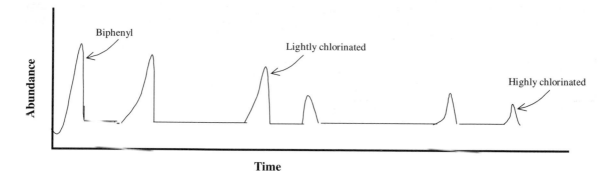

**Fig. 10.17**  High-performance liquid chromatograph output for PCBs showing the appearance of light congeners coming off earlier compared to more complex ones.

ern Europe. Much less information is available on the behavior of the compounds in tropical ecosystems and the potential of the organisms in that region still remain largely unrealized.

## References

Alexander M. (1994). Biodegradation and Bioremediation. Acad. Press, New York, NY.

Alexander M. (1967). Breakdown of pesticides in soils. *In*: N.C. Brady (ed.). Agriculture and the Quality of Our Environment. Amer. Associ. Adv. Sci. Washington DC, pp. 331-342.

Arif, J.M., W.A. Smith, and R.C. Gupta. 1999. DNA adduct formation and persistence in rat tissues following exposure to the mammary carcinogen dibenzo[*a,l*]pyrene. *Carcinogenesis* **20**: 1147-1150.

Atlas R.M. and R. Bartha (1973). Stimulated biodegradation of oil slicks using oleophilic fertilizers. *Environ. Sci. Tech.* **7**: 538-541.

Bailey V.L. and W.B. McGill (2002). Fate of $^{14}$C-labeled pyrene in a creasote and octadecane in an oil-contaminated soil. *Soil Biol. Biochem.* **34**: 423-433.

Chang B.V., L.C. Shiung, and S.Y. Yuan (2002). Anaerobic biodegradation of polyaromatic hydrocarbon in soil. *Chemosphere* **48**: 717-724.

Curtis R.F., C. Dennis, J.M. Gee, M.G. Gee, N.M. Griffiths, D.G. Land, J.L. Peel and D. Robinson (1974). Chloroanisoles as a cause of musty taint in chickens and their microbiological formation from chlorophenols in broiler house litters. *J. Agric. Food Sci.* **25**: 811-828.

Daughton C.G. and T.A. Ternes (1999). Pharmaceutical and personal care products in the environment: agents of subtle changes? *Enviro. Health Perspec.* **107**: 907-938.

Donnelly P.K. and J.S. Fletcher (1995). PCB metabolism by ectomycorrhizal fungi. *Bull. Enviro. Contam. Toxicol.* **54**: 507-513.

Fuhrman J.A. (1999). Marine viruses and their biogeochemical and ecological effects. *Nature* **399**: 541-548.

Gavalchin J. and S.E. Katz (1994). The persistence of fecal-borne antibiotics in soil. *J. AOAC Inte.* **77**: 481-485.

Glass B.L. (1972). Relation between the degradation of DDT and the iron redox system in soils. *J. of Agric. Food Chem.* **20**: 324-327.

Guenzi W.D. and W.E. Beard (1967). Anaerobic degradation of DDT and DDD in soil. *Science* **156**: 1116-1117.

Halling-Sørensen B., S.N. Nielsen, P.F. Lanzky, F. Ingerster, H.C.H. Lützhøff, and S.E. Jørgensen (1998). Occurrence, fate and effects of pharmaceutical substances in the environment—a review. *Chemosphere* **36**: 357-393.

Hatzinger P.B. and M. Alexander (1995). Effect of aging of chemicals in soil on their biodegradability and extractability. *Environ. Sci. Tech.* **29**: 537-545.

Hayes L.A. and D.R. Lovley (2002). Specific 16S rDNA sequence associated with naphthalene degradation under sulfate-reducing conditions in harbor sediments. *Microb. Ecol.* **43**: 134-145.

Holten Lützhøft H-C.H., W.H.J. Vaes, A.P. Freidig, B. Halling-Sørensen, and J.L.M. Hermens (2000). 1-Octanol/water distribution coefficient of oxolinic acid: influence of pH and its relation to the interaction with dissolved organic carbon. *Chemosphere* **40**: 711-714.

Huberer T. (2002). Tracking persistent pharmaceutical residues from municipal sewage to drinking water. *J. Hydrol.* **266**: 175-189.

Hovarth R.S. (1971). Cometabolism of the herbicide 2,3,6-trichlorobenzoate. *J. Agric. Food Chem.* **19**: 291-293.

Ingerslev F. and B. Halling-Sørensen (2000) Biodegrability Properties of Sulfonamides in activated sludge. *Environ. Toxicol. Chem.* **19**: 2467-2473.

Isaacson R.F. and M.F. Torrence (2002). The role of antibiotics in agriculture: a report from the American Academy of Microbiology, Washington DC. Available at http://www.asmusa.org.

Jjemba P.K. (2002a). The potential impact of veterinary and human therapeutic agents in manure and biosolids on vegetation: a review. *Agric., Ecosys. Environ.* **93:** 267-278.

Jjemba P.K. (2002b). The effect of chloroquine, quinacrine, and metronidazole on both soybean (*Glycine max*) plants and soil microbiota. *Chemosphere* **46:** 1019-1025.

Jjemba P.K. and B.K Robertson (2003). The fate and potential impact of pharmaceutical Compounds to non-target organisms in the environment. Third Int. Conf. Endocrine Disrupters and Pharmaceutical Compounds in Groundwater. Minneapolis, MN, pp. 184-194.

Jjemba P.K., B.K. Kinkle, and J.R. Shann (in litt.). In situ enumeration and identification of pyrene-degrading soil bacteria.

Johnson B.T., R.N. Goodman, and H.S. Goldberg (1967). Conversion of DDT to DDD by pathogenic and saprophytic bacteria associated with plants. *Science* **157:** 560-561.

Joner E.J., S.C. Corgié, N. Amellal, and C.J. Leyval (2002). Nutritional constraints to degradation of polycyclic aromatic hydrocarbons in a simulated rhizosphere. *Soil Biol. Biochem.* **34:** 859-864.

Jørgensen S.E. and B. Halling-Sørensen (2000). Drugs in the environment. *Chemosphere* **40:** 691-699.

Lau P. and V. de Lorenzo (1999). Genetic engineering: the frontier of bioremediation. *Environ. Sci. Tech.* **4:** 124A-128A.

Madsen E.L., J.L. Sinclair, and W.C. Ghiorse (1991). In situ biodegradation: microbiological patterns in a contaminated aquifer. *Science* **252:** 830-833.

Marengo J.R., R.A. Kok K. O'Brien, R.R. Velagaleti, and J.M. Stamm (1997). Aerobic degradation of ($^{14}$C)-sarafloxacin hydrochloride in soil. *Environ. Toxicol. Chem.* **16:** 462-471.

Monod J. (1949). The growth of bacterial culture. *Ann. Rev. Microbiol.* **3:** 371-394.

Neidhart F.G., J.L. Ingraham, and M. Schaechter (1990). Physiology of the Bacterial Cell: A Molecular Approach. Sinauer Assoc., Sunderland, MA.

Neilson A.H, A.S. Allard, P.A. Hynning, M. Remberger and L. Lander (1983). Bacterial methylation of chlorinated phenols and guaiacols; formation of veratroles from guaiacols and high-molecular weight chlorinated lignin. *Appli. Environ. Microbiol.* **45:** 774-783.

Parr J.E. and S. Smith (1974). Degradation of DDT in an Everglade muck as affected by lime, ferrous iron, and anaerobiosis. *Soil Sci.* **118:** 45-52.

Phale P.S., H.S. Savithri, N.A. Rao, and C.S. Vaidyanathan (1995). Production of biosurfactant "Biosur-Pm" by *Pseudomonas maltophila* CSV89: characterization and role in hydrocarbon uptake. *Arch. Microbiol.* **163:** 424-431.

Radehaus P.M. (1998). Microbiologically mediated reactions in aquatic systems. *In*: D.L. Macalady (ed.). Perspectives in Environmental Chemistry. Oxford Univ. Press, New York, NY, pp. 210- 230.

Ralebitso T.K., E. Senior, and H.W. van Verseveld (2002). Microbial aspects of atrazine degradation in natural environments. *Biodegradation* **13:** 11-19.

Ramos J.L., A. Wasserfallen, K. Rose, and K.N. Timmis (1987). Redesigning metabolic routes: manipulation of TOL plasmid pathway for catabolism of alkylbenzoates. *Science* **235(4788):** 593-596.

Richardson M.L. and J.M. Bowron (1985). The fate of pharmaceutical chemicals in the aquatic environment. *J. Pharm. Pharmacol.* **37:** 1-12.

Robertson B.K. and M. Alexander (1994). Growth-linked and cometabolic biodegradation: possible reasons for occurrence or absence of accelerated pesticide biodegradation. *Pesticide Sci.* **41:** 311-318.

Robertson B.K. and P.K. Jjemba (in litt.). Enhanced degradation of sorbed 2, 4, 6-trinitrotoluene (TNT) in the presence of a microbial consortium.

Rochkind-Dubinsky M.L., G.S. Sayler and J.W. Blackburn (1987). Microbiological Decomposition of Chlorinated Aromatic Compounds. Marcel Dekker, Inc., New York, NY.

Rosenberg A. and M. Alexander (1980a). Microbial metabolism of 2,4,5-trichlorophenoxyacetic acid in soil, soil suspensions, and axenic cultures. *J. Agric. Food Chem.* **28:** 297-302.

Rosenberg A. and M. Alexander (1980b). 2,4,5-trichlorophenoxyacetic acid (2,4,5-T) decomposition in tropical soil and its cometabolism by bacteria *in vitro*, *J. Agric. Food Chem.* **28:** 705-709.

Ruelle P. and U.W. Kesselring (1997). Aqueous solubility prediction of environmentally important chemicals from the mobile order thermodynamics. *Chemosphere* **34:** 275-298.

Sayler G.S. and S. Ripp (2000). Field applications of genetically engineered microorganisms for bioremediation processes. *Curr. Opin. Biotech.* **11:** 286-289.

Schipper L.A., B.P. Degens, G.P. Sparling, and L.C. Duncan (2001). Changes in microbial heterotrophic diversity along plant successional sequences. *Soil Biol. Biochem.* **33:** 2093-2103.

Schwarzenbach R.P., P.M. Gschwend, and D.M. Imboden (2002). Environmental Organic Chemistry. John Wiley, Hoboken, NJ.

Shannon C.E. and W. Weaver (1963). The Mathematical Theory of Communication. Univ. Illinois Press, Urbana, IL.

Stone A.T. (1998). Metal-catalyzed hydrolysis of organic compounds in aquatic environments. *In*: D.L. Macalady (ed) Perspectives in Environmental Chemistry. Oxford Univ. Press, New York, NY, pp. 75- 93.

Stoner D.L. (1994). Hazardous organic waste amenable to biological treatment. *In*: D.L. Stoner (ed.). Biotechnology for the Treatment of Hazardous Waste. Lewis Publ., Boca Raton, FL, pp. 1-25.

Stuer-Lauridsen F., M. Birkved, L.P. Hansen, H.-C.H. Lützhøft, and B. Halling-Sørensen (2000). Environmental risk assessment of human pharmaceuticals in Denmark after normal therapeutic use. *Chemosphere* **40:** 783-793.

Suflita J.M., A. Horowitz, D.R. Shelton, and J.M. Tiedje (1982). Dehalogenation: A novel pathway for the anaerobic biodegradation of haloaromatic compounds. *Science* **218:** 1115-1117.

Tang J., H-H. Liste and M. Alexander (2002). Chemical assays of availability to earthworms of polycyclic aromatic hydrocarbons in soil. *Chemosphere* **48:** 35-42.

Tate III R.L. (1987). Soil Organic Matter: Biological and Ecological Effects. John Wiley, New York, NY.

Thiele S. (2000). Adsorption of the antibiotic pharmaceutical compound sulfapyridine by a long-term differently fertilized loess Chernozem. *J. Plant Nutr. Soil Sci.* **163:** 589-594.

Tolls J. (2001). Sorption of veterinary pharmaceuticals in soils: a review. *Environ. Sci. Technol.* Ingerslev F. and B. Halling-Sørensen (2000). Biodegradability properties of sulfonamides in activated sludge. *Environ. Toxicol. Chem.* **19:** 2467-2473

Torsvik V. and L. Øvreås (2002). Microbial diversity and function in soil: from genes to ecosystems. *Curr. Opin. Microbiol.* **5:** 240-245.

Unterman R., D.L. Bedard, M.J. Brennan, L.H. Bopp, F.J. Mondello, R.E. Brooks, D.P. Mobley, J.B. McDermott, C.C. Schwartz, and D.K. Dietrich (1988). Biological approaches for PCB degradation. In: G.S. Omenn et al. (eds.). Reducing Risks from Environmental Chemicals Through Biotechnology. Plenum Press, New York, NY, pp. 253-269.

Yoshitomi K. and J.R. Shann (2001). Corn (*Zea mays* L.) root exudates and their impact on [14]C-pyrene mineralization. *Soil Biol. Biochem.* **33:** 1769-1776.

Zhou J., B. Xia, D.S. Treves, L.-Y. Wu, T.L. Marsh, R.V. O'Neill, A.V. Palumbo, and J.M. Tiedje (2002). Spatial and resource factors influencing high microbial diversity in soil. *Appl. Environ. Microbiol.* **68:** 326-334.

# 11

# Microbiology of the Atmosphere

The biology of microorganisms and other microscopic biological materials in the atmosphere as well as their deposition to new locations involves a wide range of specialities including physics, bacteriology, mycology, chemistry, meteorology, botany, pathology, epidemiology, and entomology. Biomaterials in the atmosphere are responsible for a variety of infections and health hazards to humans, including allergies, asthma, and other respiratory disorders. This chapter focuses primarily on the activity of microorganisms as they affect the composition of the atmosphere. Emphasis on biological air pollutants in this section is not intended to underestimate the importance of nonbiological sources of air pollution.

The atmosphere as an environment consists of four distinct layers termed the troposphere, stratosphere, mesosphere and thermosphere in order of increasing height from the Earth's surface (Fig. 11.1). The troposphere contains approximately 80% of the total mass of the atmosphere and virtually all the atmospheric water, i.e., water vapor, liquid water in the form of clouds and rain, hail, ice, and snow (Kim, 1994). This is a very small mass, considering the fact that the Earth has a diameter of about 12,000 km. The depth of the troposphere ranges from 8 km in the polar region to 15 km in the equatorial region and also depends on the prevailing weather conditions (Gregory, 1961; Philander, 1998). It is characteristically more turbulent than all the other layers. The stratosphere, on the other hand, is quite stratified which restricts the vertical movement of air. This attribute causes pollutants to remain for a long time in this region. Thus, pollutants ejected into the stratosphere have a very long residence time. The stratosphere also contains a layer of 12-30 km depth that is enriched with ozone (i.e., the ozone layer) which plays a crucial role in determining temperatures on the Earth's surface as will be outlined shortly. The troposphere and stratosphere combined contain about 99% of the total atmospheric mass. Above the stratosphere is the mesosphere layer with about 30 km deep. The atmosphere is characterized by a drop in temperature with increasing altitude, with the mesosphere being even colder than the stratosphere. Within the mesosphere, the vapor freezes into clouds. Whatever microbes and other inert particles in the aerosol are able to make it to that height participate in nucleation, the initial process in the formation of clouds. The thermosphere, which lies above the mesosphere, comprises the ionosphere and exosphere. It is fairly inert with no bioaerosols or other particulate material.

The atmosphere absorbs about 70% of incident sunlight. If it were nonexistent, the average temperatures all over the Earth's surface would be below freezing all year around. Naturally, under those conditions, all the water

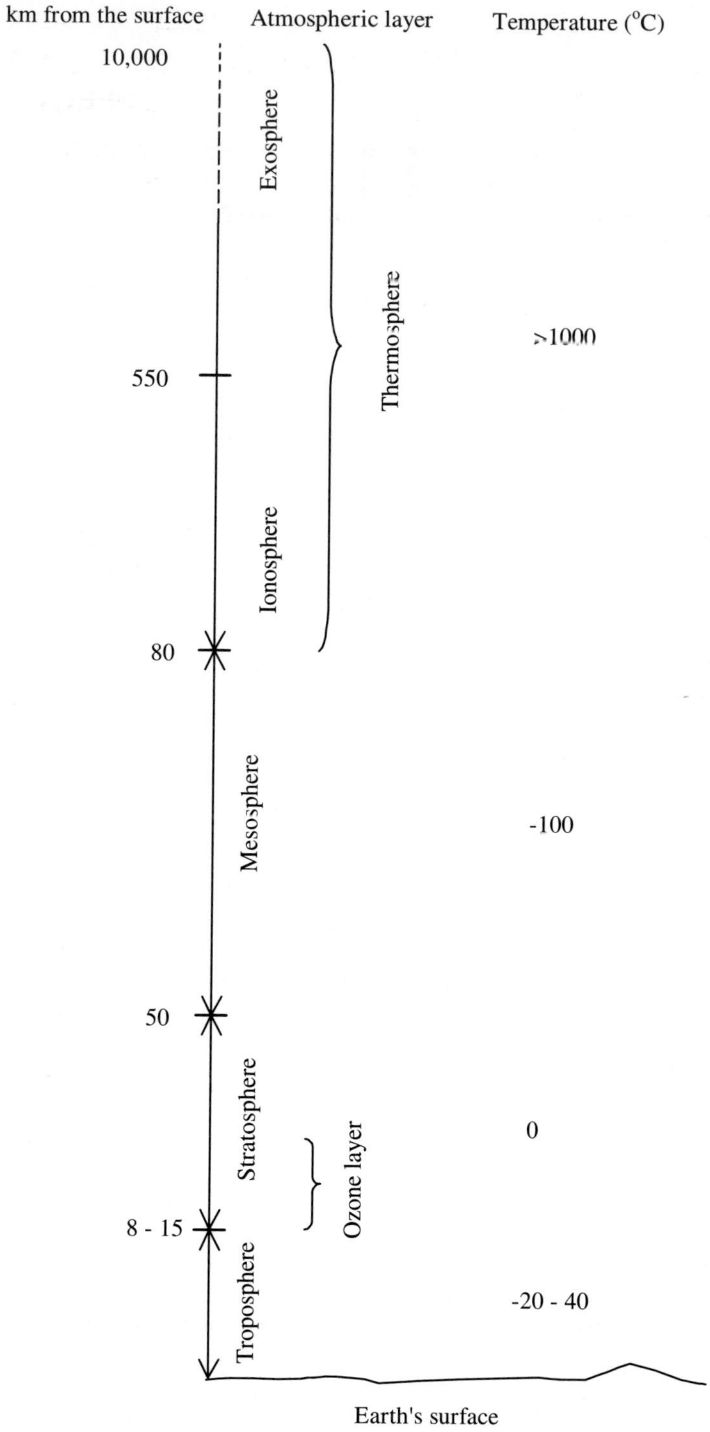

**Fig. 11.1** The three main layers that make up the atmosphere and their relative temperature. Note that the figure is not on scale.

would be frozen and life as we know it would probably be nonexistent. Compelling evidence about this contention can be drawn if we compare the conditions on Earth with those of its immediate neighbors in the universe, i.e., Mars and Venus, which are inhospitable. Mars has a very thin "atmosphere", to serve as a blanket, which is primarily composed of $CO_2$. To that effect, the temperatures on Mars are so low that all the water is frozen (Table 11.1). On the other hand, Venus has a huge greenhouse effect due to its thick "atmosphere" that is also primarily composed of $CO_2$. Temperatures are so high on Venus that its water evaporated and escaped into space long ago (Philander, 1998). Most of the absorption of the solar radiation before it reaches the Earth's surface is attributed to the ozone layer.

The Earth's atmosphere acts as a blanket that traps heat, a phenomenon referred to as the greenhouse effect, so called because the greenhouse gases, like glass on a greenhouse, allow increasing solar rays to pass through but reflects outgoing thermal radiation. The atmosphere is therefore a very delicate component of the environment [whose presence and well-being is essential to existing life on Earth.] Whereas the presence of some microorganisms is widely readily associated with airborne infections, the connection between microorganisms and other aspects of the atmosphere is rarely recognized. Microbiological processes exert a major influence on the composition of the atmosphere, affecting the environment and our well-being on Earth. It is imperative that we understand the dynamics of how microorganisms and their activities exert such an influence.

## 11.1 IMPACT OF MICROBIAL ACTIVITIES ON THE GREENHOUSE EFFECT

Although seldom directly realized, microbial activities significantly affect the greenhouse effect and other global environmental changes. Through their activities, microorganisms produce and consume atmospheric gases that affect the climate, affecting the concentration of inorganic gases in the atmosphere (King et al., 2001). The Earth's atmosphere is predominantly composed of nitrogen and oxygen which together account for 99% of the atmospheric gases. These two abundant diatomic gaseous molecules do not participate in the greenhouse effect as they cannot absorb the less energetic infrared rays reflected from the Earth because the photons of infrared rays have so little energy. However, they play a major role in absorbing the incoming energetic short wavelength UV rays at heights of 200 km and above, that is, within the stratosphere. In this form, they act as a first line of defense for life on Earth against incoming UV radiation, the second line of defense being furnished in the stratosphere by the ozone layer.

It is worth noting, therefore, that the greenhouse effect depends not on these two most abundant gases present in the atmosphere, but rather on the triatomic gas molecules which are present in minute concentrations, viz. carbon dioxide (0.035%), water vapor, methane (0.00017%) and ozone (0.000001-0.000004%). These four gases interact with the less energetic infrared rays, playing a crucial role in absorbing the incoming long infrared radiation. They are transparent to the heat from the sun

**Table 11.1** Temperatures of the planets

| Planet | Distance from sun (x$10^6$ km) | Global average temperature (°C) | Light intensity (watts/m$^2$) |
|---|---|---|---|
| Mars | 228 | −53 | 589 |
| Earth | 150 | +15 | 1,372 |
| Venus | 108 | +430 | 2,613 |

*Source*: Philander (1998) with permission from Princeton University Press, Princeton, NJ.

(short UV rays), our major source of heat, and their increasing concentrations in the atmosphere have been linked to slight increases in global temperature (Table 11.2). Whereas much of the debate about global warming has greatly focused on average global warming, more attention should be paid to regional changes in temperatures, which are predicted to substantially alter global circulations and precipitation patterns (Philander, 1998). These climatic alterations are, in turn, expected to cause a rise in sea levels, inundating some coastal areas while causing drought in others. The changes are also predicted to result in the spread of diseases into new areas, as is discussed in Chapter 15.

### 11.1.1   Ozone

Ozone in itself is not of any known microbial origin. However, considering its role in the greenhouse effect, it is important to briefly discuss how it impacts the environment once it is affected by the other triatomic gases. Ozone is a deep blue, explosive and very poisonous gas. Given these characteristics, it may seem ironical that so much interest is given to protecting its presence in the atmosphere. Interest in its preservation stems from the fact that it is a powerful greenhouse gas as it absorbs both the short wavelength energetic UV rays from the sun and also some of the less energetic infrared rays reflected from the Earth. As indicated earlier, it provides the second line of defense against the incoming UV rays that escape interception in the thermosphere by the more abundant gases, nitrogen and oxygen. Its abundance in the stratosphere is based on the fact that molecular oxygen decreases with altitude since gravity keeps most $O_2$ molecules near the Earth's surface. High in the atmosphere, UV photodissociates molecular $O_2$ into two oxygen atoms and this provides a mixture of $O_2$ molecules and atoms which react to form the unusual ozone ($O_3$) molecules

$$O_2 + UV \xrightarrow{\text{photodissociation}} O + O \qquad ...(1)$$

$$O_2 + O \longrightarrow O_3 \qquad ...(2)$$

The photodissociation of molecular oxygen into oxygen atoms (eqn. 1) occurs in the radiation band wavelength ($< 0.286$ $\mu$m). Ozone is particularly abundant in the stratosphere because at this height there is a balance between elevation and the abundance of $O_2$ molecules which form and collide with O atoms to form ozone.

### 11.1.2   Carbon dioxide

Although many of the increases in the other three triatomic gases are mainly produced as by-products of fossil fuel combustion, other activities, which include agriculture (in which microbial processes are involved), forestry clearing, and other industrial activities also contribute. Some aspects about these gases participating in the biogeochemical cycling of nutrients have been discussed in Chapter 9. In the present Chapter, we highlight their involvement in perturbations of the atmosphere, perturbations that directly deplete the ozone layer, reducing the protection that life on Earth receives from that layer. Carbon dioxide is the most important of all the greenhouse gases. It plays a contradictory role on Earth as it is the

**Table 11.2**   Sensitivity of mean global temperature to concentrations of various greenhouse gases

| Greenhouse gas | Current concentration | Change in concentration | Warming (°C) |
|---|---|---|---|
| $CO_2$ | 350 ppm | +100 ppm | 0.5 - 1.6 |
| Methane | 1.8 ppm | Doubling | 0.2 - 0.7 |
| $N_2O$ | 300 ppb | 50% increase | 0.2 - 0.6 |
| Ozone | 0.02-0.14 ppm | 50% decrease | 0.5 - 1.7 |
| Stratospheric water vapor | ? | Doubling | 0.6 |

Compiled from Ciborowski (1989), Conrad (1995), and Jaffe (1992).

medium through which the solar energy is transformed into living matter. Human activities release carbon dioxide to the atmosphere through the burning of coal, oil, wood, natural gas, and other organic compounds. Due to these activities, the concentration of carbon dioxide in the atmosphere has increased more than 7% over the last 30 years and is still rising rapidly. Thus, $CO_2$ concentrations have increased from 270 ppm in the mid-1800s to about 365 ppm today (King et al., 2001). Microorganisms play a central role in the accumulation of this gas in the atmosphere by influencing the stability or instability of its storage pools as discussed in section 9.2.

### 11.1.3 Nitrogen Oxides (NO and N$_2$O)

Nitrogen oxides (NOX), i.e., nitric oxide (NO) and nitrous oxide (N$_2$O), are produced as a result of fossil fuel combustion (40%) and biomass burning (25%); the remainder is produced by lightning and metabolic processes in various microorganisms in the soil. Production of both these nitrogen oxides in soils is attributed to nitrifying and denitrifying microorganisms as well as a variety of other microorganisms during the decomposition of nitrogen-rich materials (Table 11.3). Thus, high concentrations of these gases are encountered in milking sheds, cattle barns, and around manure piles (Mosier et al., 1998). Nitrogen oxides are more than 150 times as effective as greenhouse gases compared to $CO_2$ (Conrad, 1995). Nitric oxide, once released into the atmosphere, reacts catalytically with ozone as per eqn. (3), generating another greenhouse gas, nitrogen dioxide (NO$_2$), which can react with water droplets in the atmosphere to form nitric acid. The acid treks back to Earth as acid rain. Nitrogen dioxide also participates in photolytic reactions (eqn. 4), regenerating nitrous oxide and singulate oxygen (O) which combines with oxygen (eqn. 5), regenerating ozone to recycle into reaction (3).

$$NO + O_3 \rightarrow NO_2 + O_2 \qquad ...(3)$$

$$NO_2 + h\nu \rightarrow NO + O \qquad ...(4)$$

$$O + O_2 \rightarrow O_3 \qquad ...(5)$$

Notice that during these reactions, both NO and NO$_2$ are regenerated and participate in more ozone destruction reactions leading to chain reactions. As more NO is generated by microbial processes in the soil over time, the amount accumulated in the sources exceeds its sinks by as much as 30%. As a matter of fact, soils account for about 40% of the global NO budget. It is produced by methylotrophs under anaerobic conditions (Ren et al., 2000).

For quite a while, the biological production of nitrous oxide was attributed to nitrifiers and denitrifiers. Later, a variety of other prokaryotes

**Table 11.3** Metabolic reactions responsible for generating nitrogen oxides in the atmosphere

| Reaction | Favorable environment | Microorganisms or process |
|---|---|---|
| $NH_4^+ \rightarrow NO_2^- \rightarrow N_2O$<br>$NH_4^+ \rightarrow NO_2^- \rightarrow NO$ | Oxic | Nitrifiers and methanotrophs deriving carbon from $CO_2$ and/or organic sources |
| $NO_3^- \rightarrow NO_2^-$   → $NH_4^+$<br>  ↘ $N_2O$ | Oxic | Yeast (e.g. *Rhodotorula* sp.) and fungi (e.g. *Fusarium oxysporum*) |
| $NO_3^- \rightarrow NO_2^- \rightarrow N_2O$ | Anoxic | Denitrifiers and methanotrophs |
| $NO_3^- \rightarrow NO_2^-$   → $NH_4^+$<br>  ↘ $N_2O$ or NO | Anoxic | Dissimilatory nitrate reduction to ammonium by organisms (e.g. *E. coli*, *E. carotovora*, *S. marcescens*, *Bacillus*, *Citrobacter*, *K. pneumoniae*, *A. vinelandii*, and *Wollinella succinogenes*) |

such as *Azotobacter vinelandii, Erwinia carotovora, Serratia marcescens, Klebsiella pneumoniae, E. coli, Enterobacter aerogenes,* and fungi such as *Fusarium oxysporium* as well as some yeasts (*Rhodotorula* sp.) were also shown to produce nitrous oxide (Bleakley and Tiedje, 1983). These organisms are generally more numerous in the environment compared to nitrifiers and denitrifiers. What is generated by microbes from nitrate depends on environmental factors such as oxygen levels in the microenvironment (Table 11.3). Formation of nitrous oxide is also stimulated by high inputs of nitrogen fertilizers, especially in flooded paddy fields due to their anoxic nature, which result in enhanced nitrate reduction (Freney and Denmead, 1992). Incidentally, rice plants are also capable of emitting the gas directly into the atmosphere through their gas vesicular system (Mosier et al., 1990). Other agricultural practices such as animal manure management, nitrogen fixation, and application of various N-fertilizer carriers, as well as the fertilizer application patterns adopted, have an impact on the levels of emission. The estimated emissions of nitrous oxide attributed to agricultural soils in different parts of the world are shown in Figure 11.2.

Generally, about 70% of the nitrous oxide emitted in the atmosphere is derived from soil where it has a residence time of 110 to 150 years (Mosier et al., 1998). Thus its concentrations are annually increasing at an average of 0.3% per annum, mostly from microbial processes during decomposition (Conrad, 1995; Mosier et al., 1998). Even where increases are directly attributable to the burning of biomass, the burning may produce nitrogen and other nutrients, making them available for enhanced soil microorganisms which in turn augments emissions of both nitric oxide and nitrous oxide from the soil (Anderson et al., 1988). A 50% increase in the present levels of nitrogen oxides in the atmosphere is estimated to raise the mean global temperatures by 0.2-0.6°C over time (Table 11.2).

## 11.1.4 Methane and Water Vapor

Methane is another important greenhouse gas whose production centrally involves microorganisms. It is mainly released during anaerobic decomposition of organic matter in marshes (wetlands), landfills, marine sediments, rice paddies, and from cattle as well as other rumi-

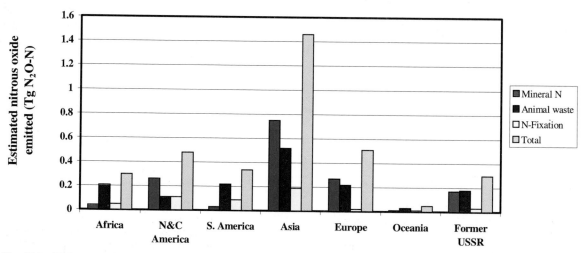

**Fig. 11.2** Estimates of the direct and indirect global emissions of nitrous oxide produced from agricultural soils that receive different N-sources (based on data from Mosier et al., 1998 with permission from Plenum/Kluwer Academic, New York).

nants, termite guts, the mammalian intestinal tract, and from other anoxic habitats (eqns. 6 and 7; Table 11.4). Most of the emissions are biogenic in origin, contributing approximately 60% of all the methane generated per annum, with only 15-20% methane emissions originating from natural gas pipes and coal mine leakages. Emissions from ruminants and paddy fields increase with increasing human populations.

$$H_2 + CO_2 \rightarrow CH_4 + H_2O \qquad ...(6)$$

$$Acetate \rightarrow CH_4 + CO_2 \qquad ...(7)$$

Soils are the most important source of atmospheric methane where it is generated by methanogens such as *Methanobacterium, Methanococcus,* and *Methanosarcina* while they decompose organic materials. All these microorganisms are sensitive to pH and have three compounds they can use, i.e., hydrogen $(H_2)$, formic acid (HCOOH), and acetic acid $(CH_3COOH)$. The process by which methane is formed in soil occurs in several steps that originate from the breakdown of polysaccharides to organic acids, alcohols, hydrogen, and $CO_2$ (Fig. 11.3). On the other hand, methane is oxidized by methylotrophs. However, methylotrophs are not that abundant in the environment, in particular anaerobic environments where methane is generated. This results in a spatial separation between the sources and sinks, leading to a net accumulation of approximately 60 million metric tons of methane in the troposphere annually. Furthermore, the gas has a life span of about 10 years in the atmosphere where it slowly decomposes photochemically through reactions involving the hydroxyl radical (OH) to yield molecular fragments that are involved in ozone destruction. A methane molecule is about 30 times more potent than $CO_2$ as a greenhouse gas (Conrad, 1995). A small fraction of methane escapes into the stratosphere where it is oxidized to produce stratospheric water vapor—a powerful greenhouse gas as it effectively absorbs infrared radiation over a broad spectrum of colors.

Water (in the form of vapor) is probably the most important greenhouse gas known. The paradox lies in the fact that, assuming a ready source of water, the warmer the atmosphere, the more water vapor becomes available. In other words, an increase in atmospheric temperature (as would occur during global warming) is likely to increase the amount of water vapor in the atmosphere which in turn further increases the temperature. Such a situation is believed to have occurred on Venus many years ago (Philander, 1998).

Other potential greenhouse gases such as

**Table 11.4** Sources of methane emissions into the atmosphere

| Source | | Total emissions | |
|---|---|---|---|
| | | Tg/year[a] | Percent |
| Biogenic | Cattle and other ruminants | 85 | 27.3 |
| | Rice paddies | 45 | 14.5 |
| | Termites | 4 | 1.3 |
| | Swamps, oceans, sediments, and lakes | 39 | 12.5 |
| | Landfills and others | 7 | 2.3 |
| Abiogenic | Biomass burning | 75 | 24.2 |
| | Coal mining and natural gas leakages | 54 | 17.4 |
| | Automobiles | 1 | 0.3 |
| | Volcanoes | 0.5 | 0.2 |

[a]*Note*: 1 Tg = $10^{12}$ g
(Source: Ciborowski (1989).

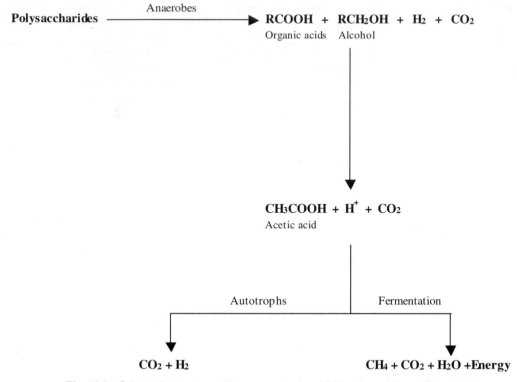

**Fig. 11.3**   Schematic summary of the processes by which methane is formed in soils.

hydrogen, carbon monoxide, and carbonyl sulfide are also controlled by microorganisms (Conrad, 1996). Chlorine, methylchloroform, methylene chloride, Halon-1301, carbon tetrachloride, peroxyacetyl nitrate (PAN), and sulfur hexafluoride are also important greenhouse gases as they are extremely powerful absorbers of infrared radiation. However, they have no known microbial origin and hence are not considered here.

## 11.2   BIOAEROSOLS

Bioaerosols are colloidal suspensions of solid particles or liquid droplets in the air containing viable pollen grains or microorganisms. Bioaerosols originate from various sources: natural (i.e., soil, aquatic, rainsplash, and vegetation) and anthropogenic (i.e., agricultural sources and other human activities which range from talking, sneezing, walking, spraying,

manufacturing, waste treatment, etc.). The concentration of viable bioaerosols usually varies inversely with altitude because the greatest potential source of bioaerosols is on the Earth's surface. The greatest reduction in vertical concentration of bioaerosols occurs in the first 50 meters or so above the Earth's surface where they are often associated with several health problems, some of which are listed in Table 11.5.

Bioaerosol particles are not always spherical in nature and their shape generally affects their trajectory. The mathematics involved in determining the effect of various shapes of bioaerosols to their transfer in the atmosphere is beyond the scope of this book. Bioaerosols range from a single bacterium, virus or spore (i.e., 0.65 μm aerodynamic diameter) to aggregations of microorganisms which may or may not be attached to soil particles or other

**Table 11.5**  Some airborne microorganisms of major concern to human health

| Type of microorganism | Causative agent | Health effects |
|---|---|---|
| Algae | *Chlorella, Chlorococcum, Anabaena,* | Allergic reactions |
| Bacteria | *Yersinia pestis* | Bubonic plague |
|  | *Bacillus* sp. | Hypersensitivity pneumositis |
|  | *Serratia marcescens* | Respiratory tract infection |
|  | *Staphylococcus, Streptococcus* sp. | Fevers, colds |
|  | *Bordetella pertussis* | Pertussis, whooping cough |
|  | *Mycobacterium tuberculosis* | Tuberculosis |
|  | *Corynebacterium diptheriae* | Diphtheria |
| Fungi | *Alternaria, Cladosporium, Penicillium, Fusarium, Aspergillus fumigatus* | Allergic reactions |
|  | *Aspergillus flavus* | Pulmonary aspergillosis |
| Protozoa | *Acanthamoeba* sp. | Eye infections and possibly, together with other protozoa, may be directly included in aerosols with *Legionella* spp. |
|  | *Naegleria fowleri* | Meningitis |
| Viruses | Poliovirus | Polio |
|  | Rhinoviruses | Common cold |
|  | Influenzavirus | Influenza |
|  | Varicella zoster | Chicken pox |
|  | Paramyxoviruses | Measles and mumps |

forms of airborne debris with sizes as large as 20 μm aerodynamic diameter. The larger bioaerosol particles tend to remain viable for longer durations in the atmosphere than do smaller ones (Lighthart, 1994).

The dispersal and movement of bioaerosols in the atmosphere greatly depend on air movement. Thus, wind speed, wind direction, and atmospheric stability play an important role in determining the local concentration of bioaerosols released into the atmosphere. Their dispersal is generally enhanced under strong winds and strong atmospheric instability as they follow the eddy currents in the airflow. Dispersal in the atmosphere occurs in a manner analogous to diffusing molecules and obeys Fick's law. According to that law, the rate of diffusion is proportional to the concentration gradient of the diffusing material. A gradient transfer model has been developed for modeling the dispersal of pathogens and pollen in cropping systems (McCartney, 1997). The model is composed of three components:

- the particles leaving a point-source horizontally by wind balanced by,
- the difference between the rate at which particles leave that source vertically due to both sedimentation and diffusion, and
- the rate at which the particles are deposited on surfaces.

In mathematical terms, the model is expressed as:

$$u[\delta C(x, z)/\delta x] = \delta z[K_z(\delta C(x, z)/\delta z] + v_s[\delta C(\delta x, z)/\delta z] + S(x, z) \quad ...(8)$$

where $C(x, z)$ is the concentration of particles at distance x and height z away from the source, u the wind speed, $K_z$ the vertical diffusion coefficient, (C/z) the concentration gradient, $v_s$ the speed at which the particle falls, and $S(x, z)$ the rate of removal of particles by deposition (McCartney, 1997). Since the size of the eddy currents are responsible for an increase in diffusion with height, the vertical diffusion coefficient ($K_z$) also increases with height. However, in more stable conditions such as within a stable vegetation canopy, the vertical diffusion co-efficient is proportional to height and can be simply derived from

$$K_z = 0.4u(z-d) \qquad ...(9)$$

where d is the zero plane displacement and is estimated to be two-thirds of the canopy (McCartney, 1997). Other more complex models have been proposed for use in instances wherein the gradient transfer model is deemed inappropriate but their use is beyond the scope of this book.

Although this model was initially developed for the spread of airborne pathogens in agricultural fields, it can be adapted, with pertinent modifications, to model the dispersal of bioaerosols around waste-treatment facilities, urban environments, and other outdoor environments where bioaerosol pollutants are of concern. For example, in urban areas the aerosols collide with surfaces and objects such as buildings, moving traffic, plants, etc. Vehicular activities also generate a large number of microorganisms rafted in dust particles.

Because of the continuous motion of the atmosphere, the sources and sinks of bioaerosols may be quite apart, causing temporal differences in bioaerosol densities even over a short period of time. Thus, local concentration of bioaerosols may become high under stable conditions with weak wind. Topography also plays a role in the dispersal of pollutants. For example, in montane regions nocturnal drainage flow formed during clear nights due to cooling of the surface carries bioaerosols from the surrounding elevated environments toward valleys, resulting into fairly high concentrations of bioaerosols in the valley at night. The time of day also has an indirect bearing on the density of bioaerosols as it influences air temperatures, causing convective differences in wind speed and direction, and intensity (Pady and Kramer, 1967).

## 11.2.1 Types of bioaerosols

### 11.2.1.1 Pollen

Pollen comprises a significant part of bioaerosols. Although not conventionally regarded among microorganisms, its inclusion in this text is based on the fact that most pollen in the air is microscopic, and from an environmental health perspective can cause considerable discomfort and misery. The incredibly high interest in pollen in bioaerosols arises from the fact that its concentration in the atmosphere consistently correlates with allergy incidences (Babu et al., 1974; Tilak, 1989; Comtois and Boucher, 1996). Most potent is pollen produced by anemophilous (as opposed to entomophilous) plants (Tilak, 1989). Anemophilous plants shed large quantities of pollen compared to entomophilous plants. The latter produce fewer large and brightly colored flowers to attract pollinators. Furthermore, in entomophilous plants, the stamens are concealed in the flower and the floral morphology restricts shedding of pollen in the atmosphere. However, broad generalizations are hard to apply in all situations because some entomophilous plants such as *Albizia, Callistemon*, and *Eucalyptus* also produce the dry, powdery pollen typical of anemophilous plants which, although less abundant in the atmosphere, can cause significant allergenic symptoms.

Comtois and Boucher (1996) studied the relationship between the ragweed plant phenology (architecture) and position of the pollen-emitting flowers on the plants and the airborne pollen cycles. They found that the highest quantities of pollen in the air corresponded to the time when the maximum number of flowers from the apex, primary branch, and secondary branches had opened. However, comparison of their data over a two-year period (1990 and 1991) also indicated that flowering is affected by the cumulative heating degree days over 5°C. Thus, pollen shading is also related to other environmental factors such as temperatures and humidity (Barnes et al., 2001). Prolonged heat can enhance the dispersal of allergens and changes in weather patterns such as the El Niño Southern oscillation (ENSO) affect the production of pollen by plants, perhaps indirectly contributing to increased incidence of asthma and other respiratory disorders (Epstein, 2001). This contention has implications in medical environmental microbiology, the focus of Chapter 14.

### 11.2.1.2 Bacteria

The most predominant types of bacteria in bioaerosols are Gram-positive cocci such as *Staphylococcus*, *Sarcina*, *Micrococcus*, and *Corynebacter* as well as spore-forming rods. Their abundance is only second to that of fungi. Gram-negative rods such as *Achromobacter, Flavobacterium*, and *Pseudomonas* are also widely encountered. The endotoxins produced from the bacterial cell walls, in particular the Gram-negative bacteria due to their lipopolysacchride membrane, are suspected to cause respiratory disorders and, in some cases, gastrointestinal disorders (Laitinen et al., 1992). Airborne concentrations of 7,200 colony forming units (CFU) m$^{-3}$ of Gram-negative bacteria are equivalent to 30 mg m$^{-3}$ of air endotoxins, an endotoxin concentration which has been proposed as the occupational exposure limit (Laitinen et al., 1992). However, Gram-negative bacteria do not appear to be carried in large numbers from their point-source compared to Gram-positive bacteria (Table 11.6); the reason for this is not too clear. A *Limulus* test was designed to determine the endotoxicity of Gram-negative bacteria and $\beta$-1,3-glucans found in many fungi. However, the *Limulus* aggregation reaction can be subjected to cross-reaction by substances that are not endotoxic and/or inhibited by these substances, giving false positive or false negative results.

Of utmost concern among airborne bacteria are *Legionella* spp., in particular *Legionella pneumophila* which causes more than 95% of all cases of Legionnaires' disease (Abu Kwaik et al., 1998). Transmission of the pathogen by person-to-person has not been documented and, although the pathogen is ubiquitous, especially in aquatic environments, transmission to humans occurs primarily by aerosols generated by power-generating cooling towers, waste-treatment plants, whirlpools, overhead showers, abattoirs, and industrial areas. In a survey of 800 cooling towers for buildings around the Baltimore-Washington DC area over a two-year period (1989-1990), Miller and Kenepp (1996) consistently isolated *Legionella* spp. using a set of selective media coupled with fluorescent antibody confirmations on suspect colonies. *Legionella pneumophila* was encountered in 75% of the towers surveyed over that time period. In some instances, the towers were found to contain *Legionella* spp. as the only culturable bacteria on buffered charcoal yeast extract (BCYE) agar, a medium for total heterotrophs similar to R2A agar and trypticase soyagar (TSA). That observation markedly suggests that some selective factor in the cooling towers might favor *Legionella* spp. over other heterotrophs.

Once inhaled, the generated aerosols containing *L. pneumophila* pose a potential danger. Levels of 1,000 *L. pneumophila* ml$^{-1}$ of water in the cooling towers or higher are reported as the threshold for potential Legionnaires' disease outbreaks (Miller and Kennep, 1996). The organism has also been reported, in some cases, to be intimately associated with various protozoan hosts which can protect it against disinfection with biocides, heating or ultraviolet light (Abu Kwaik et al., 1998). In some instances, this *Legionella* species is released surrounded by a protozoan vesicle which renders it highly resistant to disinfection (Berk et al., 1998). As a matter of fact, *L. pneumophila* does not multiply extracellularly in the environment; its multiplication occurs only within an infected host (i.e., protozoa or human). Once inside the protozoan host, the bacterium dramatically increases resistance to a whole range of environmental extremes such as high temperature, acidity, and osmotic pressure (Abu Kwaik et al., 1997), coupled with an increased infectivity of mammalian cells (Cirillo et al., 1994), and resistance to antibiotics (Barker et al., 1995). Nonculturable *L. pneumophila* is also revived after incubation within protozoa (Steinert et al., 1997). At this point, available information indicates that it is parasitic because during the processes of multiplying, increasing in infectivity, and attaining enhanced resistance to environmental extremes, *L. pneumophila* somehow kills the protozoan host through, as yet unknown mechanisms. It is not clear whether some aerosols containing *L. pneumophila* also contain aerosolized protozoa.

### 11.2.1.3 Fungi

A wide range of fungal spores which cause diseases and economic losses to various crops are airborne (Table 11.6). Some fungi, especially of genera of Deuteromycotina and Zygomycotina are pathogenic, causing allergic or hypersenstivity reactions, especially in imunocompromised individuals. Commonly associated with hypersensitivity reactions are *Alternaria alternata*, *Aspergillus niger*, *A. fumigatus*, *A. flavus*, *A. parasiticus*, *Penicillium* spp., *Paecilomyces variotii*, *Emericella nidulans*, *Aerobasidium pullulans*, *Cladosporium cladosporioides*, *Geotrichum candidum*, *Eurotium chevalieri*, *Trichoderma viride*, and *Fusarium* spp., most of them causing hypersensitivity due to the toxins (mycotoxins) they produce. This list is not exhaustive. More than 400 mycotoxins are produced as secondary metabolites (Tuomi et al., 2000) and therefore, their production does not correlate with the growth of the species that produce them. From an ecological perspective, it is not clear as to why fungi produce toxins at

any one point but probably they are generated as a protective mechanism against a broad range of competitors. Even within the same species, toxin production can vary greatly between isolates, demonstrating the wide array of toxins which can be generated by fungi in nature.

Most of the mycotoxins are nonvolatile at ambient temperature but human can be exposed to them through house dust (Angulo-Romero et al., 1996). Once inhaled, the fungal spores may cause pulmonary mycotoxicosis due to absorption of the toxins through the mucous membranes of the respiratory tract (Tobin et al., 1987). Mycotoxins have also been extracted from bulk materials within buildings even in situations where the associated fungal species that generate those toxins are not recovered on growth medium (Croft et al., 1986; Tuomi et al., 2000). Apparently aflatoxins, produced solely by fungi, are the most potent toxin known today. Other mycotoxins, such as sterigmatocystin, are also quite potent.

Of most concern currently is the Deuteromycetes fungus, *Stachybotrys chartarum*. Although this fungus and its ability to produce toxic metabolites, particularly trichothecenes such as satratoxins, have been known from as early as 1837 (Nelson 2001), it is becoming increasingly implicated in respiratory and central nervous system disorders (Johanning et al., 1996) and pulmonary hemorrhage (Vesper et al., 1999) associated with moldy conditions indoors. The fungus is fairly widespread in soil (Khallil and Abdel-Sater, 1992), and in dust for that matter, where it does a great job in the degradation of cellulose and hemicellulose debris. Its growth indoors is enriched under high moisture conditions (more than 95% relative humidity; Rowan et al., 1999) on surfaces rich in cellulose, which it uses as a carbon source under low-nitrogen conditions. It is therefore a threat to human and veterinary animal health in housing units that use cellulose material for walls (i.e., dry wall,

**Table 11.6** Some fungi of agricultural importance that are transmitted through the air

| Fungus | Organism affected by fungus | Associated disease |
|---|---|---|
| *Pyricularia oryzae* | Rice | Rice blast |
| *Puccinia penniseti* | Pearl millet | Pearl millet rust |
| *Schlerospora graminicola* | Pearl millet | Pearl millet rust |
| *Claviceps fusiformis* | Pearl millet | Pearl millet rust |
| *Pyricularia penniseti* | Pearl millet | Pearl millet rust |
| *Exserohilum furcicum* | Sorghum | Sorghum leaf blight |
| *Peronosclerospora soghi* | Sorghum | Sorghum leaf blight |
| *Ustillago scitaminea* | Sugarcane | Sugarcane smut |
| *Exobasidium vexans* | Tea | Tea blister blight |
| *Phytophthora palmivora* | Cocoa | Black pod disease |
| *Cercospora* spp. | Peanuts | Cercospora leaf spot |

wood, gypsum board, cardboard) that become moist, which is typically the case in water-damaged buildings (Johanning et al., 1996; Tuomi et al., 2000). However, the spores are produced in a slimy mass and not likely to become airborne unless the mass dries out.

### 11.2.1.4 Microalgae

The air can also contain microalgae particles, some of which have allergenic compounds (Table 11.5). Nineteen genera belonging to families Cyanophyceae and Chlorophyceae were isolated from indoor and outdoor samples in Hyderabad, India (Rao and Jadhav, 1997). *Stagonema* sp. was found exclusively in the house of a nasobronchial patient, whereas *Gloeocystis* sp. was found only in the control house. Unfortunately, no antibody tests were conducted in that study although studies elsewhere also suspect airborne microalgae as a biologic factor in respiratory symptoms (Bernstein and Safferman, 1970, 1973; Pinto, 1972).

### 11.2.1.5 Viruses

Viruses such as polio, measles, mumps, common cold and chicken pox (Table 11.5) are transmitted as aerosols from infected individuals through talking, coughing and sneezing, besides other modes of transmission. Survival of these viruses indoors is well documented

and hence mitigation of airborne viral infections understood (Akers, 1973; Marks et al., 2000). However, in outdoor environments their viability is probably greatly impacted by environmental factors such as UV, moisture stress, and unfavorable temperatures. Whereas the mechanisms for denaturation due to temperature are not known, UV acts against viruses by denaturing the nucleic acids while moisture stress is detrimental because it denatures the capsid and/or virus protein coat. Aerosolized viruses are also generated during the treatment of sewage. Evidence from around a waste-treatment plant indicated that viruses can be carried for greater distances than bacteria (Carducci et al., 1999). Irrespective of the sampling equipment adopted (see sampling in Section 11.2.3), sampling for viruses in bioaerosols requires collecting large volumes. Recovery of viruses also requires prehumidifying the air being sampled, a process that dramatically increases collection efficiency by 2 to 3 log units for most types of viruses, in particular RNA-containing viruses (Akers, 1973). Reduced recovery at low relative humidity is possibly attributable to damage in the protein coat and/or capsid.

### 11.2.1.6 Dust mites and related house allergens

The bodies, feces, and secretions of house dust mites (*Dermatophagoides farinae, D.*

*pteromyssinus, Blomia tropicalis,* and *Euroglyphus maynei*) are sources of common allergens which are airborne in house dust. Mites survive better under humid environs with a relative humidity of 60-75% and temperatures in the range of 15.6 - 24°C (O'Rourke et al., 1996). Although they are from the world of entomology, including them under environmental microbiology is relevant because they are microscopic.

## 11.2.2  Survival of bioaerosols in the environment

Disruption and evaporation of aerosol droplets as well as the survival of the microorganisms contained therein, are influenced by the prevailing atmospheric conditions such as temperature and relative humidity. Their removal from the atmosphere to the Earth's surface is facilitated by gravitational settling and scavenging by rain water and, in temperate regions, snow. In open space, movement and dispersal of bioaerosols is a physical process governed by the basic laws of physics. They are therefore subject to inertial resistance and to drag or viscosity effects of the atmosphere. Resistance due to inertia can be defined according to Newton's laws (eqn. 10) whereas drag is defined according to Stokes' law (eqn. 11).

$$F_D = C_D \pi \rho_g D_p^2 V^2/8 \qquad ...(10)$$

$$F_D = \pi \eta V D_p \qquad ...(11)$$

where $F_D$ is the atmospheric drag force, $C_D$ the drag coefficient, $\rho_g$ the density of ambient air (g $ml^{-1}$), $D_p$ the particle diameter (m), and V, the velocity of the particle (m $s^{-1}$), $\eta$ is the coefficient of viscosity of ambient air (dynes × s $cm^{-2}$). Thus, the lighter individual (as opposed to aggregated) bioaerosol particles tend to stay suspended in the atmosphere for a longer duration because a smaller drag force acts on them compared to the force that acts on larger particulates. Thus smaller particles of less than 10 μm in diameter are suspended longer in the atmosphere than larger particles,

the settlement of particulates increasing with increasing size of the particles of the same density.

Microbes in ambient air often occur as aggregates or on soil (dust) debris. Their survival increases with increment in the aerodynamic size of the particles (Tong and Lighthart, 1998). However, airborne microbes typically do not grow or multiply in ambient air due to a lack of nutrients. They can be transmitted for long distances. Desiccation is the most fundamental stress of bioaerosols as it causes alterations in the cell membrane, resulting in a leaky gel-like structure. The rate and degree of desiccation depend on the ambient temperature and relative humidity but it is difficult to separate the effects of relative humidity from the effects of temperature because the former depends on the latter. Irrespective of the prevailing relative humidity in the environment, airborne microbes undergo some desiccation. Most stressful to bioaerosols are relative humidity levels of less than 20% and above 80% (Webb, 1959).

For prokaryotes, the cell wall composition affects their resistance and hence survival in aerosols. Gram-positive bacteria are generally more resistant due to their relatively tough cell wall compared to Gram-negative bacteria. The cell wall of Gram-positive bacteria consists of approximately 80% peptidoglycans whereas that for Gram-negative bacteria has only 5-10% peptidoglycans by weight, the rest comprising polysaccharides, proteins, and phospholipids (see Fig. 2.2).

Survival of the organisms in bioaerosols is also influenced by solar radiation and the chemical composition of the atmosphere. The solar radiation per unit area hitting the Earth's surface varies significantly depending on the latitude and season but bioaerosols in the ambient region are subject to both visible (400-750 nm wavelength) and ultraviolet (290-400 nm wavelength) light, the irradiation causing either lethal or sublethal damage (Tong and Lighthart, 1997). Sublethal damage causes delays in growth of the bioaerosol microorganisms, even when they settle on suitable surfaces (Dimmick, 1960). However, damage to cells due to irradiance appears to affect the microorganisms to a differing degree. Preliminary experiments by Tong and Lighthart (1998) using terrestrial and extraterrestrial solar spectra showed that airborne microorganisms can potentially be used as a solar UV radiation dosimeter. With increase in UV solar radiation, survival of the ambient bacterial populations dramatically decreases (Fig. 11.4). The practical ramification from this observation is that with depletion of the ozone layer, the expected increase in UV in the atmosphere carries an increased lethal effect on bioaerosols. Thus, natural ambient microbial populations in the atmosphere over time can be used as an early detection/warning system for progressive increases in UV radiation. The better survival in

larger as opposed to smaller aerodynamic particles is possibly a result of better protection of the microorganisms in larger particles against UV radiation, nutrient stress, as well as moisture depletion.

The atmosphere is never free of bioaerosols but their concentrations greatly depend on the nature of the source coupled with prevailing environmental factors (temperature, relative humidity, wind speed and duration, time of day, radiation, location, etc.). Bioaerosols are in particularly high doses within and around waste-treatment plants, certain farm operations, and at compost sites where, unless appropriate measures are taken, they can pose some risk to workers and surrounding communities (Laitinen et al., 1992; Fischer et al., 1999), causing what is normally referred to as microbial air pollution (MAP). Thus, employees in these settings tend

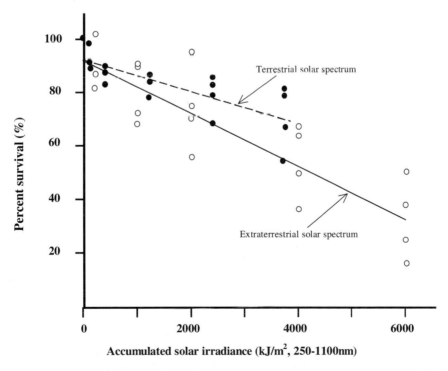

**Fig. 11.4** Survival of aerosolized bacterial populations after exposure to terrestrial and extraterrestrial solar radiation. The extraterrestrial solar radiation had a higher percent UV (i.e., 0.11, 0.60 and 4.90% UVC, UVB, and UVA, respectively) than terrestrial solar radiation with 0.01, 0.18, and 3.32% UVC, UVB, and UVA respectively (adapted from Tong and Lighthart, 1998; reprinted with permission from Elsevier).

**Table 11.7** Airborne concentrations of microorganisms (CFU $m^{-3}$) in and around the Iztapalapa (NE Mexico City) domestic waste transfer station

| Sampling station | Range of total bacteria (CFU $m^{-3}$) | Gram-positive bacteria | | Range of mesophilic fungi | Penicillin | |
| --- | --- | --- | --- | --- | --- | --- |
| | | Range | % Total | | Range | % Total |
| Unloading trucks | 2,200-14,800 | 11-3,920 | 14 | 170-14,800 | 28-14,830 | 70 |
| Loading containers | 350-14,800 | 11-4,480 | 21 | 340-14,800 | 63-14,830 | 75 |
| Lunchroom | 2,220-14,800 | 21-1,280 | 3 | 6,020-14,800 | 95-14,800 | 79 |
| 50 m downwind | 270-4,220 | 0-200 | 3 | 165-14,800 | 22-6,450 | 35 |
| 20 m upwind | 60-740 | 0-14 | 1 | 20-175 | 4-100 | 45 |

The air was sampled using two-stage Andersen samplers with orifice diameters for stages 1 and 2 of 1.5 and 0.4 mm respectively. Each sampling event lasted 5 min at 28.3 L $min^{-1}$ and was mounted on a 2 m high tower. Samplers were loaded with plastic petri dishes containing 20 ml of either trypticase soyagar (for bacteria) or malt extract agar (for fungi). The TSA and MEA plates were incubated at 35°C (2 days) and 25°C (3 days) respectively.
Source: Rosas et al. (1996) with permission from CRC Lewis.

to report a higher incidence of respiratory symptoms and abdominal disorders. Aerosols can be transmitted from source to recipient in dust, air, and by handling contaminated surfaces. The concentration of microorganisms is often higher downwind from the source than upwind (Table 11.7). This reality should ideally be considered when siting waste-processing facilities in close proximity to neighborhoods.

Besides concern over bioaerosols to human health, airborne particles greatly contribute toward the deterioration of structures, ranging from buildings, pieces of art, concrete, and metallic structures. Most of these structures contain some level of organic constituents, providing different ecological niches that are exploitable by microorganisms. These structures are often exposed to dust which settles on them, introducing the airborne microflora that subsequently grows once other environmental factors such as temperature, moisture content (humidity), light, and pH permit (Nugari et al., 1993). On such hardy structures, nutrients can be provided from the accompanying dust and/ or inorganic air pollutants such as sulfur dioxide, hydrogen sulfide, ammonia, heavy metal dusting, etc. which may be present in that environment. The resultant damage to the structures can be aesthetic or structural as a result of the metabolites and extracellular enzymes that the colonizing microflora excretes (Ciferri, 1999). Oxidation-reduction of the existing cations can also lead to the deposition of precipitates that foul the structure in question. Frequently isolated from deteriorated or deteriorating structures are soil inhabiting fungi (such as *Aspergillus, Altenaria, Aureobasidium, Cladosporium, Chaetomium,* and *Penicillium*), and bacteria (*Alcaligenes, Arthrobacter, Bacillus, Flavobacterium, Pseudomonas, Norcadia,* and *Streptomyces*), the composition of the thriving microflora greatly depending on the chemical composition of the structure in question (Ciferri, 1999). Algae, cyanobacteria, and lichens can also thrive on structures that are exposed to light, weathering the surface layers.

### 11.2.3  Sampling for bioaerosols

The sampling of aerosols requires separation and collection of particles from the airstream. To be quantitative, the air intake has to have a known flow rate (i.e., volume of air taken in per unit time). The process relies on the principles of impaction, filtration or impingement. Impaction is used to separate particles from a gas stream based on the inertia of the particles. An impactor consists of a series of nozzles through which a desired diameter size of particles, and those with a smaller diameter, pass. The particles are collected onto a target consisting of a growth medium (agar) on a petri dish, greased plate, sticky glass slide, or a filter material. The target may be incubated to allow the collected bioaerosol to grow forming visible colonies, weighed to determine the amount of sample collected, or washed and the wash solution analyzed. Alternatively, the collection facility is stained (e.g. with Fuchsin) and the collected aerosol material directly observed (or counted) under the microscope. When observed or quantified by incubating the plates, the underlying assumption for the colonies formed is that each colony represents a single aerosol particle. This assumption may inaccurately represent the bioaerosols because as the number of particles deposited on the growth medium increases, the probability that the next particle will impact a "clean" spot decreases. A correction factor to adjust for this inaccuracy was proposed by Jensen et al. (1994):

$$P_r = N\{1/N + 1/(N-1) + 1/(N-2) + \ldots + 1/(N-r+1)\} \quad \ldots(12)$$

where  r is the number of observed colonies on the plate;

$P_r$ the expected number of culturable particles to produce r colonies; and

N the total number of holes per impactor stage.

Collection of aerosol particles by filtration is a common practice in industrial hygiene. Air filters collect approximately 100% of the particles larger than the pore size. Since the

**Table 11.8** Efficiency of different media to collect aerosolized spores using the cyclone scrubber

| Medium | % recovery efficiency of various microorganisms with different media[1,2] | | | |
|---|---|---|---|---|
| | *B. subtilis* | *E. coli* | *S. marcescens* | *A. aerogenes* |
| 0.5% Gelatin milk | 81.6 | 79 | 89 | 60.8 |
| Gelatin milk | 81.9 | 70.8 | 77 | 79 |
| Glycerol-gelatin milk phosphate | 85.5 | 45 | 70 | 33.2 |
| 0.06% Tween 80 | 81.8 | –[3] | – | 69.7 |
| *S. marcescens* medium | 74 | ND[4] | 68 | ND |
| *A. aerogenes* medium | 86 | ND | ND | 87.4 |
| *E. coli* medium | 85.3 | 67 | ND | ND |

[1] *B. subtilis* as spores, *E. coli*, *S. marcescens* and *A. aerogenes* as cells.
[2] % recovery efficiency was in relation to the AGI-30 impinger sampler.
[3] *E. coli* and *S. marscesens* were killed in the AGI-30 impinger.
[4] Not determined.
Source: White et al., (1975) with permission from the American Society of Microbiology.

removal of particles occurs by collision and attachment of the particles to the membrane surface, particles of smaller pore sizes may also be efficiently collected. The sample collected on the filter can, depending on the objective of the study, be either directly examined under the microscope after staining, or washed off with 0.02% Tween and cultured.

Bioaerosols can also be sampled by impingement to collect both culturable and nonculturable bioaerosols. Most commonly used is the all-glass impinger (AGI-30) sampler (Lekme et al 1981). Its inlet tube is designed to simulate the collection of particles in the nasal passage, a design that makes it ideal to use for studying the respiratory infection potential of airborne microorganisms.

Where culture methods are relied on to determine the composition of bioaerosols, irrespective of the sampling method used to trap the sampled air, the types of organisms present are reflected by the type of medium in the trapping system. Most commonly used is Lauria Berttani (LB) agar or R2A agar (Lighthart, 1997). White et al. (1975) used different types of media to collect *Bacillus subtilis* spores, *S. marcescens, A. aerogenes* and *E. coli* aerosols. They found no difference in the percent recovery of *B. subtilis* irrespective of the type of medium used. However, differences in the

quantities of *E. coli* and *S. marcescens* recovered were noticeable depending on the medium (Table 11.8). Thus, the collection medium is critical, especially with vegetative cells as opposed to spores. On the whole, variability between samples could be substantial due to other environmental factors such as relative humidity, temperature, and wind speed. Wind speed affects the impact and possibly affects the extent of cell damage. Monitoring bioaerosols should include the viable, viable but nonculturable (VBNC), and nonviable microorganisms. Furthermore, just like other environments, the atmosphere is presumed to contain a whole range of microorganisms that have never been cultivated. Unfortunately, culture-independent techniques have not yet been used extensively in aerobiology.

## References

Abu Kwaik Y., L-Y. Gao, O.S. Harb, and B.J. Stone (1997). Transcriptional regulation of the macrophage-induced gene (*gspA*) of *Legionella pneumophila* and phenotypic characterization of a null mutant. *Molec. Microbiol.* **24:** 629-642.

Abu Kwaik Y., L-Y. Gao, B.J. Stone, C. Venkataraman, and O.S. Harb (1998). Invasion of protozoa by *Legionella pneumophila* and its role in bacterial ecology and pathogenesis. *Appl. Environ. Microbiol.* **64:** 3127-3133.

Akers T.G. (1973). Some aspects of the airborne inactivation of viruses. *In:* J.F.P. Hers and K.C. Winkler (eds.). Airborne Transmission and Airborne Infection. John Wiley and Sons, New York, NY, pp. 73-81.

Anderson I.C., J.S. Levine, M.A. Poth, and P.J. Riggan (1988). Enhanced biogenic emissions of nitric oxide and nitrous oxide following surface biomass burning. *J. Geophys. Res.* **93**: 3893-3898.

Angulo-Romero J., F.Infante-García-Pantaleón, E. Domínguez-Vilches, A. Mediavilla-Molina, and J.M. Caridad-Ocerín (1996). Pathogenic and antigenic fungi in school dust of the south of Spain. *In:* M. Muilenberg and H. Burge (eds.). Aerobiology. CRC Lewis Press, Inc., New York, NY, pp. 49-65.

Babu C.R., A.B. Singh, and D.N. Shivpuri (1974). Allergic factors and symptomatology of respiratory allergy patients. *J. Asthma Res.* **16**: 97-101.

Barker J., H. Scife, and M.R.W. Brown (1995). Intraphagocytic growth induces an antibiotic-resistant phenotype of *Legionella pneumophila. Antimicrob. Agents Chemother.* **39**: 2684-2688.

Barnes C., F. Pacheco, J. Landuyt, F. Hu, and J. Portnoy (2001). The effect of temperature, relative humidity and rainfall on airborne ragweed pollen concentrations. *Aerobiologia* **17**: 61-68.

Berk S.G., R.S. Ting, G.W. Turner, and R.J. Ashburn (1998). Production of respirable vesicles containing live *Legionella pneumophila* cells by two *Acanthamoeba* spp. *Appl. Environ. Microbiol.* **64**: 279-286.

Bernstein I.L. and R.S. Safferman (1970). Viable algae in house dust. *Nature* **227**: 851-852.

Bernstein I.L. and R.S. Safferman (1973). Clinical sensitivity to green algae demonstrated by nasal challenge and in-vitro tests of immediate hypersensitivity. *J. Allergy* **51**: 22-28.

Bleakley B.H. and J.M. Tiedje (1983). Nitrous oxide production by organisms other than nitrifiers or denitrifiers. *Appl. Environ. Microbiol.* **44**: 1342-1348.

Carducci A., C. Gemelli, L. Cantiani, B. Casini, and E. Rovini (1999). Assessment of microbial parameters as indicators of aerosol from urban sewage treatment plants. *Lett. Appl. Microbiol.* **28**: 207-210.

Ciborowski P. (1989). The greenhouse gases: Sources, sinks, trends, and opportunities. *In:* D.E. Abrahamson (ed.). The Challenge of Global Warming. Island Press, Washington DC, pp. 213-230.

Ciferri O. (1999). Microbial degradation of paintings. *Appl. Environ. Microbiol.* **65**: 879-885.

Cirillo J.D., L.S. Tompkins, and S. Falkow (1994). Growth of *Legionella pneumophila* in *Acanthamoeba* spp. *Appl. Environ. Microbiol.* **64**: 279-286.

Comtois P. and S. Boucher (1996). Phenology and aerobiology of short ragweed (*Ambrosia artemisiifolia*) pollen. *In:* M. Muilenberg and H. Burge (eds.). Aerobiology. CRC Lewis Press, Inc., New York, NY, pp.17-26.

Conrad R. (1995). Soil microbial processes involved in production and consumption of atmospheric trace gases. *Adv. Microb. Ecol.* **14**: 207-250.

Conrad R. (1996). Soil microorganisms as controllers of atmospheric trace gases ($H_2$, CO, $CH_4$, OCS, $N_2O$, and NO). *Microbiol. Rev.* **60**: 609-640.

Croft W.A., B.B. Jarvis, and C.S. Yatawara (1986). Airborne outbreak of trichothecene toxicosis. *Atm. Environ.* **20**: 549-552.

Dimmick R.L. (1960). Delayed recovery of airborne *Serratia marcescens* after short-term exposure to ultra-violet irradiation. *Nature* **19**: 251-252.

Epstein P.R. (2001). Climate change and emerging infectious diseases. *Microb. and Inf.* **3**: 747-754.

Fischer G., T. Müller, R. Ostrowski, and W. Dott (1999). Mycotoxins of *Aspergillus fumigatus* in pure culture and in native bioaerosols from compost facilities. *Chemosphere* **38**: 1745-1755.

Freney J.R. and O.T. Denmead (1992). Factors controlling ammonia and nitrous oxide emissions from flooded rice fields. *Ecol. Bull.* **42**: 188-194.

Gregory P.H. (1961). The Microbiology of the Atmosphere. Leonard Hill Books, Aylesbury, Bucks, UK.

Jaffe D.A. (1992). The nitrogen cycle. *In:* S.S. Butcher, R.J. Charlson, G.H. Orians and G.V. Wolfe (eds.). Global Biogeochemical Cycles. Acad. Press, New York, NY, pp. 262-284.

Jensen P.A., B. Lighthart, A.J. Mohr, and B.T. Shaffer (1994). Instrumentation used with microbial bioaerosols. *In:* B. Lighthart and A.J. Mohr (eds.). Atmospheric Microbial Aerosols: Theory and Applications. Chapman & Hall, New York, NY, pp. 226-284.

Johanning E., R. Biagini, D. Hull, P. Morey, B. Jarvis, and P. Landsbergis (1996). Health and immunology study following exposure to toxigenic fungi (*Stachybotrys chartarum*) in a water-damaged office environment. *Intnatl. Arch. Occup. Environ. Health* **68**: 207-218.

Khallil A.M. and A.M. Abdel-Sater (1992). Fungi from water, soil and air polluted by industrial effluents of Manquabad superphosphate factory. *Intnatl. Biodeter. Biodegrad.* **30**: 363-386.

Kim J. (1994). Atmospheric environment of bioaerosols. *In:* B. Lighthart and A.J. Mohr (eds.). Atmospheric Microbial Aerosols: Theory and Applications. Chapman & Hall, New York, NY, pp. 28-67.

King G.M., D. Kirchman, A. Salyers, W. Sclesinger, and J.M. Tiedje (2001). Global environmental change: microbial contributions, microbial solutions. Amer. Soc. Microbiol., Washington DC. Available online at http://www.asmusa.org/pasrc/pdfs/globalwarming.pdf.

Laitinen S., A. Nevalainen, M. Kotimaa, J. Lisivuori, and P.J. Matikainen (1992). Relationship between bacterial counts and endotoxin concentrations in the air of wastewater treatment plants. *Appl. Environ. Microbiol.* **58**: 3774-3776.

Lemke L.L., R.N. Kniseley, R.C. van Nostrand, and M.D. Hale (1981). Precision of the All-glass Impinger and the Anderson Microbial impactor for air sampling in solid-waste hauling facilities. *Appl. Environ. Microbiol.* **42**: 222-225.

Lighthart B. (1994). Physics of microbial aerosols. *In:* B. Lighthart and A.J. Mohr (eds.). Atmospheric Microbial Aerosols: Theory and Applications. Chapman & Hall, New York, NY, pp. 5-27.

Lighthart B. (1997). The ecology of bacteria in the alfresco atmosphere. *FEMS Microbiol. Ecol.* **23**: 263-274.

Marks P.J., I.B. Vipond, D. Carlisle, D. Deakin, and R.E. Fey (2000). Evidence of airborne transmission of Norwal-like virus (NLV) in a hotel restaurant. *Epidemiol. Infect.* **124**: 481-487.

McCartney H.A. (1997). Modelling the dispersal of fungal spores and pollen by wind. *In*: S.N. Agashe (ed.). Aerobiology. Sci. Publ. Inc., Enfield, NH, pp. 327-332.

Miller R.D. and K.A. Kenepp (1996). *Legionella* in cooling towers: Use of Legionella-total bacteria ratios. *In*: M. Muilenberg and H. Burge (eds.). Aerobiology. CRC Lewis Press, New York, NY, pp. 99-107.

Mosier A.R., S.K. Mohanty, A. Bhadrachalam, and S.P. Chakravorti (1990). Evolution of dinitrogen and nitrous oxide from soil to the atmosphere through rice plants. *Biol. Fert. Soils* **9**: 61-67.

Mosier A.R., J.M. Duxbury, J.R. Freney. O. Heinemeyer, and K. Minami (1998). Assessing and mitigating $N_2O$ emissions from agricultural soils. *Climatic Change* **40**: 7-38.

Nelson B.D. (2001). *Stachbotrys chartarum*: The toxic indoor mold. http://www.apsnet.org/online/feature/stachbotrys (last checked February 26th, 2002).

Nugari M.P., M. Realini, and A. Roccardi (1993). Contamination of mural paintings by indoor airborne fungal spores. *Aerobiologia* **9**: 131-139.

O'Rourke M.K., C.L. Moore, and L.G. Arlian (1996). Prevalence of house dust mites from homes in the Sonoran Desert, Arizona. *In*: M. Muilenberg and H. Burge (eds.). Aerobiology. CRC Lewis Press, Inc., New York, NY, pp.67-80.

Pady S.M. and C.L. Kramer (1967). Diurnal periodicity of airborne bacteria. *Mycologia* **59**: 714-716.

Philander S.G. (1998). Is the Temperature Rising?: The Uncertain Science of Global Warming. Princeton Univ. Press, Princeton, NJ.

Pinto C.B. (1972). Airborne algae as possible etiologic factor in respiratory allergy in Caracas, Venezuela. *J. Allergy Clin. Immun.* **49**: 356-358.

Rao K.S.R. and M.J. Jadhav (1997). Isolation of microalgae from the dust obtained from indoor and outdoor sources. *In*: S.N. Agashe (ed.). Aerobiology. Sci. Publ. Inc., Enfield NH, pp. 49-58.

Ren T., R. Roy, and R. Knowles (2000). Production and consumption of nitric oxide by three methanotrophic bacteria. *Appl. Environ. Microbiol.* **66**: 3891-3897.

Rosas I., C. Caldern, E. Salinas, and J. Lacey (1996). Airborne microorganisms in a domestic waste transfer station. *In*: M. Muilenberg and H. Berge (eds.). Aerobiology. CRC Lewis Press, Inc., New York, NY, pp. 89-98.

Rowan N.J., C.M. Johnstone, R.C. McLean, J.G. Anderson, and J.A. Clarke (1999). Prediction of toxigenic fungal growth in buildings by using a novel modelling system. *Appl. Environ. Microbiol.* **65**: 4814-4821.

Steinert M., L. Emody, R. Amman, and J. Hacker (1997). Resuscitation of viable but nonculturable *Legionella pneunmophila* Philadelphia JR32 by *Acanthamoeba castellani. Appl. Environ. Microbiol.* **63**: 2047-2053.

Tilak S.T. (1989). Environmental Ecology and Aerobiology. Today and Tomorrow's Printers and Publ., New Dehli, India.

Tobin R.S., E. Baronowski, A.R. Gilman, T. Kuiper-Goodman, J.D. Miller, and M. Giddings (1987). Significance of fungi in indoor air: report of working group. *Can. J. Public Health* **78**: S1-14.

Tong Y. and B. Lighthart (1997). Solar radiation has a lethal effect on natural populations of culturable outdoor atmospheric bacteria. *Atm. Environ.* **31**: 897-900.

Tong Y. and B. Lighthart (1998). Effect of simulated solar radiation on mixed outdoor atmospheric bacterial populations. *FEMS Microbiol. Ecol.* **26**: 311-316.

Tuomi T., K. Reijula, T. Johnsson, K. Hemminki, E-L. Hintikka, O. Lindroos, S. Kalso, P. Koukila-Kähkölä, H. Mussalo-Rauhamaa, and T. Haahtela (2000). Mycotoxins in crude building materials from water-damaged buildings. *Appl. Environ. Microbiol.* **66**: 1899-1904.

Vesper S.J., D.G. Dearborn, I. Yike, W.G. Sorenson, and R.A. Haughland (1999). Hemolysis, toxicity, and randomly amplified polymorphic DNA analysis of *Stachybotrys chartarum* strains. *Appl. Environ. Microbiol.* **65**: 3175-3181.

Webb S.J. (1959). Factors affecting the viability of airborne bacteria. I. Bacteria aerolized from distilled water. *Can. J. Microbiol.* **5**: 649-669.

White L.A., D.J. Hadley, D.E. Davids, and R. Naylor (1975). Improved large volume sampler for the collection of bacterial cells from aerosol. *Appl. Environ. Microbiol.* **29**: 335-339.

# 12

# Interaction of Metals and Metalloids with Microorganisms in the Environment

## 12.1 CHEMICAL AND PHYSICAL PROPERTIES OF ENVIRONMENTAL SIGNIFICANCE

Elements in nature occur as metals, metalloids, or nonmetals (Fig. 12.1). Metals are generally lustrous and can be drawn into thin wires (i.e., ductile) or hammered into various shapes (i.e., malleable). They give up electrons, becoming positively charged. They also often have the ability to conduct electricity and/or heat. Metalloids on the other hand, display some characteristics of metals and nonmetals. As we move across the periodic table (Fig. 12.1) from left to right and from the top of the column downward, the elements that behave like metals more and more display metallike characteristics. It is also noticeable from the periodic table that most elements are metals, with a small fraction behaving as non-metals and an even smaller fraction as metalloids. Within the same group (i.e., columns in Fig. 12.1), the elements have similar properties although the properties vary somewhat in a regular predictable pattern. Thus, the grouping described by chemists enables us to make comparisons and/or to predict how the respective elements are expected to affect biological molecules in the environment. In chemistry books, physical and chemical properties of metals and metalloids are often considered separately. It is important to note, however that in biological systems these properties are interrelated. In this chapter we explore some of these interrelationships and how they impact microbes in the environment.

Metals and metalloids are classified as essential or nonessential (Fig. 12.1). Essential metals and metalloids are generally involved in the survival and growth of microorganisms in the environment by stabilizing biological structures such as proteins and membranes as well as acting as catalysts in a wide array of biological processes. Thus, when in short supply, they limit the growth of organisms. It can be seen from the periodic table that a number of essential metals belong to the first three to four rows in Groups IA, IIA, as well as the first row in the transition metals. Members of both Groups IA, IIA react by ionization, forming ionic bonds either with negatively charged cross-linking ligands in cell walls or with intercellular anions. The transition metals on the other hand, bind ligands more strongly than Groups IA, IIA as they have partly filled $d$ orbital shells. It can also be seen that most of the transition metals are not biologically essential (Fig. 12.1). Only two metalloids, i.e., boron and silicon, are cardinal to biological systems. Among the essential metals and metalloids, only Si, Co, and Ni can bond with carbon—the central element in

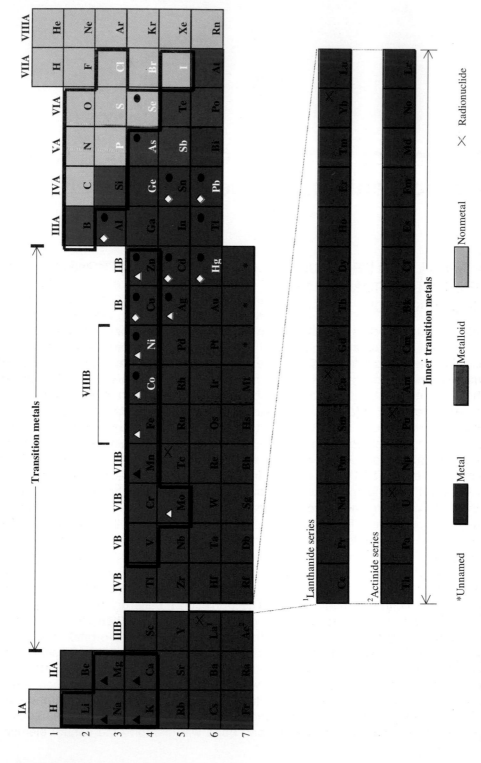

**Fig. 12.1**   Periodic table of elements (i.e. metals, metalloids, and nonmetals). In the table, toxic heavy metals (■), essential elements (■), radionuclides (✕), essential elements (✕), radionuclides (●), essential elements (■), as well as hard acids (▲), intermediate acids (▲), and soft acids (◇) are distinguished. Elements written in white form biogenic bonds with carbon in the environment.

all biological systems. However, other nonessential elements such as Ge, As, Sb, Pb, and Hg can also bond with carbon, generating molecules potentially toxic to forms of life in the environment.

A number of the essential transition metals are also referred to as heavy metals. However, it should be noted that although this term arbitrarily refers to all elements in the periodic table with metallic properties having an atomic number greater than 20, when used in the literature, it almost exclusively refers to the few toxic metals and metalloids highlighted in Figure 12.1, which are toxic to biological cells. Thus, to maintain consistency with the rest of the literature, we restrict its use in this text to those few toxic metals and metalloids highlighted. They exert toxicity by influencing biochemical activities, cell morphology, and growth. However, the individual effects of heavy metals vary greatly from microbe to microbe depending on the respective metal and its concentration. Solubility of compounds in water and/or lipids is used as an indicator of degradability in the environment (Wood, 1974). The inner transition metals are toxic but mostly rare and very insoluble in water and lipids. Hence their interaction with microorganisms is not discussed here. The interaction of uranium, an inner transition metal, has been appropriately discussed under the adaptation of microorganisms to extreme environments (Chapter 7).

Metal cations are also categorized as hard, intermediate, or soft acids. Hard acids are small, do not contain unshared pairs of electrons in their outermost (valence) shell, and usually possess a high charge. They are thus highly electronegative and not readily polarized. Most metals of fundamental cellular constituents are hard acids (Figure 12.1) but because they have fewer electrons and are not easily polarized, they cannot effectively compete with soft metals for binding sites. They are thus, readily displaced from their sites by soft cations (Hughes and Poole, 1989). Soft acids on the other hand have atoms with a low positive charge and contain unshared pairs of electrons in their valence shell. Compared to hard acids, they have a low electronegativity and are highly polarizable. Their electron shells are easily deformed and hence the name "soft" acids. They are associated with environmental pollution, forming more stable complexes with cellular components and ligands such as carboxylate groups in the cell wall, compared with hard acids. Although their toxicity to microorganisms is the primary focus of these metals in this chapter, it is worth noting that they also exert toxicity to higher organisms, including humans. For example, cadmium, lead, mercury, and tin (tributyltin and organotin compounds) have been implicated in disrupting the endocrine system (Colburn et al., 1996). Arsenic, beryllium, cadmium, and nickel have also been linked to an increased incidence of cancer in cases of occupational exposure and have been demonstrated as carcinogens in laboratory animals. When excessive in the environment, soft acids (metals) will outcompete other cations in the cytoplasm for sites on intracellular molecules (i.e., proteins, polysaccharides, and nucleic acids).

The intermediate acids, notably copper, iron, and molybdenum, are important components of enzymes, possibly because of their ability to adapt various oxidation states. From the foregoing discussion, it is apparent that addressing the problems associated with environmental and industrial pollution from toxic metals and metalloids requires us to focus on the behavior of microorganisms interacting with soft and intermediate acids in the environment. Although metals and metalloids cannot be biodegraded, unlike organic compounds, microorganisms influence the behavior of these ions in the environment by transforming the metals from one chemical state into another. Thus, it is imperative to understand the effects of metals on microorganisms and the effect of microorganisms on metals in the environment.

## 12.2 SOURCES OF METAL POLLUTION

Soft acids are generally not abundant in the biosphere. However, due to industrial activities and deliberate, as well as accidental discharge, these heavy metals are increasingly becoming concentrated in specific localities and cause pollution problems. Thus, through human activities such as mining, smelting, manufacturing, industrial discharge, breakdowns in galvanized appliances and fixtures, and from the whole range of personal care products that we use on a regular basis, we have created environments with a high concentration of heavy metals. Of major concern are lead, cadmium, zinc, copper, arsenic, aluminum, nickel, mercury, and manganese. These metals exert toxicity through five general mechanisms (Gadd, 1986):

(1) Displacing the essential metal ions and thus denying these ions access to those sites. For example, $Cd^{2+}$ ion has an ionic radius similar to $Ca^{2+}$ (0.95 Å) and because of its position in the periodic table (Group IIB), also resembles $Zn^{2+}$. It will therefore bind to predominantly sulfur-containing sites, such as cysteine, which are normally occupied by $Zn^{2+}$, and to the oxygen-containing ligand sites preferred by $Ca^{2+}$ (e.g. in aspartate, glutamate).

(2) Binding more strongly to functional sites normally occupied by essential metals, thus blocking the essential functional groups of biological molecules such as enzymes and disrupting the essential metals.

(3) Changing the conformation of biological molecules (i.e. proteins and nucleic acids), thus disrupting the integrity of entire cells and/or their membrane, rendering them inactive.

(4) Decomposing essential metabolites.

(5) Changing the osmotic balance around the cells.

Thus, pollution by metals affects every aspect of microbial metabolism and activity, including respiration, membrane transport, and the synthesis and activity of ribosomes, in some cases resulting in death of the cells. As indicated earlier, some of the heavy metals are cardinal but only in relatively minute quantities, functioning as metalloenzymes (Table 12.1). For example, copper at 0.6 - 2.7 µg $L^{-1}$ and zinc at 1.5 -10 µg $L^{-1}$ are essential for metabolism but can be toxic to living cells at higher concentrations (Giller et al., 1998).

Because of a large surface-to-volume ratio, microorganisms provide a large contact area which can interact with metals in the surrounding environment. They accumulate metals through various mechanisms including complexation, adsorption, precipitation, and active transport into the cell. The uptake of metals by some microorganisms is biphasic. In such instances, the first phase is metabolism independent (biosorption), the cell being the major site of uptake. This phase is rapid and may account for as much as 90% of the total amount of heavy metals taken up by microorganisms in some systems (Fehrmann and Pohl, 1993). Biosorption is unaffected by temperature, at least within the range of 4 - 30°C, resulting from an electrostatic interaction of the metal ions with the exposed reactive groups on the surface of the microorganisms. However, it is reduced by low pH as well as the presence of other cations which compete for the same sites, and anions (such as $PO_4^{2-}$, $SO_4^{2-}$, $OH^-$) which effect precipitation. In soils and sediments, the availability of metals to microorganisms in general, and heavy metals in particular can be influenced by several factors such as pH, soil type, and the levels of organic matter. Thus, the availability and subsequent toxicity of heavy metals is reduced under acidic conditions. However, if the concentration of metals is extremely low (i.e., less than $10^{-8}$ to $10^{-4}$ M), its availability can be less dependent on the pH (Krantz-Rülcker et al., 1996). Clay minerals,

**Table 12.1**   Some common metalloenzymes and the enzyme functions with which they are associated.

| Heavy metal | Enzyme | Enzyme function |
|---|---|---|
| Cobalt (Co) | Methyl malonyl-CoA transferase | Transfer of methyl groups |
| Copper (Cu) | Cytochrome oxidase (cytochrome $a_3$) | Reduction of oxygen during respiration |
| Iron (Fe) | Cytochromes a, b, and c | Electron transport system (ETS) during respiration |
|  | Nitrate reductase | $NO_3^- + 2H^+ + 2e \rightarrow NO_2^- + H_2O$ |
| Manganese (Mn) | Superoxide dismutase | $2O_2^- + 2H^+ \rightarrow H_2O_2 + O_2$ |
| Molybdenum (Mo) | Nitrogenase | Nitrogen fixation |
| Nickel (Ni) | Carbon monoxide dehydrogenase | $CO + H_2O \rightarrow CO_2 + H_2$ |
| Zinc (Zn) | Carbonic anhydrase | $H_2CO_3 \rightarrow H_2O + CO_2$ |

*Source*: Ehrlich (1986) with permission from the author.

particularly vermiculite and montmorrillonite organic components such as humic substances, amino acids, proteins, polysaccharides and chelating agents, as well as cations and anions can also reduce toxicity. Both vermiculite and montmorrillonite are chelating agents and act by forming complexes. Thus toxicity to metals depends to some extent on the prevailing environmental conditions. The second phase is metabolism dependent, consisting of actively transporting the metals across the membrane into the cell.

## 12.3   EFFECTS OF METAL POLLUTION ON MICROBES

Microorganisms, through metabolic activities, play an important role in the mobility of metals in the environment. They exert their influence through the enzymes and other compounds they release, enhancing the solubility of the metals. For example, extracellular polymers of microbial origin can enhance the mobility of metals by desorbing the metals off contaminated soil colloids (Chen et al., 1995). However, extracellular polysaccharides are quite diverse in nature, to the extent that they differ even between microbial species. The composition of extracellular polysaccharides can also vary greatly for the same species under different environmental conditions. Under similar environments *Pseudomonas cepacia, P. putida*, and another unidentified soil isolate 9702-M4 differed tremendously in the ability to desorb cadmium and lead (Table 12.2). In Chen's study, the presence of extracellular polymers from isolate 9702-M4 reduced the distribution coefficient of cadmium and lead by 60% and 87% respectively. On the other hand, polymers from *P. putida* reduced the distribution coefficient of these two metals by 12% and 93% respectively (Table 12.2). Reduction in the distribution coefficient of the metal in the presence of extracellular polymers was attributed to an increase in release of the adsorbed metal by complexation and subsequent dissolution with

**Table 12.2**   Effect of bacterial extracellular polymers on the distribution coefficients of cadmium and lead

| Organism | Gram reaction | Catalase | Percent reduction in distribution coefficient[a] | |
|---|---|---|---|---|
|  |  |  | Lead | Cadmium |
| *Pseudomonas cepacia* 249-100 | — | % | 85 | 94.4 |
| *Pseudomonas putida* | — | % | 12 | 92.8 |
| Soil isolate 9702-M4 | — | — | 60 | 87 |

[a]The distribution coefficient was determined by fitting the data to a linear isotherm $\Gamma - K_d C_s$ in which $\Gamma$ is the amount of metal (i.e., Cd or Pb) adsorbed per g of the sand, $K_d$ is the metal distribution coefficient and $C_s$ the equilibrium concentration of the metal in the solution phase. All results are based on a polymer concentration of 10 mg/l (Adapted from Chen et al., 1995; reprinted with permission from Elsevier).

the polymer. Enhanced mobility of the metals was further demonstrated in leaching sand columns that received extracellular polymers compared to columns where no polymers were present (Chen et al., 1995). However, information about what quantities of extracellular polymers are produced by microorganisms in their natural environments is limited. If they are of any consequence in enhancing mobility of metals in soils and sediments, the naturally generated polymers must not be highly sorbable to soil colloids. The use of microbes in controlling pollution from metals and in economically extracting metals ores (biomining) are discussed in Sections 12.5 and 12.7 respectively.

Metabolic processes as reflected by basal respiration can also be negatively affected by heavy metals. The significantly reduced metabolic activity subsequently leads to a decreased microbial biomass (Table 12.3) because when under stress from metals, microorganisms may divert energy from growth to maintenance functions. In most cases, pollution occurs from a multiplicity of metals. Despite this reality, most studies have only investigated the effects of individual metals on microorganisms. Most of the studies on the effects of metals to microorganisms have also used culturing techniques with constituents such as peptone, yeast extract, silica gel, or gelatin as part of the test medium. However, these constituents readily complex with most metals, a shortcoming that impacts most of the results obtained, especially in situations wherein micromolar concentrations of the metal being tested are the norm (Chaudri et al., 1993; Giller et al., 1998). Furthermore, the toxic forms of most metals are free ions, a fact that makes it difficult to adequately study heavy toxicity using conventional culturing techniques because, as indicated above, metals tend to form complexes and precipitates in the conventional growth media, including agar. This reality can explain, at least in part, why the effects of metals on microorganisms reported by various research groups are sometimes contradictory.

Similar to the situation in a growth medium, speciation of metals occurs in the environment, mostly due to the prevailing pH. Thus the pH of the environment plays an important role in determining the toxicity of metals in the environment. Generally, acidic pH ranges favor the existence of metals in their ionic form. As the pH increases, the functional sites on the biological surface become deprotonated, binding more cations. However, under alkaline pH conditions precipitation occurs for most metals, which generally reduces their availability and subsequent toxicity to microorganisms in the environment (Babich and Stotzky, 1983). In soils and sediments, metal pollution is often associated with other adverse conditions besides pH, such as low organic matter content, low nutrient availability, and especially in industrial pollution, a high concentration of recalcitrant organic compounds, conditions per se harsh for microorganisms. However, the uptake and subsequent toxicity of the metal in question can, in some instances, also greatly depend on the physiological status of the cells in the environment since dormant cells are less likely to take up metal pollutants (Insam et al., 1996; Krantz-Rülcker et al., 1996). For some metals, adsorption and subsequent uptake can be passive, at least in part, occurring to the same extent in both starved and nonstarved cells.

How microorganisms in particular, and biological systems in general, deal with metal pollution is of primary interest in environmental microbiology. Microbes can be susceptible, tolerant, or resistant to heavy metals. According to Gadd (1992), tolerance and resistance are distinct, the former referring to the ability of the organism in question to cope with the toxicity based on its intrinsic properties while the latter refers to the ability of the organism to survive metal stress by using some detoxification mechanism induced in direct response to the metal. Selective pressures from metal-containing environments have led to the development of resistance systems to virtually all toxic metals. Resistance to metals in prokaryotes is

**Table 12.3** Potential effect of heavy metal contamination on basal respiration and microbial biomass in *Deschampsia cespitosa* (L. Beauv.) rhizosphere soils from six sites in Canada[1].

| Site | Description | Soil pH (KCl) | % Organic matter | Metal concentration ($\mu$g g$^{-1}$ soil) | | Basal respiration ($\mu$g CO$_2$ soil h$^{-1}$) | Microbial biomass ($\mu$g Cmic g$^{-1}$ soil) |
| --- | --- | --- | --- | --- | --- | --- | --- |
| | | | | Total Cu | Total Ni | | |
| Cypress Lake | CaCO$_3$-clay, roots partially submerged in water | 7.8 | 22.43 a | <20 | <30 | 16.19 a | 3120 a |
| Little Current | Calcareous bedrock | 8.2 | 5.71 b | <20 | <30 | 1.62 b | 347 b |
| Chutes | Sandy soil, outwash from a waterfall, patchy vegetation | 6.5 | 0.54 d | <20 | <30 | 1.30 b | 156 c |
| Falconbridge | Sandy soil, rich in Cu, Ni; smelter active; contaminated | 5.3 | 1.28 c | 417 - 984 | 374 - 1341 | 0.53 c | 104 d |
| Roastbed | Tailings rich in Cu, Ni (i.e., contaminated) | 6.9 | 0.54 d | 417 - 984 | 374 - 1341 | 0.75 c | 113 d |
| Coniston | Glacial till, sand and clay; contaminated | 5.5 | 0.40 d | 70 - 454 | 78 - 412 | 0.56 c | 125 cd |

[1]Numbers accompanied by the same letter within the column are not significantly different; $p < 0.05$ n = 20).
*Source:* Insam et al. (1996) with permission from Elsevier.

mostly mediated by plasmids and is highly specific for a particular metal, the plasmid DNA inserting itself autonomously even between prokaryotes of unrelated genera (Summers, 1986). This allows the resistance plasmids to become widely disseminated through the prokaryote population of a given environ. Metal-resistance systems have been found on plasmids of every bacterial group tested (Summers, 1986; Silver and Phung, 1996). However, some resistance, particularly for essential metals, is chromosome mediated. Plasmid-mediated resistance to metals can be moved rapidly between cells within a population. In this manner, microorganisms reduce the burden of carrying the genes for resistance to nonessential metals since they are only needed on certain occasions (Bruins et al., 2000). Some of the known operons for various heavy metals are listed in Table 12.4. Thus resistance to metals is found in a wide range of bacteria and fungi and these mechanisms may have developed shortly after prokaryotes evolved (Bruins et al., 2000). It should be noted that there is no general mechanism for resistance to all heavy metals. In bacteria, resistance to metals is usually closely associated with resistance to antibiotics (Timoney et al., 1978; Colwell et al., 1986), both types of resistance being mediated by the same plasmid and existing on the same locus. This observation has implications in using microorganisms to prospect for metals (see Section 12.7).

Generally, Gram-negative bacteria display greater tolerance to metal toxicity than Gram-positive bacteria because the former have a more complex three-layered cell wall (Fig. 2.2) that more effectively immobilizes metals. Other traits of enhanced metal uptake and/or immobilization include the presence of an extracellular gelatinous matrix and especially in fungi, the production of extracellular melanin. For example, *Zoogloea ramigera* strains with a gelatinous matrix were demonstrated to bind more metals during the treatment of waste water than strains lacking such a matrix (Norberg and Persson, 1984).

There are six known mechanisms of resistance and tolerance to metals by microorganisms (Table 12.5). Through these mechanisms microorganisms generally strive to modify the metals to less toxic forms. One interesting mechanism involves the microorganisms in question taking up the metal in large quantities, sequestering it, and thus rendering it insoluble. For example, in studies by Bianchi et al. (1981), $Mn^{2+}$-resistant *Saccharomyces cerevisiae* mutants were found to take up more $Mn^{2+}$ ions than the sensitive wild-type, sequestering the ions in the vacuole. Manganese competes with magnesium, an essential nutrient, inhibiting the uptake of the latter. However, sequestration of metals can occur within or outside the cell (i.e., intracellular or extracellular respectively). When intracellular, metallothionein or a cysteine-rich protein is produced, production of the former currently only known to occur in cyanobacteria, e.g., *Synechococcus* spp. (Cook et al., 1998). Intracellular sequestration is more widespread among microorganisms.

Some microorganisms can exclude (efflux) heavy metals as a mechanism of tolerance or resistance. Exclusion may be active (ATPase-linked) or passive (non-ATPase-linked). For example, *Penicillium ochro-chloron* can grow at very high concentrations of $Cu^{2+}$ without accumulating much of this ion whereas *Thiobacillus ferroxidans* can accumulate uranium on its cell wall with very little of this element finding its way inside the cell (Gadd et al., 1984). Many Gram-negative bacteria release

**Table 12.4** Some of the known operons for resistance to heavy metals in microorganisms

| Heavy metal | Operon | Reference |
|---|---|---|
| Arsenic | *ars* operon | Sato and Koboyashi (1998) |
| Copper | *cop* operon | Bruins et al. (2000) |
| Mercury | *mer* operon | Narita and Endo (2001); Smith et al. (1998) |
| Zinc | *smt* operon | Cook et al. (1998) |
| Cadmium | *cad* operon | Silver and Phung (1996) |

**Table 12.5** Currently known mechanisms of tolerance and resistance to metals by microorganisms

| Mechanism | Description and examples | Reference |
|---|---|---|
| Compartmentalization | Accumulation of the metal inside the cells but leaving the essential functions protected from the toxic metal | Bruins et al., 2000 |
| Efflux | Enhanced ability of the organism to export the metal from the cell. For example, arsenate, which is readily taken up as a phosphate analog, can be effluxed by *Bacillus subtilis* during a period of phosphate abundance. | Sato and Koboyashi, 1998 |
| Biosorption | The metal becoming bound by proteins, the cell wall or by extracellular polymers, e.g. *Klebsiella aerogenes* strains with protective polypeptide layer accumulates more Cd(II) but survives better under excessive Cd than do strains without the polypeptide layer. | Scott and Palmer, 1990 |
| Volatilization | The organism playing a role in transforming the metal into a volatile product e.g. methylmercury demethylated by Gram-positive *Bacillus* sp.; *Staphylococcus aureus*) and Gram-negative (*E. coli, Pseudomonas aeruginosa, Thiobacillus ferroxidans, Serretia mercens*) bacteria to generate mercury ($Hg^0$) which is volatilized. | Bruins et al., 2000 |
| Exclusion | A decreased uptake of the metal in question. For example, Cu(II) is excluded by *E. coli* due to alternations in the membrane chemical protein porins. | Rouch et al., 1995 |
| Sequestration | Formation of insoluble products e.g. *Saccharomyces cerevisiae* excretes large amounts of glutathione which complexes with heavy metals that precipitate inside the cell. | Bianchi et al., 1981; Cook et al., 1998 |

polysaccharides that coat the cell surface, giving it a net negative charge that attracts (adsorbs) cations. Biosorption also occurs in bacteria that naturally form an extracellular polysaccharide coating, which provides sites for attachment of metal cations.

The effects of metals on viruses have not been well researched. However, preliminary information indicates that some viruses, such as poliovirus, can be inactivated by metal ions, in particular divalent and trivalent cations (Abad et al., 1994). However, others such as hepatitis A, rotavirus, adenovirus, and viccinia are resistant to metals, possibly because they have an inherently more stable molecular structure.

## 12.4 METAL CORROSION

Metals can undergo continuous oxidation and reduction through both chemical and microbiological processes, leading to what is referred to as corrosion. Corrosion involves the dissolution of a metal from anodic sites with the subsequent acceptance of electrons at cathodic sites. It occurs both under oxic and anoxic environments. During corrosion consumption of electrons varies, depending on the redox potential of the surface. Under oxic environments, oxygen serves as the electron acceptor, forming a variety of oxides and hydroxides. At a low redox potential, protons become the electron acceptors yielding $H_2$ and other reduced products. Corrosion is accelerated by removal of the end products. Thus, in the presence of bacterial biofilms on the metal surfaces, uptake of oxygen is enhanced, creating localized zones of differential aeration. This in turn produces cathodic areas where electrons are continuously accepted, leading to reduction of the metal, and anodic areas where the oxidized metal dissolves, resulting in a corrosion current and dissolution of the metal in question.

For economic reasons, corrosion of iron and steel has been more widely studied than in other metals. However, it is important to recog-

nize that a variety of other metals such as zinc, lead, copper, and aluminum can also become corroded. For example, the fungus *Cladosporium resinae* has been widely shown to corrode aluminum alloys in jet fuel tanks (Miller, 1981). Its success in colonizing the tanks is linked to its ability to grow on JP-4 fuel and other straight-chain organic compounds as a sole C source. Its spores are fairly abundant in the atmosphere and are thought to insinuate in the tanks at high altitudes during the intake of air as the jet fuel level drops. Some spores are also believed to enter the tank during routine refueling. Thus no metal is totally immune from colonization by microorganisms which facilitate the biological processes, involving a variety of bacteria, fungi, and algae (Table 12.6). Corrosion of metals is therefore a widespread problem attributable to chemical and biological processes.

Because corrosion can also occur at substantial rates in the absence of bacteria, the role of bacteria in metal corrosion was overlooked for some time. Understanding the electrochemistry of microbiologically influenced corrosion (MIC) was also hampered earlier by separation among disciplines of solid physics, microbiology, electrochemistry, and material engineering. However, with forged alliances, it is beginning to be understood. The strict differentiation between chemical and microbial corrosion is often difficult especially under both aerobic and anaerobic conditions because the two processes enhance each other. Some of the novel approaches to sort out

**Table 12.6** Types of microorganisms associated with corrosion of metals

| Microorganism | Examples |
|---|---|
| Bacteria | *Aerobacter, Bacillus, Flavobacterium, Pseudomonas, Thiobacillus* |
| Algae | *Chlorococcus, Oscillatoria, Crenotilus, Navicula, Scenedesmus, Ulothrix* |
| Fungi | *Arnorphotheca, Cladosporium, Hormodendrum* |

the role of microorganisms have involved monitoring the cell density in biofilms, measuring the accumulating ATP, and the content of lipopolysaccharides on the surface of corroding metals in relation to dissolved oxygen and electric potential over time (Lee and Charaklis, 1993; Chen et al., 1997). Thus, microscopic examination also reveals that under MIC, the extent of microbial colonization of deteriorating surfaces corresponds to distinct anodic and cathodic areas associated with gradients of oxygen and pH (Fig. 12.2). The presence of bacteria in pitted areas of metals has also been demonstrated using scanning electron microscopy (Lee and Charaklis, 1993). Growth of the microbial colony actually sets up differential reactions, the continued growth keeping the concentration of oxygen diminished toward the center of the colony and thus establishing an electrochemical gradient. Even when some aerobic microorganisms die off at the center of the colony due to oxygen stress with continued anaerobiosis, the resultant debris from the lysed morbid cells maintains the electrochemical gradient. This enables corrosion to continue even in the absence of microbial growth on the surface. It is important to note from the onset that unlike the case in pure culture laboratory situations, under natural environmental conditions the colony may not

be solely due to a single culture but rather to a variety of microbial species. If the anaerobic zone is formed at a microsite where an anaerobic microorganism was already present but could not thrive initially due to unfavorable anoxic conditions, its growth will be enhanced under the newly established conditions. Corrosion by anaerobes, in particular the sulfur-reducing bacteria (SRBs) will now be discussed.

It is important to note that although MIC destroys metals, it also provides nutrition to the H-consuming microorganisms. It can occur in almost any industrial situation in which microbial growth takes place, including anaerobic conditions (e.g. anaerobic iron pipes). In aqueous environments, the effects of corrosion by fungi are often outcompeted by bacteria. Under anaerobic environments, anaerobes, in particular SRBs such as *Desulfovibrio* sp. play a major role in facilitating corrosion. Corrosion under anaerobic conditions also occurs mainly by electrochemical anodic-cathodic reactions which involve the removal of hydrogen from polarized cathodic areas, oxidizing it to $H^+$ and generating some electrons in turn (eqn 1). Then hydrogen reduces sulfates to sulfide corrosion products (eqn 3).

Anodic reaction $4Fe \rightarrow 4Fe^{2+} + 8e$ ...(1)

$$8H_2O \rightarrow 8H^+ + 8OH^- \quad ...(2)$$

Cathodic reaction $8H^+ + 8e \rightarrow 4H_2$ ...(3)

Microbial action $4H_2 + SO_4^{2-} \rightarrow S^{2-} + 4H_2O$
...(4)

Corrosion product $S^{2-} + Fe^{2+} \rightarrow FeS$ ...(5)

The SRBs are known to cause cathodic hydrogen depolarization, the resultant $H_2S$ damaging such passive metals as stainless steel by accelerating anodic interactions (Chen et al., 1997). It is obvious that the above reactions require no oxygen. The process accounts for heavy economic losses in sewage pipes, oil pipes, etc. whereby iron is reduced by protons as follows

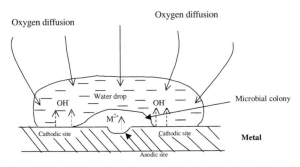

**Fig. 12.2** Schematic diagram showing the formation of cathodic and anodic sites on a metal surface during microbiologically influenced corrosion (MIC). Oxygen diffuses into the liquid but conditions toward the surface of the metal remain more anoxic favoring the activity of anaerobes, particular sulfur-reducing bacteria.

$$Fe^0 + 2H^+ \rightarrow Fe^{2+} + H_2 \; \Delta G^0 \simeq 0 \quad ...(6)$$

This reaction occurs on the metallic pipe surface and the $H_2$ produced tends to stay on the surface and is continuously sucked off by the SRBs for their respiratory requirement, a process that continuously dissolves the metallic pipe (eqns. 3 and 5). Without the hydrogen uptake by the SRBs, the corrosion process is terminated.

The rate of corrosion has been found to correlate with hydrogenase activity (Bryant et al., 1991). SRBs are quite widespread in various environments ranging from sediments, permafrost, deep seas, waste water, etc., throughout the world. High population densities are usually detected at anaerobic corrosion sites but it is important to note that not all anaerobic corrosion is attributable to SRBs as hydrogenase negative *Desulfovibrio* sp. (as opposed to their positive counterpart) can also cause some limited corrosion (Miller, 1981).

Metals are often covered with microorganisms in the form of a biofilm that also comprises the deposits and exopolymers generated by such microorganisms. The deposition and exopolymers bind the microorganisms to the metal and are also directly involved in the dissolution of the metal and its corrosion (Chen et al., 1997). The microbial exopolymers are quite diverse and include hexoses, teichoic acids, amino acids, *O*-methyl and *O*-acetyl residues, with functional groups that readily bind metal ions. Growth conditions also directly affect the quantity of exopolymers produced. Corrosion can also be enhanced by enzymatic products released during the growth of certain microorganisms. In some instances, corrosion is due to the metabolic products excreted during normal growth and/or under nutrient stress. Such metabolites include citric, isocitric, $\alpha$-aconistic, and *cis*-glutonic acids. The extent to which the microbes adhere to the metal depends on environmental conditions, in particular the pH and redox potential.

## 12.5 BIOTRANSFORMATION AND BIOREMOVAL OF HEAVY METALS FROM POLLUTED ENVIRONMENTS

Microbes can reduce or oxidize metals, thus removing them from specific environments such as waste water, activated sludge, sediments, soils, and biosolids. The transformation processes initially involve either absorption (uptake) or adsorption of the metal. Because of their small size, microorganisms exhibit a large surface-area-to-volume ratio, a surface that may interact with dissolved constituents in the environment. Thus, many microorganisms have the ability to selectively accumulate metals to the extent that microbial biomass is estimated to accumulate more than 30% of the metals in the terrestrial environment despite the fact that it comprises less than 1% of the terrestrial solid component (Ledin et al., 1999; Krantz-Rülcker et al., 1996). Absorption is particularly more pronounced at neutral pH (for bacteria) to low pH (for fungi). However, in terrestrial environments, the amounts accumulated by microorganisms also depend on the metal in question, as well as the amount and portions of other sorbents such as organic matter and clay content. In marine environments, algae are better suited to remove metals than other microorganisms because of their extensive biomass and their higher propensity to tolerate other environmental extremes such as acidity, which are often associated with heavy metal content in these environments.

On an industrial scale, biotransformation and bioremoval of metals in polluted mining and milling discharge waste water have been achieved through meanders, impoundments, and wetlands (Brierley et al., 1989; Ehrlich, 1995). Practically, the removal and transformation of heavy metals in these systems are attained through interactions among both microflora (algae, bacteria, fungi, and protozoa) and macroflora. The meanders are designed to favor a substantial growth of single and multicellular algae such as *Chlorella, Cladophora, Oscillatoria, Spirogyra, Hydrodictyon*, and *Rhizoclonium* sp. When the algae die and sink, most of the metals they accumulate are incorporated into anoxic sediments where they are transformed by sulfur-reducing bacteria, usually forming FeS, ZnS, CuS, CdS, etc., and end up sequestered. Some systems have been reported to remove Pb, Zn, Fe, Cu, Ni, and Cd from polluted waterways at efficiencies as high as 99% (Brierley et al., 1989). Recent studies by Terry and Stone (2002) with live and dead alga *Scenedesmus abundans* showed that bioremoval occurs rapidly (within the first 24 h), removal being pronounced when live versus dead alga was used.

The use of live vis-à-vis dead microbial cells can be advantageous in situations wherein the metals are not at detrimental concentrations because it provides for continuous growth and metabolic uptake. However, with heavy pollution it is difficult to maintain viable microflora and also ensure sufficiently adequate contact between the biomass and the pollutants. Furthermore, bioremoval is greatly dependent on seasons which vary in temperatures and microbial activity. In the presence of excessive heavy metals, therefore, it could be more cost effective to use nonliving microbial biomass to avoid interruption in the uptake of metals due to toxicity. The functional groups on nonliving microbial cell walls can also adsorb metals (Fehrmann and Pohl, 1993; Terry and Stone, 2002). As a matter of fact, dead biomass can, in some instances, provide a higher capacity for uptake of metals than living biomass or other commercially available adsorbents. Table 12.7 shows the concentration factors of cadmium by a variety of nonliving cyanobacteria and algae compared to activated carbon, silica gel and siliceous earth. All of the dead species, except *Porphyridium purpureum*, had more adsorption capabilities than activated carbon, silica gel and siliceous earth. A comparison of the surface quality of the species tested, activated carbon, silica gel, and siliceous earth (reflected by adsorption per surface unit) showed that both *Porphyridium purpureum* and Chlorophyceae have a low surface quality. Phaeophyceae have the highest surface quality, possibly due to the munuronic and polyguluronic acids in the cell wall (Fehrmann and Pohl, 1993).

Metals and metalloids may also be transformed by microorganisms to some organic forms (i.e., methylation). The methylated products can then be further converted and volatilized. In the environment, sediments are a major location in which most pollutants are biomethylated. The methylated products can be of increased or reduced toxicity compared to the respective parent cation in the environment, its toxicity being reflected by its lipophilicity (Hughes and Poole, 1989; Wood, 1974). Absorption processes may be less effective in situations wherein metal ion concentrations increase with time, such as in waste-damping sites. The transformation and removal of metals through biotic mechanisms is discussed below for mercury, arsenic, and selenium, three heavy metals of significance in the environment.

Several microorganisms have the capability of tolerating and absorbing high doses of radioactive metals (Fredrickson et al., 2000). The tolerance of microorganisms to radioactivity was discussed in Chapter 7 but it is important to point out that the concepts of bioremoval and biotransformation of metals are also being used to clean up radioactive sites that were either left behind after the arms race of the twentieth century or have naturally occurring radioactive metals. The ability of protozoa and other phagocytosic feeders to ingest metal precipitates directly has not been well studied. Nilsson (1979) fed *Tetrahymena* spp. with lead acetate in axenic culture and observed the metal to precipitate and bind onto the membrane spaces and vacuoles. However, it is not clear as to what extent phagocytosis of heavy metals in the environment by protozoa and other phagocytosic microorganisms contribute to immobilizing metals. An appreciation of the role

**Table 12.7** Concentration factors of algal biomass after addition of 10 ml cadmium solution and the related cadmium adsorption surface

| Taxonomic group | Species | Concentration factor at different concentrations of cadmium ($\mu$g ml$^{-1}$)[a] | | | | Cd adsorption per surface unit ($\mu$g Cd m$^{-2}$ cm$^3$)[b] |
|---|---|---|---|---|---|---|
| | | 0.5 | 1.0 | 2.0 | 3.5 | |
| Cyanobacteria | *Anabaena inaequalis* | 167 | 167 | 168 | 168 | 36.2 |
| | Anabaena lutea | 172 | 172 | 172 | 172 | 33.2 |
| | *Nodularia harveyana* | 144 | 148 | 150 | 151 | 12.5 |
| | *Nostoc commune* | 168 | 170 | 171 | 171 | 14.1 |
| Chlorophyceae | *Chlamydomonas sp.* | 188 | 188 | 188 | 187 | 14.5 |
| | *Klebsormidium sp.* | 131 | 139 | 143 | 145 | 17.1 |
| | *Ulothrix gigas* | 117 | 120 | 122 | 123 | 17.2 |
| Eustigmatophyceae | *Eustigmatos vischeri* | 177 | 182 | 184 | 185 | 25.5 |
| Phaeophyceae | *Ectocarpus siliculosus* | 190 | 192 | 193 | 194 | 47.7 |
| | *Halopteris scoparia* | 176 | 176 | 175 | 175 | 51.8 |
| | *Spermatochnus paradoxus* | 186 | 187 | 188 | 188 | 42.0 |
| | *Sphacelaria rigida* | 181 | 181 | 182 | 182 | 42.5 |
| Rhosphyceae | *Porphyridium purpureum* | 98 | 86 | 80 | 78 | 13.8 |
| Tribophyceae | *Bumilieriopsis filiformis* | 173 | 170 | 168 | 168 | 24.2 |
| | *Ophiocytium maius* | 180 | 180 | 180 | 180 | 27.5 |
| Other materials | Activated carbon | 161 | 137 | 125 | 120 | 20.6 |
| | Silica gel | 186 | 161 | 149 | 144 | 33.3 |
| | Siliceous earth | 200 | 154 | 131 | 121 | 18.7 |

[a]Concentration factors with different concentrations of cadmium were calculated from the amount of cadmium adsorbed on the surface of the biomass divided by the amount of cadmium remaining in the ambient solution, i.e. $\mu$g Cd g$^{-1}$ biomass/$\mu$g Cd ml$^{-1}$ solution). Due to the experimental conditions (10 ml Cd solution, 50 mg biomass), the maximum possible concentration factor was 200.
[b]The specific surface area was determined by means of laser diffractometry. The surface area was then used to calculate the cadmium adsorption per surface unit ($\mu$g Cd m$^{-2}$ cm$^3$) based on data from the 2.0 $\mu$g ml$^{-1}$ cadmium concentration.
*Source:* Fehrmann and Pohl (1993) with permission from Kluwer Academic Publishers.

of protozoa in removing metals from contaminated environments is essential considering the abundance and role of protozoa in sewage where heavy metals are usually abundant.

## 12.6 BIOLOGICAL CYCLING OF SPECIFIC HEAVY METAL POLLUTANTS

### 12.6.1 Mercury

Mercury can exist in nature in three states: metallic ($Hg^0$), mercuric ($Hg^{2+}$) and mercurous ($Hg^+$). Metallic mercury is relatively nontoxic because it is highly insoluble. In the absence of microbes, mercury can last in the environment indefinitely. However, it rarely occurs as a pure metal in nature as it is easily converted by

peroxidase and catalase enzymes to highly toxic organic alkyl and aryl forms. More often encountered is the mercuric form often found as mercury sulfide (cinnabar, HgS) at concentrations of approximately 0.5 ppm in the Earth's crust (Weast, 1984). Microorganisms are quite central in the transformation and subsequent transport of mercury in the environment as they are squarely involved in methylating mercuric cations, generating far more toxic products than the $Hg^{2+}$ parent ion. Methylation of inorganic Hg(II) is a fairly widespread process in a variety of habitats (soil, water, sediments, and even the human gut). Methylation processes derive the methyl group from methylocabolamine ($CH_3B_{12}$), a compound present in many microbial species and implicated in the methylation of a number of other metals (Pb, Sn, Pd, Pt, Au and Tl) and metalloids (As, Se, and Te) (Ridley et al., 1977). In fact, bacteria that lack this compound show a high sensitivity to methylmercury (Hughes and Poole, 1989), which occurs under both aerobic and anaerobic environments. Freshwater levels of methylmercury stand at about 3.96 ng $l^{-1}$, whereas in ocean waters the concentration is about 0.088 ng $l^{-1}$, the actual concentration varying depending on the pH and season (Thayer, 1995).

Most of the studies to understand this process were done under aerobic environments in which bacteria, yeast (*Cryptococcus* sp.), algae (e.g. *Chlamydomonas* sp.), and protozoa were associated with methylation and dimethylation of mercury (Thayer, 1995: Fig. 12.3). However, as can be seen in the biological cycling of mercury, large amounts of mercuric compounds end up in anaerobic sediments. Most sediments

are anaerobic. More attention is being devoted to studying this process under anaerobic conditions. Preliminary information has implicated SRBs, notably *Desulfovibrio desufricans* (Pak and Bartha, 1998) and *Clostridium* sp. (Narita and Endo, 2001) in methylation of mercury under anaerobic environments. The methylating SRBs appear to be facilitated by some interspecific hydrogen acetate transfer from the methanogens to them (Pak and Bartha, 1998). The rate of methylation greatly depends on environmental conditions.

Because methylmercury is produced at a rate faster than it can be degraded by microorganisms, is very soluble in water, and diffuses readily through biological membranes, it may accumulate in seafood and subsequently be consumed by humans. Even small amounts of methylmercury can be so devastating that it is unsafe to eat fish containing more than 0.5 ppm. In humans, mercury toxicity is manifested

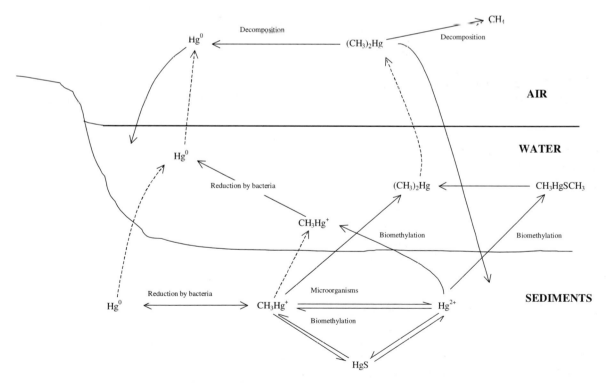

**Fig. 12.3**  Cycling of mercury in the environment.

in nephrotic syndrome, acrodynia ("pink disease"), erethism, and both respiratory and renal failure. In Minamata Bay and Nigata (Japan), methylmercury levels as a result of industrial waste dumping reached levels as high as 908 mg kg$^{-1}$ (Thayer, 1995) and was responsible for several deaths in both areas in the 1950s. Cases of methylmercury poisoning in both locales continue to the present day despite cleanup operations over more than half a century (Narita and Endo, 2001).

Mercury is highly toxic because of its affinity for thiols (such as -SH and S-S functional groups) which are abundant in all living cells. Mercury can also bind to nitrogen in nucleic acids, interfering with protein synthesis and gene action. The methylated compound flows more readily in the environment than Hg and is far more soluble in lipids (i.e., lipophilic) with a partition coefficient of 191.3 (Thayer, 1995). Although the original contamination at Minamata Bay and Nigata was caused by human activities, microbial activity worsened the problem by methylating the mercuric compounds resulting in more disaster.

Although microorganisms are involved in generating this toxic product, they are also directly involved in the detoxification of methylmercuric compounds, providing volatile mercury (Hg$^0$). Detoxification in microorganisms requires two enzymes, i.e., mercury lyse and mercury(II) reductase and occurs in two distinct steps (eqns 7 and 8). The resultant Hg0 can then be readily volatilized through the cellular membrane, thus removing them from the contaminated environment (see Fig. 12.3). The HX is recycled to detoxify more organomercury. The reduction process is not linked to respiration.

$$RHgX + HX \xrightarrow{\text{organomercurial lyse}} RH + HgX_2 \quad ...(7)$$

$$HgX_2 + NADPH + H^+ \xrightarrow{\text{mercury (II) reductase}} Hg^0 + NADP + 2HX \quad ...(8)$$

Some microorganisms use a different detoxification mechanism by converting the mercuric ion (Hg$^{2+}$) to methylmercury (CH$_3$Hg$^+$) or dimethylmercury ((CH$_3$)$_2$Hg) (see Fig. 12.3). If not reduced to volatile mercury, the methylmercuric compounds can enter the food chain of larger aquatic and terrestrial organisms. To emphasize the role of microorganisms in the cycling of this element in the environment, Smith et al. (1998) recently showed that *Bacillus subtilis* and *Streptomyces vonezuela* can readily oxidize volatile mercury (Hg$^0$) vapor to the soluble mercuric (Hg$^{2+}$) form. This oxidation is believed to be widespread among prokaryotes but its significance in the global mercury cycle is not yet fully assessed.

Mercuric compounds can also be detoxified when some microorganisms intracellularly accumulate or extracellularly bind and either complex or precipitate them. Toxicity of mercury (and other heavy metals such as Pb, Cd, Zn, and Cu for that matter) toward the oxidizing organisms, can also be reduced by the production of some extracellular thiols such as thiosulfate (S$_2$O$_3^{2-}$) and tetrathionate (S$_4$O$_6^{2-}$) (Walnright and Grayston, 1983).

The resistance of microorganisms to mercury has been quite well studied, especially in aerobes. It is encoded by the *mer* operon (Table 12.4) which is induced at approximately 0.01 µM Hg (II) concentration in the cytoplasm (Smith et al., 1998). A wide range of bacteria genera such as *Vibrio, Pseudomonas, Acinetobacter, Achromobacter, Corynebacteria, Bacillus, Cytophaga*, and *Flavobacterium* spp. show resistance to mercury (Colwell et al., 1986). Preliminary evidence indicates that the same *merA* gene confers resistance in anaerobes as well (Narita and Endo, 2001) but the resistance to this metal under anaerobic conditions has not yet received the attention it deserves. As indicated earlier, most of the mercury that contaminates waterways ends up concentrated in anaerobic sediments. Mercury-resistant organ-

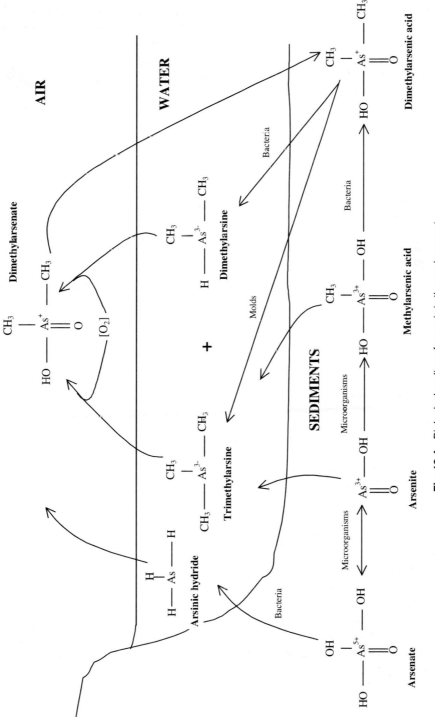

**Fig. 12.4**  Biological cycling of arsenic in the environment.

isms are categorized into either narrow- or broad-spectrum organisms (Bruins et al., 2000). The former lack organomercurial lyse and are therefore not resistant to most organomercurial compounds. On the other hand, the latter category has both enzymes and is resistant to most organomercurial compounds. Organomercurial lyse has a broad specificity for a variety of organomercurial substrates.

## 12.6.2   Cadmium

Cadmium is widely used by industry in the stabilization of plastics, as a coating of other metals to protect them from corrosion, fertilizers, pesticides, and for electroplating. Inside the cell, it can bind to sulfihydryl groups, interfering with important cellular functions. It is one of the most toxic metals known. Just like mercury, cadmium exerts toxicity by binding with thiol groups. It has ionic attributes that are similar to those of zinc (Fernandez-Leborans and Herrero, 2000) and calcium (Hughes and Poole, 1989). For example, the ionic radius of Cd is $0.97 \times 10^{-10}$ whereas the radius for Ca is $0.99 \times 10^{-10}$. Thus, its uptake competes with that of $Ca^{2+}$ and $Zn^{2+}$. It can also effectively displace these two cations in the cell. A recent report by Bruins et al. (2000) indicated that Cd can also enter the cell through the Mn(II) uptake system. Thus, cadmium has a variety of avenues to enter microbial cells.

Resistance to Cd is more prevalent among Gram-negative than Gram-positive bacteria. Resistance is possibly attributable to the presence of metallothionein-like proteins which are rich in cysteine, that bind and detoxify a range of heavy metals. The cadmium *cad* operon is induced by a variety of divalent cations (Silver and Phung, 1996). Resistance to cadmium in microorganisms is mostly attained by the microbial cells actively expelling the cation from the cell (i.e. efflux).

## 12.6.3   Arsenic

Pollution by arsenic is mainly derived from mine drainage and from combustion of fossil fuels. Anthropogenic sources of arsenic in the environment include the usage of arsenic-based wallpaper, semiconductors, fossil fuels such as coal, and from mining and smelting activities. Arsenic compounds also enter the environment as components of herbicides, fungicides and fertilizers. Its chemistry is similar to that of phosphorus, its immediate neighbor in the periodic table (Fig. 12.1). Arsenic can be transformed into three states (i.e., arsenate [As(V)], arsenite [As(III)] and arsine [As(-III)], the dynamics between each state being achieved by both chemical and microbiological means. In the past, much emphasis was given to the chemical reactivity of arsenic, with the role of microorganisms in its speciation and mobility in the environment underrated. The biological cycle of arsenic is shown in Figure. 12.4 and clearly reveals that it is greatly mediated by microorganisms. Thus, a number of bacteria (e.g. *Micrococcus*) and algae (e.g. *Chlorella*) can, under anaerobic conditions, transform arsenate (V) to a more mobile arsenite(III) in anaerobic environments. *Desulfotomaculum auripigmentum*; *Sulfurospirillum arsenophilum*, and *Chrysiogenes arsenatis* have also been shown to reduce As(V) to As(III) with sulfide as the electron donor (Newman et al., 1997; Oremland and Stolz, 2000). Studies by Dowdle et al. (1996) show that this process is completely inhibited if microorganisms are absent in the system (Fig. 12.5). Furthermore, no transformation was detected when the sediments were incubated under aerobic conditions. Reduction to As(III) is stimulated by the electron donor $H_2$ (Fig. 12.5). These observations clearly demonstrate that the reduction of As(V) to As(III) is an anaerobic microbial proc-

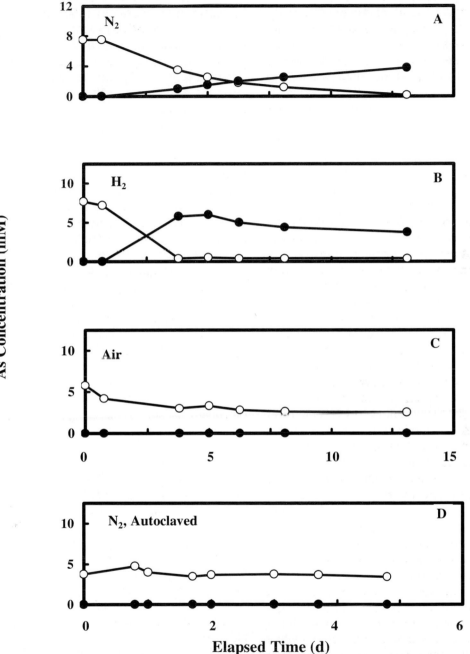

**Fig. 12.5** Reduction of As(V) (○) to As(III)(●) in sediments incubated under anaerobic (A and B) compared to aerobic (C) conditions. Panel D shows the absence of reduction when microorganisms are absent indicating that the process is biological in nature (adapted from Dowdle et al., 1996 with permission from the American Society for Microbiology).

ess. The highly toxic product (As(III)) can inhibit other ecologically important processes such as denitrification, sulfate reduction, and methanogenesis.

The reverse process, i.e., oxidation of arsenite (As(III)) to arsenate (As(V)), is also facilitated by a variety of bacteria such as *Achromobacter* sp., *Alcaligenes fecealis*, *Xanthomonas* sp. and *Pseudomonas* sp., as well as yeasts (Fig. 12.4). This transformation is very similar to the oxidation of nitrite to nitrate and is mostly favored under marine environment where P levels are depleted. Alternatively, arsenite can be methylated by a variety of microorganisms, to dimethylarsine [$(CH_3)_2AsH$] or trimethylarsine [$(CH_3)_3As$], products that are readily soluble in water. The biomethylation of arsenic can also occur in eukaryotes, includant humans (possibly facilitated by the microbes in the gastrointestinal system). The resultant compounds after methylation show appreciably less toxicity compared to the parent compound, i.e., arsenite. This is contrary to most other metals and metalloids which on methylation form products that are more toxic (e.g. see methylation of mercury above). The methylated arsenic products are also readily volatilized. In fact, biomethylation and subsequent volatilization has been used to remediate arsenic contamination and is a major mechanism by which microorganisms can detoxify arsenic(III)-based pollutants.

Arsenate can also be readily reduced under anaerobic conditions such as anoxic sediments to arsenic hydride ($AsH_3$) by microorganisms, e.g. *Pseudomonas* and *Alcaligenes* sp., the resultant hydride being readily volatilized. This reduction has not been reported in any fungus and may be solely performed by prokaryotes. It is believed to be a major mechanism by which arsenic compounds are lost from soils (Thayer, 1995).

As is clearly obvious from the foregoing account, the toxicity of arsenic compounds is related to the oxidation state of the arsenic ion.

The nontoxic arsenate (As(V)) is an analog of inorganic phosphate. In fact, in the environment and in metabolic processes, the similarity between phosphates and arsenate results in the competition for surface sites in sediments (Dowdle et al., 1996) and in metabolic reactions (Hughes and Poole, 1989). Under periods of phosphate abundance, arsenate enters the cell through the *Pit* system, a constitutive nonspecific P system. Under P-stressful environments, however, the more specific *Pst* system is induced (Bruins et al., 2000). Some microorganisms have the ability to tolerate As(V) by inactivating nonspecific *Pst* system (Nies and Silver, 1995). Thus, where P is depleted, by photosynthesis, the arsenate and phosphorus concentrations may be coequal, competing with the phosphates, leading to the uptake of arsenate. The arsenate is later excreted as methylarsenic acid and trimethylarsenic acid. The acids are further reduced to diethyl- and triethylarsine but these can be oxidized again during tides and current flow, to the acid that is volatilized. The importance of organisms in general, and microbes in particular in the cycling of arsenic in the environment is further displayed by seasonal changes in arsenics in phytoplanktons (Michel et al., 2000).

### 12.6.4 Selenium

Selenium is a metalloid and thus has chemical and physical properties intermediate between those of metals and nonmetals. It is widely distributed in the Earth's crust at low concentrations ranging between 0.1-2 mg kg$^{-1}$ soil (Frankenberger and Karlson, 1994). It can become an environmental contaminant due to human activities such as drainage from mines, combustion of fossil fuels (particularly coal), application of pesticides, use as a pigment in plastics and paint, inks, lubricants, and in the manufacturing of semiconductors (Haygarth, 1994). Thus, selenium toxicity problems are more widespread in areas where irrigation drainage water accumulates. Selenium has oxidation states of zero (elemental Se), −2 in combination with metals or hydrogen, and +4 or +6 in combination with oxygen. It can thus occur in different forms in the environment. It is required only in minute quantities in biological systems and is very toxic at high concentrations.

Just like arsenate, microbes can use oxidized forms of Se, for example Se(VI) as terminal electron acceptors to support growth. A variety of bacteria, algae, and fungi can transform selenium compounds (Table 12.8), thus playing a central role in its cycling in the environment and affecting its availability as well as toxicity. Transformation of selenium is more extensive under aerobic but can also occur under anaerobic conditions. Reduction of selenate (Se(VI)) leads to a more stable and less toxic selenite (Se(IV)), particularly in the presence of high $SO_4^{2-}$ levels (Oremland 1994). Selenite can then be reduced further to solid, relatively unreactive elemental selenium (Se(0)).

Selenium can also be methylated, just like arsenic and mercury. On a global scale, the biomethylation of volatile selenium on the terrestrial surface and oceans contributes 1,200 and 5,000-8,000 tones of methylated selenium compounds respectively, in the atmosphere per annum (Haygarth, 1994). Thus biomethylation, principally effected by microbes, is an important source of selenium in the atmosphere. In soil, volatilization is greatly affected by the pH and soil porosity as these attributes also affect microbial activity. Volatilization is maximum in neutral, slightly alkaline, or slightly acidic pH soils, and is greatly diminished under extremely acidic conditions (Tan et al., 1994). Several lines of evidence have unequivocally linked microbes to the methylation process. For example, the volatilization of selenium was greatly reduced by 80-90% when prokaryotic antibiotics were added to a plant nutrient solution in the presence of plant roots (Zayed and Terry, 1994). The small amounts of methylation under these conditions may have arisen due to fungi or some prokaryotes that remained unaffected by the antibiotics used. A comparison of methylation in sterile versus nonsterile soils, sewage, and sediments also demonstrates the absence of methylation in sterile conditions, clearly indicating the biological nature of this process (Doran and Alexander, 1977). The methylation process is often limited by nutrients, in particular carbon, in oligotrophic environments such as soil and sediments. Thus, organic amendments have been reported to increase the volatilization of alkyl selenides (Frankenberger and Karlson, 1994), making the addition of limiting nutrients a remediation strategy in cleaning up Se-contaminated sites. Other factors which can influence the volatilization process include moisture content, prevailing temperatures, concentration of Se ions, and type of Se species. The methylated form can also be methylated further, generating volatile dimethyl forms such as dimethyl diselenide [$(CH_3)_2SeO_2$], methane selenol [$CH_3SeH$], dimethyl selenone [$(CH_3)_2SeO_2$], and dimethyl selenenyl sulfide [$CH_3SeSCH_3$] (Table 12.8).

Microbes can also reduce selenium to hydrogen selenide ($H_2Se$), one of the most toxic forms of Se known (Doran and Alexander, 1977), or under rare conditions, oxidize selenium to selenite ($SeO_3^{2-}$) and selenate ($SeO_4^{2-}$) (Sarathchandra and Watkinson, 1981). Both of these possibilities have only been investigated under laboratory conditions and their significance in the environment remains largely unknown.

At high concentrations, selenium can inhibit the growth of a variety of microorganisms (Oremland, 1994). It can exert its toxicity by mimicking sulfur. The biochemistry of Se is very similar to that of sulfur, to the extent that some biological systems, e.g. legumes and sulfate-reducing bacteria, incorporate Se instead of S into proteins. Selenium can also undergo assimilatory reduction using enzymes and mechanisms that apply to sulfur. The two metalloids are neighbors in the periodic table (Group VIA; Fig. 12.1).

**Table 12.8** Some microbes which have been demonstrated to grow on selenium compounds

| Microorganisms | Selenium compound | Aeration status | Products | Reference |
|---|---|---|---|---|
| **Bacteria** | | | | |
| *Pseudomonas* sp. | Selenomethyionine | Aerobic | Dimethyl diselenide | Doran and Alexander, 1977 |
| *Thauera selenatis* | Selenate ($SeO_4^{2-}$) and selenite ($SeO_3^{2-}$) | Anaerobic | Elemental Se | Macy, 1994 |
| *Wolinella succinogenes* | Selenate and selenite | Anaerobic | Elemental Se | Oremland, 1994 |
| *Bacillus megaterium* | Elemental Selenium | Aerobic | Selenite ($SeO_3^{2-}$) | Sarathchandra and Watkinson, 1981 |
| *Agrobacterium* sp. | Selenocystine | Aerobic | Elemental Se | Doran and Alexander, 1977 |
| *Xanthomonas* sp. | Trimethylselenonium | Aerobic | Dimethylselenide | " |
| *Corynebacterium* sp. | Elemental selenium, selenocystine, selenomethionine, methaneselenate, methanoselenonate, selinite, selenate, or trimethylselenonium | Aerobic | Dimethylselenide | " |
| *Micrococcus lactilyticus* | Elemental Selenium | Anaerobic | Hydrogen selenide ($H_2Se$) | Oremland, 1994 |
| **Fungi** | | | | |
| *Candida humicola* | Selenecious soil and drainage water | Aerobic/Anaerobic | Volatile (possibly DMSe) | Frankenberger and Karlson, 1994 |
| *Aspergillus* sp. | Selenocystine | Aerobic | Elemental Se | Doran and Alexander, 1977 |
| *Penicillium* sp. | Selenocystine | Aerobic | Elemental Se | " |
| **Algae** | | | | |
| *Gracilaria blodgettii* | Se(IV) ions | Aerobic | Accumulation by the algae | Wang and Dei, 1999 |
| *Chlorella vulgaris* | Se(IV) and Se(VI) | Aerobic | Uncharacterized organic Se | Hu et al., 1997 |

## 12.6.5 Biotransformation of other heavy metals

Other heavy metals such as tin, lead, and nickel can also be transformed by microorganisms in a manner generally similar to those described above. Most of their sources of pollution in the environment are associated with human activities. For example, organotin compounds are widely used as fungicides because of their high toxicity. They exert toxicity by blocking ATPase and thus ATP synthesis. Lead in terrestrial environments and in the atmosphere is derived from combustion of gasoline. Just like organotin, organolead compounds are more toxic than inorganic lead due to their high solubility in lipoproteins and thus greater permeability through membranes (Thayer, 1995). As a matter of fact, tetraethyllead [$(C_2H_5)_4Pb$] is the most serious organometal pollutant generated through the activities of mankind as an additive of gasoline. It volatilizes into the atmosphere. While the use of lead-containing gasoline is almost phased out in most developed countries, its use in developing countries still continues. For quite a while it was not clear whether lead is biomethylated through microbial processes. However, work by various groups has implicated *Aeromonas, Acinobacter, Flavobacterium,* and *Pseudomonas* in its methylation to trimethyllead and tetramethyllead (Summers and Silver, 1978). A microbial consortium has also been shown to methylate lead (Thayer, 1995). The addition of methyl- or ethyllead compounds to soil can inhibit important biological processes such as the transformation of nitrogen.

Microorganisms resistant to inorganic tin have been isolated from polluted sites. Tin can be methylated, the process occurring predominantly in sediments after Sn(IV) is reduced to Sn(II) (Hallas et al., 1982). Tributyltin decomposition occurs by both biotic and abiotic processes depending on the prevailing environmental conditions, in particular light, and nutrients. Various organotin compounds are detrimental to soil microorganisms and have been widely used as active ingredients in pesticides. However, tolerance to tin can also develop in microorganisms although the mechanisms have not been widely studied (Suzuki et al., 1992).

## 12.7 USING MICROBES TO LEACH METALS

Metal-resistant microorganisms are being increasingly utilized to solubilize and recover metals from their mineral rocks and metal-polluted environments such as soils, water, and sewage. In Section 12.6 we briefly discussed the possibility of using microorganisms to decontaminate polluted sites such as sediments, soils, and waterways. In this section we focus on use of microorganisms to economically extract precious metals from their native rocks, mineral waste piles, and mine drainage systems, a process popularly termed **biomining** or **bioleaching**. Such recovery processes usually involve chemolithotrophs that derive their energy from oxidizing the ore, leading to acidic by-products. They obtain their carbon from $CO_2$, nitrogen from inorganic and organic compounds (e.g. dinitrogen and amino acids respectively) and trace element requirements from impurities in the ore. Their leaching capability can be greatly inhibited if organic matter is excessive but since they generate acids during the leaching process, they eventually create acidic environments that can subsequently eliminate other microorganisms, favoring the chemolithotrophs.

The principles of bioleaching were initially researched using copper sulfide and pyrite (FeS) ores. They involve both biochemical and chemical processes as shown in the reactions given below (eqns 10 to 13). Thus, the microorganisms oxidize the sulfide ores to meet their energy requirements. This oxidation process requires a great deal of $O_2$ and produces sulfuric acid, which is an environmental problem. However, the aforesaid reaction is really a summary of these four reactions:

$$FeS_2 + 7O_2 + 2H_2O \rightarrow 2FeSO_4 + 2H_2SO_4$$
$$...(9)$$

$$4FeSO_4 + O_2 + 2H_2SO_2 \rightarrow 2Fe_2(SO_4)_3 + 2H_2O$$
$$...(10)$$

$$FeS_2 + Fe_2(SO_4)_3 \rightarrow 3FeSO_2 + 2S \qquad ...(11)$$

$$2S + 3O_2 + 2H_2O \rightarrow 2H_2SO_4 \qquad ...(12)$$

The initial reaction (eqn 9) is facilitated by microorganisms, especially *Thiobacillus* spp. Reactions (10) and (12) also involve bacterial activity whereas reaction (11) is of a purely chemical (i.e., nonbiological) nature. During reactions (10) and (12) the sulfate or sulfide moiety is generally oxidized to provide energy for the chemolithotrophs. The $Fe_2(SO_4)_3$ formed in reaction (10) is soluble and can react with other elements in soil, precipitating into a mineral-gelosite. A more simplified version of the above reactions which results into a condition known as "acid...drainage" is:

$$4FeS_2 + 15O_2 + 2H_2O \xrightarrow{\text{Bacteria}} 2Fe(SO_4)_3$$
Pyrite

$$+ \quad 2H_2SO_4 \qquad ...(13)$$
(Acid mine drainage)

Other sulfide ores containing precious metals such as chalcopyrite ($CuFeS_2$), silver, gold, platinum, and uranium have also been commercially extracted from their natural ores or waste piles using microorganisms and involving similar oxidation (Ehrlich, 1995). The acid generated can depress the pH to as low as 0.7, thus creating an environment that can only be withstood by extremely acid-tolerant species, notably *Thiobacillus thioxidans, T. ferroxidans. Leptospirrillum ferroxidans, Gallionella* sp. and *Metallogenium* (Miller, 1981). Some fungi such as *Aspergillus* spp., and *Fusarium* spp., also oxidize iron and sulfur under these conditions but preliminary information suggests that the rates of oxidation of S, sulfates, and sulfides by eukaryotes are far less comparable to those of prokaryotes, possibly because of apparent differences in the accessibility of the ores to be solubilized (see discussion of direct and indi-

rect mechanisms below).

In typical bioleaching field operations, controlling the pH is usually not necessary as the system tends to be self-regulating. However, under anaerobic conditions, some heterotrophs can also leach metals. When leaching is heterotroph-based, the ore constituents solubilized are usually reduced by serving as a terminal electron acceptor. Heterotrophic leaching requires a near neutral pH, a condition that also favors many fungi. No selective advantage is deliberately imposed by pH in this instance as most microbes can thrive at this pH. However, selective pressure can be imposed by the prevailing oxygen tension. Most of the microorganisms which have been documented in bioleaching are mesophiles. However, oxidation of metal ores is an exothermic process and, in mineral waste piles, causes the temperature to rise to 60°C or higher. For the process to sustain itself, some of the participating microorganisms have to be thermophilic. This aspect of biomining has currently not been adequately researched to identify the major microbial players which can continue the process above the mesophilic temperature range.

The specific engineering systems used in bioleaching are similar to those elucidated for bioremoval as outlined in Section 12.5 and aim at optimizing the conditions under which microbial flora attack the ore to promote bioleaching. The biochemical mechanisms of solubilizing the constituents of the ores can be direct or indirect. Direct mechanisms involve catalysis by microbial enzymes in the oxidation (i.e., removal of electrons) or reduction (i.e., addition of electrons) and require that the microorganisms be in physical contact with the metal. In most instances, this involves the solubilized metal to be bound onto the biological cell surface. Direct mechanisms are generally associated with prokaryotes as their respiratory system (electron transport system, ETS) is located in the plasma membrane and cell envelope as opposed to eukaryotes where the ETS is associated with mitochondria (an internal organelle). Thus, prokaryotes are more able to directly attack insoluble ores such as FeS, CuS, $MnO_2$, geothite (FeOOH), because the necessary enzymes located in the plasma membrane can readily come into contact with the ores (Ehrlich, 1981, 1995). Work by Beveridge et al. (1982) indicated that both teichoic and teichuronic acid are the primary binding sites of metals in the cell wall surface. Accumulation of metals due to exuded polysaccharides was demonstrated with *Zoogloea ramigera* enabling it to adsorb 0.3g and 0.1g copper and cadmium respectively (Brooks, 1995). Indirect mechanisms, on the other hand, involve the production of metabolic products such as organic acids (e.g. citric, oxalic, acetic, propionic, lactic) and inorganic acids (e.g. sulfuric, nitric), as well as ligands (e.g. extracellular polysacharides) and alkali (e.g. ammonia and amines) by the microorganisms. The products generated then chemically react with the ore, solubilizing the metals.

*In situ*, the processes adopted on an industrial scale utilize microorganisms which are native to the ore, stimulating their activity by either providing nutrients such as nitrogen or phosphorus, and manipulating other environmental factors (pH, temperature, redox potential, oxygen content, etc.). In this manner, the microbial weathering processes normally, very slow in nature, are greatly enhanced. During mining the earth is usually opened to extract the ore. Low-grade ore is usually uneconomical to extract by conventional means and so is piled and thereby subjected to oxidation over time. Present advances in bioleaching have enabled the mining industry to economically extract metals from such piles, taking advantage of leaching using microorganisms such as *Thiobacillus*. In some instances, impure ores are piled and water run through them. The mineral-rich runoff is then subjected to electrophoresis to collect the metal.

Microbes are also increasingly being used in prospecting for mineral deposits. For this purpose, the microorganisms must distinctly differ from those in the surrounding areas that do not contain the metal and the conditions in the vicinity of the mineral deposits must differ both chemically and physically from adjacent areas (Parduhn, 1995). Use of microorganisms in mineral prospecting was initiated in the 1950s by scientists of the former Soviet Union while studying specific microflora associated with mineral deposits. Based on those studies it was realized, for example, that the presence of large ratios of *Thiobacillus denitrificans* to *T. thiooxidans* delineated the presence of sulfide ores. Since then, other microbial traits such as the presence of resistance to antibiotics have been linked to the presence of metals and used in prospecting for them (Parduhn, 1995). Colwell et al. (1986) identified many bacteria taxa that were tolerant to various heavy metals (Co, Hg, Mo, and Pb) from polluted waters and sediments in the Chesapeake Bay. Some of the taxa were also resistant to various antibiotics. Studies by Watterson et al. (1986) compared the abundance of *Bacillus cereus* resistant to penicillin in various soils. They found a high correlation between the abundance of penicillin-resistant *B. cereus* with metal content in a copper deposit soil. The soil also had a high population of *Penicillium* sp., indicating an ecological relationship between penicillin-producing fungi and penicillin-resistant bacteria in this naturally metalliferous soil. In this particular instance, the penicillin produced by the fungus was possibly chelating the heavy metals (due to the penicillamine present), detoxifying the metals, enabling the penicillin-resistant bacteria to thrive. Thus, growth medium that contains penicillin as a selective marker has been used to effectively complement prospecting for minerals over environmentally diverse terrains in Mexico and North America (Parduhn, 1995). This aspect of environmental microbiology is still in its infancy however, more than half a century after it was pioneered.

## References

Abad F.X., R.M. Pintó, J.M. Diez, and A. Bosch (1994). Disinfection of human enteric viruses in water by copper and silver in combination with low levels of chlorine. *Appl. Environ. Microbiol.* **60:** 2377-2383

Babich H. and G. Stotzky (1983). Synergism between nickel and copper in their toxicity to microbes: mediation by pH. *Ecotoxicol. Environ. Saf.* **6:** 577-589.

Beveridge T.J., C.W. Forsberg, and R.J. Doyle (1982). Major sites of metal binding in *Bacillus lichenformis* walls. *J. Bacteriol.* **150:** 1438-1448.

Bianchi M.E., M.L. Carbone, G. Lucchini, and G.E. Magni (1981). Mutants resistant to manganese in *Saccharomyces cerevisiae. Curr. Genet.* **4:** 215-220

Brierley C.L., J.A. Brierley, and M.S. Davidson (1989). Applied microbial processes for metals recovery and removal from wastewater. *In:* T.J. Beveridge and R.J. Doyle (eds.) Metal Ions and Bacteria. John Wiley & Sons, New York, NY, pp. 359-382.

Brooks R.R. (1995). The nature, structure, and environment of microorganisms. *In:* R.R., Brooks, C.E. Dunn, and G.E.M. Hall (eds.). Biological Systems in Mineral Exploration and Processing. Ellis Horwood, New York, NY, pp. 159-175.

Bruins M.R., S. Kapil, and F.W. Oehme (2000). Microbial resistance to metals in the environment. *Ecotoxicol. Environ. Safety* **45:** 198-207.

Bryant R.D., W. Jansen, J. Boivin, E.J. Laishley, and J.W. Costerton (1991). Effect of hydrogenase and mixed sulfate-reducing bacterial populations on the corrosion of steel. *Appl. Environ. Microbiol.* **57:** 2804-2809.

Chaudri A.M., S.P. McGrath, K.E. Giller, J.S. Angle, and R.L. Chaney (1993). Screening of isolates and strains of *Rhizobium leguminosarum* biovar *trifolii* for heavy metal resistance using buffered media. *Environ. Toxicol. Chem.* **12:** 1643-1651.

Chen G., R.J. Palmer, and D.C. White (1997). Instrumental analysis of microbiologically influenced corrosion. *Biodegradation* **8:** 189-200.

Chen J.-H., L. Lion, W. Ghiorse, and M. Shuler (1995). Mobilization of adsorbed cadmium and lead in aquifer material by bacterial extracellular polymers. *Water Res.* **29:** 421-430.

Colburn T., D. Dimanoski, and J.P. Myers (1996). Our Stolen Future. Penguin Books, New York, NY.

Colwell R.R., D. Allen-Austin, T. Barkay, J. Barja, and J.D. Nelson Jr. (1986). Antibiotic-resistant bacteria associated with environmental heavy metal concentrations. *In:* D. Carisle, W.L. Berry, and I.R. Kaplan (eds.). Mineral Exploration: Biological Systems and Organic Matter. Prentice Hall, Englewood Cliff, NJ, pp. 282-300.

Cook W.J., S.R. Taylor, and L.M. Hall (1998). Crystal structure of the cyanobacterial metallothionein repressor smtB: a model for metalloregulatory proteins. *J. Molec. Biol.* **275:** 337-346.

Doran J.W. and M. Alexander (1977). Microbial transformation of selenium. *Appl. Environ. Microbiol.* **33:** 31-37.

Dowdle P.R., A.M. Lavernman, and R.S. Oremland (1996). Bacterial dissimilatory reduction of arsenic (V) to arsenic (III) in anoxic sediments. *Appl. Environ. Microbiol.* **62:** 1664-1669.

Ehrlich H.L. (1981). Geomicrobiology. Marcel Dekker, New York, NY.

Ehrlich H.L. (1986). Interactions of heavy metals and microorganisms. *In:* D. Carlisle, W.L. Berry, I.R. Kaplan and J.R. Walterson (eds.). Mineral Exploration: Biological Systems and Organic Matter. Rubev Volume V. Prentice Hall, Englewood Cliff, NJ, pp. 221-237.

Ehrlich H.L. (1995). Microbial leaching of ores. *In:* R.R. Brooks, C.E. Dunn, and G.E.M. Hall (eds.). Biological Systems in Mineral Exploration and Processing. Ellis Horwood, New York, NY, pp. 207-231.

Fehrmann C. and P. Pohl (1993). Cadmium adsorption by the non-living biomass of microalgae grown in axenic mass culture. *J. Appl. Phycol.* **5:** 555-562.

Fernandez-Leborans G. and Y.O. Herrero (2000). Toxicity and bioaccumulation of lead and cadmium in marine protozoan communities. *Ecotoxicol. Environ. Saf.* **47:** 266-276.

Frankenberger, Jr. W.T. and U. Karlson (1994). Microbial volatilization of selenium from soils and sediments. *In:* W.T. Frankenberger Jr. and S. Benson (eds.). Selenium in the Environment. Marcel Dekker, Inc., New York, NY, pp. 369-387.

Fredrickson J.K., H.M. Kostandarithes, S.W. Li, A.E. Plymale, and M.J. Daly (2000). Reduction of Fe(III), Cr(VI), U(VI), and Tc(VII) by *Deinococcus radiodurans* R1. *Appl. Environ. Microbiol.* **66:** 2006-2011.

Gadd G.M. (1986). Fungal response towards heavy metals. *In:* R.A. Herbert and G.A. Codd (eds) Microbes in Extreme Environments. Acad. Press, New York, NY, pp. 83-110.

Gadd G.M. (1992). Metals and microorganisms: a problem of definition. *FEMS Microbiol. Lett.* **100:** 197-203.

Gadd G.M., J.A. Chudek, R. Foster, and R.H. Reed (1984). The osmotic response of *Penicillium ochro-chloron* changes in internal solute levels in response to copper and salt stress. *J. Gen. Microbiol.* **130:** 1969-1975.

Giller K.E., E. Witter, and S.P. McGrath (1998). Toxicity of heavy metals to microorganisms and microbial processes in agricultural soils: A review. *Soil Biol. Biochem.* **30:** 1389-1414.

Hallas L.E., J.C. Means, and J.J. Cooney (1982). Methylation of tin by estuarine microorganisms. *Science* **215:** 1505-1507.

Haygarth P.M. (1994). Global importance and global cycling of selenium. *In:* W.T. Frankenberger Jr. and S. Benson (eds.). Selenium in the Environment. Marcel Dekker, Inc., New York, NY, pp. 1-27.

Hu M., Y. Yang, J.M. Martin, K. Yin, and P.J. Harrison (1997). Preferential uptake of Se(IV) over Se(VI) and the production of dissolved organic Se by marine phytoplankton. *Mar. Environ. Res.* **44:** 225-231.

Hughes M.N. and R.K. Poole (1989). Metals and Microorganisms. Chapman and Hall, New York, NY.

Insam H., T.C. Hutchinson, and H.H. Reber (1996). Effects of heavy metal stress on the metabolic quotient of the soil microflora. *Soil Biol. Biochem.* **28:** 691-694.

Krantz-Rülcker C., B. Allard, and J. Schnürer (1996). Adsorption of IIB-metals by three common soil fungi: comparison and assessment of importance for metal distribution in natural soil systems. *Soil Biol. Biochem.* **28:** 967-975.

Ledin M. (2000). Accumulation of metals by microorganisms—processes and importance for soil systems. *Earth-Sci. Rev.* **51:** 1-31.

Ledin M., C. Krantz-Rülcker, and B. Allard (1999). Microorganisms as metal sorbents: comparison with other soil constituents in multi-compartment systems. *Soil Biol. Biochem.* **31:** 1639-1648.

Lee W. and W.G. Charaklis (1993). Corrosion of mild steel under anaerobic biofilm. *Corrosion* **49:** 186-199.

Macy J.M. (1994). Biochemistry of selenium metabolism by *Thauera selenatis* gen. nov. sp. nov. and use of the organism for bioremediation of selenium oxyanions in San Joaquin Valley drainage water. *In*: W.T. Frankenberger Jr. and S. Benson (eds.). Selenium in the Environment. Marcel Dekker, Inc., New York, NY, pp. 421-444.

Michel P., B. Boutier, and J.-F. Chiffoleau (2000). Net fluxes of dissolved arsenic, cadmium, copper, zinc, nitrogen and phosphorus from the Gironde Estuary (France): seasonal variations and trends. *Estuarine Coastal and Shelf Science* **51:** 451-462.

Miller J.D.A. (1981). Metals. *In*: A.H. Rose (ed.). Economic Microbiology Vol. 6. Microbial Biodeterioration. Acad. Press, New York, NY, pp. 149-202.

Narita M. and G. Endo (2001). Characteristics and mechanisms of mercury resistance of anaerobic bacteria isolated from mercury polluted sea bottom sediments. *In*: M. Healy, D.L. Wise, and M. Moo-Young (eds.). Environmental Monitoring and Biodiagnostics of Hazardous Contaminants. Kluwer Acad. Publi., Boston, MA, pp. 155-165.

Newman D.K., T.J. Beveridge, and F.M.M. Morel (1997). Precipitation of arsenic trisulfide by *Desulfotomaculum auripigmentum*. *Appl. Environ. Microbiol.* **63:** 2022-2028.

Nies D.H. and S. Silver (1995). Ion efflux systems involved in bacterial metal resistances. *J. Indus. Microbiol.* **14:** 189-199.

Nilsson J.R. (1979). Intracellular distribution of lead in *Tetrahymena* during continuous exposure to the metal. *J. Cell Sci.* **39:** 383-396.

Norberg A. and H. Persson (1984). Accumulation of heavy metals by *Zoogloea ramigera*. *Biotechnol. Bioengi.* **26:** 239-246.

Oremland R.S. (1994). Biogeochemical transformations of selenium in anoxic environments. *In*: W.T. Frankenberger Jr. and S. Benson (eds.). Selenium in the Environment. Marcel Dekker, Inc., New York, NY, pp. 389-419.

Oremland R.S. and J. Stolz (2000). Dissimilatory reduction of selenate and arsenate in nature. *In*: D.R. Lovley (ed.). Environmental Microbe-Metal Interactions. ASM Press, Washington DC, pp. 199-224.

Pak K.-R. and R. Bartha (1998). Mercury methylation by interspecies hydrogen and acetate transfer between sulfidogens and methanogens. *Appl. Environ. Microbiol.* **64:** 1987-1990.

Parduhn N.L. (1995). Geomicrobiological prospecting for petroleum and minerals. *In*: R.R., Brooks, C.E. Dunn, and G.E.M. Hall (eds.). Biological Systems in Mineral Exploration and Processing. Ellis Horwood, New York, NY, pp. 177-206.

Ridley W.P., L.J. Dizikes, and J.M. Wood (1977). Biomethylation of toxic elements in the environment. *Science* **197:** 329-332.

Rouch D.A., B.T.D. Lee, and A.P. Morby (1995). Understanding cellular responses to toxic agents: a model for mechanism choice in bacterial metal resistance. *J. Indus. Microbiol.* **14:** 132-141.

Sarathchandra S.U. and J.H. Watkinson (1981). Oxidation of elemental selenium to selenite by *Bacillus megaterium*. *Science* **211:** 600.

Sato T. and Y. Kobayashi (1998). The *ars* operon in the skin element of *Bacillus subtilis* confers resistance to arsenate and arsenite. *J. Bacteriol.* **180:** 1655-1661.

Scott J.A. and S.J. and S.J. Palmer (1990). Sites of cadmium uptake in bacteria used for biosorption. *Appl. Microbiol. Biotechnol.* **33:** 221-225

Silver S. and L.T. Phung (1996). Bacterial heavy metal resistance: new surprises. *Ann. Rev. Microbiol.* **50:** 753-789.

Smith T., K. Pitts, J.A. McGarvey, and A.O. Summers (1998). Bacterial oxidation of mercury metal vapor, Hg(0). *Appl. Environ. Microbiol.* **64:** 1328-1332.

Summers A.O. (1986). Genetic mechanisms of heavy-metal and antibiotic resistances. *In*: D. Carisle, W.L. Berry, and I.R. Kaplan (eds.). Mineral Exploration: Biological Systems and Organic Matter. Prentice Hall, Englewood Cliff, NJ, pp. 265-281.

Summers A.O. and S. Silver (1978). Microbial transformation of metals. *Ann. Rev. Microbiol.* **32:** 637-672.

Suzuki S., T. Fukagawa, and K. Takama (1992). Occurrence of tributyltin- tolerant bacteria in tributyltin- or cadmium-containing seawater. *Appl. Environ. Microbiol.* **58:** 3410-3412.

Tan J.A., W.Y. Wang, D.C. Wang, and S.F. Hou (1994). Adsorption, volatilization, and speciation of selenium in different types of soils in China. *In:* W.T. Frankenberger Jr. and S. Benson (eds.). Selenium in the Environment. Marcel Dekker, Inc., New York, NY, pp. 47-67.

Terry P.A. and W. Stone (2002). Biosorption of cadmium and copper contaminated water by *Scenedesmus abundans*. *Chemosphere* **47:** 249-255.

Thayer J.S. (1995). Environmental Chemistry of the Heavy Elements: Hydrido and Organo Compounds. VCH Publishers, New York, NY.

Timoney J.F., J. Port, J. Giles, and J. Spanier (1978). Heavy metals and antibiotic resistance in the bacterial flora of sediments of New York Bight. *App. Environ. Microbiol.* **36:** 465-472.

Wainright M. and S.J. Grayston (1983). Reduction in heavy metal toxicity towards fungi by addition to media of sodium thiosulfate and sodium tetrathionate. *Trans. Brit. Mycolog. Soc.* **81**: 541-546.

Wang W-X. and R.C.H. Dei (1999). Kinetic measurement of metal accumulation in two marine macroalgae. Mar. Biol. (Berlin) **135**: 11-23.

Watterson J.R., L.A. Nagy, and D.M. Updegraff (1986). Penicillin resistance in soil bacteria is an index of soil metal content noar a porphyry copper deposit and near a concealed massive sulfide deposit. *In*: D. Carisle, W.L. Berry, and I.R. Kaplan (eds.). Mineral Exploration: Biological Systems and Organic Matter. Prentice Hall, Englewood Cliff, NJ, pp. 328-350.

Weast R.C. (1984). CRC Handbook of Chemistry and Physics. CRC Press Inc., Boca Raton, FL (65th ed.).

Wood J.M. (1974). Biological cycles for toxic elements in the environment. *Science* **183**: 1049-1052.

Zayed A.M. and N. Terry (1994). Selenium volatilization in roots and shoots: effects of shoot removal and sulfate levels. *J. Plant Physiol.* **143**: 8-14.

# 13

# Water and Biosolids Microbiology

## 13.1  WATER SUPPLY

Water supports a chain of life from the small planktons to the fish, birds and their predators, thus showing an endless cyclic transfer of materials from life to life. Approximately 70% of the Earth's surface is covered by water, the nonoceanic surface area covering only 148 million $km^2$ (Butcher et al., 1992). Amidst this abundance of water, however, there is a shortage as most of the water is unuseable for human consumption, industry or agriculture because it is either polluted or contains high levels of sea salts. The problem of water pollution by human activities further compounds the shortages. Such pollution mostly emanates from deposition of domestic wastes from towns and cities, fertilizer, pesticides and herbicide leachate and runoff from farmland; radioactive waste from reactors; as well as chemical waste from industries, laboratories and military installations. In various cities, numerous domestic waste treatment facilities provide for outgrowth pipes that extend for a few kilometers offshore, damping the generated waste in open waters. However, with growing populations, particularly around coastal areas, our waterways cannot be indefinitely used as a sink for untreated global refuse.

Pollution of water can introduce the pollutant into the water cycle, affecting a wide range of organisms. For any community to develop and sustain itself, water supply is essential. At the other end of this prerequisite, is the need for water treatment. The problems associated with water pollution by chemicals have been discussed in both Chapters 10 (organic compounds) and 12 (heavy metals). The present chapter focuses on pollutants of a biological nature and their treatment. The approach adapted is just for convenience and should not be interpreted to imply that these two types of contaminants are exclusive of each other. In fact, both types of contaminants do occur together in the same body of water or biosolids in time and space. Pollution by biological materials occurs when feces, particularly of human origin, enter the water supply system. Thus, in this chapter, we focus on minimizing the introduction of such contamination of the environment in the first place, by looking at the treatment of wastewater and sewage sludge. Waste water and sewage sludge are collectively referred to as biosolids.

## 13.2  WATERBORNE DISEASES

Excretions from humans and livestock comprise the major source of biological materials that contaminate our waterways. The amount of urine and feces excreted by the adult human ranges between 0.2-0.9 $kg^{-1}$day depending on climate, diet, water consumption, and occupation (Franceys et al., 1992). Excretions by various types of livestock, such as cows and pigs, are even larger (Spaepen et al., 1997). Inadequate and unsanitary disposal of these feces, particularly human feces, leads to the contamination of water sources. Such contamination exposes the population to both communicable and noncommunicable diseases, some of which are listed in Table 13.1. Under normal circumstances, human urine is sterile and the presence of pathogens indicates either an infection of the host or pollution by feces. Human feces, on the other hand, always contain a variety of microorganisms which pose a health hazard to humans. The socioeconomic impact of the waterborne diseases associated with these wastes cannot be underestimated. Fortunately, with proper sanitation, these waterborne diseases are completely preventable. Most susceptible to these infections are children less than 5 years of age because their immune systems are less developed. In developing countries, the immune system in children tends to be further impaired by malnutrition. Also equally susceptible are individuals whose immune system is compromised, such as people under chemotherapy, on dialysis or transplant patients as well as people with AIDS.

Gastroenteritis occasionally occurs due to enteric bacteria, in particular *E. coli*, *Salmonella* sp. and *Shigella* sp. Typhoid fever (by *Salmonella typhii*) is still a major problem in developing countries. Outbreaks of typhoid were also quite prevalent at the advent of the industrial revolution as industrialization tends to congregate people, making it easier for the pathogen to spread through contact. A substantial

**Table 13.1** Occurrence of some representative pathogens in urine, feces and sullage

| Pathogen | Associated disease | Present in | | |
|---|---|---|---|---|
| **Bacteria** | | Urine | Feces | Sullage |
| *Escherichia coli* | Diarrhea | x | x | x |
| *Leptospira interrogans* | Leptospirosis | x | | |
| *Salmonella typi* | Typhoid fevers | x | x | x |
| *Shigella* spp | Shigellosis | | x | |
| *Vibrio cholerae* | Cholera | | x | |
| Viruses | | | | |
| Poliovirus | Poliomyelitis | | x | x |
| Coxsackie viruses | Meningitis, pneumonia, fever | | x | |
| Rotaviruses | Gastroenteritis | | x | |
| Hepatitis virus A | Infectious hepatitis | | x | |
| Protozoa | | | | |
| *Entamoeba histolytica* | Amoebiasis | | x | |
| *Giardia lamblia* | Giardiasis (diarrhea, weight loss) | | x | x |
| *Cryptosporidium* spp. | Cryptosporidiosis | | x | x |
| Helminths (vegetative forms and ova) | | | | |
| *Ancylostoma duodenale* | Hookworm | | x | x |
| *Ascaris lumbricoides* | Digestive/nutritional disorders | | x | x |
| *Toxicara* spp. | Fever, abdominal pains, neurological symptoms | | x | |
| *Fasciola hepatica* | Liver fluke | | x | |
| *Necator americarus* | Hookworm disease | | x | x |
| *Schistosoma* spp. | Schistosomiasis (Bilharzia) | x | x | x |
| *Taenia* spp. (tapeworm) | Nervousness, insomnia, anorexia, and abdominal pains | | x | x |
| *Trichuris trichiura* (whipworm) | Abdominal pains, diarrhea, anemia, weight loss | x | x | |

*Source*: Franceys et al. (1992); US EPA (1992); Smith (1996).

amount of information about viruses and protozoa in fecal material is available. Fungi are rare in fecal material, although isolated cases of *Aspergillus fumigatus* in sludge material have been reported (Angle, 1994). Evidence of fungal infections due to potable water at an epidemiological level is rare. Most of the problems associated with algae in water relate to palatability, odor, and turbidity concerns (Paralkar and Edzwald, 1996), although a few types of algae are also toxic. Most notable are *Anabaena flosaquae, Aphinizomenon flos-aquae,* and *Microcystis aeruginosa* (LeChevallier et al., 1999a). The first two species produce neurotoxins whereas toxins from *M. aeruginosa* affect the liver. Algae can be removed by filtering and chemical treatment with chlorine or ozone. The odors are probably due to extracellular polymers by algae. Between 7-50% of the C assimilated by algae is released into the water in the form of extracellular polymers (Fogg, 1966). The excreted polymers include amino acids, lipids, sugars, glycolic acids, organic acids and polysaccharides.

As indicated in Chapter 5, viruses of major concern in water include poliovirus, Coxsackie virus, echoviruses, and other enteroviruses such as adenoviruses, rhenoviruses, rotoviruses, heptatis A virus (infectious hepatitis), and Norwalk virus (diarrhea). Since viruses only multiply within a living host, their numbers cannot increase in water unless their host cells are also present. Thus, any treatment, dilution and natural inactivation of the water, waste water and sewage reduces the number of viruses, minimizing the chances of viral outbreaks where water treatment is enforced. However, some viruses may survive treatment and, depending on the type of virus, even minimal quantities are capable of producing infections. Protozoa of major concern include *Entamoeba histolytica, Giardia lamblia,* and *Cryptosporidium* sp., causing dysentery, giardiasis, and cryptosporidiosis respectively. Unlike the first two diseases, there is no effective treatment against cryptosporidiosis. *Cryptosporidium parvum* is an endoparasite which forms an oocyte because it has four sporozoites inside it. Once the oocyte is ingested, it releases all the four sporozoites (see Chapter 14). The sporozoites cause a bout of intermittent diarrhea; an individual with a sound immune system can overcome the infection. However, if the immune system is weakened, for example through infection by HIV, the subject may not recover (Kelly et al., 1997).

Currently, cryptosporidiosis is diagnosed better through active surveillance by using molecular biological techniques. Surveillance is either active or passive. The former involves surveying the blood within the population for the presence of antibodies against *Cryptosporidium* sp. or testing water for the presence of the parasite using fluorescent microscopy. However, this testing approach, together with the test for viruses in water, is expensive. Furthermore, both tests require a good level of scientific skill to execute (see Section 13.4) and are not readily available, especially in developing countries. Thus, for both protozoan and viral infections, most water treatment and health agencies have adapted passive surveillance whereby they wait for an outbreak to happen and then adopt a course of action. Infections by *Cyclospora cayetanensis* are also of increasing concern, particularly among immunocompromised individuals. Most of the water and biosolids-related disease outbreaks can be avoided by filtration, chlorination, and provision of adequate protection to our watersheds.

## 13.3  WATER PURIFICATION AND TREATMENT

Some water treatment agencies routinely use both filtration and chemical treatment in various modifications. In general, depending on the source of water, the purification and treatment regimen aims at:

(1) pretreating the water with chemicals such as alum or polymers to destroy some microorganisms and control taste and odors;

(2) mixing the water rapidly to distribute the chemicals evenly;

(3) coagulating and flocculating the impurities;

(4) sedimenting the flocculated particles;

(5) filtrating the water to remove the extraneous material that may not have sedimented; and

(6) post-treating with chlorine to inhibit the growth of microbes in the distribution system.

Thus the first step involves collection in a reservoir and allowing the existing particulate matter to settle (Fig. 13.1). Settlement is enhanced by adding compounds such as a combination of aluminum and iron salts commercially marketed as alum or with ferric chloride which forms aluminum phosphate (eqn 1) and ferric phosphate (eqn 2) precipitates respectively.

$$Al^{3+} + PO_4^{3-} \longrightarrow AlPO_4 \quad ...(1)$$
Alum

$$FeCl_3 + PO_4^{3-} \longrightarrow FePO_4 + 3Cl^- \quad ...(2)$$

After flocculating the particulate matter, the water is filtered through beds of sand and/or charcoal to remove or at least reduce protozoan cysts and oocysts, as well as nematode and helminth eggs. The charcoal also adsorbs some of the organic compounds including some pharmaceutical and personal care products that may be present. As a final step, the water is disinfected prior to pumping to the various distribution systems.

Chlorination is usually aimed at bacteria and viruses. It does not effectively affect most helminth eggs or protozoan cysts and oocysts (Feachem et al., 1983; Mara and Cairncross, 1989). In fact, some recent outbreaks of protozoa-related infections, notably *Gardia lamblia* and *Cryptosporidium parvum*, have been attributed to water despite its chlorination. Viral in-

fections also remain a concern in water. Their potential impact on public health under different water treatment regimens is discussed in the next chapter.

If the effluents are rich in nutrients, chlorination can also be ideal for the regrowth of some excreted bacteria which survive it, e.g. coliforms which take advantage of the low microbiological activity and plentiful nutrients (Feachem et al., 1983). Chlorination also generates toxic chemicals, e.g. trihalomethane (THM), which affect aquatic life and have adverse health effects in humans (Craun et al., 1994; Boorman et al., 1999). Generation of monochloramine ensures a good disinfectant and eliminates the production of both ammonia and nitrogen gas, thus reducing ammonia accumulation problems in the water.

Chlorination has traditionally been used to destroy or inactivate bacteria, but has also proven effective against viruses but not against protozoan cysts and helminth eggs. Its germicidal action is caused by hypochlorous acid. The disinfection chemistry is as follows:

$$Cl_2 + H_2O \longrightarrow \underset{\text{Hypochlorous acid}}{HOCl} + \underset{\text{Hydrochloric acid}}{HCl}$$
$$...(3)$$

$$HOCl + NH_3 \longrightarrow \underset{\text{Monochloramine}}{NH_2Cl} + H_2O \quad ...(4)$$

$$NOCl + NH_2Cl \longrightarrow \underset{\text{Dichloramine}}{NHCl_2} + H_2O \quad ...(5)$$

$$2NHCl_2 + H_2O \longrightarrow N_2 + HOCl + H^+ + 3Cl^-$$
$$...(6)$$

The hypochlorous acid (HOCl) produced in the first reaction is a strong oxidant that prevents much of the cellular enzyme system from functioning. It is neutral (i.e., without charge) and rapidly diffuses through the cell wall just like water. Hypochlorous acid reacts with inorganic nitrogenous metabolites such as ammonia, proteins, and amino acids. Ammonia is also usually mixed with chlorine in municipal

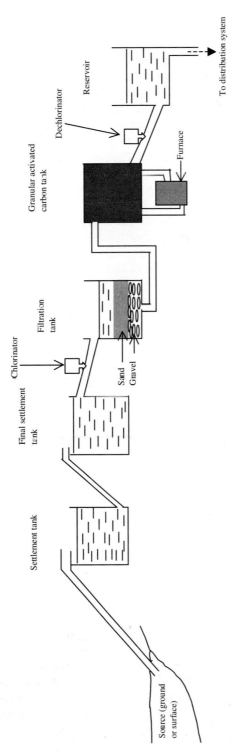

**Fig. 13.1**  Purification and treatment of drinking water.

water-treatment systems to form chloramines which are very stable compounds. Monochloramine, produced in the second step, is a good disinfectant. However, chloramines are toxic to fish and when chlorinated water is used in aquariums, the chloramines should be oxidized to counter toxicity. The final reaction occurs spontaneously, leading to the breakpoint in chlorination.

Ultraviolet light is used in some situations to disinfect water supplies, especially where the volumes being disinfected are relatively small and the water has low turbidity. In this case it successfully eliminates any possibility of posttreatment chemical residues.

Ozone has also been used by some water utilities to inactivate microorganisms but its use can cause the formation of undesirable organic by-products.

However, these processes in themselves offer no guarantee against posttreatment contamination, especially during the flow of water through the distribution system. For example, *Mycobacterium* spp. were found to grow in disinfected water and in distribution systems comprising existing biofilms (Tsintzou et al, 2000; Falkinham et al., 2001). A plethora of infections in individuals whose immune system is compromised have been traced to mycobacteria whose source is probably aquatic in origin. Their presence in the system is noticeably more frequent in water that has a high assimilable organic carbon and/or biodegradable organic carbon (Falkinham et al., 2001). Once established in a distribution system, *Mycobacterium* spp. are very hard to eliminate as they are highly resistant to ozone and chlorine-based disinfectants. Results by Taylor et al. (2000) also showed that the water-grown *Mycobacterium* cells are less susceptible to disinfection with chlorine, monochloramine, chlorine dioxide, and ozone (Table 13.2), possibly because their rate of growth in water is much slower compared to the growth rate in conventional laboratory media. *Mycobacterium* spp. were also found in distribution systems that

contained no detectable fecal indicators, i.e., total coliforms, fecal coliforms, and streptococci (Tsintzou et al, 2000), reinforcing the likelihood that absence of fecal coliforms does not accurately predict the absence of this pathogen in drinking water. As a matter of fact, preliminary evidence suggests that there is a negative association between *Mycobacterium* spp. and enterobacteria. Concerns about *Mycobacterium* spp. and a variety of other emerging pathogens such as *Listeria* sp., enteropathogenic *E. coli* (O157 and related serotypes such as O55:H7 and O111:NM), *Helicobacter pyroli, Aeromonas hydrophila* and *Campylobacter jejuni, Cyclospora cayetanensis,* Heptatis A (HAV), Norwalk and Norwalk-like viruses (NLV) (LeChevallier et al., 1999a, b), raises new challenges not only from a water quality testing perspective, but also in devising disinfection strategies that ensure delivering safe drinking water to the final consumers.

## 13.4  WATER QUALITY TESTING

It would be impractical to run tests for all pathogens in sewage sludge and waste water. Thus, monitoring for representative pathogens and nonpathogenic indicator organisms is done. More specifically monitored are fecal coliform bacteria which are themselves usually nonpathogenic and always abundant in untreated sewage sludge as indicators of the potential presence of fecal waste in general and pathogens in particular. Monitoring of fecal coliforms is a widely adapted approach because it is easy and inexpensive.

Considering the wide range of waterborne pathogens and the skills associated with successfully working with some of them, many monitoring programs assay for indicator organisms instead. Indicator organisms widely used to indicate fecal contamination are members of the Enterobacteriaceae family commonly referred to as coliforms of which fecal coliforms and fecal streptococci are a subset. Although

**Table 13.2**  Mean calculated disinfection contact time ($CT_{99.9\%}$) for *Mycobacterium avium* strains and *E. coli* with various disinfectants[1]

| Disinfectant | Mean disinfection contact time ($CT_{99.9\%}$) for | | | | | | Ratio of $CT_{99.9\%}$ (*M. avium*/*E. coli*) |
| --- | --- | --- | --- | --- | --- | --- | --- |
| | *M. avium* A5 | *M. avium* 1060 | *M. avium* 1508 | *M. avium* 5002 | *M. avium* 5502 | *E. coli* | |
| Chlorine | | | | | | | |
| Grown on media | 106 | 204 | 134 | 126 | 51 | 0.09 | 580-2300 |
| Grown on water | 1552 | 1,445 | 596 | 962 | 551 | ND | ND |
| Monochloramine | 97 | 458 | 548 | 1,710 | 91 | 73 | 1.2-23 |
| Chlorine dioxide | ND | 8 | ND | 11 | 2 | 0.02 | 100-550 |
| Ozone | ND | 0.17 | ND | 0.12 | 0.10 | 0.002 | 50-85 |

[1]Disinfectans were at a concentration of 1 ppm. Each mean is based on three replicates.
*Source:* Taylor et al., 2000 with permission from the American Society for Microbiology.

fecal coliforms are usually nonpathogenic, both the nonpathogenic and pathogenic bacteria enter freshwater bodies through human and other warm-blooded animal sewage. The presence of fecal coliforms thus indicates the presence of pathogenic organisms. The coliform test is itself suspect as it only detects both coliforms of fecal origin (i.e., enteric bacteria of *E. coli* category and their relatives) and nonfecal origin that grow naturally in soil (Perez-Rosas and Hazen, 1989). The coliform test basically involves the biodegradation (fermentation) of lactose, producing gases ($CO_2$, $H_2$ and $N_2$) as well as an acid (APHA et al., 1995). The gases produced are very easily detected if a lactose medium in a tube containing another smaller inverted airtight tube is fermented when spiked with a quantity of the test sample, and incubated at 37°C. If coliforms are present, they break down lactose, generating gases which in turn accumulate in the inverted tube. Lactose is used as the test compound because enterics are the most likely microorganisms to ferment it. On confirming the presence of coliforms, more specific and detailed tests are then conducted (APHA, AWWA, WEF, 1995). The test will not tell you about the presence or absence of viruses, protozoa such as *Cryptosporidium*, or other bacteria not related to *E. coli*.

The presence of viruses is determined using cell culture. Several gallons of the water to be tested are pumped through a charged filter onto which viruses, if any, are trapped. The filter is eluted using beef extract which neutralizes the charge on the filter. The eluted sample is then concentrated by adjusting the beef extract pH. The concentrated sample is inoculated onto Buffalo green monkey (BGM) kidney cells and incubated at 37°C over a two-week period (Fout et al., 1996). The entire assay takes a month or longer to finalize, a duration that reduces the usefulness of the results. Furthermore, some important waterborne viruses, e.g., Norwalk, rotoviruses, and Hepatitis A are not detectable with the BGM cell line. Molecular techniques to detect these viruses have been developed (Atmar et al., 1995, 1996). However, these methods are still only qualitative and lack the capability to determine whether any detected virus-related nucleic acid corresponds to viable, and therefore infectious virus particles. Thus these molecular techniques tend to be useful as a supplement to conventional cell culture techniques.

*Giardia* and *Cryptosporidium* spp. are detected using epifluorescence (Fout et al., 1996; Rose, 1997). A large volume of water is run through a yarn-wound filter, just as for viruses. The retained particulates are eluted and then concentrated by centrifugation. *Giardia* cysts and *Cryptosporidium* oocysts are separated from other particulates by flotation on percoll sucrose solution (sp. gr. 1.1) and indirectly stained with fluorescent antibodies (APHA et al., 1995). Based on size, internal structures, shape and morphology, the parasites are enumerated. However, some of the cysts and oocysts may be lost during concentration of the sample or be trapped with organic detritus and not float. Thus, in standard interlaboratory performance evaluations in the United States, an average of only 35-43% of the oocysts in spiked samples were recovered (Schaefer, 2001). An ideal method of detection should preferably be able to detect low numbers of these pathogens in large volumes of water. Furthermore, the detection methods currently used involve lengthy processing and microscop time and are unsuitable for determining whether the cysts or oocysts are viable.

Considering the complexity of the tests for most pathogens of major concern, some tests that use surrogates are under development (Baker, 1998; Leclerc et al., 2000). Recent waterborne disease outbreaks have prompted increased interest in the validity of various water-quality parameters for assessing treatment plant efficiency. Direct monitoring of various pathogens such as encysted protozoa is problematic because of the random occurrence of such pathogens, poor recovery, and lack of specificity for some pathogens. The density of parasites correlates with coliforms and coliform bacteria have traditionally been used as indicators of fecal pollution (APHA et al., 1995; Orangui et al., 1987). This correlation works well with raw water but does not extend to finished drinking water. Monitoring of indigenous spores of aerobic spore-forming bacteria represents a viable method for determining the performance of treatment plants. The method compares the concentration of spores in source water and effluents, thus providing an indication of the efficiency of the respective plants in removing biological particles. Their removal closely parallels the removal of other particles. Unlike most microbiological methods, spores do not propagate in the various unit processes, are very resistant to disinfection, and can be detected throughout the treatment process.

Use of the method allows concentration of volumes using a membrane filter and permits the analyst to increase the volume of water examined. Turbid water can also be conveniently subdivided into smaller portions, filtered, and the microbial counts combined so as to obtain an aggregate spore concentration. However, no single surrogate appears to be consistently representative for a wide range of pathogens (Fig.13.2). There are no reliable indicators of excreted viruses, helminths, and protozoa. Furthermore, the density of microorganisms in marine waters can also be influenced by the direction and speed of the wind since these affect the movement of currents and the spread and distribution of suspended microorganisms (Smith et al., 1999).

## 13.5  SEWAGE TREATMENT

Sewage refers to all the water from the household that is used for washing and toilet facilities as well as runoff that drains into the sewer lines. Prior to treatment, it can be composed of as much as 99.9% water and 0.1% suspended solids. To date, a number of cities in the world have only rudimentary sewage treatment

systems discharging untreated or nearly raw sewage into the waterways. Depending on the processing regimen adapted, the final product can end up with more than 50% solids (referred to as sludge or biosolids), and a liquid effluent (wastewater) that is free of pathogens. Whether treated or not, the final product is rich in a variety of essential plant nutrients such as N, P and K, and trace nutrients but unfortunately often contains substantial quantities of heavy metals originating from synthetic detergents that are toxic to both plants and animals. Nitrogen, phosphorus, and potassium in untreated sewage are in concentration ranges between 10-100, 5-25 and 10-40 mg $L^{-1}$ respectively. The phosphorus mainly comes from detergents and is mostly in the form of phosphates ($PO_4^{3-}$), polyphosphates, and organic P. Phosphorus is normally most limiting in lake and sea water. So if P is not adequately reduced in the final liquid biosolids effluent prior to discharge into waterways, it causes algal growth, a condition known as eutrophication. Treatment of sewage is mainly a biological and more specifically a microbiological process but also includes some chemical and physical processes. Thus treated waste water and biosolids will contain less N and P compared to untreated sewage but will contain approximately the same amount of K as stipulated above, depending on the treatment process. The goals of treating sewage are therefore to:

(1) reduce organic material;
(2) reduce the amount of nutrients, especially N and P;
(3) sequester or degrade toxic chemicals such as solvents and heavy metals; and
(4) remove pathogens, i.e., disinfect the water prior to discharge into open waterways.

Wastewater and sewage treatment basically aims at reducing the biochemical oxygen demand (BOD) as described below. Prior to treatment, sewage has about $10^8$ viable bacteria $ml^{-1}$ and $10^4$ protozoa $ml^{-1}$, the bacterial population greatly depending on the type of

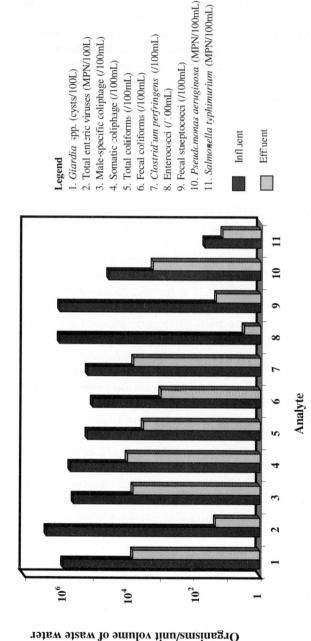

**Fig. 13.2** Reduction of pathogens (*Giardia* sp., enteric viruses, and *Salmonella typhimurium*) in waste water during treatment in relation to a variety of typically used indicators (Jjemba, unpubl.) Note that the density of microrganisms (y-axis) is on a logarithmic scale.

**Legend**
1. *Giardia* spp. (cysts/100L)
2. Total enteric viruses (MPN/100L)
3. Male-specific coliphage (/100mL)
4. Somatic coliphage (/100mL)
5. Total coliforms (/100mL)
6. Fecal coliforms (/100mL)
7. *Clostridium perfringens* (/100mL)
8. Enterococci (/100mL)
9. Fecal streptococci (/100mL)
10. *Pseudomonas aeruginosa* (MPN/100mL)
11. *Salmonella typhimurium* (MPN/100mL)

Influent
Effluent

growth medium used. Among the protozoa, ciliates are most dominant in sludge but flagellates and amoeba may also occur in significant numbers. Protozoa play a significant role in reducing the bacteria, BOD, and amount of suspended solids (Table 13.3). Some of the bacterial genera routinely found in sewage that are not necessarily pathogenic but play an important role in treatment are listed in Table 13.4, although the list is by no means exhaustive. For example, a variety of nitrifiers are always present but because they grow slowly compared to other bacteria, are rarely isolated, and their presence not widely reported. The conventional treatment process involves both primary and secondary processes (Fig. 13.3). The stages of treatment and how they impact BOD are discussed below.

**Table 13.3** Impact of ciliates on the quality of waste water

| Parameter | With ciliates | Without ciliates |
|---|---|---|
| Viable bacteria (ml$^{-1}$) | $1$-$9 \times 10^6$ | $1.6 \times 10^8$ |
| Suspended solids (mg l$^{-1}$) | 26 - 34 | 86 - 118 |
| Total BOD (mg l$^{-1}$) | 7 - 24 | 53 - 70 |
| Soluble BOD (mg l$^{-1}$) | 3 - 9 | 30 -35 |

*Source*: Curds (1982) with permission from the Annual Reviews of Microbiology.

## 13.5.1 Primary treatment

The primary treatment process is designed to remove the large floating and suspended solids, thus reducing the risk of damaging or blocking the pumps and valves used in subsequent sewage treatment stages. If not removed, the large and suspended solids would also take up volume and exert BOD during subsequent treatment. However, even before the primary treatment process starts, processes such as screening and grit removal are conducted. Although primary treatment is often wrongfully regarded as nonmicrobiological, some microbiological changes do occur during this phase. Notably, the large microbial particles and biofilms attached to existing suspended solids settle during this stage together with about 50% of the suspended solid materials. Furthermore, about 25% of the BOD is removed. BOD is a measure of the biologically degradable organic matter and reflects the amount of oxygen that is required by bacteria to metabolize the existing organic matter. BOD is normally determined in a dark sealed container at 20°C over a 5-day period (BOD$_5$) with an active microflora under aerobic conditions by providing dissolved oxygen. Incubation in darkness ensures that the oxygen utilized by respiration is measured. If incubated in light, some algae, if present, will liberate oxygen due to photosynthesis, giving an erroneous measurement.

During the treatment of sewage, the degradable organic matter is the major donor of electrons to facilitate the biological processes. It is desirable that the treatment regimen followed in a particular environmental setup reduce the BOD as fast as possible. At 20°C, the concentration of dissolved oxygen at saturation is only 9.1 mg L$^{-1}$. Prior to determining the BOD, the fecal sludge samples have to be properly diluted, or else the oxygen supply will be depleted before oxidation is completed. Dilution is done with a sterile phosphate buffer and without leaving any air space (or bubbles) in the bottle, which is then sealed to avoid entry of atmospheric air during the assay. In practice, since the appropriate dilution is not certain at the onset, several different dilutions are made. The BOD in one set of the diluted bottles is determined immediately as to provide the initial measurement (i.e., BOD at day zero). The assay uses a 5-day duration because during this period the majority of the BOD will be due to reduced organic matter. Nitrifiers grow more slowly than heterotrophs. If BOD attributable to nitrifiers is also required, the assay can be conducted for a longer period of time. The assay closely represents the natural processes which occur when the material being tested (sewage or waste water in this case) is discharged in natural waterways and, therefore, gives a

**Table 13.4** Some microbial genera of environmental importance encountered in sewage

| Category | Representative organisms | Remarks |
|---|---|---|
| Aerobic prokaryotes | *Achromobacter, Alcaligenes, Acinetobacter, Aeromonas, Aerobacter, Bravibacterium, Flavobacterium, Corynebacterium, Spirillum, Comamonas* spp. | A few, such as *Aeromonas* spp. are of medical concern. This range of organisms plays a role in the aerobic treatment of sewage. |
| Facultative aerobes | *Escherichia coli, Cytophaga, Micrococcus, Pseudomonas, Hypomicrobium, Cyanobacterium, Vibrio cholera, Streptococci, Lactobacilli, Salmonella* spp. | *E. coli* is the main indicator of fecal contamination. *V. cholera* is a lethal pathogen. *Cyanobacterium* sp. may contribute to the undesirable process of bulking. *Pseudomonas* sp. can use some of the nitrate as a terminal electron acceptor in the absence of oxygen. |
| Anaerobes | *Clostridium perfringens, Sphaerotilus, Butyribacterium, Megasphera* spp., *Methanogens*, Nitrifers (*Nitrobacter* and *Nitrosomonas*), *Bacteroides, Geotrichum, Propiobacterium* spp. | *Sphaerotilus, Beggiatoa,* and *Geotrichum* spp. facilitate floc formation which is necessary for secondary sludge to settle by gravity. However, when they become too abundant in activated sludge, they cause bulking as they interfere with the flocs causing them to compact loosely and thus less likely to settle. Most methanogens remain uncultured. |
| Protozoa | Ciliates (e.g. *Aspidisca, Chilodonella, Opercularia,* and *Vorticella*), Flagellates (e.g. *Giardia* spp.), Sarcodina (e.g. *Naegleria, Acathamoeba* spp.), Sporozoans, mainly *Cryptosporidium* spp. | Ciliates are most dominant and play a major role in the treatment process. |
| Viruses | Poliovirus, Coxsackie virus, Norwalk, Hepatitis A, Rotaviruses | Rotaviruses responsible for untold infant mortality, especially in developing countries. Norwalk and Hepatitis A are hard to detect in sewage without molecular probes because they do not infect the typically used cell lines in culture. |

somewhat accurate indication of the production potential of the material in question. The primary sludge sedimented during this phase provides the bulk of biosolids and is removed from the primary sedimentation tank periodically into an anaerobic digester to undergo further treatment (see secondary treatment-anaerobic phase below).

## 13.5.2 Secondary treatment

Secondary treatment is done in two phases— **aerobic** and **anaerobic**—and further reduces BOD (Fig. 13.3). The **aerobic phase** is designed to encourage the growth of aerobic microorganisms. Appropriate aeration is achieved by using either trickling filters or aeration tanks to stimulate aerobic heterotrophic bacteria as well as protozoa. Free-living bacteria are consumed and recycled in the system. During this process most of the remaining suspended solids are metabolized by existing aerobes such as *Zoogloea ramigera, Comomonas, Bacillus, Flavobacterium* spp., and a variety of others (Table 13.4) deposited as activated sludge, so called because the sludge is "activated" by being enriched with active bacteria captured by sedimentation and recycled to cause more flocculation. During this stage the bacterial cells not associated with the floc, and therefore freely suspended in solution, are more readily available for predation by protozoa (particularly the ciliates; Table 13.3). Continuing predation of the free-swimming bacteria thus tends to select for floc-forming species, augmenting the build-up of flocs. Flocculated solids eventually settle out, leaving a fairly clear effluent that has a diminished amount of suspended solids and BOD (Table 13.4). The effluent is then disinfected, for example, by chlorinating or ozonation as described for drinking water in Section 13.3 above. The disinfected effluent is dechlorinated (if chlorine has been used) and then either safely discharged into the waterway or used for other purposes such as irrigation (see Section 13.6 below). With waste water, sufficient disinfection is usu-

ally attained with 10-30 mg chlorine per liter of treated waste water. Low levels of microorganisms left in the treated waste water, once added to natural waterways are even further reduced depending on the resultant dilution, flow rate of the water in the waterway, and prevailing temperature. The secondary sludge is transferred to an anaerobic tank and combined with primary sludge to process further into biosolids.

If still loaded with excessive nutrients, the effluent undergoes tertiary treatment (see Section 13.5.3). If ciliates are absent in the system, this favors the growth of sheathed bacteria, e.g. *Sphaerotilus natans* and *Streptothrix* spp., which with growth form filaments, commonly referred to as sewage fungus; because of its sheath, this fungus is ingested by the various types of protozoa with difficulty. The predominance of these types of bacteria in secondary sludge causes the latter to float rather than sediment out, leading to a phenomenon called foaming or bulking. When this happens, the bulked material flows out together with the wastewater effluent which, if discharged into waterways, causes serious pollution problems. Microscopic examination of foams also reveals an abundance of branched or unbranched mycolic actinomycetes (mycolata) assigned to genera *Corynebacterium, Dietza, Gordonia, Mycobacterium, Norcadia, Rhodococcus, Skermania, Tsukamurell,* and *Williamsia* (Davenport et al., 2000). Davenport et al. (2000) observed a distinct relationship between the concentration of mycolata and the presence of foams in treatment plants. However, their investigations did not consider protozoa, indicating instead that for foaming to occur the mycolata have to attain a certain threshold.

The **anaerobic phase** seeks to reduce the BOD of the biosolid matter further. Under anaerobic conditions, the matter is slowly hydrolyzed and fermented, reducing the organic solids (polysaccharides, fats, and proteins) content by about 50%. This process also greatly reduces the offensive odors associated with sludge. It is often necessary to supply

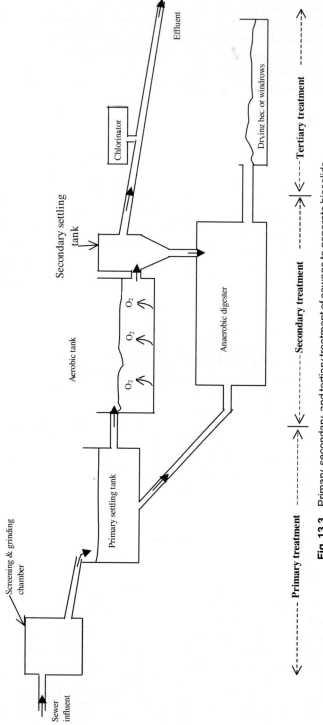

**Fig. 13.3** Primary, secondary, and tertiary treatment of sewage to generate biosolids.

some external energy in the form of heat to maintain a temperature that accelerates the biochemical reactions involved. In cases where heat is supplied, the mesophilic temperature range of 25-40°C is targeted to enhance the naturally occurring mesophiles. In some rare cases, thermophilic temperatures of 50-70°C are used. The actual digestion in the anaerobic systems is mostly attributable to three loosely classified groups of bacteria, i.e., hydrolytic, methanogenic, and fermentative. Considerable overlap is seen between the hydrolytic and fermentative groups. Hydrolytic bacteria hydrolyze insoluble complex polymers such as cellulose, hemicelluolose, and starch. Because these molecules are large and cannot be readily transported into the cell, their hydrolysis occurs extracellularly. The range of microorganisms able to hydrolyze these compounds narrows as the molecular size and complexity of the molecules increases.

Fungi and protozoa play no significant role in this anaerobic system and are only present in low numbers. The most dominant are bacteria of genus *Bacteroides* but fermentative organisms such as *Clostridium* spp., *Propiobacterium, Butyribacterium,* and *Megasphora* spp. are also found in substantial numbers. They ferment a wide range of existing substrates—from sugars, amino acids, purines, pyrimidines, and alkanoic acid generating alcohols, to a variety of volatile fatty acids (VFAs). The existing methanogens are difficult to culture and therefore not widely characterized. However, their diversity is beginning to unfold based on recent advances in molecular probes. They are slow growing and hence limit the rate in anaerobic digesters. Their nutritional capability appears to be limited to one- and two-carbon compounds. Thus complex polymers such as cellulose must be broken down into simpler compounds such as short fatty acids and acetate, before they can be converted by methanogens to methane and $CO_2$ (eqn. 7).

$$CH_3COOH \rightharpoonup CH_4 + CO_2 \qquad ...(7)$$

Some methanogens can also use the carbon dioxide generated together with hydrogen to generate more methane (eqn. 8). The methane generated during the entire process can actually be used as a fuel for heating the digester.

$$CO_2 + 4H_2 \rightharpoonup CH_4 + 2H_2O \qquad ...(8)$$

The removal of pathogens during anaerobic digestion depends on temperature, retention time, and type of process involved. Retention times vary depending on the anaerobic system used (Table 13.5). Typical retention time (RT) within the anaerobic digester is 14-30 days; the actual time is determined according to the mathematical relationship:

$$RT = \frac{Volume}{Flow\,rate} \qquad ...(9)$$

A longer RT is usually required because of the slow growth of fatty acid syntrophs and methanogens which would otherwise be washed out (Table 13.6).

Considering the fast growth rate of fermentative bacteria, which are acid-produc-

**Table 13.5** Mean sludge retention times with different anaerobic digestion temperature

| Type of anaerobic digestion | Temperature range (°C) | Retention time (days) |
|---|---|---|
| Cold digestion | <20 | >100 |
| Mesophilic digestion | 25-40 | 12-15 |
| Thermophilic digestion | 50-70 | 25-35 |

Compiled from Sterritt and Lester (1988).

**Table 13.6** Generation time for different groups of organisms in anaerobic digesters

| Group of bacteria | Doubling time (h) |
|---|---|
| Fermentative bacteria | 1-5 |
| $H_2/CO_2$ methanogens | 5-10 |
| Fatty acid oxidizers | 4-8 |
| Acetate-using bacteria | 1-9 |

ing, compared to the slow-growing methanogens, acetate-producing and fatty acid oxidizers, the system is likely to become quite acidic. Thus a base (calcium hydroxide) is added periodically to keep the pH in balance. There is growing interest in treating other wastes, e.g. agricultural and industrial, including highly chlorinated organic compounds such as PCB and TCE (see Chapter 11) in a similar fashion, rather than incinerating them since incineration generates more undesirable refuse in the form of gases. Because the anaerobic process depends on the interaction of several bacterial populations, the biggest problem is getting these populations established and properly balanced in the anaerobic tank. However, once established, the system runs quite well.

The process leaves a stable and inert sludge that still has the potential to provide regrowth for the few existing pathogens. Furthermore, this matter can still contain viable helminth eggs and infectious viruses. As mentioned before, primary and secondary treatment of sewage are both generally ineffective against helminth eggs and may not meet the Engelberg guidelines of less than 1 viable intestinal nematode egg per 100 g. As a matter of fact, helminths and nematodes are of major concern in the use of biosolids in agricultural systems. Not only are they a potential health hazard, but are also a potential source of parasites for some crops such as potatoes (Lewis-Jones and Winkler, 1991). For example, *Heterodera* sp. are a major parasites in the potato industry and can survive these treatment regimens. Once introduced in the agricultural field, they are very hard to eliminate. Their eggs can remain viable for many years. Other *Heterodera* species attack a wide range of crops (Table 13.7). To inactivate helminth eggs and viruses as well as to minimize the chance of regrowth of bacterial pathogens, biosolids are treated with lime to raise the pH to 12 and maintained at that pH for at least 2 hours (US EPA, 1992). This process also greatly reduces the moisture content. Moisture can be reduced further in drying beds or windrows. Dehydration of the biosolids also reduces their potential to attract vectors such as insects, birds and rodents, which can otherwise reintroduce pathogens. After this final set of procedures, the biosolids should be free from pathogens and can be safely used as a fertilizer or disposed of in a landfill. Where used as a fertilizer, some issues may arise because of the normally prevalent high content of heavy metals (McBride, 1995; Brown et al., 1996). With proper treatment (see Chapter 12), removal of heavy metals from activated may range from 1-83% (Brierley et al., 1989).

### 13.5.3 Tertiary treatment

Tertiary treatment is not always necessary and the need to conduct it depends on the intended use of the waste water in question. It is intended to remove nonbiodegradables such as $PO_4^{2-}$, $NO_3^-$. Domestic sewage has a BOD:N:P ratio of 100:5:1, with nitrogen mostly occurring in sludge in the ammonium form and comprising about 30 mg $L^{-1}$ (Lester and Birkett, 1999). It is

**Table 13.7**  Some common *Heterodera* sp. and their associated host plants

| *Heterodera* species | Common name | Host crops |
|---|---|---|
| *H. pallida* | Pale potato cyst nematode | Potato, tomato |
| *H. rostochiensis* | Potato cyst nematode (PCN) | Potato, tomato |
| *H. schachtii* | Beet cyst nematode | Sugar beet |
| *H. cruciferae* | Cabbage cyst nematode | Brassica |
| *H. carotae* | Carrot cyst nematode | Carrot |
| *H. punctata* | Grass root nematode | Grasses |
| *H. trifolii* | Clover cyst nematode | Grasses |

desirable to reduce the nitrogen in the final effluent to less than 10 mg L$^{-1}$. While some of the nitrogen will be removed during treatment by heterotrophs since they assimilate it to satisfy their nitrogen requirement, the amounts removed through this process are usually inconsequential. Nitrifying bacteria, through a two-step process, convert the ammonium to nitrite and then to nitrate (eqns. 10 and 11). The first stage is performed by *Nitrosomonas* and the second stage by *Nitrobacter*.

$$2NH_4^+ + 3O_2 \rightleftharpoons 2NO_2^- + 2H_2O + 4H^+ \quad ...(10)$$

$$2NO_2^- + O_2 \rightleftharpoons 2NO_3^- \quad\quad\quad ...(11)$$

Nitrifying bacteria are slow growing compared to other bacteria in activated sludge. It is essential, therefore, to retain the treated sludge in the tank for several days to preclude greatly diminishing their populations in an established treatment system. A period of about 8 days is recommended.

It is worth noting that from the reactions above, most of the nitrogen is not removed but rather only stoichiometrically converted from ammonium to nitrate. Both nitrite and nitrate are undesirable in the effluent because high nitrate concentrations can cause methanoglobinaemia (blue baby disease) in young babies fed on bottled milk made with such polluted water. Methanoglobinaemia can be fatal if not treated. Phosphorous levels in the product can also be of concern. These nutrients can be reduced either by chemical (using alum) or biological means prior to disinfection and discharge, to preclude eutrophication in the receiving waters. From a biological perspective, the normal microbial cell requires about 1-3% of its dry weight as P (Bitton, 1994). Thus the biological approach of removing P relies on several organisms, e.g. *Moraxella, E. coli, Aerobacter, Acinetobacter, Pseudomonas, Beggiotoa,* and *Mycobacterium* which can accumulate P in excess of their normal cell requirement. The excess P is typically stored in polyphosphate granules and serves as a P and

energy source during stress. Toxic chemicals, e.g. organics, if of concern, are also removed by activated carbon filtration, though this process tends to be expensive. In a few cases, some pathogens in the sludge can also be reduced by disinfection or incineration. However, successful disinfection of sewage using chemical disinfection is less predictable compared to its success in disinfecting drinking water, and may only be adopted where the highest level of quality control and management is assured.

### 13.5.4 Stabilization ponds

Where the infrastructure is limiting but land is abundant, or for smaller communities, stabilization ponds (also called lagoons or retention ponds) are used to achieve what can be achieved with the primary and secondary treatments outlined above. These are advantageous in that they require minimal cost, less intensive maintenance, and low technical know-how, but still retain the ability to remove a high degree of the excreted pathogens. Furthermore, they require no energy expenditure other than solar and have a high ability to absorb organic and inorganic loads. They are usually of choice in warm climates if land is available at a reasonable cost. Best results demand that several ponds be connected together in a series as this consistently gives a better pathogen removal than a single pond with the same retention time. Under tropical conditions, fecal coliform concentrations of less than $10^3$ per 100 ml are attainable in a series of 4-6 waste stabilization ponds with an overall retention time of 20 days or longer (Orangui et al., 1987). These populations correspond to removals of 99.9-100% of fecal coliforms (Table 13.8). Inactivation of pathogens in stabilization ponds is attributable to the effects of UV since daylight is the major driving force of the process. Other factors that affect inactivation include changes in temperature, pH, conductivity, and chemical oxygen demand (COD).

**Table 13.8**   Mean populations of viruses and bacteria in raw waste water (RW) and effluents in a series of stabilization ponds at 27±1°C

| Organism | Mean populations 100 ml$^{-1}$ (for bacteria) or 10 L$^{-1}$ (for viruses)[a] | | | | | | Removed | |
| --- | --- | --- | --- | --- | --- | --- | --- | --- |
| | RW | AP | FP | MP1 | MP2 | MP3 | Percent (%) | Log$_{10}$ units |
| Bacteria | | | | | | | | |
| Fecal coliforms | $2 \times 10^7$ | $4 \times 10^6$ | $8 \times 10^5$ | $2 \times 10^5$ | $3 \times 10^4$ | $7 \times 10^3$ | 99.97 | 4 |
| Fecal streptococci | $3 \times 10^6$ | $9 \times 10^5$ | $1 \times 10^5$ | $1 \times 10^4$ | $2 \times 10^3$ | 300 | 99.99 | 4 |
| *Clostridium perfringens* | $5 \times 10^4$ | $2 \times 10^4$ | $6 \times 10^3$ | $2 \times 10^3$ | $1 \times 10^3$ | 300 | 99.40 | 2 |
| Total bifidobacteria | $1 \times 10^7$ | $3 \times 10^6$ | $5 \times 10^4$ | 100 | 0 | 0 | 100 | 7 |
| Sorbitol-positive bifids | $2 \times 10^6$ | $5 \times 10^5$ | $2 \times 10^3$ | 40 | 0 | 0 | 100 | 6 |
| Campylobacters | 70 | 20 | 0.2 | 0 | 0 | 0 | 100 | 1 |
| *Salmonella* sp. | 20 | 8 | 0.1 | 0.02 | 0.01 | 0 | 100 | 1 |
| Viruses | | | | | | | | |
| Enteroviruses | $1 \times 10^4$ | $6 \times 10^3$ | $1 \times 10^3$ | 400 | 50 | 9 | 99.91 | 3 |
| Rotaviruses | 800 | 200 | 70 | 30 | 10 | 3 | 99.63 | 1 |
| BOD (mg L$^{-1}$) | 215 | 36 | 41 | 21 | 21 | 18 | 91.6 | |

[a]AP an anaerobic pond with mean retention time of 1 day; FP a secondary facultative pond while MP1-MP3 were maturation ponds. FP and MP1-MP3 had a retention time of 5 days each.

*Source:* Orangui et al. (1987) with permission from the International Water Association.

By nature of design and lack of mechanical mixing, the pond naturally partitions into an aerobic and an anaerobic section (Fig. 13.4). Aerobic heterotrophic bacteria dominate in the aerobic zone just as in the case of aerobic treatment reactors but rotifers and copepods also play a role by grazing on free-swimming bacteria, thus improving turbidity. Algae also, if present, play a major role as they supply some oxygen to the system through their photosynthetic activity. Algae also utilize some of the ammonia and phosphorous, thus reducing the undesirable presence of these excessive nutrients. Protozoa do not seem to feature prominently in the microbiological processes in stabilization ponds. The degree of reduction of bacteria in a pond can be estimated from the formula

$$R = 1 + Kt \qquad ...(12)$$

where R is the ratio of concentration of fecal coliforms in incoming and outflowing water; t the retention time of the waste in the pond in days; and K the rate of die-off of fecal bacteria which depends on the mean temperature (T) in °C. K is estimated from the formula $K = 2.6(1.19)^{T-20}$. For design purposes, the mean monthly temperature of the coldest month is used (Mara and Cairncross, 1989). Processes in the anaerobic zone are fairly sensitive to temperature and are greatly diminished below 17°C (Lester and Birkett, 1999).

Stabilization ponds are effective in removing pathogens in wastes. However, in considering the removal of pathogens from wastes, the number of pathogens that survive is more important than the number removed or killed. As much as 99.9% of the pathogens are removed in some processes but the remaining 0.1% can still be significant in view of the high concentration of pathogens initially present in the wastes. To put this in perspective, in Table 13.8, the raw waste water had $2 \times 10^7$ fecal coliforms per 100 ml, a population that is typical. Removal of 99.97% of these pathogens still left $7 \times 10^3$ coliforms per ml, a population that is still substantial and does not meet the Engelberg guidelines of <1,000 fecal coliforms per 100 ml. The degree of pathogen removal is thus conveniently expressed in terms of $\log_{10}$ units rather than percentages.

Unlike sewage which is usually deliberately treated to reduce pathogens, nutrients, and heavy metals, animal manure is usually piled, composted, or stored as a slurry in manure tanks, lagoons or pits without deliberate treatment. Runoff from livestock farms also pollutes our waterways. Although in the majority of cases livestock dung does not pose as high a danger as human feces, some cases of bacterial pathogens such as *E. coli* O157:H7 infiltrating into surface and groundwater areas by runoff from livestock farms or unincorporated manure have been reported (Gagliardi and

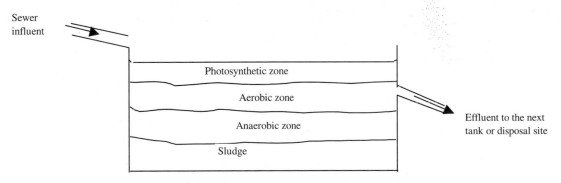

**Fig. 13.4**   Design of a typical stabilization pond showing the photosynthetic, aerobic, and anaerobic zones respectively.

Karns, 2000). Densities as high as $10^9$ *Cryptosporium* oocysts per gram feces have been reported in diarrhetic calves (Angus, 1986) and a variety of other animals (such as beavers, cats, dogs, etc.). Several environmentally stable types of viruses are also found in animal wastes used as manure (Pell, 1997). None of these types of viruses is able to survive aerobic composting of manure when temperatures reach 60°C or higher in the compost. These pathogens can pose a danger to human health.

### 13.5.5 Other methods of treating sewage

Other treatments include composting, physically heating the sludge using hot gases or heat exchangers, radiation, pasteurization, and incineration. The effectiveness of heating processes depends on the duration of heating and the temperature to which the sludge is subjected. Both radiation (depending on the dosage) and high temperature can reduce all types of pathogens to below detectable levels. Helminth ova are fairly resistant to high temperatures, however, and viruses quite resistant to radiation. The principles outlined for composting in Chapter 7 also apply to sewage treatment. Other processes are generally not widely practiced and only briefly discussed.

## 13.6 SELF-PURIFICATION CAPACITY OF NATURAL WATERS

More than half the world's population live in developing and newly industrializing countries with no access to piped water or a reliable water supply. Under these circumstances, human waste material is predominantly deposited on-site in latrines or their modification (Franceys et al., 1992). Furthermore, a good proportion of the world population also uses on-site septic tanks, some of which leak fecal material into the groundwater (Scandura and Sobsey, 1997). Under these natural conditions, the decomposition of fecal material is carried out by the native

bacteria although fungi, protozoa, and other organisms may also participate. Decomposition in these settings involves:

- breakdown of complex compounds, e.g. proteins into simpler more stable forms;
- formation and release of various gases ($NH_3$, $CH_4$, $CO_2$ and $N_2$) into the atmosphere;
- formation of soluble materials which filter into the surrounding soil and may leak into the groundwater; and
- destruction of most pathogens which are unable to survive the prevailing high temperatures of 70°C or higher and/or nutrient stress.

If oxygen is present in these environments, aerobes will be the major decomposers. As the oxygen decreases, facultative aerobes still continue the decomposition process.

There is insufficient documentation as to whether infiltration of effluents from latrines into soil occurs through the bottom or through the sidewalls of the pit. One argument indicates that the flow is through the base of the pit and drainage trenches, with lateral (sidewall) infiltration only negligible. Another possibility is that the soil pores at the base become rapidly blocked by sludge, making sidewall infiltration more predominant (Francheys et al., 1992). Either of these possibilities is more likely applicable to a specific environment, depending on soil texture, porosity characteristics of the soil profile in question, and height of the water table. Predominantly clayey and silty soils have smaller pores than sandy soils in which the pores are large and hence less likely to be clogged by effluents. Fecal-related nutrients, in particular nitrates (Fourie and Van Ryneveld, 1995; Girard and Hillaire, 1997) and fecal pathogens (Scandura and Sobsey, 1997), have been reported to leak into aquifer and groundwater, especially in urban areas if homemade latrines are not properly designed, and in rural areas with on-site septic tanks.

The level of the water table changes with seasons. In the unsaturated zone, the flow of liquid is induced by gravity as well as by cohesive and adhesive forces in the soil. In a saturated zone, all the pores are filled with water and drainage depends on their size and the difference in levels between water table and liquid in the pit. In situations wherein the soil pores become clogged, clogging can be caused by:

- growth of microorganisms and their wastes (i.e., formation of biofilms that eventually block the flow channels);
- direct blockage by solids as they try to filter through;
- precipitation of insoluble salts; and
- swelling of clay minerals such as montmorrillonite/kaolinites.

However, clogging is not instant. When the liquids first infiltrate into the unsaturated soil, aerobic bacteria decompose much of the organic material filtered into the soil, keeping the pores cleared. As the organic material builds up, conditions become more and more anaerobic and this slows the decomposition rate, leading to an accumulation of insoluble salts.

Mostly encountered at substantial distances in the effluent from pit latrines and on-site septic tanks are viruses, bacteria, and chemicals which could contaminate the groundwater. However, because of their relatively large size, protozoa and helminths are rapidly strained out by the soil. Phosphorus, because of its capacity to adsorb to soils, particularly clayey or acidic soils, is also rarely a problem in such effluents (Fourie and Van Ryneveld, 1995). Of most concern is the potential contamination of groundwater by nitrates. Nitrates are not adsorbed or physically retarded by soil and can travel large distances in soils with high hydraulic conductivities. Because of their small size, viruses are little affected by filtration and their removal is almost entirely by adsorption onto soil particles. Compared to bacteria, viruses reach greater distances (see Chapter 5). Their adsorption is greatest in soils with a high clay content and is favored by the low flow rates typical in an unsaturated zone (Angle et al.,

1994). Adsorbed bacteria and viruses survive for longer periods of time in moist than in dry soil and thus their ultimate fate depends on the texture, organic matter content, and water-holding capacity of the soil. Generally, there is some risk of latrine and leaky septic tanks effluent matter polluting the groundwater if the water table is low, since the presence of fissures or holes caused either by burrowing animals or roots can allow a rapid flow of the effluents to the underlying groundwater with little removal of microorganisms (Scandura and Sobsey, 1997). Bacteria and viruses can travel distances similar to those traveled by groundwater in about 10 days, a distance of about 15 m, before being destroyed (Lewis-Jones and Winkler, 1991).

## 13.7 REUSE OF SEWAGE AND OTHER WASTES

In many countries where the supply of water is lower than the demand, inevitably waste water is used for irrigation of crops or consumption by animals (including humans). This is particularly the case in arid and semiarid regions where a major conflict between domestic and industrial uses competes with agricultural demand for water. This conflict is often resolved by domestic and industrial requirements being met primarily with fresh water. The waste water generated from these two sources is then treated and used for irrigation purposes. Furthermore, treated or untreated excreta is used for a variety of purposes, namely, fertilizing soil, production of biogas, aquiculture (fertilization of fish ponds and production of algae) as well as irrigation purposes (Mara and Cairncross, 1989). For example, more than 6.5 Mt. (dry weight basis) of sewage sludge are generated annually in the European Union, of which more than 37% is utilized for agriculture (Smith, 1996). Quantities in other developed countries and in densely populated developing countries in Asia are also considerably large. Application of waste water to cropping fields can therefore reduce or even totally eliminate the need for applying conventional fertilizers. Animal wastes from intensive and semi-intensive farming are often applied to agricultural crops and/or discharged into waterways, finding their way into the environment. Where used for supplying nutrients in soil, both biosolids and animal manures provide additional benefits of increased humus content, which in turn significantly improves the soil structure and water-retention capacity.

With such practices, the degree of hazard depends on the level of treatment of the waste prior to reuse, nature of the crop, method of application, condition of the soil, air, and water as well as, in the case of piped sewage systems, the concentration of heavy metals. If present in sufficient levels, bacteria, viruses, and helminth eggs can cause diseases in humans and other animals when the sewage sludge or waste water are used as soil conditioners or fertilizers. Thus, despite all its benefits, untreated excreta used for agricultural purposes is a potent source for transmission of excreta-related diseases. Reported survival times of various fecal-related pathogens in soils and on crops vary widely (Table 13.9). These variations are due to differing climatic factors, strains, and the various assay methods used. What is apparent is that almost all excreted pathogens, in particular bacteria, viruses, and helminth eggs can survive in soil and on crops for a sufficiently long time to become potentially hazardous to humans. Viruses survive for longer periods of time because they exhibit few metabolic functions. Parasites (protozoa and helminths) form environmentally resistant structures such as ova and cysts which can survive for several months in soil. Protozoan cysts are relatively more susceptible to environmental factors such as desiccation, heat, and sunlight and thus pose less threat to public health in reused waste water and biosolids. Unlike viruses, protozoa, and helminths, even very small populations of bacteria can proliferate rapidly under the right environmental conditions. Thus care must be taken to ensure that the biosolids remain fairly

dehydrated to minimize the resurgence of bacterial growth.

US EPA (1992) categorized biosolids as either class A or B. In class A biosolids, the pathogens (i.e., enteric viruses, pathogenic bacteria, and viable helminth ova) have to be reduced to below detectable levels. For class B biosolids, the goal is to reduce the pathogens to levels unlikely to pose a threat to public health and the environment if specific restrictions are imposed on the site where the biosolids have been applied. Thus, class B biosolids may contain reduced but still significant densities of pathogenic bacteria, viruses, protozoa and viable helminth ova. Site restrictions are designed to minimize the potential for human and animal contact with the material for a period of time after application until environmental factors have further reduced the pathogens. These restrictions include setting a lapse period before crops are harvested and ensuring restricted access to grazing animals. The methods currently used to test the decline of pathogens other than bacteria in biosolids are still by and large time consuming, expensive and, to some extent, inadequate. More specifically, helminths are tested by incubating samples for 2-4 weeks and examining them for embryonation. Likewise, viruses are tested using BGM kidney cells and looking for formation of plaques. My experience with this method showed that plaque formation could be inhibited in the samples tested using a total culturable virus assay similar to the one described for drinking water (Section 13.4). Furthermore, the recovery of viruses in spiked biosolid samples remained variable.

Besides pathogens and heavy metals, another aspect of emerging concern is the presence of pharmaceutical and personal care products in treated waste water, biosolids, and animal manure (Ternes, 1998; Ingerslev and Halling-Sørensen, 2000, 2001). Their potential impact on micro- and macroorganisms in the environment is currently under study. At present, the degradability of these compounds

**Table 13.9**  Survival times of some excreted pathogens in soil and on crops (20-30°C)

| Pathogen | Survival time (days)[a] | |
|---|---|---|
| | In soil | On crops |
| Viruses | | |
| Enteroviruses (poliovirus, Coxsackie virus and echovirus) | 100 (<20) | 60 (<15) |
| Bacteria | | |
| Fecal coliforms | 70 (<20) | 30 (<15) |
| *Salmonella* spp. | 70 (<20) | 30 (<15) |
| *Vibrio cholerae* | 20 (<10) | 5 (<2) |
| Protozoa | | |
| *Entamoeba histolytica* (cysts) | 20 (<10) | 10 (<2) |
| Helminths | | |
| *Ascaris lumbricoides* (eggs) | Months | 60 (<30) |
| Hookworm (larvae) | 90 (<30) | 30 (<10) |
| *Taenia saginata* (eggs) | Months | 60 (<30) |
| *Trichuris trichiura* (eggs) | Months | 60 (<30) |

*Source:* Feachem et al. (1983) with permission from John Wiley & Sons Limited.
[a]Usual survival time shown in parentheses.

in waste water and biosolids is largely not known but *in-vitro* studies have shown that several therapeutic agents do exert adverse effects on both micro- and macroorganisms in the environment (Ariese et al., 2001; Jjemba, 2002a, b). Once introduced into the environment, the mobility and sorption of these compounds also greatly influence their availability for uptake by organisms. Although information about the behavior of these compounds in the environment is still lacking, several parallels can be predicted based on what is known from other well-researched organic compounds (i.e., pesticides, herbicides and fungicides).

# References

Angle S.S. (1994). Sewage sludge: Pathogen considerations. *In:* C.E. Clapp, W.E. Larson, and R. H. Dowdy (eds.). Sewage Sludge: Land Utilization and the Environment. SSSA Miscellaneous Publ. Madison WI.

Angus K.W. (1986). Survival of *Cryptosporidium* in excreta: Zoonotic aspects of infections and possible implications for spread by agricultural use of sewage sludge. *In:* J.C. Block, A.H. Havelaar, and P. L' Hermite (eds.). Epidemiological Studies of Risks Associated with the Agricultural Use of Sewage Sludge: Knowledge and Needs. Elsevier Sci. Publ., New York, NY, pp. 98-101.

APHA, AWWA, WEF (1995). Standard Methods for the Examination of Water and Wastewater. Amer. Public Health Assoc., Amer. Water Works Assoc., and Water Environ. Federation, Washington DC (19th ed.).

Ariese F., W.H.O. Ernst, and D.T.H.M. Sijm (2001). Natural and synthetic organic compounds in the environment-a symposium report. *Environ. Toxicol. Pharmacol.* **10:** 65-80.

Atmar R.L., F.H. Neill, J.L. Romalde, F. Le Guyader, C.M. Woodley, T.G. Metcalf, and M.K. Estes (1995). Detection of Norwalk virus and hepatitis A virus in shellfish tissues with the PCR. *Appl. Environ. Microbiol.* **61:** 3014-3018.

Atmar R.L., F.H. Neill, C.M. Woodley, R. Manger, G.S. Fout, W. Burkhart, L. Leja, E.R. McGovern, F. Le Guyader, T.G. Metcalf, and M.K. Estes (1996). Collaborative evaluation of a method for the detection of Norwalk virus in shellfish tissues by PCR. *Appl. Environ. Microbiol.* **62:** 254-258.

Baker K.H. (1998). Detection and occurrence of indicator organisms and pathogens. *Water Environ. Res.* **70:** 405-418.

Bitton G. (1994). Wastewater Microbiology. Wiley-Liss, Inc., New York, NY.

Boorman G.A., V. Dellarco, J.K. Dunnick, R.E. Chapi, S. Hunter, F. Hauchman, H. Gardner, M. Cox, and R.C. Sills (1999). Drinking water disinfection byproducts: review and approach to toxicity evaluation. *Environ. Health Persp.* **107(Suppl. 1):** 207-218.

Brierley C.L., J.A. Brierley and M.S. Davidson (1989). Applied microbial processes for metals recovery and removal from wastewater. *In:* T.J. Beveridge and R.J. Doyle (eds.). Metal Ions and Bacteria. Wiley Intersci, New York, NY, pp. 359-382.

Brown S.L., R.L. Chaney, C.A. Loyd, J.S. Angle, and J.A. Ryan (1996). Relative uptake of cadmium by garden vegetables and fruits grown on long-term biosolid-amended soils. *Environ. Sci. Techn.* **30:** 3508-3511.

Butcher S.S., R.J. Charlson, G.H. Orians, and G.V. Wolfe (eds.) (1992). Introduction. Global Biogeochemical Cycles. Acad. Press, New York, NY, pp. 1-7.

Craun G.F., S. Regli, R.M. Clark, R.J. Bull, J. Doull, W. Grabow, G.M. Marsh, D.A. Okun, M.D. Sobsey, and J.M. Symons (1994). Balancing chemical and microbial risks of drinking water disinfection, Part II. Managing the risk. *J. Water SRT-Aqua* **43:** 207-218.

Curds C.R. (1982). The ecology and role of protozoa in aerobic sewage-treatment processes. *Ann. Rev. Microbiol.* **36:** 27-46.

Davenport R.J., T.P. Curtis, M. Goodfellow, F.M. Stainsby, and M. Bingley (2000). Quantitative use of fluorescent in situ hybridization to examine relationships between mycolic acid-containing actinomycetes and foaming in activated sludge plants. *Appl. Environ. Microbiol.* **66:**1158-1166.

Falkinham III J.O., C.D. Norton, and M.W. LeChevallier (2001). Factors influencing numbers of *Mycobacterium avium, Mycobacterium intrecellulare,* and other mycobacteria in drinking water distibution systems. *Appl. Environ. Microbiol.* **67:** 1225-1231.

Feachem R.G., D.G. Bradely, H. Garelick, and D.D. Mara (1983). Sanitation and Disease: Health Aspects of Excreta and Wastewater Management. World Bank Studies in Water Supply and Sanitation, 3. John Wiley, Chichester, UK.

Fogg G.E. (1966). The extracellular products of algae. *Oceanog. Mar. Biol. Ann. Rev.* **4:** 195-?

Fourie A.B. and M.B. Van Ryneveld (1995). The fate in the subsurface of contaminants associated with on-site sanitation: A review. *Water S A (Pretoria)* **21:** 101-111.

Fout G.S., F.W. Schaefer III, J.W. Messer, D.R. Dahling, and R.E. Stetler (1996). ICR Microbial Laboratory Manual. US EPA Office Res. Develop., Cincinnati, OH.

Franceys R., J. Pickford, and R. Reed (1992). A Guide to the Development of On-site Sanitation. WHO, Geneva, Switzerland.

Gagliardi J.V. and J.S. Karns (2000). Leaching of *Escherichia coli* O157:H7 in diverse soils under various agricultural management practices. *Appl. Environ. Microbiol.* **66:** 877-883.

Girard P. and Hillaire M.C. (1997). Determining the source of nitrate pollution in the Niger discontinuous aquifers using the natural $^{15}$N/$^{14}$N ratios. *J. Hydrol. (Amsterdam)* **199:** 239-251.

Ingerslev F. and B. Halling-Sørensen (2000). Biodegradability properties of sulfonamides in activated sludge. *Environ. Toxicol. Chem.* **19:** 2467-2473.

Ingerslev F. and B. Halling-Sørensen (2001). Biodegradability of metronidazole, olaquindox, and tylosin and formation of tylosin degradation products in aerobic soil-manure slurries. *Ecotoxicol. Environ. Safety* **48:** 311-320.

Jjemba P.K. (2002a). The effect of chloroquine, quinacrine, and metronidazole on both soybean (*Glycine max*) plants and soil microbiota. *Chemosphere* **46:** 1019-1025.

Jjemba P.K. (2002b). The potential impact of veterinary and human therapeutic agents in manure and biosolids on vegetation: a review. *Agric., Ecosys. Environ.* **93:** 267-278.

Kelly P., S.K. Baboo, P. Ndubani, M. Nchito, N.P. Okeowo, N.P. Luo, R.A. Feldman, and M.J.G. Fathing (1997). Cryptosporidiosis in adults in Lusaka, Zambia, and its relationship to oocsyt contamination of drinking water. *J. Infect. Dis.* **176:** 1120-1123.

LeChevallier M.W., M. Abbaszadegan, A.K. Camper, C.J. Hurst, G. Izaguirre, M.M. Marshall, D. Naumoritz, P. Payment, E.W. Rice, J. Rose, S. Schaub, T.R. Slifko, B.D. Smith, H.W. Smith, C.R. Sterling, and M. Stewart (1999a). Emerging pathogens—viruses, protozoa and algal toxins. *J. AWWA* **91:** 110-121.

LeChevallier M.W., M. Abbaszadegan, A.K. Camper, C.J. Hurst, G. Izaguirre, M.M. Marshall, D. Naumoritz, P. Payment, E.W. Rice, J. Rose, S. Schaub, T.R. Slifko, B.D. Smith, H.W. Smith, C.R. Sterling, and M. Stewart (1999b). Emerging pathogens—bacteria. *J. AWWA* **91:** 101-109.

Leclerc H., S. Edberg, V. Pierzo, and J.M. Delattre (2000). Bacteriophages as indicators of enteric viruses and public health risk in groundwaters. *J. Appl. Microbiol.* **88:** 5-21.

Lester J.N. and J.W. Birkett (1999). Microbiology and Chemistry for Environmental Scientists and Engineers. E & FN Spon, London, UK.

Lewis-Jones R. and M. Winkler (1991). Sludge Parasites and other Pathogens. Ellis Horwood, New York, NY.

Mara D. and S. Cairncross (1989). Guidelines for the Safe Use of Wastewater and Excreta in Agriculture and Aquiculture. WHO, Geneva, Switzerland.

McBride M.B. (1995). Toxic metal accumulation from agricultural use of sludge: Are US EPA regulations protective? *J. Environ. Qual.* **24:** 5-18.

Orangui J.I., T.P. Curtis, S.A. Silva, and D.D. Mara (1987). The removal of excreted bacteria and viruses in deep waste stabilization ponds in northeast Brazil. *Water Sci. Techn.* **19:** 569-573.

Paralkar A. and J.K. Edzwald (1996). Effect of ozone on EOM and coagulation. *J. AWWA* **88:** 143-154.

Pell A. (1997). Manure and microbes: public and animal health problem? *Diary Sci.* **80:** 2673-2681.

Perez-Rosas N. and T.C. Hazen (1989). In situ survival of *Vibrio cholerae* and *Escherichia coli* in a tropical rain forest watershed. *Appl. Environ. Microbiol.* **55:** 495-499.

Rose J.B. (1997). Environmental ecology of *Cryptosporidium* and public health implications. *Ann. Rev. Public Health* **18:** 135-161.

Scandura J.E. and M.D. Sobsey (1997). Viral and bacterial contamination of groundwater from on-site sewage treatment systems. *Water Sci. Techn.* **35:** 141-146.

Schaefer F.W. (2001). Can we believe our results? *In:* M. Smith and K.C. Thompson (eds.). *Cryptosporidium*: The

Analytical Challenge. Roy. Soc. Chem., Cambridge, UK, pp. 155-161.

Smith P., C. Carroll, B. Wilkins, P. Johnson, S.N. Gabhainn, and L.P. Smith (1999). The effect of wind speed and direction on the distribution of sewage-associated bacteria. *Lett. Appl. Microbiol.* **28:** 184-188.

Smith S.R. (1996). Agricultural Recycling of Sewage Sludge and the Environment. CAB Intnatl., Wallingford, UK, 382 pp.

Spaepen, K.R.I., L.J.J. Van Leemput, P.G. Wislocki, and C. Verschueren (1997). A uniform procedure to estimate the predicted environmental concentration of the residues of veterinary medicines in soil. *Environ. Toxicol. Chem.* **16:** 1977-1082.

Sterritt R.M. and J.N. Lester (1988). Microbiology for Environmental and Public Health Engineers. Spon, London, UK.

Taylor R.H., J.O. Falkinham III, C.D. Norton, and M.W. LeChaveallier (2000). Chlorine, chloramine, chlorine dioxide, and ozone susceptibility of *Mycobacterium avium. Appl. Environ. Microbiol.* **66:** 1702-1705.

Ternes T.A. (1998). Occurrence of drugs in German sewage treatment plants and rivers. *Water Res.* **32:** 3245-3260.

Tsintozou A., A. Vantarakis, O. Pagonopoulou, A. Athanassiadou, and M. Papapetropoulou (2000). Environmental mycobacteria in drinking water before and after replacement of the water distribution network. *Water, Air, Soil Pollut.* **120:** 272-282.

US EPA (1992). Environmental Regulations and Technology. Control of Pathogens and Vector Attraction in Sewage Sludge. Report No. EPA/625/R-92/013. Office Res. Develop., Washington DC.

# 14

# Medical Environmental Microbiology

## 14.1 ECOLOGY OF INFECTIOUS AGENTS AND PROGRESSION OF DISEASES

Within the microbial world, relatively few species are pathogens. Most microorganisms carry out essential biogeochemical cycling processes (Chapter 9) or are closely associated with plants and animals in stable, beneficial relationships (Chapter 8). Many infectious disease agents live in the natural environment where part or all of their life cycle occurs. The natural environment in this instance also includes various parts of the human body (both internal and external) where resident microflora, once established, remains quite stable and is routinely found. This stability is referred to as microbial homeostasis and arises from various intermicrobial and microbial-host interactions. The environmental phase in which the pathogen exists dictates the mode of transmission that the pathogen exploits to complete its life cycle.

In cases of pathogenic microorganisms, it is important to understand the causative agent(s) and the symptoms associated with the disease they cause. Conventional medical microbiology has, for generations, focused on the identification of causative agents, occurrence, prevalence, and treatment of diseases in subjects with a limited focus on the ecology of the disease and how the disease-causing agents behave in relation to prevailing environments. Some infectious diseases occur in a cyclic manner, suggesting that environmental factors play a role in their incidence. Furthermore, seasonal trends of disease infection have also been documented in folk medicine for centuries but, until recently, not much scientific attention was paid to this fact and so systematic understanding of the prevalence and progression of diseases was lacking. Now it is known that changes in environmental conditions affect the timing and severity of disease outbreaks. Scientists from various disciplines, including meteorology, epidemiology, microbiology, medicine, and oceanography have recently collaborated to study how environmental patterns affect the occurrence and prevalence of pathogenic microorganisms as well as vectors (NRC, 2001). This approach is generating more and more opportunities for understanding infectious diseases and their environmental links.

Realization that all infections involve a pathogen, host(s), and the environment is of paramount importance. Both emergence and survival of pathogens in the environment are influenced by several environmental factors, including salinity, type of microorganism, temperature, sediments, nutrients, antagonistic factors, light, and dissolved oxygen. Effective control strategies require a clear understanding of the life cycle and mode of transmission of the disease. Some pathogens require a vector to complete their life cycle. For pathogens to ultimately cause disease, interactions between other existing microorganisms and the host are cardinal.

This chapter highlights the relationships between some microorganisms of health significance and environmental factors that facilitate their prevalence, intensity of disease outbreaks, and survival of the pathogen between outbreaks. They enable integration of the principles of microbiology with environmental factors for better understanding the occurrence and progression of infectious diseases. This implicitly necessitates a clear understanding of the life cycle of the pathogen and the how and why those pathogens survive and proliferate in particular ecosystems. These life cycles are directly or indirectly affected by environmental factors, which in turn requires an assessment of the mode of transmission and the alternative hosts of the pathogen in question, including host-parasite interactions, nutritional patterns of the pathogen, as well as its replication and growth. This further demands knowledge of the biochemistry and immunological aspects of disease development. Once clearly identified, the key environmental factors can be manipulated to provide prophylactic and curative strategies. Environmental (and climatic) changes at the local, regional, and global scale are also under scrutiny. How do these changes affect pathogens and their vectors? Without doubt, some of these changes enable pathogens or vectors to proliferate in new areas.

Based on mode of transmission, the relevant infectious diseases in environmental microbiology can be categorized in three distinct groups, viz., airborne diseases that infect through the respiratory tract, vector-transmitted, and food and waterborne diseases transmitted through the mouth. Examples of typical diseases of environmental consequence from each category or mode of transmission and the associated causative agents are listed in Table 14.1. The mode of transmission enables us to examine the common attributes that influence the occurrence and spread of the infection. However, categorization based on the mode of transmission and/or the portals of entry is blurred because some diseases are transmitted through more than one mode. For example, tuberculosis and poliomyelitis are airborne but also transmitted through tainted water. Furthermore, certain microorganisms considered vector-borne often rely on an intimate relationship with another organism. For example, *Schistosoma* and *Vibrio cholera* are intimately linked with snails and copepods respectively, although they are considered waterborne. Thus, categorization based on mode of transmission should be considered arbitrary.

Interest in climate-disease links has likewise increased because the patterns are naturally affected by changes in temperature, moisture, humidity, and a variety of other environmental conditions. For example, in some tropical countries the incidence of malaria and other infectious diseases such as Rift Valley fever (Linthicum et al., 1999), cholera (Pascual et al., 2000), and encephalitis (Reisen et al., 1993) appear to peak right after the El Niño weather phenomenon due to associated flooding. The use of climate and weather patterns can increase the sensitivity of surveillance programs by lowering the threshold of disease detection and increasing the intervention response. Thus, in some instances diseases have been predicted by combining both the long- and short-range environmental components such as changes in vegetation and vector populations. To be certain of the impending disease prevalence, the vectors have to be tested for the presence of the infectious agent (Eldridge, 1987). Some of these diseases and their environmental linkages are further discussed below.

## 14.2 VECTOR-BORNE DISEASES

Vector-borne diseases are those in which a host or vector carries the pathogen for part of the

**Table 14.1** Some infectious diseases and the environmental conditions with which they are mostly associated

| Category | Disease | Causative agent | Favorable environmental condition and other remarks |
|---|---|---|---|
| Airborne | Pneumonia | *Streptococcus pneumoniae* (G+) | Cold; damp |
| | Tuberculosis | *Mycobacterium tuberculosis* (G+) | A reemerging disease of major concern |
| | Influenza | Influenza virus | Cold and humid that favor survival of the aerosolized virus |
| | Meningococcal meningitis | *Nesseira meningitidis* | Dry/dust storms |
| | Pneumocystis carinii pneumonia | *Pneumocystis carinii* | Of major concern in immunocompromized patients |
| Vector-borne | Hantavirus pulmonary disease (HPS) | Sin Nombre virus | Damp conditions favorable to the deer mouse (*Peromyscus maniculatus*) vector |
| | Lyme disease | *Borrelia burgdcrferi* | Warm and moist favorable to tick (*Ixodes dammini*) vectors |
| | Bubonic plague | *Yersinia pestis* (G-) | Warm, dusty conditions favorable to flea vectors |
| | Malaria | *Plasmodium vivax* | Warm and wet coupled with stagnant water that serves as a breeding ground for the *Anopheles* mosquito |
| | Dengue fever | Dengue virus | Warm, humid and wet conditions favorable for the vector *Aedes aegypti* mosquito. The eggs of the vector are killed by freezing environmental conditions whereas adults are restricted below 10°C (Epstein, 2001). |
| | Rift valley fever | Rift valley fever virus (RVF) | Warm and damp conditions favorable for the *Culex* mosquito |
| | St. Louis encephalitis | St. Louis encephalitis virus (SLE) | Flooding followed by dry warm conditions (stagnant water) that favors *Culex* mosquitoes |
| | Trypanasomiasis | *Trypanasoma gambiense* | Warm and wet conditions favorable to Tsetse flies |
| Food and water-borne | Staphylococcus food poisoning | *Staphylococcus aures* (G+) | Their detection using culture-based techniques is difficult because they sometimes form tiny variant colonies which escape detection. They are a major concern in antibiotic resistance transfer causing nosocomial infections. |
| | Aflotoxin poisoning | *Aspergillus flavus* (a fungus) | The most potent toxin known to humans |
| | Amoebic dysentery | *Entamoeba histolytica* | Present in aquatic environments exposed to fecal material. |
| | Cholera | *Vibrio cholera* (G–) | Flooding |
| | Cryptosporidiosis | *Cryptosporidium* sp. | Wet/flooding. Oocysts can survive for long durations in moist environment but survival is threatened by elevated temperature (e.g. under composting), as well as pH and temperature extremes. |
| | Giardiasis | *Giardia lamblia* | Wet/flooding. Forms cysts whose survival in the environment is assumed to be similar to that of *Cryptosporidium* oocysts. |

*Table 14.1 contd.*

*Table 14.1 contd.*

| | | |
|---|---|---|
| Hepatitis A | Hepatitis RNA virus | Persistent in the environment and stable at pH extremes (pH 1 to 11.5). Hepatitis A infections in temperate regions are associated with seasonal variations which have not yet been fully explained. |
| Poliomyelitis | Poliovirus | Routinely encountered in waste water and sewage. It can survive long periods of desiccation but appears less persistent than Norwalk and Hepatitis A. |
| Gastroenteritis | Norwalk and Norwalk-like viruses | Predominantly introduced in the environment in sewage and waste water. Persistent in the environment. |
| Shigellosis | *Shigella* sp. (G-) | Usually occurs in brackish waters but can undergoes a viable but nonculturable state that is still infectious. |
| Typhoid fever | *Salmonella typhinurum* (G-) | Occurs in environments under the influence of fecal matter. Persistent in the environment and can undergo viable but nonculturable state that is still infectious. |

pathogen's life cycle. Typical vectors are arthropods and rodents and although macro in themselves, their coexistence with microscopic infectious agents comes into the limelight in environmental microbiology. Specific examples of these diseases are listed in Table 14.1. The rate at which vector-borne pathogens multiply, as with most other pathogens, depends on ambient temperatures. Thus, if the vector is cold blooded, replication within it can be greatly inhibited when ambient temperatures drop below a certain threshold, dramatically reducing disease incidence (Gillet, 1974).

## 14.2.1 Malaria

In human history, malaria is responsible for the highest mortality rate. It is estimated that of all the people who have died, half have died of malaria. Current cases of malaria worldwide are estimated at about 300-500 million clinical cases per annum, of which about one million each year translates into approximately 1% of deaths per year (NRC, 2001). Malarial ailments go back to ancient times and their symptoms have been recorded in ancient Roman and Greek writings in which, at such an early time, malaria was associated with swamps and marshes. In 1880, a French military surgeon, Charles Laveran, while stationed in Algiers discovered its causative agent to be *Plasmodium vivax,* a pathogenic protozoan. Nine years later, another military surgeon, Ronald Ross, while stationed in Myanmar (formerly Burma) discovered that the protozoan parasite is carried by *Anopheles* mosquitoes. For their discoveries, each won the Nobel Prize in 1902 and 1907 respectively. Since then, other species of *Plasmodium* such as *P. falciparum, P. malarie* and *P. ovale* have been found to cause malaria of different severity. The infection cycle of the parasite and subsequent progression of the disease are shown in Fig. 14.1 and involves distinct events in both the mosquito vector and in humans. Specifically, both fertilization of the plasmodium and growth of sporozoites occur within the mosquito. On biting a subject (hu-

man), the mosquito transmits the sporozoites into the bloodstream. These accumulate in the liver, multiply asexually, and are released into the bloodstream again where they infect the red blood cells (rbcs). In the rbcs, they reproduce further, rupture the cells, and invade even more cells. The periodic rupture and release of merozoites into the bloodstream cause a cycle of symptoms in the host ranging from chills, headache, fever, and sweats. When the mosquito bites the infected person again, the gametes of the pathogen enter its stomach and its sexual cycle begins again. Symptoms include fever, chills during repeated release from the rbcs, anemia, and an enlarged spleen. Quinine and other drugs are often ineffective because the sporozoites escape treatment within the liver.

The previous malaria eradication programs of the 1950s to 1970s focused on two approaches: (i) use of chloroquine to eliminate human bloodstream infection and (ii) spraying DDT to kill the mosquitos. The eradication program succeeded only in North America, Europe, and some parts of the near East and Far East Asia mostly because of economic development and better housing as well as better access to medical care. The eradication program totally ignored sub-Saharan Africa where malaria is most endemic. Furthermore, in such places as Latin America and parts of Asia, application of both chloroquine and DDT selected for chloroquine-resistant parasites and DDT-resistant mosquitoes. In some of these areas, malaria has surged dramatically. Eradication of *Plasmodium* vectors from their natural habitats (forests and grasslands) in the tropics, is seemingly impracticable.

A conventional approach has been to eradicate the mosquito vectors by destroying their breeding grounds and adults so as to disrupt the cycle. However, this approach has been of limited success since in wet areas it is impossible to eliminate stagnant water, a potential breeding ground for mosquitoes. Stagnant water occurs in a variety of areas, especially in the tropics that receive frequent rain. Even small amounts of water such as on plant litter, can act as a breeding ground for mosquitoes. Canals used in the irrigation of dry areas also provide breeding grounds for the mosquitoes. Measures to control the vectors have in the past particularly involved intensive spraying in endemic areas with fairly persistent compounds such as DDT. Over time, mosquitoes resistant to these pesticides of choice have emerged. A more recent approach has been to develop a vaccine against malaria (Belperron et al., 1999).

Successful control of the disease requires a clear understanding of the growth cycle of the causative agent. The protozoan parasite, *P. vivax*, produces sexual stages in human blood but fertilization occurs only after the gametes are taken up by the blood-feeding mosquito (*Anopheles* sp.). Thus, the mosquito serves as an intermediate host and this adaptation in itself avoids hyper-infestation as it precludes completion of the entire generation in a single host. The stages occurring in man are independent of the mosquito and vice versa. After fertilization, the sporozoites develop inside the mosquito intestine and lyse. Transmission to other humans is effected when the sporozoite-containing mosquito bites healthy individuals for a blood meal (Fig. 14.1). Transmission of the disease is effectively optimized between 20-30°C and relative humidity greater than 60%; *Anopheles* mosquito vectors occur when temperatures exceed 22°C (Epstein, 2001). These environmental conditions are ideal for mosquitoes and favor amplification of the infectious agents within the mosquito host (Reeves et al., 1994). Thus mosquito population sizes which depend on the availability of water and temperature can be predicted to minimize human infections (Rose et al., 2001).

The sporozoites of the female anopheles are injected in humans through mosquito bites. They travel through the bloodstream to the liver where they mature into schizonts. Schizonts are released within 8 - 25 days into the bloodstream as merozoites and produce symptomatic infections as they invade and destroy the red blood cells. Each merozoite is capable of infecting an rbc and initiating a new cycle of asexual reproduction. Some merozoites re-invade the liver and cause relapse fevers 6-11 months later when they mature and release more merozoites. Some merozoites, on entering the rbcs, mature even further into erythrocytic schizonts and lyse their host rbcs as they complete maturation, releasing the next generation of merozoites which invade previously uninfected rbcs. Within the rbcs, some of the secondary merozoites differentiate to sexual forms, i.e., microgametocytes (♂) and macrogametocytes (♀). What factors trigger sexual differentiation are not known. When the female anopheles bites an infected individual, it takes up both macro- and microgametocytes which fertilize each other to form a zygote. This is the only diploid stage during the cycle. Shortly after fertilization, the zygote undergoes meiotic division to form a haploid, invades the gut of the mosquito, and develops an oocyst (ookinate) on the outer layer of the stomach wall. The ookinate penetrates the wall and matures to produce sporozoites which migrate to the salivary gland of the mosquito and the cycle is repeated. It is important to note that the absence of the right type of mosquito vector, i.e., a female anopheles mosquito, prevents transmission of the gametocytes. However, transmission from an infected human to a healthy individual can still occur, for example, through blood transfusion from infected donors or through the use of nonsterile contaminated needles (Tsai and Krogstad, 1998).

Most of the work to understand malaria has focused on preventing infection and not on preventing the disease from advancing once an individual is infected. Furthermore, resistance of malarial parasites to the drugs of choice is increasing (Tsai and Krogstad, 1998). How is the incidence of this disease going to change in light of increasing resistance of its vectors to

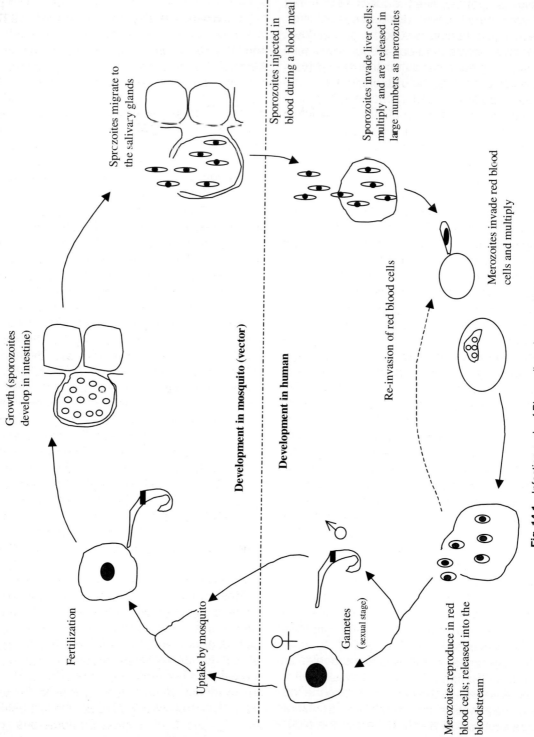

**Fig. 14.1** Infection cycle of *Plasmodium vivax* causing malaria in humans.

pesticides? Development of *Plasmodium* is regulated by the host's body temperature which normally fluctuates over a 24-hour period. Thus the parasite seems to release its merozoites in 24 hour multiples, a trend that seems to display a classical but intriguing biological clock. The parasite seems to carefully time its gametocytes to mature at night, a time when mosquitoes are on a feeding spree and hence are in a better position to transmit the parasite to a new host. Development of an *in-vitro* culture system for *P. falciparum* permitted the application of advances in molecular and cell biology to research its control. With the current advances in molecular biology, possibly it is worth attempting to change some of the physiological aspects of the mosquito to prevent fertilization of the gametes.

The linkage between mosquito-borne diseases and environmental conditions has also been long known and pertains to both increase in number of vectors (i.e., mosquitoes) and amplification of the infectious agent within the vector during wet warm months (Reisen et al., 1993; Reeves et al., 1994). *Plasmodium falciparum* matures in 26 days at 20°C (68°F) compared to just 13 days at 25°C (77°F). Furthermore, the mosquito vectors tend to bite more frequently at higher temperatures (Epstein, 2000). After amplification of the pathogen in the vector exceeds a threshold, transmission into humans occurs. Thus, environmental (climatic) data are increasingly being used to predict disease outbreaks, lowering the threshold, and increasing the time available to put intervention plans in place (Linthicum et al., 1999; Rose et al., 2001).

## 14.2.2 Plagues

Other infectious diseases which have been of significance in world history include bubonic plague caused by *Yersinia pestis*, a Gram-negative, nonspore-forming, nonmotile facultative anaerobic intracellular pathogen. The pathogen belongs to the Enterobacteriaceae family and grows optimally at 28-30°C and pH 7.2-7.6, although it can tolerate 4-40°C and pH 5-9.6 extremes (Perry and Fetherston, 1997). Two other *Yersinia* spp., *Y. pseudotuberculosis* and *Y. enterocolitica*, are also pathogenic to humans, causing pneumonic plague. From a historical perspective, bubonic plague can be traced as far back as A.D. 541 to A.D. 544 in the Middle East (Perry and Fetherston, 1997). The infection spread to various parts of North Africa, Europe, as well as central and southern Asia, amounting to the first known plague pandemic. It caused massive depopulation, although the exact numbers are not known with certitude. The second pandemic is believed to have started from central Asia in the fourteenth century, spreading westward along trade routes into Europe. It climaxed around the period A.D. 1347-1351 and is estimated to have claimed 17-28 million people in Europe alone. This pandemic is also the first well-documented report of biological warfare by humans. The army of Genghis Khan was besieged by The Tartars in the Crimean city of Kaffa (presently called Feodosiya in southern Ukraine) on the shores of the Black Sea. The siege created crowding within the city and exposed the troops to the plague. To break loose, Khan's army catapulted plague-ridden cadavers to their enemy over the city walls and managed to escape to Italy but also inadvertently spread the *Y. pestis*-infected *fleas.* This movement brought the fleas, and therefore the disease, into contact with one of the most dangerous rodent reservoirs, the black rat (*Rattus rattus*), already living in fairly large numbers close to humans in cities, thereby accelerating the spread of the plague into several cities in western Europe. Over a two-year period (1348-1350), the disease killed approximately 25% of the inhabitants of Europe at that time (Odum, 1989). In the second half of the fourteenth century, many cities in Europe lost 50% or more of their inhabitants due to the disease and the later population explosion of that continent can largely be traced to the successful control of the plague. On a positive note, this pandemic stimulated numerous policies in medical education, public health regulations, and the development of hospitals with the notion of treating the sick rather than just isolating them (NRC, 2001). The third pandemic is believed to have started in 1855 and, although declining, still continues, posing a danger to public health (Garrett, 2000; Perry and Fetherston, 1997).

Bubonic plague may seem a disease of the past. However, it raged through the Indian subcontinent as recently as the late 1980s to the mid-1990s (Garrett, 2000). It is still one of the most feared infectious diseases and has reservoirs on every continent. Most of the rodents die of the plague after infection. However, a low proportion of them develop and maintain a balanced chronic state after infection, acting as reservoirs for disease transmission to humans via fleas after the pathogen has multiplied in the latter's midgut (Fig. 14.2). More than 1,500 species of fleas have been identified as vectors but some of them have a slightly acidic (pH 6-7) midgut that precludes multiplication of the pathogen. Most epidemics have been associated with the oriental rat flea (*Xenopsylla cheopis*) but several other species have been implicated (Table 14.2). During a normal blood-feeding episode, the flea is able to prevent regurgitation of its blood meal in a structure that separates the midgut and the esophagus called the proventriculus. However, if the blood meal contains the pathogen, the flea midgut exhibits cohesive dark brown masses that contain the pathogens which extend into the proventriculus within two days after infection. The dark brown mass subsequently blocks the proventriculus and prevents further blood meals from reaching the stomach. In a frantic effort to survive from starvation and dehydration, the now infected feeding flea shifts its diet to warm-blooded animals such as rats (*Rattus rattus*) and humans. Through this feeding strategy, the flea spreads the pathogen to human beings.

On infecting of the human, *Y. pestis* travels in the blood to the lymph nodes where it causes inflammatory swellings (buboes). Inside the monocytes, the pathogen forms a distinct capsule around itself to prevent phagocytosis, enabling it to grow intracellularly. Capsule formation has been reported to occur at 37°C but not at 26°C (Perry and Fertherston, 1997), indicating tremendous adaptation of the pathogen to warm-blooded animals for its survival and disease progression. These two temperatures represent those encountered within warm-blooded mammals and fleas respectively. Indeed, experiments with mutants unable to form capsules have shown a reduced virulence of *Y. pestis* in some animal models. The pathogen eventually spreads into the bloodstream causing septicemia (an exotoxin is produced which inhibits respiration in the mitochondria). As the disease progresses, secondary buboes are formed in the peripheral lymph nodes and essential organs such as the liver, lungs, and spleen, become infected. Secondary infection also leads to the development of pneumonic plague which can be spread in respiratory droplets (aerosols). Symptoms include pain, delirium, shock and then death in five days.

The pathogen can survive in a dormant state in soil for an extended period. It can also hibernate in the guts of fleas without causing them any apparent damage, i.e., as a benign commensal. This relationship changes, however, when the bacterium switches on its hemin-storage (*hms*) gene, secreting some proteins that infect and block the vector flea's foregut (Hinnebusch et al., 1996). The physiological mechanism causing the blockage is not known but in response to this attack, as mentioned above, the infected flea, in what appears to be a last effort to avoid starvation, aggressively bites warm-blooded animals. It has been speculated that switching on the hms proteins enables the pathogen to use the stored hemin as a nutritional source of iron (Perry and Fetherston, 1997). The bacteria directly deliver their toxic proteins (Yop for Yersinia outer proteins) into the eukaryotic cell cytosol, a mechanism referred to as "type III". The type III secretion protein, which is also present in other

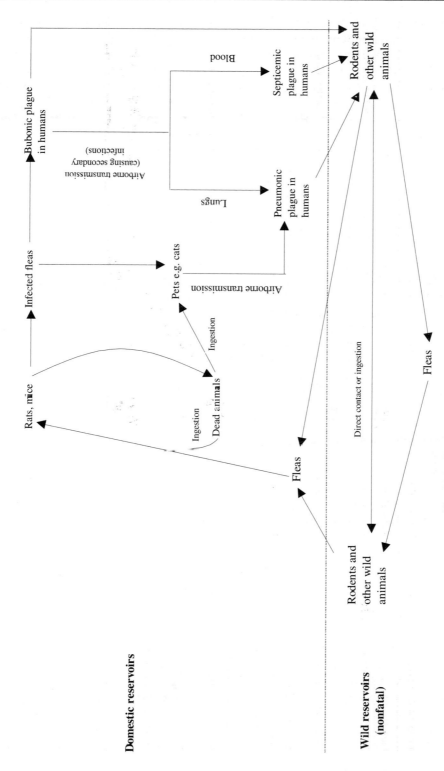

**Domestic reservoirs**

**Wild reservoirs (nonfatal)**

**Fig. 14.2** Transmission of *Yersinia pestis* to humans.

**Table 14.2** Several fleas species that have been implicated in the transmission of *Yersinia pestis*

| Flea species | Typical host | Common habitat and other remarks |
|---|---|---|
| *Xenopsylla cheopis* (Oriental rat flea) | *Rattus rattus* | Prefers moderately warm, moist climates but is distributed worldwide. Generally infests burrow-dwelling rats in urban areas; ingests 0.C3-0.5 µL blood. It is the most important vector for *Y. pestis*. |
| *Oropsylla montanus* (Rock squirrel flea) | Rock squirrels | Their role as vectors is variable. Active during sumrrer and fall months which correlates with subsidence of squirrel epizootics. |
| *Malaraeus telchinum* | Mice | Found in California meadows. Not very significant vectors. |
| *Nosopsyllus fasciatus* | Domestic rats | Adapted to damp, cool conditions. |
| *Xenopsylla brasiliensis* | ? | Occurs in Africa, India, and S. America |
| *Xenopsylla astia* | ? | In Indonesia and SE Asia |
| *Xenopsylla vexabilis* | ? | Pacific Islands |
| *Pulex irritans* (Human flea) | ? | Erroneously referred to as the human flea but has ε wide host range. It is a moderately poor vector. |
| *Ctenocephalides tesquorum* | ? | Important in the former USSR |
| *Oropsylla silantiewi* | ? | Important in the former USSR |
| *Rhadinopsylla ventricosa* | ? | important in the former USSR |
| *Ctenocephalides felis* (Cat flea) | Cats | Very poor vectors but isolated cases of transmission do occur. Very similar in appearance to the dog flea (*C. cauis*). |

Compiled from Perry and Fetherston, (1997).

pathogens such as *Salmonella* spp., *Shigella* spp., *Pseudomonas aeruginosa*, *Chlamydia psittaci*, and enteropathogenic *E. coli* is specifically triggered by contact with host cells and delivers its toxin directly into the host cell, thus avoiding the latter's defense mechanism. Once inside the cells of warm-blooded animals, the proteins injected by the pathogen incapacitate the cell and block the signaling and phagocytotic capacity of the immune system (Barinaga, 1996). The environmental factors that trigger off the *hms* genes are not known, but once identified, they could serve as an early warning system for plague in endemic areas. *In-vitro* studies have shown that deprivation of $Ca^{2+}$ ions leads to cessation of growth in *Yersinia* and triggers secretion of the Yop proteins (Cornelis, 1998). The protein is only secreted when the bacterium adheres to living target eukaryotic cells.

On the whole, the host-parasite relationship requires the parasite (*Yersinia* spp.) to adapt to minimize disease symptoms in order to ensure its survival and continued dissemination. To attain this, *Yersinia* spp. have alternative hosts such as lice, ticks, and enzootic mammalian hosts (e.g. domestic dogs, ferrets, coyotes, skunks, racoons, bears, goats, camels, etc.) that are more resistant to plague (Perry and Fetherston, 1997). More than 200 mammalian species have been reported to be naturally infected with *Y. pestis* as hosts or carriers but rodents are the most important hosts for the long-term survival of this pathogen. This gives *Yersinia* spp. the capacity to survive in the environment. Currently, the role of extrinsic environmental factors such as humidity and temperature on the transmission of *Y. pestis* is not clearly understood and still remains a challenge for environmental microbiologists and public health professionals. However, plague epidemics have been observed to subside under conditions of low humidity and high temperatures, possibly due to a disruption in the life cycle of the fleas. Blockage of the proventriculus is substantially reduced and can, in some instances

clear at higher temperatures (>28-30°C) (Perry and Fetherston, 1997). This indicates that environmental temperatures have an important role in plague outbreaks. Transmission of the pathogen from infected fleas to mammals is rare in the absence of a blocked proventriculus.

In the Northern hemisphere temperate regions, the disease typically occurs in late spring through the summer months (i.e., May to September). Diagnosis is mainly by serology, with the presence of anti-F1 antibody being the gold standard for routine identification of *Yersinia* spp. (Perry and Fetherston, 1997). From the clinical perspective, prevention and control are attained by prophylactic doses of antibiotics such as streptomycin, gentamicin, tetracycline, oxytetracycline, chloramphenicol and/or doxycycline (Perry and Fetherston, 1997). With quick diagnosis, these antibiotics are quite effective and antibiotic-resistant *Yersinia* strains are still rare. Vaccination with either a live attenuated (Pgm⁻) or formalin-fixed virulent *Y. pestis* strain is also effective for high risk areas. However, the antibodies wear off quickly, requiring periodic booster shots at 1-2 year intervals. Furthermore, vaccination does not offer protection against the pneumonic form of plague. From an environmental perspective, measures that make residential areas less appealing to rodents significantly reduce the incidence of plague in humans. This approach, coupled with active surveillance in endemic areas greatly control potential outbreaks. It is difficult to differentiate between pathogenic and nonpathogenic strains of *Yersinia* spp., and no specific procedure has been described for detecting this pathogen in water. Methods used for detection in food might prove applicable to water analysis.

## 14.3 FOOD- AND WATERBORNE DISEASES

Major advances in the treatment of water commenced at the turn of the twentieth century mostly through the disinfection of water with chlorine. This measure dramatically reduced waterborne diseases such as cholera and typhoid, particularly in developed countries. In the mid-1940s, primary treatment of sewage emerged followed by secondary treatment in the 1970s. Despite these advances, waterborne diseases are still of major concern in public health and environmental microbiology, affecting food supplies through the contamination of seafood and irrigated plants. Thus, the majority of waterborne diseases originate from use of water contaminated with human and animal feces. Such uses may range from directly drinking the tainted water to consuming plants that have been washed or irrigated with water that has not been adequately treated. Other uses include recreational activities. So many disease-causing microorganisms are waterborne and waterborne infections are global and comprise over half the world's patients at any given time. Inactivation or die-off rates of pathogens are higher in waters of greater salinity as coliforms survive poorly in marine waters, which contributes to their being inadequate indicators of overall pathogens. In assessing the efficacy of water and wastewater treatment regimens the focus has usually been on monitoring culturable indicator organisms. However, a range of microorganisms associated with diseases, e.g. *Cryptosporidium, Giardia, Mycobacterium*, hepatitis A virus, and *Legionella*, are resistant to routine disinfection (Rose, 1997; Smith and Rose, 1998). This paradox has important implications in the long-term survival and persistence of these pathogens in the environment, which translates into a continued threat to public health.

The survival of bacteria and viruses has historically been measured by cultivation techniques. However, as outlined in Chapter 6, this approach indubitably underestimates the abundance of microorganisms in the environment. As outlined in Chapter 13, fecal coliforms, especially *E. coli*, are widely used as indicator organisms in environmental microbiology. However, the die-off rates for many pathogens

in the environment are slower than for *E. coli*. Furthermore, emerging pathogens such as enteropathogenic *E. coli* (i.e., O157, O55:H7, and O111:NM), two decades after their initial characterization and identification, have been associated with severe diarrhetic infections with increasing frequency (Powell et al., 2000). No biochemical marker is available for differentiation between pathogenic and nonpathogenic strains of *E. coli*. *E. coli* O157:H7 was able to survive on vegetables after the vegetables were fertilized with cattle manure, and was fatally transmitted to a 2-year-old and his sibling in Maine. *E. coli* O157:H7 outbreaks linked to apple cider have also been suspected to have originated from the application of contaminated manure in the apple orchard (Pell, 1997). This strain causes hemolytic uremic syndrome (HUS), a rare kidney disorder whose symptoms include hematic diarrhea, followed by renal failure. The organism is particularly dangerous for children whose immune system is not fully developed, as well as the elderly and immunocompromised individuals. From an ecological perspective, the organism is able to proliferate in the rumen and is prevalent in about 5% of dairy cows in temperate regions (Pell, 1997). It can survive adverse environmental conditions such as low temperatures (<8°C), pH lower than 4 and high levels of salt (Clavero and Beuchat, 1996).

Of increasing concern is the prevalence in rivers/marine environments of bacteria resistant to the various antibiotics in use. Whereas a number of bacteria may not necessarily be of direct medical concern, there is fear of them transferring the drug-resistance properties to germs that can still be cured with antibiotics (Isaacson and Torrence, 2002). When water containing these drug-resistant nonpathogenic microorganisms is used for recreational purposes, the resistant bacteria can get into the human gut and transfer such resistance to pathogens of human concern. For example, the increased prevalence of vancomycin-resistant enterococci in humans has been linked to the widespread use of avoparcin, a vancomycin analogue widely used in livestock (poultry) production as a growth promoter. Alternatively, resistance can be acquired directly by opportunistic pathogens present in (waste) water, rendering the infections difficult to treat. Vancomycin resistance has been shown to persist in exposed livestock (poultry) and individuals for more than three years after terminating its use, indicating that resistance to some antibiotics is quite stable even in a nonselective environment (Borgen et al., 2000). Vancomycin-resistant enterococci are an important cause of nosocomial infections worldwide (Harbarth et al., 2001) and have been isolated from a variety of sources, including farm animals, contaminated water, and sewage. A recent study in Norway isolated vancomycin-resistant enterococci in 98.6% and 17.8% of the poultry- and farmer-derived fecal samples respectively from avoparcin-exposed farms (Table 14.3). All of the isolated vancomycin-resistant strains exhibited a high level of vancomycin resistance [MIC $\geq$ 256 mg L$^{-1}$] and possessed the *vanA* gene for resistance (Borgen et al., 2000). Resistance to this antibiotic was found in only 1.4% and 11.8% of farmer- and poultry-derived fecal material respectively from unexposed farms. The *vanA* gene is often located in a transposon readily transferred to other enterococci and even across other species, e.g., *Staphylococcus* sp. (Acar et al., 2000). Staphylococcus infections constitute the majority of nosocomial infections in hospitals. Other resistance genes, notably *vanB* and *vanD*, have also been reported and their mode of transfer is somewhat similar to *vanA*.

In an integrated livestock-fishing pond system in Thailand, resistance to ciprofloxacin was 80% and that of oxytetracycline and sulfamethoxazole 100% in the opportunistic pathogen *Acinetobacter* spp. recovered from water and sediment samples (Petersen et al., 2002). This result is significant because in clinical settings, antibiotics such as ciprofloxacin are considered a last line of defense as a treatment regimen. Exposure of microorganisms to antibiotics, in particular man-made antibiotics that do not occur naturally such as ceftazidine and cefotaxime, was initially thought to originate from the fact that a number of these compounds pass through the patient's system without being metabolized and hence end up in waste water. Most pharmaceutical compounds are excreted in fairly substantial quantities (Table 14.4), ending up in biosolids, manure, and waterways. They are commanding increasing scrutiny in the environment (Halling-Sørensen et al., 1998; Heberer, 2002; Jjemba, 2002). However, there are wide disparities between cultures and countries on the use of antibiotics (Harbarth et al., 2001, 2002). For example, it is estimated that only about 0.01-10% of certain antibiotics are used for human therapeutic purposes in the United States, depending on the particular antibiotic (Garrett, 2000), the rest largely fed to livestock at subtherapeutic levels as a growth promoter. The connection between use of antibiotics as

**Table 14.3** Percentage of vancomycin-resistant enterococci positive and negative fecal samples from farms and farmers in Norway in 1998[a]

| Vancomycin-resistant enterococci | Farmers Exposed | Unexposed | Poultry Exposed | Unexposed |
|---|---|---|---|---|
| Positive | 17.8% | 1.4% | 98.6% | 10.8% |
| Negative | 82.2% | 98.6% | 1.4% | 89.2% |
|  | n=73 | n=74 | n=73 | n=74 |

[a]Exposed refers to farms where avoparcin an analogue for vancomycin, was previously used as a growth promoter; unexposed refers to farms in which avoparcin had never been used.
*Source:* Based on data from Borgen et al. (2000).

**Table 14.4**  Excretion and biodegradability of some of the commonly used therapeutic agents[a]

| Compound | Excretion rates (%) | | Biodegradability[b] |
| --- | --- | --- | --- |
| | Unchanged | Metabolites | |
| Amoxicillin | 80-90 | 10-20 | No data |
| Ampicillin | 30-60 | 20-30 | No data |
| Carbamazepine | 1-2 | No data | Persistent |
| Chlorophenicol | 5-10 | No data | No data |
| Chlortetracycline | >70 | No data | $T_{1/2}$ = 20 days (soil) but can be more than 64 days |
| Cyclophosphamide | 14-53 | No data | Persistent |
| Estrogen | No data | No data | Persistent |
| Ibuprofen | 1-8 | No data | Inherently degraded |
| Ifosfomide | 14-53 | No data | Persistent |
| Ivermectin | 40-75 | 25-60 | Persistent |
| Metronidazole | 40 | No data | Persistent |
| Oxytetracycline | >80 | No data | $T_{1/2}$ = 20 days |
| Penicillin G | 50-70 | 30-70 | Partially degraded |
| Phenazone | No data | No data | Persistent |
| Sarafloxacin | No data | No data | Persistent |
| Sulphamethoxine | .15 | No data | Persistent |
| Sulphadimethoxine | No data | No data | Persistent |
| Streptomycin | >66 | No data | Persistent |
| Tetracycline | 80-90 | No data | Persistent |
| Tylosin | 28-76 | No data | $T_{1/2}$ = 3-8 d (slurry) |
| Virginiamycin | 0-31 | No data | Persistent |

[a]Reproduced from Jjemba (2002) with permission from Elsevier.
[b]$T_{1/2}$ is the duration it takes for half the compound to be degraded. Biodegradability is reported as persistent by the cited reference, which, in the present interpretation, suggests that the compound is not readily degradable but quantitative information, such as half-life in the environment, is not available.

growth promoters and antibiotic resistance transfer to humans and nontargeted animals has been demonstrated (Gilliver et al., 1999; Thiele-Bruhn, 2003). Resistance can spread from one host to another or by horizontal transfer from a common reservoir.

## 14.3.1  Cholera

Cholera has swept the globe several times in the past and is endemic in Asia, particularly in India and Bangladesh due to seasonal flooding of the Ganges and the Yangtse Rivers during the monsoon rains. The disease is virtually absent in developed countries but is still common where sewage treatment is absent or poorly implemented. Thus, in Asia and Africa cholera

cases are estimated to exceed 5 million each year (Chiang and Mekalnos, 1999). It is caused by *Vibrio cholerae,* a Gram-negative, nonsporulating rod with a single polar flagellum, which is transmitted almost exclusively via contaminated water. A relatively large inoculum ($10^8$ - $10^{11}$ organisms) is required for successful infection because *V. cholerae* is acid sensitive and cannot readily withstand low pH environments of the stomach (Chiang and Mekalanos, 1999). The causative organism grows in the small intestine and produces an enterotoxin (choleragen) which induces the secretion of chlorides, bicarbonates, and water. These electrolytes and the excess water together with the epithelial wall cells are excreted, taking on the appearance of a "bloody rice watery stool". Some fluids can also be lost through violent vomiting. The loss of fluids through vomiting and diarrhea acts in the dissemination of the vibrio back into the environment. Such a sudden loss of tremendous amounts of body fluids (12-20 liters day$^{-1}$) can lead to an instant loss of up to 10% of the victim's body weight in a matter of hours. This causes shock, which in turn causes the blood to become viscous so that vital organs are disabled, a condition that subsequently leads to death. With cholera infections, it is common to learn of people who wake up apparently healthy and are dead by sundown due to dehydration. Recovery from the disease provides effective immunity based on the antigenic activity of both the enterotoxin and the cells. However, the same person can still be susceptible to other strains of *V. cholerae*.

The story of the great outbreak of cholera in London in 1854-62 which caused 20,000 deaths is known to almost every microbiologist and epidemiologist today. A London physician, Dr. John Snow, mapped the occurrence of cholera cases and found that 500 cases lived within a few blocks and all of the inhabitants drew water from one public water pump located on Broad Street. Dr. Snow reasoned that whatever was causing this malady was clearly centered on the Broad Street pump. In a decisive move to practice preventive medicine, Dr. Snow removed the handle from the pump, a move which brought the epidemic under control within a few days. Later, it was discovered that the pump was in fact drawing water from a source that was near a leaky sewage pipe, allowing *Vibrio cholerae* to seep into the water supply. In areas such as Bangladesh where cholera is endemic, the incidence of the disease is known to follow a seasonal pattern (Pascual et al., 2000) with environmental parameters such as temperature, salinity, etc., implicated in the seasonal distribution of toxigenic *Vibrio cholerae* and the occurrence of clinical cases (Colwell, 1996: Lobitz et al., 2000). The distribution of toxigenic

*Vibrio cholerae* in the environment and the occurrence of clinical cases of cholera appears to be dependent on sea surface temperatures and sea surface height (Lobitz et al., 2000). Besides temperature and height, other parameters such as salinity, precipitation, and the abundance of planktons affect the incidence and dynamics of the disease in an intricate relationship presented in Figure 14.3. This Figure clearly shows how atmospheric composition and related human activities that lead to environmental changes can influence the emergence and persistence of some diseases. Such intricate relationships are not limited to cholera only but to varying degrees can also affect the emergence and/or spread of other diseases such as Hantavirus pulmonary infection, dengue fever, West Nile virus infections, etc., as discussed later in this chapter.

Cholera develops after ingestion of a suitable inoculum which can range between $10^4$-$10^9$ cells. The cells colonize the small intestines and attach firmly to the epithelium wall where they grow, releasing a toxin. Excretion of cholera vibrios by convalescent carriers may occur for more than one year while asymptomatic carriers can excrete them for 1-2 weeks. Mortality from cholera can be as high as 60%. Treatment involves replacing fluids and nutrients and administering streptomycin or tetracycline-type antibiotics. Control requires improved sanitation and the elimination of alternative hosts as its spread is mostly attributed to polluted water and fecal contamination of food. Elimination of shellfish, considered an alternative host, from the coast of Bangladesh, is showing some signs of control. *Vibrio cholerae* is free living in water and associates with planktons, in particular copepods (Colwell, 1996; Chowdhury et al., 1997).

Although more than 150 serogroups of *V. cholerae* are known, only the O1 and O139 serovars are known to cause cholera (Chiang and Mehalanos, 1999) but other pathogenic species such as *V. vulnificus*, *V. parahaemmolyticus*, and *V. alginolyticus* are

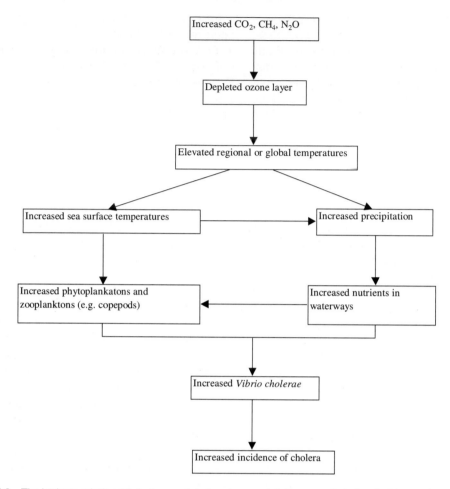

**Fig. 14.3**   The intricate relationship between global environmental change and cholera incidences in endemic areas.

also sources of concern in public health, causing gastroenteritis infections. The O1 serovar is further divided into two biotypes, classical and E1 Tor. These two biotypes are assigned according to the unique ability of the latter to produce hemolysin and mannose-sensitive hemagglutinin. The *V. cholera* genome has high plasticity, forming new strains (Rubin et al., 1998). It is speculated that through the horizontal transfer of genetic elements, conjugation and transduction, virulent *V. cholera* strains evolved (Chiang and Mekalanos, 1999). The *toxR* regulates the cholera toxin expression in response to various environmental conditions

such as temperature, pH, and oxygen tension (Otterman and Mekalanos, 1994). The filamentous appendages, referred to as the toxin-coagulated pilus (TCP), that is required for auto-agglutination of bacterial *V. cholera* cells required in the colonization of the mucosal surface was probably acquired by transduction. Toxin-coagulated pili are required for successful colonization of the human intestine and also serve as receptors to the cholera toxin bacteriophage (CTXØ). All toxigenic *V. cholerae* strains possess toxic-coagulated pili. The O139 serovar which first surfaced in India and Bangladesh in 1992 may have resulted from acquisition of its toxicity genes from other serovars by horizontal transfer. This serovar carries a self-transmissible element (SXT) by conjugation that confers resistance to multiple antibiotics and contributes positively to its environmental fitness. Thus, transmission of genetic elements has played a significant role in the emergence of new pathogenic strains. The environmental conditions that favor these process need to be elucidated in order to control future cholera outbreaks and emergence of even more lethal strains.

As outlined in Chapter 6, bacteria undergo viable but nonculturable (VBNC) states depending on a variety of environmental factors such as temperature, salinity, aeration, cell concentration, and nutrient concentration. VBNC *Vibrio cholerae* produces the toxin (Hasan et al., 1994) that can cause cholera in human volunteers and/or revert to the culturable infectious state following human passage (Oliver, 2000). Thus, loss of culturability does not guarantee loss of pathogenicity and nonculturable pathogens can pose some public health threats and are important in epidemiology. In areas such as Bangladesh where cholera is endemic, the pathogen exists in its VBNC state between epidemics, thereby continuing to pose a potential threat to public health. It is worthy note that the ability of these intestinal pathogens to multiply within the intestinal mucosa has been found to be more important in the manifestation of disease than just toxin production (Rahman et al., 1996). The VBNC state in *V. vulnificus* has been shown to be induced by temperature changes, with temperatures below 5°C being favorable for this state (Linder and Oliver, 1989; Wolf and Oliver, 1992). The abundance of this pathogen has also been found to be directly related to water temperatures and salinity (Motes et al., 1998). Thus, when maintained at 10, 15, 20, and 30°C, *V. vulnificus* remained culturable throughout the test period (40 days) compared to those main-

tained at 5°C (Wolfe and Oliver, 1992). The plasmids essential for virulence are also maintained as the cells enter the VBNC state. These observations have implications for the long-term survival and cycling of such pathogens in water. This observation likewise has implications for the detection (or lack of detection) of some pathogens in environmental samples during colder periods. However, high temperatures, differences in UV illumination, presence/absence of plasmids, and salinity have also been associated with loss of culturability in some bacterial species (Barcina et al., 1989, Oliver, 2000), indicating that the mechanisms of loss of culturability vary greatly from species to species, and depend on a variety of environmental and genetic factors.

In cholera endemic areas, survival of the pathogen has also been found to be remarkably promoted by its attachment to copepods. As a matter of fact, the presence of copepods enables *V. cholerae* to survive routine treatment such as chlorination (Table 14.5). Copepod-associated *V. cholerae* is somehow induced to a VNBC state after treatment with alum for 4 hours, indicating that copepods play an important role in the survival of *V. cholerae* in the environment. *Vibrio* cells survive by possibly being directly ingested in the copepod gut where they can escape the harsh water treatment procedures (Chowdhury et al., 1997). In endemic areas, the removal of copepods by filtration is therefore highly recommended prior to chlorination or treating with alum. The filtration step would also remove the need for high concentrations of chemicals such as hypochlorite and alum during water treatment, reducing the accumulation of potentially carcinogenic byproducts. Thus, in endemic areas combining both chlorination and filtration can effectively eliminate cholera from drinking-water supplies.

Other bacteria such as *V. vulnificus*, *Salmonella*, *E. coli*, *Shigella*, *Campylobacter*, *Legionella*, *Listeria* sp., and *Helicobacter* also

**Table 14.5** Survival of free-living and copepod-associated *V. cholerae* treated with sodium hypochlorite or alum[a]

| Concentration (%) | | Type of *V. cholerae* | DFA-DVC ($\times 10^2$ cells ml$^{-1}$)[b] |
|---|---|---|---|
| Alum | Na-hypochlorite | | |
| None | None | Free-living | 1,700 |
| | | Copepod-associated | 84 |
| 0.50 | 1.00 | Free-living | 583 (34.29) |
| | | Copepod-associated | 6 (7.14) |
| 0.50 | 2.50 | Free-living | 913 (53.71) |
| | | Copepod-associated | 4 (4.76) |

[a]The cells were treated with alum or Na-hypochlorite for 4 hours. In all instances where Na-hypochlorite alone or in combination with alum were used, no *Vibrio cholerae* were detected (detection limit of 10 cells ml$^{-1}$) on Luria-Bertani (LB) agar or thiosulfate-citrate-bile salt-sucrose (TCBS) agar. TCBS agar is a highly selective medium routinely used for the enumeration and isolation of *V. cholerae*.
[b]Numbers in (parentheses) are the percentage of cells surviving treatment as determined by the direct fluorescent-antibody and direct viable count (DFA-DVC) method as a fraction of those in the control treatment.
*Source*: Modified from Chowdhury et al. (1997) with permission from the American Society for Microbiology.

undergo the VBNC state, suggesting that this state is part of the normal life cycle of a variety of bacterial strains with consequences to public health. VBNC *Shigella* also maintain a biologically active shiga toxin and remain capable of adhering to the intestinal epithelium (Fig. 14.4). Humans appear to be the major reservoir although other primates also carry the pathogen (Powell et al., 2000). The fact that *E. coli* likewise undergoes culturability changes has public health ramifications because *E. coli* is widely used as the gold standard for indicating fecal pollution in both water and food. Thus, the presence of VBNC *E. coli* would give a false impression about the safety of these foodstuffs in some instances. This further makes the usefulness of the coliform assay for evaluating water quality questionable.

## 14.3.2 Typhoid and other Salmonella-related enteric diseases

The best-known member of the *Salmonella* genus is *Salmonella typhi*, the most virulent serovar in this genus. The organism responsible for typhoid fever and the disease is specific to humans but can be transmitted by other animals, such as cattle, sheep, pigs, and chicken (Pell, 1997) through food or water contaminated with fecal material. These animal carriers, however, do not suffer from the disease.

Thus, it is often associated with contaminated poultry and poultry products. Typhoid fever has an incubation period of about two weeks and starts off as a high fever and a continual headache. Subsequently, diarrhea appears in the third week. In the gastrointestinal tract, *S. typhi* penetrates intracellularly and multiplies rapidly. Evidence using mutants shows that *Salmonella* sp. need lipopolysaccharides to bind onto the epithelial cell wall (Licht et al., 1996). They eventually enter the bloodstream and are widely distributed throughout the body. The pathogen heavily infects the bile duct, spleen, bone marrow, and lungs from where reinfection occurs. Symptoms include diarrhea, headache, vomiting, fever, and nausea. It can be treated with chlorophenicol and if left untreated, can be fatal within a few weeks. It is still a significant cause of illness or even death in some countries with poor sanitation. Immunization with a killed vaccine (injectable) is a common practice in developed countries. Live, orally ingested vaccines are now available. Other *Salmonella* spp. such as *S. typhimurium, S. paratyphi,* and *S. enteritidis,* cause severe diarrhea and other enteric diseases and are gaining notoriety as formidable pathogens too. Unlike *S. typhi,* these other species have nonhuman hosts.

Excretion of the pathogen in feces can continue in infected persons for 2-12 months. It is

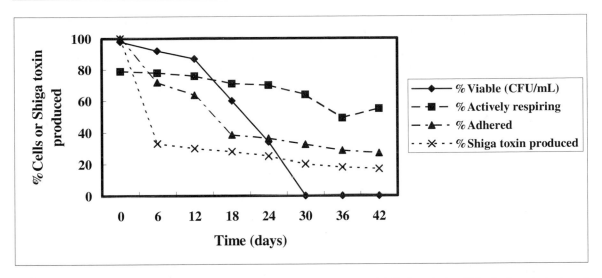

**Fig. 14.4** Production of shiga toxin with various changes in culturability and viability in a virulent *Shigella dysenteriae* type 1 over a 42-day period. The initial inoculum of $10^7$ cfu ml$^{-1}$ was incubated at 30°C with agitation (100 rpm) in sterile water. One hundred ml were drawn at each sampling interval in triplicates and plated on Trypticase soy agar to determine cfu. Active respiration was determined using INT (2-*p*-iodophenyl-3-nitrophenyl-5-phenyl tetrazolium chloride) and/or elongation due to a gyrase inhibitor. It is a fraction of direct viable counts. The percent cells adhered were determined with Henle 407 cells mantained in basal Eagle medium containing 10% calf serum. The day 0 datum for this parameter was a positive control with no shiga toxin (i.e., 100% cells adhered). Figure based on data from Rahman et al. (1996) with permission from the American Society for Microbiology.

estimated that about 1% of the world population excretes *Salmonella* at any one time. *Salmonella* is commonly detected in waters which receive agricultural-based influents, storm water, sewage, streams, and stabilization ponds. *Salmonella* is inactivated to the VBNC state, the rate of inactivation being inversely proportional to temperature (Roszak et al., 1984; Chmielewski and Frank, 1995) and nutrients. On addition of nutrients, Roszak et al. (1984) resuscitated *S. enteriditis* almost to the original culturable cell density. However, if the cells underwent dormancy for an extended period of time (>21 days), resuscitation through provision of nutrients was not successful under laboratory conditions. It is not clear how frequently resuscitation occurs in the environment. The problem of *Salmonella* infections has been exacerbated by the increasing resistance of *Salmonella* spp. to antibiotics. This resistance has resulted in an increase in *Salmonella* spp. in environmental matrices compared to other competing microbes, thus increasing the risk of infection. However, there are procedural limitations in the detection of *Salmonella* (and *Shigella*) in environmental samples using conventional culture techniques that involve selective enrichment and plating, followed by fluorescent antibody staining. DNA probes appear promising, especially under aquatic environments.

### 14.3.3 Viruses

The effects of viral infections can range from overt disease to acute self-limiting respiratory, skin, gastrointestinal, and ear infections, to extreme, gastrointestinal and liver disorders or even death. Of more than 30,000 hospitalizations due to diarrhea in ten sentinel hospitals in the United States in 1990-1992, more than 94% no definite causative agent was diagnosed although viruses were suspected

(Glass et al., 2000). Prior to that period, diagnostic tools for detection of viruses in clinical samples were not yet robust but even in what was classified under unknown etiologic sources (Fig. 14.5), viral infection, in particular through potable water, was suspected.

Much emphasis has been given in environmental microbiology to the elimination (or at least reduction) of enteric viruses from (waste)water during routine treatment as a means of controlling the spread of viral diseases such as poliomyelitis, Hepatitis A, and rotaviruses. Hence the linkage between viral infections and the environment is an area of increasing research and interest. The influence of the environment (i.e., seasonal changes) on viral diseases such as influenza is well documented, with most outbreaks occurring in late fall, winter, and early spring in temperate regions (NRC, 2001). Seasonal variations in the spread of influenza are also facilitated in part by the indoor life style during the cold seasons. However, humidity also plays a role in the survival of aerosolized viral droplets released during coughing and sneezing (Schulman and Kilbourne, 1963). Recent evidence likewise shows a seasonal trend in the occurrence of several other viral infections such as rotaviruses, rheoviruses, and caliciviruses (i.e., Norwalk and Norwalk-like viruses), clearly indicating the interplay of environmental factors in

their spread and persistence.

Considerable progress in understanding the survival of viruses in the environment has greatly depended on bacteriophages (especially coliphages) and enteroviruses (i.e., poliovirus, echovirus, and Coxsackie virus). This approach has helped us to sort out the basic science of viral infection and transmission but does not answer some of the basic questions of how other viruses of public health concern, in particular caliciviruses, hepatitis A, and rotaviruses persist and survive in the environment. Coliphages are not of consequence to public health as they only attack bacteria. Nor do their prevalence and survival adequately mimic those of viruses of concern to public health. Enteroviruses are commonly found in many healthy individuals and often cause mild or symptomless diseases, except in rare cases where they can develop into full-blown infections such as poliomyelitis. This "less harmful coexistence" between humans and poliovirus is possibly attributable to immunity introduced through routine immunization a few days after birth.

Like NLVs, neither rotaviruses nor Hepatitis A virus could be adequately inactivated by chlorination alone or chlorination combined with copper and/or silver disinfectants. In contrast poliovirus was inactivited whereas adenoviruses were inactivated quite slowly by

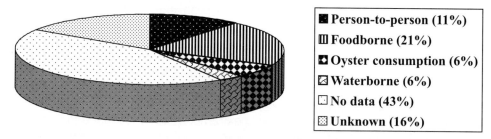

**Fig. 14.5** The suspected mode of transmission of gastroenteritis for 90 outbreaks in the United States during January 1996 to June 1997. It is plausible that the majority of outbreaks when the mode of transmission is not known or where no data exist are most likely attributable to water but remained unresolved because the detection methods were insufficiently developed at the time (Figure based on data from Fankhauser et al., 1998 with permission from the University of Chicago Press, Chicago, IL)

these treatments (Abad et al., 1994). Electron microscopic observations have shown that in the presence of free chlorine together with heavy metals (copper and silver), viruses form aggregates, a strategy to enhance their survival in the environment (Abad et al., 1994). These results have important ramifications in the treatment and disinfection of waste water and biosolids as these materials typically have high cation concentrations (Brown et al., 1996; McBride, 1995). The foregoing results likewise indicate that poliovirus is not an adequate model viral strain for use in disinfection studies.

Considering the rate of infection and subsequent death, particularly of the young, immunocompromised, and the elderly from rotaviruses as well as other viral infections, a concerted effort to understand the dynamics of these viruses in the environment is needed. Rotavirus infections are responsible for about one-third of the children hospitalized for diarrhea in the United States (Glass et al., 1996) and thousands of child deaths in developing countries. Furthermore, healthy adults may be reinfected throughout life despite prior infection since protective immunity against rotaviruses is short lived given the fact that there are many serotypes (Gerba et al., 1996). Based on nucleic acids alone, rotaviruses have RNA strands and are less prone to errors and mutations since the DNA polymerase has proof-reading mechanisms. They also have a double envelope that neutralizes antibodies following natural infection, thus conferring immunity (Unicomb et al., 1999). This physical barrier may also protect these viruses during disinfection of water systems using chlorine.

Elimination of smallpox from the face of the Earth following vaccination campaigns was successful because the smallpox virus has no host other than humans. Since the viruses of current concern to public health have alternative hosts, exterminating them is not a viable option. Thus our focus should include gaining insights into how they survive under various environmental conditions or identifying specific environmental linkages to viral disease infections. Lessons learned from the poliovirus suggest that immunization coupled with adequate sanitation is essential in controlling viral diseases. In a recent study by Ford-Jones et al. (2000) in the Greater Toronto Metropolitan area, 77% of the children hospitalized with rotavirus infections were cared for in their own home or by a close relative compared to only 13% of the infected who were going to family day-care homes and 8% going to child-care centers. This evidence suggests an in-home causative agent, most probably water, despite its treatment. In that same study, rotavirus infection was also more prevalent in winter and spring (February- May) than in autumn, clearly indicating a seasonal linkage to rotavirus infections. Rotavirus infection in the United States has also been linked to season and geographic locations, peaking in the Southwest USA in November and moving across the country toward the Northeast where peak activity occurs in March and April, hospitalizing over 55,000 children per year (Glass et al., 1996; Torok et al., 1996). Rotaviruses can survive in water for more than a week depending on environmental conditions, although due to difficulties in the rotavirus culture system, most of the information about its survival and elimination from water (e.g. Rao and colleagues' data as well as Berman and Hoff's summarized in Table 14.6) has been obtained using simian rotavirus which, although similar in morphology to human rotaviruses, is noninfectious to humans. The limited number of studies on rotaviruses infectious to humans have shown that these viruses are more persistent in the environment than simian rotaviruses and polioviruses (Raphael et al., 1985; Pancorbo et al., 1987).

With such a consistent infection cycle, coupled with the fact that differing strains of the virus are encountered in different parts of the USA, mere migration of the virus from the Southwest to the Northeast and influence of temperature and humidity were ruled out by

**Table 14.6** Various disinfection techniques and their efficacy on HAV, NLVs, and rotaviruses during the treatment of drinking or raw water

| Disinfection | Efficacy and other remarks |
|---|---|
| UV irradiation | Rotaviruses resistant to UV in some instances (Gerba et al., 1996). HAV more resistant to UV than poliovirus but an initial infectious titre of 4425 $TCID_{50}ml^{-1}$ was reduced by 99.9% in 7.8 min irradiation at 254 nm with a 15 W lamp (Lévéque et al., 1995). By comparison, poliovirus was reduced to a similar titre under the same conditions in 2.9 min. |
| Thermal disinfection | NLVs stable with freezing and at 60°C (Glass et al., 2000). |
| Filtration | Removed 99.98% simian rotavirus SA-11 and 98.85% HAV compared to 98.6% poliovirus when combined with alum coagulation, flocculation, and sedimentation with 1 mg $mL^{-1}$ ferric chloride ($FeCl_3 \cdot 6H_2O$) and aluminum sulfate (alum, $Al_2(SO_4)_3 \cdot 18H_2O$) in pilot plant studies (Rao et al., 1988). |
| Metals | |
| Copper and Silver | Combination of $Cu2+$ (700 g $L^{-1}$) and/or Ag+ (70 g $L^{-1}$) with chlorine is effective against poliovirus but ineffective against HAV and rotavirus (Abad et al., 1994). |
| Ferric chloride | Used as a coagulant, removed >99% rotaviruses and 71-79% HAV compared to 74% poliovirus, depending on the turbidity of the water in batch studies (Rao et al., 1988). |
| Magnesium | Typically used to soften drinking water. $MgCl_2$ (100 mg $L^{-1}$) removed >90% rotavirus and >95% HAV compared to 99.6% poliovirus in batch studies (Rao et al., 1988). |
| Calcium | $CaCO_3$ (100 mg $L^{-1}$) as a softener removed 13% rotavirus and only 9% HAV compared to 11% poliovirus in raw water (pH 5.9). However, on raising the pH to 9.5, $CaCO_3$ removed 67% rotavirus and 96% HAV compared to >99% poliovirus in batch studies (Rao et al., 1988). |
| Calcium and magnesium combined | Removed total hardness from raw water and also removed >99% rotavirus and >97% HAV compared to 96% poliovirus in batch studies (Rao et al., 1988) |
| Oxidizing agents | |
| Chlorine | 0.75 mg chlorine $L^{-1}$ was ineffective on rotaviruses in tap water (Raphael et al., 1985). Levels as high as 10 ppm did not inactivate NLVs (Glass et al., 2000). Less than 2.6 $log_{10}$ reduction of HAV and rotavirus by 1 mg free chlorine $L^{-1}$ compared to more than 4 $log_{10}$ reduction of poliovirus (Abad et al., 1994). |
| Chlorine dioxide | Inactivation of simian rotavirus SA11 by 0.5 mg chlorine dioxide $L^{-1}$ was slower at pH 6 than at pH 10 (Berman and Hoff, 1984). |
| Chloramines | Rotaviruses were more resistant to chloramines than poliovirus. For example, simian rotavirus SA11 was resistant to a high concentration of 10 mg monochloramine $L^{-1}$ at pH 8, inactivation of the pure virus requiring more than 6 h whereas the cell-associated was inactivated in more than 10 h (Berman and Hoff, 1984). |
| Ozone | Rotaviruses were more resistant to ozone than poliovirus (Gerba et al., 1996) |
| Non-oxidizing agents | |
| Liming | Raises the pH. In batch studies by Rao et al. (Rao et al., 1988) after 30 min. contact time, 100% HAV remained viable at pH 10.8-11.5 whereas 70%, 53%, and 33% rotaviruses were viable at pH 10.5, 10.8, and 11.0 respectively. |
| Alum coagulation (i.e., $Al_2(SO_4)_3 \cdot 18H_2O$) | Reduced turbidity and also moved 95% simian rotavirus SA11 and 97% HAV depending on turbidity, compared to 97% poliovirus when combined with flocculation, and sedimentation with 15 mg $Al^{3+} L^{-1}$ in pilot plant studies (Rao et al., 1988). |

Glass et al. (1996). Glass and coworkers (Glass et al., 1996; Torok et al., 1996) hypothesized that the infection pattern is dependent on an unknown local common reservoir. Such a reservoir could be water, some animals, and/or some aerosol. Whatever the reservoir, it seems logical that temperature and possibly moisture/humidity play a role in its activity. Although the role of temperature in the proliferation of rotaviruses when temperatures are low is still not known, these viruses are definitely capable of long-time survival, especially at low temperature. Thus, over a 65-day period, survival of rotavirus was higher at 4°C than at 20°C (two temperature extremes typically encountered in water treatment, storage, and distribution systems in temperate regions), possibly as a result of the reduced growth of bacteria and other microorganisms at low temperatures (Raphael et al., 1985). Under high microbial activity (typical of high-temperature conditions), the existing microorganisms may degrade the viral protein coat (Raphael et al., 1985). Furthermore, increased viral infectivity at low temperatures cannot be ruled out as an attribute to increased viral incidences. Studies by Pancorbo et al. (1987) showed that the antigenicity of human rotavirus Wa decreased to a different extent in waters differing in pH, organic carbon content, conductivity, and suspended solids, with viral survival being greater in more polluted waters, albeit they remained more infectious for a longer period of time in nonpolluted waters (e.g. lake water) than in polluted (creek and secondary effluent) waters.

Some rotavirus serotypes that infect humans have recently been described in livestock (cattle, horses, and pigs), suggesting that these animals may serve as secondary hosts and/or reservoirs (Cunliffe et al., 1999; Unicomb et al., 1999). Seasonal patterns in other viral diseases such as Hepatitis A (Cuthbert, 2001) and NLV (Fankhauser et al., 1998; Lodder et al., 1999) have also been reported, but in most cases remain unexplained. In the tropics, infections occur throughout the year but no proper studies have been done to determine whether some seasons have a higher disease incidence than others.

Although extensively researched in the case of prokaryotic pathogens, the role of biofilms in protecting viruses against disinfection and treatment in the environment has still to be exhaustively researched. Viruses have been detected within biofilms in outlets of drinking water treatment and distribution systems, their accumulation in such films accelerated in the presence of particulate materials, bacterial cells, and extracellular polymeric substances. The deposition of viruses into biofilms appears to be desirable in some instances. For example, in studies by Ueda and Horan (2000) using a 400 nm pore size membrane filtration bioreactor, increasing amounts of 200 nm bacteriophage were deposited in the biofilm, as evidenced from reduced detection in the effluent. The phenomena that allow for accumulation of viruses in biofilms are not yet well understood. Although no studies are available, the consistent seasonal infection cycle of some viral diseases may, at least in part, be explained by the dynamics of biofilms with regard to viruses in water distribution systems over time. Viruses adsorbed on biofilms can still be infectious. Thus, if conditions such as differences in temperature, pH, electrical potential, and light (Ista et al., 1999) within the distribution systems dislodge the viruses from the biofilms, they can end up recirculating with otherwise treated water, infecting the end users.

Deaths due to some of these viral diseases have dramatically decreased in developed countries due mostly to effective early therapeutic intervention (Cuthbert, 2001; Glass et al., 1996). However, the total cost of gastrointestinal illnesses in developed countries, most of which are of viral origin, still remains high (Payment, 1993). The death toll from viral diseases in less developed countries likewise remains quite high. Furthermore, symptomless infections may be imposing combined subtle challenges to the body's immune system, leading to low life expectancy in populations in less developed countries. The thriving and subsequent contact of these vectors with humans can be greatly influenced by environmental conditions and how we maintain our surroundings.

Viruses have been reported in drinking water that meets bacterial standards (Beuret et al., 2002; Gratacap-Cavallier et al., 2000; Keswick et al., 1984), even after routine treatment methods such as chlorination have been used (Abad et al., 1994). This reality challenges us to systematically evaluate the efficacy of various purification and disinfection techniques, especially when we consider the fact that viruses are inactivated to a different degree in different types of water (Pancorbo et al., 1987). It can be seen from Table 14.6 that information about the efficacy of a variety of disinfection techniques, particularly for NLVs, is limited. An important distinction between viruses and prokaryotes is that the former cannot replicate in the environment unless they are inside host cells. Thus, controlling virus-potential host associations is a key in controlling the proliferation of viruses in the environment. As outlined above, contact is mostly attained through consumption of contaminated water and food. NLV, rotaviruses and HAV have also been detected in treated waste water (Cuthbert, 2001; Gerba et al., 1996), biosolids (Lodder et al., 1999), and manure. Compared to enteroviruses, rotaviruses bind very poorly to soil particles which, together with better survival in soil, enables them to easily contaminate groundwater through farm runoff, on-site latrines, and leaking septic tanks (Scandura and Sobsey, 1997). Ensuring adequate treatment of waste water, biosolids, and manure to attain virus-free products is a first line of defense against introducing viruses into the public health arena, since today's waste water is tomorrow's drinking water. In the words of Nobel laureate Peter Medawar, no virus is known to do good and is a piece of bad news wrapped up in a protein (Medawar and Medawar, 1983).

Some important viruses, in particular Norwalk-like viruses, cannot be propagated in cell culture to this day. Rotaviruses that infect humans also do not grow well in cell culture (Gerba et al., 1996). To be able to determine the protein constituents of these viruses, assess their full impact in the environment, and determine their infectivity to the detection method currently used (i.e., RT-PCR), there is also a need to develop culture techniques for the rotavirus, considered a major agent of acute gastroenteritis worldwide (Glass et al., 2000). For practical purposes, it is also desirable that the developed cell line have the ability to support several types of viruses as environmental matrices usually contain more than a single viral strain.

Viral diseases such as West Nile virus (WNV), human immunodeficiency virus (HIV), Ebola, Marburg, and Lassa fever are among the two-and-half dozen new diseases that have emerged in the last quarter of the twentieth century. They bear some relevance to environmental microbiology and continue to challenge modern medicine. The vector-borne WNV is steadily capturing new ground, particularly in temperate regions. Originally isolated in the tropics in the 1940s (Smithburn et al., 1940), this virus has been reported in the Middle East, Europe, and more recently North America (Goddard et al., 2002). Since its sighting in New York in 1999, WNL has been diagnosed in more than 1,900 human cases (94 fatalities) and more than 6,000 horses (Goddard et al., 2002). The 1999 outbreak was possibly favored by the preceding warm winter, accompanied by a hot dry summer (Epstein, 2001). These environmental conditions can favor the breeding of the Culex mosquito (*Culex pipiens*) vector and the dry conditions reduce the potential predators of these vectors (e.g., dragonflies and amphibians). Furthermore, the relatively high temperatures enhance the rate at which the virus matures. What makes this virus so potentially infectious to humans in temperate regions, compared to the tropics where clinical cases are almost unheard of, is not clear but could be due to differences in resistance in the respective populations.

Acquired immunodeficiency syndrome (AIDS) was first recognized as a clinical syndrome in 1981 and is currently one of the most prevalent infectious human diseases. The current analogy is that all people who contrast the HIV virus eventually progress to AIDS. It is worthy noting that the fatalities from this disease are not caused by the infectious agent (HIV) but rather by a series of conventionally known infectious diseases caused by other opportunistic pathogens such as *Mycobacterium* and *Pneumocystis carinii*. These pathogens prior to the AIDS epidemic were almost under control but once the body's immune system has been debilitated, they invade and overwhelm the host, with fatal consequences. Controlling the virus through vaccinations and chemotherapy has proven difficult thus far because the virus has very high genetic variability. These otherwise controllable secondary infections thrive as a result of the suppressed human immune system as HIV prevents the normal division and function of T lymphocytes which are a key component of immunity. Both HIV and Ebola are transmitted primarily by direct interpersonal contact. Their emergence, just like hantavirus pulmonary infections, is suspected to be a result of our interference with the environment which eventually brought us into contact with their natural reservoirs in the wild (Grifo, 2001; Mayer, 2000).

HIV viruses are shed in fecal material (Yolken et al., 1991), arousing suspicion that the virus can be transmitted into the environment through waste water leached into the groundwater. However, the available evidence indicates that the virus is able to survive for only a short period of time in an open environment (Ansari et al., 1992).

### 14.3.4 Parasites

Protozoa (*Giardia lamblia, Cryptosporidium parvum*, and *Entamoeba histolytica*) and helminths (hookworms, roundworms, tapeworms, and whipworms) are the parasites of primary concern to environmental microbiology and public health. *Entamoeba histolytica* infects only humans, causing amoebic dysentery (amoebiasis). It is an important cause of morbidity worldwide. It is particularly prevalent in areas without proper sewage treatment where its infections may be as high as 72% of enteric protozoan infections (Rose et al., 1996). Unlike vegetative protozoan cells which are destroyed by the stomach acid (HCl), in protozoan cysts, only the protective wall is digested, thus releasing the vegetative forms. The vegetative forms then multiply in the epithelial cells of the wall in the large intestine causing severe dysentery with characteristically hematic mucous-coated feces. The vegetative forms feed on red blood cells and destroy the tissue in the gastrointestinal tract. The infection is worsened by intestinal bacteria infecting the lacerated intestinal wall and sometimes further invading other organs such as the liver. Diagnosis involves recovery and identification of the organism in the stool or the latex-agglutination and fluorescent-antibody tests. Treatment with metronidazole and iodoquinol is effective.

*Giardia lamblia* infection in the US is estimated at 60,000 cases of illness per annum with approximately 60% of the infection suspected to be waterborne, whereas cryptosporidiosis accounts for just 0.1 - 0.9% incidences of acute diarrhea. *Cryptosporidium* has been responsible for close to 0.5 million documented cases in the UK and the USA combined since 1984 and is estimated to cause between 250 and 500 million infections per annum in Africa, Asia and Latin America (Smith and Rose, 1998). Both giardiasis and cryptosporidiosis are almost exclusively associated with fecal contamination. Exposure to these parasites is also particularly endemic in areas associated with inadequate hygiene.

Similar to viruses, ingestion of small numbers of 1-10 cysts or oocysts can initiate infection. Thus, much lower levels of these pathogens are of greater public health concern than low levels of bacteria. Pathogenic enteric protozoans generally cause acute and chronic diarrhea. Eggs (cysts and oocysts) are excreted in feces in an infective form and are able to survive in the environment. They tend to be quite resistant to water disinfection compared to bacteria. Thus they are more effectively eliminated from water by filtration, reducing the oocysts by $10^{2.3}$-$10^{5.7}$ organisms per unit volume (Smith and Rose, 1998). A substantial number of organisms ($10^{1.55}$-$10^{3}$ per unit volume) can also be removed by coagulation and sedimentation. Water is the major route of exposure of both these pathogens to humans. The occurrence of *Cryptosporidium* sp. in treated waste water has been documented less often than *Giardia* sp. Studies suggest that domestic sewage discharges are a larger source of *Giardia* sp. while animals may be the major source of *Cryptosporidium* sp. In healthy individuals, both *Giardia* and *Cryptosporidium* infections are self-limiting, although they cause serious discomfort. Fifty percent of the healthy volunteers tested needed 132 *Cryptosporidium parvum* oocysts to cause illness (DuPont et al., 1995) but the pathogen can prove fatal in immunocompromised individuals. In two of the recent major *Cryptosporidium* outbreaks in the United States, mortality rates in immunocompromised individuals were 86% and 53% in Milwaukee and Las Vegas, respectively (Rose, 1997).

Giardiasis and cryptosporidiosis are rapidly increasing in importance as emerging parasitic protozoa-associated diseases (Horsburgh and Nelson, 1997) and hence merit some in-depth analysis. Even though they belong to different orders (Fig. 3.2), the life cycles of *Giardia* sp. and *Cryptosporidium* sp. are fairly complicated and somewhat similar. Once ingested, the *Cryptosporidium* oocysts or *Giardia* cysts excyst in the intestines, releasing the infectious organism. Several changes occur within the gastrointestinal tract where the

*Cryptosporidium* sp. completes its entire complex life cycle (Fig. 14.6) before the reconstituted oocysts or cysts are excreted back into the environment in very large numbers of more than $10^{9}$ oocysts per infected individual. The oocysts and cysts are very hardy and resistant to many environmental extremes, including acidity, temperature extremes, desiccation, higher atmospheric pressure, and ultraviolet radiation (Rose, 1997; Rose et al., 2000). Both *Giardia* and *Cryptosporidium* comprise several species but *C. parvum* and *G. lamblia* are the respective primary species responsible for clinical diseases in humans. Both parasites likewise have a reduced host specificity, which increases their potential for contaminating and spreading in the environment. They occupy three major reservoirs—humans, wildlife, and domestic livestock—from which they are excreted, usually in large numbers. The large numbers of pathogens excreted favor waterborne transmission. Both cryptosporidiosis and giardiasis outbreaks are associated with heavy rainfall because under these conditions, they are easily washed into reservoirs of drinking water (Wuhib et al., 1994; Alterholt et al., 1998; Rose et al., 2000). Thus, control measures have to aim at minimizing fecal material from all three sources into water and foodstuff. The introduction of these parasites from livestock into waterways can be greatly limited if farm runoff is directed into filter strips so as to reduce the risk of contaminating streams (Pell, 1997).

Very little information about the survival and inactivation of *Giardia* and *Cryptosporidium* in the environment is available. *Giardia* cysts maintain the ability to encyst at the same rate for up to 12 days in fresh waters at temperatures below 10°C. In studies by Fayer (1994) with an inoculum of 150,000 *Cryptosporidium* oocysts, 100% of the mice exposed to the inoculum heated at 67.5°C for one minute or 60°C for five minutes were infected. The implication from this study is that *Cryptosporidium* oocysts can remain infectious even under hot

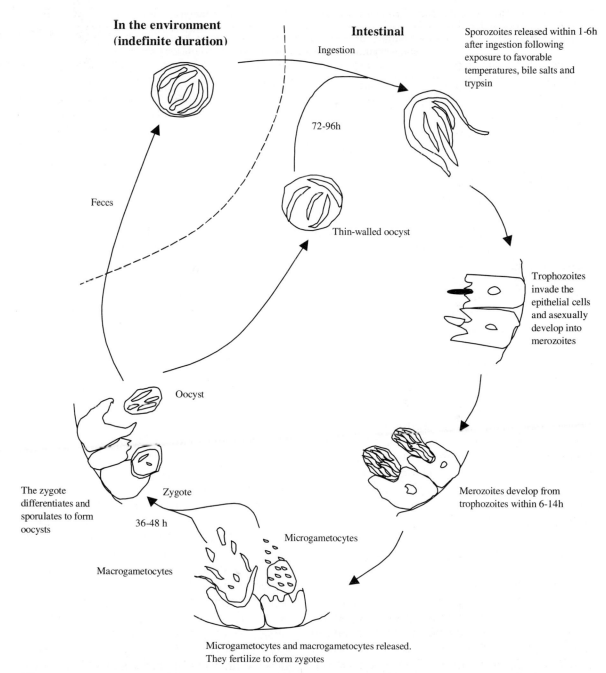

**In the environment (indefinite duration)**

**Intestinal**

Ingestion

Sporozoites released within 1-6h after ingestion following exposure to favorable temperatures, bile salts and trypsin

72-96h

Feces

Thin-walled oocyst

Trophozoites invade the epithelial cells and asexually develop into merozoites

Oocyst

The zygote differentiates and sporulates to form oocysts

Zygote

36-48 h

Merozoites develop from trophozoites within 6-14h

Microgametocytes

Macrogametocytes

Microgametocytes and macrogametocytes released. They fertilize to form zygotes

**Fig. 14.6** Ecological cycle of *Cryptosporidium* sp. Note that most of the life cycle occurs in the intestines. The only portion that occurs in the external environment can be excluded by the parasite. However, the parasite typically uses the environmental loop to spread from one host to the other. Once in the environment, the oocysts can survive indefinitely until an ideal condition (a host) appears (Figure modified from Smith and Rose, 1998 with permission from Elsevier.

**Table 14.7** Viability of *Cryptosporidium parvum* oocysts stored at various temperatures

| Storage time (h) | Percent number of mice (n = 6 unless specified) harboring developmental stage cryptosporidia after the oocyst inoculum was stored at: | | | | |
|---|---|---|---|---|---|
| | 5°C | −10°C | −15°C | −20°C | −70°C |
| 1 | ND | ND | ND | 100 | 0 (n=5) |
| 3 | ND | ND | ND | 100 (n=7) | ND |
| 5 | ND | ND | ND | 100 | ND |
| 8 | ND | 100 | 100 | 16.7 | 0 |
| 24 | ND | 100 | 100 | 0 | 0 |
| 168 | 100 | 100 | 0 | 0 | ND |

ND: Not determined
*Source*: Fayer and Nerad (1996) with permission from the American Society for Microbiology.

environmental conditions. A similar study was done to determine the impact of low (close to freezing and at subzero) temperatures on oocysts. With an inoculum kept at 5°C or −10, and −15°C, for 24 h, *Cryptosporidium* oocysts were still infectious to 100% of the mice tested (Table 14.7). However, the oocysts rapidly lost viability on freezing at −15°C and −20°C for one week. On the contrary, the oocysts were still viable even after freezing at −10°C for one week. All these observations indicate that temperature is a key environmental factor in determining how long *Cryptosporidium* remains viable and infectious in the environment. It can be further inferred from these results that water containing *C. parvum* may not be rendered completely noninfectious by freezing, e.g. by making ice cubes, for a short period of time before consumption.

## 14.4 AIRBORNE INFECTIONS

### 14.4.1 Tuberculosis

Tuberculosis caused by *Mycobacterium tuberculosis* was first described in 1882 by Robert Koch although it is believed to have emerged about 15,000 to 35,000 years ago (Navin et al., 2002). It causes about 3 million deaths worldwide each year. Currently in the US, it is the third leading cause of death among infectious diseases after influenza and pneumonia. It is an important reemerging disease, the increased incidence worldwide attributable to increased susceptibility in immunocompromised individuals. Recent statistics showed about 1.9 billion people worldwide to be infected with *Mycobacterium tuberculosis* (Navin et al., 2002). Primary infection is by inhalation of droplets or dust particles containing the viable bacteria. It settles and grows in the lungs. The immune system macrophages attack the foreign cells but the latter survive and multiply within the former, demonstrating a high level of adaptation by the pathogen. Postprimary infection involves a delayed hypersensitivity reaction, resulting in the formation of aggregates of activated macrophages called tubercles. Bacterial growth is not controlled, causing extensive damage to the lung tissue. Eventually, the pathogen colonizes other parts of the body, resulting in death of the patient but not before he/she has coughed, sneezed, and otherwise transmitted the agent to others. This trend has ecological implications because as is well known, a pathogen which kills its host does not succeed very well unless it modifies its spread. Infection can be prevented by immunization with a bacillus Calmette-Guerin (BCG) vaccine consisting of a live attenuated strain of *M. bovis*. However, the efficacy of the vaccine is quite variable and its protective efficiency varies from 0 - 80% (Bercovier, 1999).

A cure is by a compound that specifically inhibits *M. tuberculosis* cell wall synthesis. However, multidrug-resistant tuberculosis threatens efforts to control *M. tuberculosis*. The drug-resistant strains may be easily communicated to others through the air. More attention is also being paid to the infection of *Mycobacterium* through water (see Chapter 13). From a detection perspective, *Mycobacterium* species are very hard to culture. Nucleic acid probes are increasingly used in clinical and environmental microbiology to detect *Mycobacterium* spp. Thus, the detection of this pathogen in clinical samples (smears) has been greatly improved using peptide nucleic acid (PNA) probes (Stender et al., 1999a, b). This approach is an improvement over FISH in the sense that PNAs are analogs to DNA with an uncharged polyamide backbone instead of sugar phosphates. Unlike rDNA probes which cannot readily penetrate Gram-positive bacteria, PNAs diffuse through hydrophobic cell walls, easily reaching the targeted ribosomes.

### 14.4.2 Fungal infections

It may appear that bacteria, protozoa and viruses are the only microorganisms of medical environmental microbiology concern. Some fungi are also a medical nuisance, particularly to transplant, HIV, and other groups of immunocompromised individuals. Although fungi are present everywhere, they predominate in the tropics owing to the warm temperatures and humid environment. A few fungi are pathogenic whereas some are opportunitistic pathogens. In most cases, the fungal infection is airborne, infecting the lungs and subsequently spreading via the circulating blood. Of most concern in medical environmental microbiology are blastomytic infections, coccidioidomycosis, histoplasma infections, and infections by *Pneumocystis carinii*.

## 14.5  RISK ASSESSMENT

It is obvious that the absence of pathogens in a particular environ may not be practically attainable all the time. Thus, infectious diseases will always be with us and assessing the potential to infect and spread of a particular infection in relation to the prevailing environmental factors has to have some level of risk assessment. Microbial risk assessment has been used to assess the risk of infection, illness, or mortality from microbial pathogens. It has been applied more widely in the control of waterborne diseases than in any other environmentally transmitted infectious diseases. Risk assessment should primarily initially identify or define the health effects associated with a particular hazard (in this instance, pathogen). Based on the ecological principle of disease, both the host and pathogen must come into contact in time and space (i.e., exposure) for the infectious disease to occur. Exposure to a particular pathogen depends on the initial concentration of that pathogen in a particular medium (or environmental matrix) and the likelihood of the pathogen coming into contact with the protected target population (e.g., humans). Risk assessment also involves characterizing the dosage of the pathogen in question (i.e., infectious dose), assessing the size and nature of the population at risk, and establishing the potential routes of exposure. All these three components are then combined in order to estimate the magnitude of the problem to public health. In practical terms, therefore, we should ask ourselves what concentrations of a particular pathogen pose a risk (i.e., meet a minimum infectious dose) and how can such a risk be minimized? What environmental manipulations (e.g., water treatment strategies, air purification methods, hospital disinfection techniques, etc.) would minimize the number and spread of a pathogen? What environmental conditions would influence microbial survival and the final population densities of the pathogen in question? Is the presence of other nonpathogenic microorganisms of any benefit in minimizing disease and, if so, how can they be used to our advantage?

### 14.5.1  Infectious dose

As an initial step in the disease process, the infectious dose will cause colonization of the pathogen in the body. This initial step often displays no symptoms (i.e., is asymptomatic). The infectious dose is normally expressed as $ID_{50}$ which basically refers to the dose required to cause infection in 50% of the exposed population. Ideally, the $ID_{50}$ is established based on a dose-response curve. Dose-response relations have been determined for a number of pathogens especially those that are waterborne. The dose-response and related $ID_{50}$ for a wide range of pathogens that are not waterborne are not known. Based on dose-response studies in human volunteers, two models have emerged, i.e.,

   (i)   the exponential model $P_i = 1 - exp(-rN)$
         and
   (ii)  the β-Poisson model $P_i = 1 - (1+N/)^{-\alpha}$

where $P_i$ is the probability of infection (i.e., the ability of the organism to reproduce in the host); N the exposure (i.e., colony-forming units of the pathogen ingested), r the fraction of organisms that survive to initiate the infection after a host-organism interaction, and $\alpha$, and β are constants for specific organisms that define the dose-response model (Table 14.8). The exponential model assumes that even a single microorganism is capable of multiplying (exponentially) and, in a sense, interacting with the host, causing infection in an individual. Experimental evidence shows that for some pathogens, there is a more gradual response to increasing dosages of the pathogen, making determination of r difficult. Thus, another model, the β-Poisson model, assumes that the value of r is not constant but rather is described by a probability distribution (β). In other words, under the β-Poisson model, the subjects are assumed to be exposed to the infectious dose of the pathogen in a random manner. As $\alpha$ increases, the β-Poisson model comes closer to the exponential (Regli et al., 1991). The two models are derived based on data from human feeding experiments, with amounts of the respective pathogen increased (i.e., response curve). Inherent in these feeding trials is the fact that the feeding subjects are usually

healthy adults. It is also important to note that when monitoring for pathogens in the environment, most methods recover only about 40% of the organisms present. Therefore, exposure may be underestimated in a majority of instances. Development of better detection techniques, coupled with additional epidemiological information, and therefore improved risk assessment methodologies would allow for more certain determinations of the risks associated and the subsequent development of better strategies to guard against infections.

## 14.5.2 Nutrients

Both clinical and experimental evidence shows that the supply of iron to a host is an important determinant of bacterial virulence and the general dynamics of many diseases, including those transmitted by *Yersinia* spp., Enterobacteriaceae, and *Plasmodium* sp., *Entamoeba histolytica*, fungal infections, as well as a host of viral infections, e.g., hepatitis B, HIV, and herpes simplex virus (Murray et al., 1978; Weinberg, 1978; Patrura and Hörl, 1999). Thus, excessive iron in the host may increase the risk of infections. When provided with an abnormally high supply of iron, even normally healthy individuals lose their ability to repel invading pathogens (Weinberg, 1978; Bullen et al., 1991; Pickett et al., 1992). On the other hand, iron can be chelated by mammalian proteins, making it less available for invading pathogens. Thus, deprivation of iron inhibits bacterial growth and iron-chelation therapy has been reported to significantly reduce malarial protozoan parasites from the bloodstream, increasing the rate of recovery from this infection (Mohanty et al., 2002). Other iron-binding proteins include lactoferrin, deferrioxamine, and transferrin. However, the validity of manipulating iron levels is still challenged by several scholars (e.g., Oppenheimer, 2001) who attribute the results to confounding factors. These studies have, in most cases lacked a holistic consideration of the ecology of the vectors and/or pathogens, the condition of the host,

**Table 14.8** Risk assessment for various pathogens and *E. coli* in water

| Organism | Risk assessment model and parameters[1] | Remarks | Reference |
|---|---|---|---|
| *Cryptosporidium* | Exponential; r = 0.00467 | Resistant to routine water treatment in the absence of filtration. The minimum infectious dose is 1 oocyst $100 L^{-1}$ and the $ID_{50}$ is 180 oocysts $100 L^{-1}$ in water. Required to achieve the 1 in a 10,000 annual risk level. The overall risk also depends on the immune status of the individual. Concentrations of 10-30 oocysts $100 L^{-1}$ should prompt water treatment modifications. | Rose (1997); Smith and Rose (1998) |
| *Giardia* | Exponential; r = 0.0198 | Resistant to routine water disinfection in the absence of filtration. Risk from *Giardia* infection depends on the immune status of the individual but a 1 in 10,000 infections per annum is considered acceptable. This translates into a concentration of $6.75 \times 10^{-6}$ giardia $L^{-1}$. | Regli et al. (1991) |
| *Salmonella* | Exponential; r = 0.00752 | Substantial ($10^5$-$10^6$ cfu) quantities of low and moderately pathogenic strains have a 1 in 10 probability of causing illness whereas small quantities ($\simeq$ 10 cfu) of highly pathogenic strains have the same probability of causing illness. | Latimer et al. (2001) |
| Poliovirus | Exponential; r = 0.009102 | A 1 in 10,000 infections per annum is considered acceptable which translates into a concentration of $1.51 \times 10^{-5}$ poliovirus $L^{-1}$. | Regli et al. (1991) |
| *E. coli* | Poisson; $\alpha$ = 0.1097; $\beta$ = 1524 Poisson; $\alpha$ = 0.1705; $\beta$ = 1,610,000 | Feeding studies with enteropathogenic *E. coli* (EPEC) show that $10^6$ organisms have a 1 in 10 chance of causing illness. | Powell et al. (2000) |
| *Shigella* | Poisson; $\alpha$ = 0.248; $\beta$ = 3.45 | Genetically related to *Salmonella* and is very virulent. Low doses of 10 organisms (cfu) have a 1 in 10 probability of causing illness. Primates are the only known reservoir. | Powell et al. (2000); Latimer et al. (2001) |
| Rotavirus | Poisson; $\alpha$ = 0.26; $\beta$ = 0.42 | Temperature and moisture govern the survival of this and a host of other viruses in the environment. A 1 in 10,000 infections per annual is considered acceptable. This translates into a concentration of $2.22 \times 10^{-7}$ rotaviruses $L^{-1}$. | Regli et al., 1991; Gerba et al. (2002) |

[1]Risk assessment model and parameters modified from Regli et al. (1991) and Rose et al. (1996).

and the environment. Thus, although the attention in these studies and clinical trials has focused exclusively on iron, we have to bear in mind that the pathogens need a variety of other nutrients and the direct competition for nutrients between pathogens and host cells does play a pivotal role in determining the outcome of the disease process. This is best explained when we revisit the resource-ratio theory which was extensively discussed in Chapter 8. In practice, lowering the iron concentration to control the pathogen can, below a certain concentration, also negatively affect the eukaryotic host cells. Furthermore, with a diverse array of metabolic functions, the iron limitation is likely to be counteracted by other metabolites such as glutamine. A high supply of glutamine ameliorates a variety of infections. Thus, if the concentrations of both iron and glutamine are compared on the same isocline, the competitive equilibrium for these two nutrients can be deliberately managed, resulting in one of three possible outcomes (Fig. 14.7). Because of the high glutamine requirements of the host, the pathogen is assumed to be the superior competitor for it. The iron/glutamine supply ratio can also be lowered by increasing the supply of glutamine (Rombeau, 1990). Glutamine is likely to be higher in individuals with better nutritional status as is typical of people living in developed countries. This theory may, at least in part, explain why iron-chelating as a strategy of controlling malarial infections gives seemingly contradictory results in malarious and nonmalarious regions. Most malarious regions are in developing countries with less adequate nutrition, and thus a high likelihood of glutamine deficiency. Going by the same analogy, the competitive resource ratio theory may also be applicable in the control of a variety of infectious diseases. For example, deficiencies of some vitamin-dependent plasma protein correlate strongly with the proportion of AIDS (Bissuel et al., 1992) and glutathione (Staal et al., 1992).

### 14.5.3 Other interacting factors

Besides dosage, development of clinical symptoms depends on the immune status of the host, age, virulence of microorganism, and type, strain, and route of infection. Furthermore, there are many complex interactions among microorganisms that protect us from diseases, either directly by keeping the infectious agent out of contact with us or by suppressing pathogenic reactions even when the pathogen is in contact with its host. Such interactions include antibiosis and predation but can, in some instances, be quite complex, resulting in the breakdown of microbial homeostasis. Such a breakdown can also be due to the metabolites generated by a particular taxon or a group of taxa which in turn changes the environment, favoring the pathogenic organisms in turn. Using the example of dental caries development, under healthy conditions teeth are typically colonized by a range of bacteria. However, a drop in pH due to the consumption and metabolism of sugars favors the dominance of acidogenic and aciduric (acid-tolerating) bacteria such as *Streptococcus sobrinus*, *S. mutans*, *S. mitis*, and lactobacilli which mineralize the enamel, causing dental caries. In this instance, the low pH (i.e., acidity) acts as the selective factor that disrupts the homeostasis (Marsh, 2003). Other environmental factors such as temperature, oxygen tension, and moisture content can also disrupt microbial homeostasis, generating various disease incidences.

The complexity of interacting factors that may lead to disease is also exemplified by the hantavirus pulmonary disease which emerged at Four Corners in southwest USA, as briefly mentioned earlier. In that outbreak, changes in land-use patterns and environmental conditions greatly reduced coyotes, owls, eagles, hawks, kestrels, and prairie falcons that eat rodents. This series of events, together with an abundance of piñon nuts subsequently led to a tenfold increase in the hantavirus-carrying deer

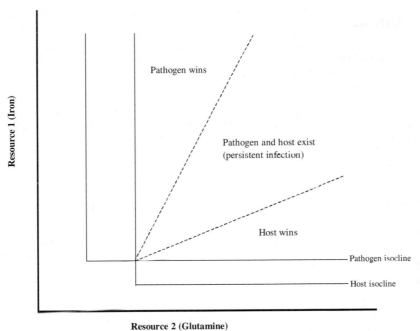

**Fig. 14.7** Hypothetical isoclines for the host and pathogen competing for two resources, glutamine and iron (adapted from Smith, 1993 with permission from Elservier).

mice populations which exposed humans to the virus (Grifo, 2001; Epstein, 2001). Deer mice feed on piñon nuts and had an unusually abundant food supply that year. The virus causes hemorrhagic fevers in humans and can be fatal. The virus may have been present in the Four Corners locality prior to these synergistic events but dormant, its transmission becoming amplified thereafter.

## 14.6 CONCLUDING REMARKS

Changes in environmental conditions, mainly as a result of local, regional or even global changes in weather, affect the geographic range of infectious diseases. These, coupled with human behavioral adaptations and public health measures impact the success or failure of such emerging and resurgent infectious diseases. Research in the vital area of linking infectious diseases to environmental changes

is still in its infancy but will, over time, prove invaluable in predicting disease outbreaks based on risk-based guidelines leading to the design of intervention measures that fall within realistic expectations. Monitoring of pathogens and environmental changes can be greatly aided by combining recent advances in remote sensing, molecular biology, geographic information systems and modeling. Last, but not least, public health programs and medical microbiology need to incorporate environmental factors in the incidence of infectious diseases in the context of selecting the best preventive and/or treatment regimes.

### References

Abad F.X., R.M. Pintó, J.M. Diez, and A. Bosch (1994). Disinfection of human enteric viruses in water by copper and silver in combination with low levels of chlorine. *Appl. and Environ. Microbiol.* **60:** 2377-83.

Acar J., M. Casewell, J. Freeman, C. Friis, and H. Goossens (2000). Avoparcin and virginiamycin as animal growth promoters: a plea for science in decision-making. *Clin. Microbiol. Infect.* **6:** 477-482.

Alterholt T.B., M.W. LeChevalier, W.D. Norton, and J.S. Rosen (1998). Effect of rainfall on giardia and crypto. *J. Amer. Water Works Assoc.* **90:**66-80.

Ansari S.A., S.R. Farrah, and G.R. Chaudhry (1992). Presence of human immunodeficiency virus nucleic acids in wastewater and their detection by polymerase chain reaction. *Appl. Environ. Microbiol.* **58:** 3984-3990.

Barcina I., J.M. González, J. Iriberi, and L. Egea (1989). Effect of visible light on progressive dormancy of *Escherichia coli* cells during the survival process in natural fresh water. *Appl. Environ. Microbiol.* **55:** 246-251.

Barinaga M. (1996). A shared strategy for virulence. *Science* **272:** 1261-1263.

Belperron A.A., D. Feltquate, B.A. Fox, T. Horii, and D.J. Bzik (1999). Immune response induced by gene gun or intramuscular infection of DNA vaccines that express immunogenic regions of the serine repeat antigen from *Plasmodium falciparum*. *Infect. Immun.* **67:** 5163-5169.

Bercovier H. (1999). Is tuberculosis really an emerging or a reemerging disease? *In*: E. Rosenberg (ed.). Microbial Ecology and Infectious Disease. ASM, Washington DC, pp. 205-212.

Berman D. and J.C. Hoff (1984). Inactivation of simian rotavirus SA11 by chlorine, chlorine dioxide, and monochloramine. *Appl. Environ. Microbiol.* **48:** 317-23.

Beuret C., D. Kohler, A. Baumgartner and T.M. Lüthi (2002). Norwalk-like virus sequences in mineral waters: one-year monitoring of three brands. *Appl. Environ. Microbiol.* **68:** 1925-1931

Bissuel F. et al. (1992). Acquired proteins deficiency: correlation with advanced disease in HIV-1-infected patients. *J. AIDS* **5:** 484-489.

Borgen K., G.S. Simonsen, A. Sundsfjord, Y. Wasteson, Ø. Olsvik, and H. Kruse (2000). Continuing high prevalence of VanA-type vancomycin-resistant enterococci on Norwegian poultry farms three years after avoparcin was banned. *J. Appl. Microbiol.* **89:** 478-485.

Brown S.L., R.L. Chaney, C.A. Loyd, J.S. Angle, and J.A. Ryan (1996). Relative uptake of cadmium by garden vegetables and fruits grown on long-term biosolid-amended soils. *Environ.l Sci. Technol* **30:** 3508-3511.

Bullen J.J., C.G. Ward, and H.J. Rogers (1991). The critical role of iron in some clinical infections. *Europ. J. Clin. Microbiol. Infect. Dis.* **10:** 613-617.

Chiang S.L. and J.J. Mekalanos (1999). Horizontal gene transfer in the emergence of virulent *Vibrio cholerae*. *In*: E. Rosenberg (ed.). Microbial Ecology and Infectious Disease. ASM, Washington DC, pp. 156-169.

Chmielewski R.A.W. and J.F. Frank (1995). Formation of viable but nonculturable *Salmonella* during starvation in chemically defined solutions. *Lett. Appl. Microbiol.* **20:** 380-384.

Chowdhury M.A.R., A. Huq, B. Xu, F.J.B. Madeira, and R.R. Colwell (1997). Effect of alum on free-living and copepod-associated *Vibrio cholerae* O1 and O139. *Appl. Environ. Microbiol.* **63:** 3323-3326.

Clavero M.R.S. and L.R. Beuchat (1996). Survival of *Escherichia coli* O157:H7 in broth and processed salami as influenced by pH, water activity, and temperature and suitability of media for its recovery. *Appl. Environ. Microbiol.* **62:** 2735-2740.

Colwell R.R. (1996). Global climate and infectious disease: the cholera paradigm. *Science* **274:** 2025-2031.

Cornelis G.R. (1998). The *Yersinia* deadly kiss. *J. Bacteriol.* **180:** 5495-5504.

Cunliffe N.A., J.S. Gondwe, R.L. Broadhead, M.E. Molyneux, P.A. Woods, J.S. Bresee, R.I. Glass, J.R. Gentsch, and C.A. Hart (1999). Rotavirus G and P types in children with acute diarrhoa in Blantyre, Malawi, from 1997 to 1998: predominance of novel P[6]G8 strains. *J. Med. Virol.* **57:** 308-12.

Cuthbert J.A. (2001). Hepatitis A: old and new. *Clin. Microbiol. Rev.* **14:** 38-58.

DuPont H.L., C.L. Chappell, C.R. Sterling, P.C. Okhuysen, J.B. Rose, and W. Jakubowski (1995). The infectivity of *Cryptosporidium parvum* in healthy volunteers. *New England J. Med.* **332:** 855-859.

Eldridge B.F. (1987). Strategies for surveillance, prevention and control of arbovirus diseases in western North America. *Amer. J. Trop. Med. Hyg.* **37:** 77S-86S.

Epstein P.R. (2000). Is global warming harmful to health? *Sci. Amer.* August:36-43.

Epstein P.R. (2001). Climate change and emerging infectious diseases. *Microbes Infect.* **3:** 747-754.

Fankhauser R.L., J.S. Noel, S.S. Monroe, T. Ando, and R.I. Glass (1998). Molecular epidemiology of "Norwalk-like viruses" in outbreaks of gastroenteritis in the United States. *J. Infect. Dis.* **178:** 1571-8.

Fayer R. (1994). Effect of high temperature on infectivity of *Cryptosporidium parvum* oocysts in water. *Appl. Environ. Microbiol.* **60:** 2732-2735.

Fayer R. and T. Nerad (1996). Effects of low temperatures on viability of *Cryptosporidium parvum* oocysts. *Appl. Environ. Microbiol.* **62:** 1431-1433.

Ford-Jones E.L., E. Wang, M. Petric, P. Corey, R. Moineddin, and M. Fearon (2000). Hospitalization for community-acquired, rotavirus-associated diarrhea. *Arch. Ped. Adol. Med.* **154:** 578-5.

Garrett L. (2000). Betrayal of Trust: The Collapse of Global Public Health. Hyperion, New York, NY.

Gerba C.P., J.B. Rose, C.N. Haas, and K.D. Crabtree (1996). Waterborne rotavirus: a risk assessment. *Water Res.* **30:** 2929-40.

Gerba C.P., I.L. Pepper, and L.F. Whitehead III (2002). A risk assessment of emerging pathogens of concern in the land application of biosolids. *Water Sci. Tech.* **46:** 225-230.

Gillet J.D. (1974). Direct and indirect influences of temperature on the transmission of parasites from insects to man. *In*: A.E.R. Taylor and R. Muller (eds.). The Effects of Meteorological Factors upon Parasites. Blackwell Sci. Publ., Oxford, UK, pp. 79-95.

Gilliver M.A., M. Bennett, M. Begon, S.M. Feore, and C.A. Hurt (1999). Antibiotic resistance found in wild rodents. *Nature* **401:** 233-234.

Glass R.I., P.E. Kilgore, R.C. Holman, S. Jin, J.C. Smith, P.A. Woods, J.C. Matthew, M. Shang, and J.R. Gentsch (1996). The epidemiology of rotavirus diarrhea in the United States: surveillance and estimates of disease burden. *J. Infect. Dis.* **174**(suppl): S5-S11.

Glass R.I., J. Noel, T. Ando, R. Fankhauser, G. Belliot, A. Mounts, U.D. Parashar, J.S. Breses, and S.S. Monroe (2000). The epidemiology of enteric caliciviruses from humans: a reassessment using new diagnostics. *J. Infect. Dis.* **181**(Suppl):S254-61.

Goddard L.B., A.E. Roth, W.K. Reisen, and T.W. Scott (2002). Vector competence of California mosquitoes for *West Nile virus. Emerg. Infect. Dis.* **8:** 1385-1391.

Gratacap-Cavallier B., O. Genoulaz, K. Brengel-Pesce, H. Soule, P. Innoenti-Fracillard, M. Bost, L. Gofti, D. Zmirou and J.M. Seigneurin (2000). Derection of human and animal rotavirus sequences in drinking water. *Appl. Environ. Microbiol.* **66:** 2690-2692

Grifo F.T. (2001). Biodiversity and human health. *In:* M.J. Novacek (ed.). The Biodiversity Crisis: Losing What Counts. New Press, New York, NY, pp. 40-43.

Halling-Sørensen, B., S.N. Nielsen, P.F. Lanzky, F. Ingerster, H.C.H. Lützhøff, and S.E. Jørgensen, (1998). Occurrence, fate and effects of pharmaceutical substances in the environment-a review. *Chemosphere* **36:** 357-393.

Harbarth S., W. Albrich, D.A. Goldmann, and J. Huebner (2001). Control of multiple resistant cocci: do international comparisons help? *Lancet Infect. Dis.* **1:** 251-261.

Harbarth S., W. Albrich, and C. Brun-Buisson (2002). Outpatient antibiotic use and prevalence of antibiotic-resistant pneumococci in France and Germany: a sociocultural perspective. *Emer. Infect. Dis.* **8:** 1460-1467.

Hasan J.A.K., M.A. R. Chowdhury, M. Shahabuddin, A.Huq, L. Loomis, and R.R. Colwell (1994). Polymerase chain reaction for detection of cholera toxin gene from viable but nonculturable *Vibrio cholerae* O1. *World J. Microbiol. Biotech.* **10:** 568-571.

Heberer T. (2002). Tracking persistent pharmaceutical residues from municipal sewage to drinking water. *J. Hydrol.* **266:** 175-189.

Hinnebusch B.J., R.D. Perry, and T.G. Schwan (1996). Role of the *Yersinia pestis* hemin storage (*hms*) locus in the transmission of plague by fleas. *Science* **273:** 367-370.

Horsburg, Jr. C.R. and A.M. Nelson (1997). Pathology of Emerging Infections. ASM , Washington DC.

Isaacson R.E. and M.E. Torrence (2002). The role of antibiotics in agriculture: a report from the American Academy of Microbiology, Washington DC. Available at http://www.asmusa.org.

Ista L.K., V.H. Pérez-Luna, and G.P. López (1999). Surface-grafted, environmentally sensitive polymers for biofilm release. *Appl. Environ. Microbiol.* **65:** 1603-1609.

Jjemba P.K. (2002). The potential impact of veterinary and human therapeutic agents in manure and biosolids on vegetation: a review. *Agric., Ecosyst. Environ.* **93:** 267-278.

Latimer H.K., L-A. Jaykus, R.A. Morales, P. Cowan, and D. Crawford-Brown (2001). A weighted composite dose-response model for human salmonellosis. *Risk Analysis* **21:** 295-305.

Lèvèque F., J.M. Crance, C. Beril, and L. Schwartzbrod (1995). Virucidal effect of UV light on hepatitis A virus in sea water: evaluation with cell culture and TR-PCR. *Water Sci. Tech.* **31:** 157-60.

Licht T.R. et al. (1996). Role of lipopolysaccharide in colonization of mouse intestine by *Salmonella typhimurium* studied by in situ hybridization. *Infect. Immun.* **64:** 3811-3817.

Linder K. and J.D. Oliver (1989). Membrane fatty acid and virulence changes in the viable but nonculturable state of *Vibrio vulnificus. Appl. Environ. Microbiol.* **55:** 2837-2842.

Linthicum K.J., A. Ayamba, C.J. Tucker, P.W. Kelley, M.F. Meyers, and C.J. Peters (1999). Climate and satellite indicators to forecast Rift Valley fever epidemics in Kenya. *Science* **285:** 397-400.

Lobitz B.L., L. Beck, A. Huq, B. Wood, G. Fuchs, A.S.G. Foruque, and R.R. Colwell (2000). Climate and infectious diseases: use of remote sensing for detection of *V. cholerae* by indirect measurement. *Proc. Natl. Acad. Sci. (USA)* **97:** 1438-1443.

Lodder W.J., J. Vinjé, H. van de Heide, A.M. de Roda Husman, E.J.T.M. Leenen, and M.P.G. Koopmans (1999). Molecular detection of Norwalk-like caliciviruses in sewage. *Appl. Environ. Microbiol.* **65:** 5624-7.

Marsh P.D. (2003). Are dental diseases examples of ecological catastrophes? *Microbiol.* **149:** 279-294.

Mayer J.D. (2000). Geography, ecology and emerging infectious diseases. *Soc. Sci. Med.* **50:** 937-952.

McBride M.B. (1995). Toxic metal accumulation from agricultural use of sludge: Are USEPA regulations protective? *J. Environ. Qual.* **24:** 5-18.

Medawar P.B. and J.S. Medawar (1983). Viruses. In Medawar PB and J.S. Medawar (ed.) Aristotle to Zoos: A Philosophical Dictionary of Biology. Harvard University Press, Cambridge, Massachusetts.

Mohanty D., K. Ghosh, A.V. Pathare, and D. Karnad (2002). Deferiprone (L1) as an adjuvant therapy for *Plasmodium falciparum* malaria. *Indian J. Med. Res.* **115:** 17-21.

Motes M.L., A. DePaola, D.W. Cook, J.E. Veazey, J.C. Hunsucker, W.E. Garthright, R.J. Blodgett, and J. Chirtel (1998). Influence of water temperature and salinity on *Vibrio vulnificus* in Northern Gulf and Atlantic Coast oysters (*Crassostrea virginica*). *Appl. Environ. Microbiol.* **64:** 1459-1465.

Murray M.J., A.B. Murray, M.B. Murray, and C.J. Murray (1978). The adverse effect of iron repletion on the course of certain infections. *Bri. Med. J.* **ii:** 1113-1115.

Navin T.R., S.J.N. McNabb, and J.T. Crawford (2002). The continued threat of tuberculosis. *Emerg. Infect. Dis.* **81:** 1187.

NRC (2001). Under the Weather: Climate, Ecosystems, and Infectious Disease. Natl. Acad. Press, Washington DC.

Odum E.P. (1989). Ecology and Our Endangered Life-Support Systems. Sinauer Assoc., Inc. Publ., Sunderland, MA.

Oliver J.D. (2000). The Public Health Significance of Viable but Nonculturable Microorganisms in the Environment. ASM Press, Washington DC, pp. 277-300.

Oppenheimer S.J. (2001). Iron and its relation to immunity and infectious disease. *J. Nutr.* **131** (Suppl. 2)**:** 616S-633S.

Ottermann K.M. and J.J. Mekalanos (1994). Regulation of cholera toxin expression. *In:* I.K. Wachsmuth, P.A. Blake, and Ø. Olsvik (eds.). *Vibrio cholerae:* Molecular to Global Perspectives. ASM, Washington DC, pp. 177-185.

Pancorbo O.C., B.G. Evanshen, W.F. Campbell, S. Lambert, S.K. Cutis, T.W. Woolley (1987). Infectivity and antigenicity reduction rates of human rotavirus strain Wa in fresh waters. *Appl. Environ. Microbiol.* **53:** 1803-11.

Pascual M., X. Rodo, S.P. Ellner, R. Colwell, and M.J. Bourma (2000). Cholera dynamics and El Niño-southern oscillation. *Science* **289:** 1766-1769.

Patrura S.I. and W.H. Hörl (1999). Iron and infection. *Kidney Intnatl.* **55** (Suppl. 69)**:** S125-S130.

Payment P. (1993). Viruses: prevalence of disease, levels, and sources. *In:* C.F. Craun (ed.). Safety of Water Disinfection: Balancing Chemical and Microbial Risks. ILSI Press, Washington DC, pp. 99-113.

Pell A. (1997). Manure and microbes: public and animal health problems? *J. Dairy Sci.* **80:** 2673-2681.

Perry R.D. and J.D. Fetherston (1997). *Yersinia pestis*—etiologic agent of plague. *Clin. Microbiol. Rev.* **10:** 35-66.

Petersen A., J.S. Andersen, T. Kaewmak, T. Somsiri, and A. Dalsgaard (2002). Impact of integrated fish farming on antimicrobial resistance in a pond environment. *Appl. Environ. Microbiol.* **68:** 6036-6042.

Pickett C.L. et al. (1992). Iron acquisition and hemolysin production by *Campylobacter jejuni. Infect. Immun.* **60:** 3872-3877.

Powell M.R., E. Ebel, W. Schlosser, M. Walderhaug, and J. Kause (2000). Dose-response envelope for *Escherichia coli* O157:H7. *Quant. Microbiol.* **2:** 141-163.

Rahman I., M. Shahamat, M.A.R. Chowdhury, and R.R. Colwell (1996). Potential virulence of viable but nonculturable *Shigella dysenteriae* type 1. *Appl. Environ. Microbiol.* **62:** 115-120.

Rao V.C., J.M. Symons, A. Ling, P. Wang, T.G. Metcalf, J.C. Hoff, and J.L. Melnick (1988). Removal of hepatitis A virus and rotavirus by drinking water treatment. *J. Amer. Water Works Assoc.* **80:** 59-67.

Raphael R.A., S.A. Sattar, and V.S. Springthorpe (1985). Long-term survival of human rotavirus in raw and treated river water. *Can. J. Microbiol.* **31:** 124-8.

Reeves W.C., J.L. Hardy, W.K. Reisen, and M.M. Milby (1994). Potential effect of global warming on mosquito-borne arboviruses. *J. Med. Entomol.* **31:** 323-332.

Regli S., J.B. Rose, C.N. Haas, and C.P. Gerba (1991). Modeling the risk from Giardia and viruses in drinking water. *J. AWWA* **83:** 76-885.

Reisen W.K., S.B. Presser, and J.L. Hardy (1993). Effect of temperature on the transmission of western equine encephalomyelitis and St. Louis encephalitis virus by *Culex tersalis* (Diptera: Culiculae). *J. Med. Entomol.* **30:** 151-160.

Rochelle P.A. and R. De Leon (2001). A review of methods for assessing the infectivity of *Cryptosporidium parvum* using in-vitro cell culture. *In:* M. Smith and K.C. Thompson (eds) *Cryptosporidium:* The Analytical Challenge. Roy. Soc. Chem., Cambridge, UK, pp. 88-95.

Rombeau J.L. (1990). A review of the effects of glutamine-enriched diets on experimentally induced enterocolitis. *J. Parent. Ent. Nutr.* **14** (suppl.)**:** 100-105.

Rose J.B. (1997). Environmental ecology of *Cryptosporidium* and public health infections. *Ann. Rev. Public Health* **18:** 135-161.

Rose J.B., J.T. Lisle, and C.N. Haas (1996). The role of pathogen monitoring in microbial risk Assessment. In: C.J. Hurst (ed.). Modeling Disease Transmission and Its Prevention by Disinfection. Cambridge Univ, Press, Cambridge, UK, pp. 75-98.

Rose J.B., A. Huq, and E.K. Lipp (2001). Health, climate and infectious disease: a global perspective. Available at http://www.asmusa.org/acasrc/pdfs/climate2.pdf. (downloaded 4/18/2001).

Rose J.B., S. Daeschner, D.R. Easterling, F.C. Curriero, S. Lele, and J.A. Patz (2000). Climate and waterborne disease outbreaks. *J. Amer. Water Works Assoc.* **9:** 77-87.

Roszak D.B., D.J. Grimes, and R.R. Colwell (1984). Viable but nonrecoverable stage of *Salmonella enteritidis* in aquatic systems. *Can. J. Microbiol.* **30:** 334-338.

Rubin E.J., M.K. Waldor, and J.J. Mekalanos (1998). Mobile genetic elements and the evolution of new epidemic strains of *Vibrio cholerae. In:* R.A. Krause (ed.). Emerging Infections. Acad. Press, New York, NY, pp. 147-161.

Scandura J.E. and M.D. Sobsey (1997). Viral and bacterial contamination of groundwater from on-site sewage treatment systems. *Water Sci. Technol.* **35:** 141-146

Schulman J.L. and E.D. Kilbourne (1963). Experimental transmission of influenza virus in mice: some factors affecting the incidence of transmitted infection. *J. Experim. Med.* **118:** 267-275.

Smith H.V. and J.B. Rose (1998). Waterborne cryptosporidiosis: current status. *Parasitol. Today* **14:** 14-22.

Smith V.H. (1993). Implications of the resource-ratio theory for microbial ecology. *Adv. Microb. Ecol.* **13:** 1-37.

Smithburn K.C., T.P. Hughes, A.W. Burke, and J.H. Paul (1940). A nuerotropic virus isolated from the blood of a native of Uganda. *Amer. J. Med. Hyg.* **20:** 471-492.

Staal F.J., S.W. Fla, M. Roederer, M.T. Anderson, L.A. Herzenberg, and L.A. Herzenberg (1992). Glutathione deficiency and human immunodeficiency virus infection. *Lancet* **339(8798):** 909-912.

Stender H., K. Lund, K.H. Petersen, O.F. Rasmussen, P. Hongmanee, H. Miorner, and S.E. Godtfresden (1999a). Fluorescence in situ hybridization assay using peptide nucleic acid probes for differentiation between tuberculous and nontuberculous *Mycobacterium* species in smears of mycobacterium cultures. *J. Clin. Microbiol.* **37:** 2760-2765.

Stender H., T.A. Mollerup, K. Lund, K.H. Petersen, P. Hongmanee, and S.E. Godtfresden (1999b). Direct detection and identification of *Mycobacterium tuberculosis* in smear-positive sputum samples by fluorescence in situ hybridization (FISH) using peptide nucleic acid (PNA) probes. *Intnatl. J. Tuberc. Lung. Dis.* **3:** 830-837.

Thiele-Bruhn S. (2003). Pharmaceutical antibiotic compounds in soils—a review. *J. Nutr. Soil Sci.* **166:** 145-167.

Torok T.J., P.E. Kilgore, M.J. Clarke, R.C. Holman, J.S. Bresee, and R.I. Glass (1996). Visualizing geographic and temporal trends in rotavirus activity in the United States. *Ped. Infect. Dis. J.* **16:** 941-6.

Tsai Y-L. and D.J. Krogstad (1998). The resurgence of malaria. *In*: W.M. Scheld, W.A. Craig, and J.M. Hughes (eds.). Emerging Infections. ASM Press, Washington DC, vol. 2, pp. 195-211.

Ueda T. and N.J. Horan (2000). Fate of indigenous bacteriophage in a membrane bioreactor. *Water Res.* **34:** 2151-2159

Unicomb L.E., G. Podder, J.R. Gentsch, P.A. Woods, K.Z. Hassan, A.S.G. Faruque, M.J. Albert, and R.I. Glass (1999). Evidence of high-frequency genomic reassortment of group A rotavirus strains in Bangladesh: emergence of type G9 in 1995. *J. Clin. Microbiol.* **37:** 1885-91.

Weinberg E.D. (1978). Iron and infection. *Microbiol. Rev.* **42:** 45-66.

Wolf P.W. and J.D. Oliver (1992). Temperature effects on the viable but nonculturable state of *Vibrio vulnificus*. *FEMS Microbiol. Ecol.* **101:** 33-39.

Wuhib T., T.M. Silva, R.D. Newman, L.S. Garcia., M.L. Pereira, C.S. Chaves, S.P. Wahlquist, R.T. Bryan, and R.L. Guerrant (1994). Cryptosporidial and microsporidial infections in human immunodeficiency virus-infected patients in northeastern Brazil. *J. Infect. Dis.* **170:** 494-497.

Yolkan R.H., S. Li, J. Perman, and R. Viscidi (1991). Persistent diarrhea and fecal shedding of retroviral nucleic acids in children infected with huma immunodeficiency virus. *J. Infect. Dis.* **164:** 61-66.

# 15

# Environmental Biotechnology and Biological Control

Biotechnology refers to the integrated use of biochemistry, molecular biology, genetics, microbiology and chemical engineering, to achieve industrial goods and services. It emerged as a result of breakthroughs in recombinant DNA. Recombinant DNA technology involves the splicing of genes and inserting genetic material taken from one location in another. It may also involve removing, adding, or transferring such genetic material from one organism to another. The gene sequences are cut out using restriction enzymes. Environmental biotechnology specifically refers to the application of biotechnology to manage environmental problems. It has traditionally included a wide array of processes such as wastewater treatment, composting, agriculture (tilling, sowing and harvesting of crops), improving agriculture through breeding, machinery, seed storage and maintenance, inoculating legumes with rhizobium and mycorrhizae and other yield enhancing microorganisms, aquiculture, silviculture, microbial mining, etc. Over the past two decades, however, environmental microbiology has explored fairly new areas which involve genetically engineering microorganisms to perform specific tasks in the environment, such as bioremediating sites that have been contaminated with explosives, pesticides, metals, treat-ment of toxic wastes using plant-microbe systems (phytoremediation), producing biodegradable plastics, bioprocessing of fossil fuels, reducing frost damage to crops, and using ultramicrobacteria in petroleum recovery and in in-situ bioremediation.

The specific protocols followed in biotechnology are eloquently outlined in a growing range of molecular biology books and manuals such as Sambrook and Russell (2001) and generally, involve preparing a "library" of foreign nucleic acid in a cloning vector which may be a plasmid or a virus. The cloned DNA is then transferred to a bacterial host cell (E. coli has been very widely used for this purpose but other bacteria can be used depending on the objective of the study). Bacterial cells are grown and selected for desirable trait(s), characterized, and used in their entirety or the desirable protein(s) extracted for the target purpose. Extraction of DNA from the cells follows a fairly simple procedure whereby cells are centrifuged to form a pellet, lysed, and the DNA then solubilized. Further purification of the DNA is obtained by adding sodium chloride and hexadecyltrimethylammonium bromide (CTAB) to remove polysaccharides and other macromolecules, and then adding phenol or chloroform. The addition of chloroform stabilizes the boundary layer between DNA and the aqueous layer and is followed by precipitating the now pure DNA with ethanol. An almost similar guideline is used to extract RNA from soils and sediments (Moran et al., 1993; Hurt et al., 2001).

In a number of instances, the extracted DNA is digested with restriction enzymes which cuts the DNA at specific sites. Key features of restriction enzymes are that they recognize specific sequences of single length and cut the DNA strand at specific locations within the recognition sequence. The probability of finding a binding sequence is inversely proportional to the sequence length. Thus for a 4-base cutting restriction enzyme, the probability is $1/(4)^4$ or 1/256; a 6-base cutting restriction enzyme (e.g. BamH1), is $1/(4)^6$ or 1/4096; an 8-base cutting restriction enzyme, is $1/(4)^8$ or 65,536; a 10-base cutting restriction enzyme, is $1/(4)^{10}$ or 1/1,048,576, and so on. The various size fragments can be visualized on a gel, with the rate of migration of the DNA through the gel depending on both the molecular weight and molecular shape of the DNA since small molecules (i.e., with a low molecular weight) move more rapidly than their large (high molecular weight) counterpart. Using *E. coli*, which has a genome of 4.7 million base pairs, a 10-base cutting restriction enzyme should find approximately five sites, thus creating 6 pieces on the gel. The DNA is then sequenced to obtain a clear analysis of the chemical structure of the genes involved (Sambrook and Russell, 2001).

## 15.1 PROBLEMS AND OPPORTUNITIES

### 15.1.1 Public concerns

There is growing interest in the area of genetic exchange in the environment because of the increasing practice of introducing genetically engineered microorganisms (GEMs) to perform particular tasks in natural systems. Some biotechnological advances have, however, inevitably raised long-term safety and environmental concerns (Marx, 1987; Henderson and Gatewood, 1998; Letourneau and Burrows, 2002). Thus, questions such as whether the engineered microorganisms will combine with the wild type or even jump across species forming "new" pathogens, the possibility of transferring (antibiotic-resistance) genetic markers from the GEMs to other organisms (including humans) in the environment, the possibility of the engineered organisms expressing unintended or undesired characteristics, how stable and/or sustainable the engineered microorganisms can be in the environment, etc., are continuously agonized over. Other concerns include uncertainties about the possible production of more toxic metabolites and to what extent we can extrapolate the behavior of GEMs from one environment to another. Five ecological parameters have been proposed to study the biosafety of GEMs in the environment, viz., their viability or ability to survive, their ability to spread, their population dynamics, their competition with other microorganisms in the environment, and their effect on biocoenoses in the environment (Tappeser et al., 2002).

Some of these questions can be addressed from a larger context by considering the fact that microorganisms are continually evolving in the environment through naturally occurring lateral and horizontal gene transfer. Thus, once in the environment, GEMs may transfer their recombinant genes to the native microorganisms, possibly creating potentially harmful hybrids. However, the limited research that has been conducted to analyze sequences shows that the integration of such genetic material into the chromosome is most successful when the donor and recipient are closely related (Ochman et al., 2000). The transferred genetic material can be associated with antibiotic resistance, changes in metabolic pathways, acquisition of virulence to benign microorganisms, growth and/or regrowth of the GEM, thus displacing the indigenous microbial population. Regrowth of GEMs can occur after environmental conditions become favorable (Ripp et al., 2000) whereas antibiotic resistance genes are associated with highly mobile genetic elements (Ochman et al., 2000).

Once released into the environment, the probability of GEM impacting nontarget organisms cannot be reduced to zero. Thus, prior to their release, proper studies have to be conducted to assess potential problems. It is important to realize that their release into the environment, even in small quantities, is difficult to contain and impossible to retrieve (Wrubel et al., 1997). In the United States, the release of such organisms is rightfully regulated under the Toxic Substances Control Act, releases only being granted after extensive scrutiny to assess the risks with a delicate balance but not to unnecessarily stifle technological advancement (Henderson and Gatewood, 1998). Similarly stringent regulations about GEMs and genetically modified organisms (GMOs) in general, have also been imposed in other countries (Braun, 2002; Gottweis, 1998). Such assessments are usually lengthy and bureaucratic and are more likely to be adapted more rapidly in bioreactor settings than the open environment. In general, biotechnological approaches that modify microorganisms by deleting genes, rather than adding new genes, seem to gain approval over a short duration through the regulatory process (Marx, 1987; Wrubel et al., 1997).

## 15.1.2 Promising opportunities

### 15.1.2.1 Bacterial artificial chromosomes (BAC)

Biotechnology also offers opportunities to exploit the functional diversity of microorganisms in nature. As pointed out in previous chapters, a wide range of eukaryotic and prokaryotic microbes are just now being explored. Tapping their full potential requires development of methods to evaluate and distinguish the biosynthetic and physiological diversity of these organisms in nature. Cloning approaches have been proposed by Rondon et al. (1999,

2000) whereby microbial DNA (>100 kb) segments were extracted directly from soil and then inserted into a surrogate *E. coli* genome using a bacterial artificial chromosome (BAC) vector. The resultant clones were then analyzed for novel phenotypic expressions in the *E. coli* clones. Considering the large size of the inserts, entire pathways in one BAC plasmid can be transferred and successfully expressed in the recipient organism. Through this approach, Rondon et al. (2000) have found clones that display novel capabilities. Further research along these lines could lead to the discovery of new natural products of commercial value, such as novel metabolic pathways, gene regulatory systems, drug-resistance genes, pharmaceutical products, and genes for virulence even from phylotypes such as *Acidobacterium, Holophaga, Verrucomicrobia,* candidate divisions TM6, TM7, WS1, and *Planctomycetes,* whose relatives have never been cultured or only rarely so.

### 15.1.2.2 Biosensors

Biosensors such as luciferase (*lux*) gene (Willardson et al., 1998; Boyd et al., 2001; Ripp et al., 2000), β-galactosidase (*gus*) gene (Sessitsch et al., 1997), and green fluorescent protein (Stiner and Halverson, 2002) have been explored as new analytical tools for monitoring microbial activity in the environment. They analyze the expression of genes by typically creating transcriptional fusions between a promoter of interest and a reporter gene. Studies using various pollutants have shown that the extent of expression of the reporter gene corresponds to the available concentration of the pollutant or its derivatives and the duration of exposure to the pollutant (Willardson et al., 1998; Stiner and Halverson, 2002). Thus, biosensors are increasingly used for measuring the bioavailability of pollutants instantly and cheaply, compared to some conventional (and often expensive) techniques (Table 15.1). With increasing concern about chemical and biological warfare, advances in biosensors as a tool

**Table 15.1**   Comparison of the concentration of toluene and its derivatives in well water using standard gas chromatography-mass spectroscopy L (GC-MS) and the pGLTUR-based bacterial biosensor

| Method | Sampling date | Compounds | Concentration | |
| --- | --- | --- | --- | --- |
| | | | By direct assay (ppm) | Toluene equivalents (ppm)[a] |
| GC-MS | 3/6/1996 | Benzene | 12 | 4.3 |
| | | Toluene | 7.4 | 7.4 |
| | | Ethylbenzene | 0.35 | 0 |
| | | Xylenes | 3.4 | 9.2 |
| | | | | **Total = 20.9** |
| pGLTUR-based biosensor | 6/4/1996 | Toluene or equivalent | 13.0 ± 1.6 | 19.8 ± 2.1 |
| GC-MS | 10/7/1996 | Benzene | 10.5 | 3.8 |
| | | Toluene | 5.7 | 5.7 |
| | | Ethylbenzene | 0.6 | 0 |
| | | Xylenes | 3.2 | 8.6 |
| | | | | **Total = 18.1** |

[a] The toluene equivalent value was calculated by multiplying the ppm concentration of each compound by the ratio of the corresponding half-maximal induction ($K_{1/2}$) to the $K_{1/2}$ of toluene. $K_{1/2}$ (μM) values for benzene, toluene, ethylbenzene, and xylenes were 468, 169, 0, and 62.4 respectively (Willardson et al., 1998) with permission from the American Society for Microbiology.

will also play a crucial role in the detection of undesirable chemical and biological agents in the environment.

### 15.1.2.3 Microarrays

Microarray analysis is also emerging as a potentially useful source of information to answer relevant questions in environmental microbiology such as pathway analysis, species identification, and investigating the mode of action of toxic compounds (Cho and Tiedje, 2001; Ye et al., 2001). The approach involves immobilizing deoxyribonucleotide probes on a gel matrix bound to a small glass surface (1.2 cm × 1.2 cm) or membrane. Microarray technology is covered in this chapter rather than the chapter concerning methods, because it displays a perfect "marriage" between biology and engineering (microchip array fabrication). It provides great potential for identifying microorganisms in their natural environments, as well as analyzing their function and dynamics simultaneously in a single assay. Its potential has yet to be rigorously tested and validated in a variety of environments, however. For one thing, the target and probe sequences are highly diverse in the environment, unlike pure culture settings where microarray technology has been more extensively used (Zhou and Thompson, 2002). Furthermore, environmental matrices such as soil and sediments often contain organic compounds, humic substances, and heavy metals, which greatly interfere with hybridization of DNA on microarrays. Using fluorescence, multiple colors can be used to label different sequences, enabling *in situ* measurements based on fluorescence patterns. Furthermore, the nucleotides can be readily adapted to detection using different colors (i.e., multicolor detection). Microarrays can be advantageous compared to gel electrophoresis because they enable rapid determination of the sequences of an unknown test sample and the array microchips are reusable.

## 15.2 SOME CASE STUDIES OF SUCCESSFUL GENE EXPRESSION

### 15.2.1 The β-glucuronidase (GUS) gene

The *gusA* gene which encodes for β-glucuronidase was initially used as a reporter gene in plants (Jefferson et al., 1987; Martin et al., 1992). It was later adapted for use as a marker for ecological studies in Gram-negative bacteria (Wilson et al., 1995) and has been specifically used to study nodule occupancy and microbial competition in the rhizosphere (Sessitsch et al., 1997; Yuhashi et al., 2000; Bloem and Law, 2001). Once the *gusA* gene is introduced into a *Rhizobium* strain, the marked bacterium is detectable because it turns a distinct blue on incubation of the bacteria in 5-bromo-4-chloro-3-indolyl β-D-glucuronide (X-GlcA). The X-GlcA imparts the blue coloration to the nodules which contain the *gus*-marked *Rhizobium* sp. as well (Fig. 15.1). It is crucial that the *gusA* marked transconjugants, just like the mutants developed by the conventional and widely used antibiotic resistance technique, possess similar effectiveness and competitiveness as the wild type. To determine whether genetically modifying the *Rhizobium leguminosurum* bv. *phaseoli* by inserting the *gusA* gene significantly affects the nodulation and nitrogen fixation potential of the bacterium, a *gusA* gene-marked transconjugant was compared with the wild type and also with a mutant obtained using the conventional antibiotic-resistance marker technique. Nodulation did not differ significantly between CIAT 899 wild type, CIAT 899::*gusA10A,* and CIAT 899 Str$^R$C (Table 15.2). The only differences seen on nodulation were attributable to differences between the two bean cultivars used, with Hondurus 35 CF 480 consistently giving more nodule biomass than Riz 44. However, this difference was only significant when the plants were inoculated with the wild-type strain. This significant difference in nodule biomass did not

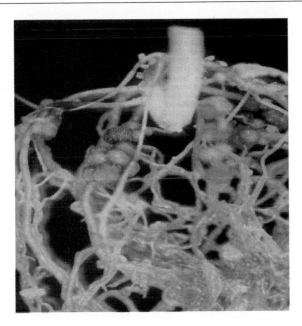

**Fig. 15.1**   Stained roots showing nodule occupancy by CIAT899 (unstained nodules) and CIAT 899::gusA10 (blue nodules). (*Source:* Sessitsch et al., 1997; reprinted with permission from Elsevier).

result into any significant difference in shoot biomass between the plants inoculated with the wild-type strain or any of the mutants, suggesting that the amount of nitrogen fixed did not differ from the wild-type and the two mutants, irrespective of the mutation strategy adapted.

**Table 15.2**   Effectiveness of CIAT899::*gusA10A* and CIAT 899 Str[R] C mutants (Jjemba, unpubl.).

| Mutant | Cultivar | Nodules (plant$^{-1}$)[a] | Nodule mass (mg plant$^{-1}$)[a] | Shoot biomass (g plant$^{-1}$)[a] |
|---|---|---|---|---|
| Uninoculated | Riz 44 | 45 b | 93.4 d | 0.48 b |
| | Hondurus 35 CF480 | 0 b | 0 d | 0.52 b |
| CIAT 899 Wild type | Riz 44 | 164 a | 156.5 c | 1.10 a |
| | Hondurus 35 CF480 | 191 a | 265.3 a | 1.32 a |
| CIAT 899::*gusA10*A | Riz 44 | 185 a | 172 bc | 1.76 a |
| | Hondurus 35 CF480 | 203 a | 220.1 ac | 1.17 a |
| CIAT 899 St[R] C | Riz 44 | 196 a | 183.4 ac | 1.17 a |
| | Hondurus 35 CF480 | 201 a | 246.1 ab | 1.39 a |
| Coefficient of variation (%) | | 29.4 | 35.0 | 28.5 |
| ANOVA | | | | |
| Cultivar | | NS[b] | *[c] | NS |
| Mutant | | ***[c] | NS | *** |
| Cultivar × mutant | | NS | NS | NS |

[a] Means accompanied by the same letter within a column do not differ significantly (p < 0.05)
[a] NS: Not significantly different at 0.001 level (***), 0.01 level (**) or 0.05 level (*).

The uninoculated plants had significantly lower biomass indicating that the plants responded to inoculation with CIAT 899 or any of its derivatives. Owing to the widespread concern about introduction of GEMs in the environment, however, its use can only be restricted to controlled conditions until proper risk assessment studies have been conducted.

## 15.2.2 *Bacillus thuringiensis* (Bt) toxin gene

*Bacillus thuringiensis*, a Gram-positive bacterium originally discovered in Thuringia province (Germany) in 1911, is a classic example of biological control of plant pests. This bacterium and its spores produce a toxin (Bt-toxin) highly poisonous to certain insects, especially the larvae of Lepidoptera, but has no effect on birds, mammals, fish and other wildlife. *B. thuringiensis* is thus an effective biological control agent already used commercially. The toxins interfere with the digestive system of insects. Shortly after eating foliage coated with this toxin, the larvae suffer paralysis, stop feeding, and soon die. A conventional practice in agriculture has been to apply a biological fungicide containing spores of *B. thuringiensis*.

The Bt toxin gene is present on a plasmid. Using genetic engineering, two improvements have been made, notably pest control on plants and in the rhizosphere. Recent advances in molecular biology and biotechnology have enabled successful cloning of the Bt-toxin gene into plants such as tomatoes, tobacco, cotton, corn, *Brassica* sp., and others, whereby the transformed plants inherently produce the toxin (Vaeck et al., 1988; Letourneau and Burrows, 2002). The process by which such transgenic plants are generated is beyond the scope of this book but from an environmental microbiology perspective, these plants aroused interest in microbiologists to acertain whether once harvested, the plant residues ploughed back into the soil would still contain the toxin and its biodegradation kinetics (Koskella and Stotzky,

1997; Saxena and Stotzky, 2001). Recent studies have shown that the Bt-toxin can bind to clays and humic substances in some soils and remain resistant to biodegradation and inactivation; in the absence of adsorption to the soil colloids, the toxins are readily utilized as a substrate by microbial consortia (Table 15.3). Whether free or bound, the toxins remain insecticidal to the larvae of various insects (Crecchio and Stotzky, 1998; Tapp and Stotzky, 1995a and b) and can persist in soil for more than 72 months (Palm et al., 1996; Tapp and Stozsky, 1998). This could constitute a hazard to nontarget organisms and raises the possibility of the target insects becoming resistant to the toxin in a manner similar to the development of antibiotic-resistant microorganisms as a result of selection and enrichment. This possibility should be factored into the assessment of potential risks associated with releasing genetically modified Bt-plants.

Clearly discernible from Table 15.3 is the fact that the toxins are utilized as a source of N when they are free. They are utilized to a much lesser extent when bound on clay or soil colloids. It can also be seen that the accessibility of the adhered toxin was slightly enhanced by alternate wetting and drying of the soil, resulting in somewhat improved degradation. Another approach involved the successful transfer of the pesticide production gene into a rhizosphere-colonizing bacterium. The Bt-toxin plasmid was moved from *Bacillus* to *Pseudomonas* sp., the pesticide-producing *Pseudomonas* inoculum protected the germinating seeds under greenhouse conditions.

## 15.2.3 GEMs for metabolizing environmental pollutants and monitoring bioremediation

Although most organic pollutants in the environment are ultimately biodegraded by existing microorganisms, the process can take years for some compounds. This reality, coupled with the fact that the number and quantities of com-

**Table 15.3** Utilization of *B. thuringiensis*-derived toxin by microorganisms as a substrate when free or bound onto clay minerals[a]

| Clay or soil fraction binding medium | Utilization of toxins as a source of | | | | | |
|---|---|---|---|---|---|---|
| | Carbon | | Nitrogen | | Carbon and nitrogen | |
| | Free | Bound | Free | Bound | Free | Bound |
| Montimorillonite homionic to sodium | + | - | + | - | + | - |
| Montimorillonite homionic to calcium | + | - | ND | ND | ND | ND |
| Montimorillonite homionic to aluminum | + | - | ND | ND | ND | ND |
| Kaolinite homionic to sodium | + | - | + | + | + | - |
| Kaolnite homionic to calcium | + | - | ND | ND | ND | ND |
| Clay-size fraction from soil | + | - | + | + | + | - |
| Clay-size fraction from a sandy loam soil amended with kaolinite (6% v/v) | + | - | + | + | + | - |
| Clay-size fraction from a sandy loam soil amended with montimorillonite | + | - | + | + | + | - |

[a]Microbial consortium initially obtained by enriching a soil slurry with citrate minimal medium and incubating at 37°C for 50 h.

+: utilization; -: no utilization of toxin as a substrate as determined by measuring $O_2$ uptake by the direct Warburg method.

ND: Not determined

*Source:* Koskella and Stotzky (1997) with permission from Elsevier.

pounds introduced in the environment have tremendously increased since the Industrial Revolution, means the more recalcitrant compounds have built up in some environments because they are degraded very slowly or incompletely. Thus, although microorganisms have a great potential to evolve under selection pressure and develop pathways that enable them to degrade these compounds, such potential can be slow in nature. Through biotechnology and genetic engineering, the evolution and development of new metabolic pathways has in some instances been initiated or accelerated with the ultimate goal of using such engineered microorganisms in bioremediation of contaminated environments. Such modifications have either genetically manipulated rate-limiting steps in known metabolic pathways or designed completely new pathways. For example, the former approach has been used to enable *Pseudomonas putida,* capable of growing on 4-methylbenzoate (4MB), to degrade 4-ethylbenzoate (4EB) (Ramos et al., 1987). This work initially developed four hypotheses to explain the lack of catabolism of 4EB by this *Pseudomonas* sp.: (a) the compound is toxic to the organism; (b) transport of the compounds into the bacterial cells is limited; (c) the failure of the compound (4EB) to effect the expression of the degradation gene(s); and/or (d) the possibility that 4EB is not a substrate for enzymatic attack. Experiments by Ramos et al. (1987) revealed that 4-ethylbenzoate is nontoxic, is not limited by transport into the cells, and is metabolized by the existing enzymes in the benzoate pathway because the first metabolite 4-methylcatechol, is produced when *P. putida* is provided with glucose and 4-ethylbenzoate. However, the next enzyme fails to act on the 4-methylcatechol, a scenario which implies enzyme inhibition. Ramos and colleague subsequently revealed that 4-ethylbenzoate fails to activate catechol 2,3,-diogenase, a key

enzyme for cleaving the aromatic ring. Mutants which selectively grow on 4-ethylbenzoate were generated by initially growing *P. putida* constitutively on a benzoate analog in the presence of 4-ethylbenzoate and then mutagenizing with ethyl methane sulfonate. The resultant mutant was cloned with a 7.7-Kb Eco RI fragment containing promoterless *lacZ* and *lacY* genes that were able to grow on 4EB as the sole carbon source, generating 4-ethylcatechol. Although subsequent studies showed that these transconjugants survive well in soil rich in organic matter, they were unable to effectively withstand contaminated soils (Huertas et al., 1998).

Genetic engineering has also been used to sense the presence of specific organic contaminants in the environment and through bioluminescence to monitor the bioremediation process *in situ* (Sayler and Ripp, 2000). Thus, *Pseudomonas fluorescens* HK44 engineered with a *lux* gene fused within the naphthalene degradative pathway (Fig. 15.2) bioluminesces as it degrades some polyaromatic hydrocarbons, enabling monitoring of degradative activity that is detected as light (Ripp et al., 2000). Under these field conditions, the *lux*-gene-tagged *P. fluorescens* displayed a similar survival pattern as the heterotrophic indigenous bacterial population in both polyaromatic hydrocarbon contaminated and noncontaminated soil over a 2-year duration. Over that period, the tagged bacteria decayed at a rate of 0.017 day$^{-1}$ and 0.014 day$^{-1}$ in contaminated and noncontaminated soil respectively. However, as numerous studies with bacterial inoculation in agricultural systems show, the survival of introduced bacterial strains varies greatly depending on several biotic and abiotic factors in the ecosystem (Jjemba and Alexander, 1999). Thus, introduced strains were recovered six years after release (Hirsch and Spokes, 1994) or became undetectable one year after introduction in soil (De Leij et al., 1995). Given this, engineering of genetically modified organisms that remain viable only in the presence of specific environmental contaminants was proposed (Keasling and Bang, 1998). Such organisms would contain self-destructive (i.e., suicidal) vectors or genes that become activated once the specific contaminant has been depleted from the environment. However, this approach offers no guarantee against potential transfer of genetic material to other microorganisms in the environment.

### 15.2.4 Other potential benefits from environmental biotechnology

Other improvements via genetic engineering have included the movement of microbial genes coding for herbicide biodegradation into crops, enabling the application of herbicides to fields, killing the weeds but sparing the crops, and the development of Ice-minus bacteria that reduce frost damage to crops (Lindow, 1990). *Pseudomonas fluorescens* was engineered to protect plants from frost damage by deleting

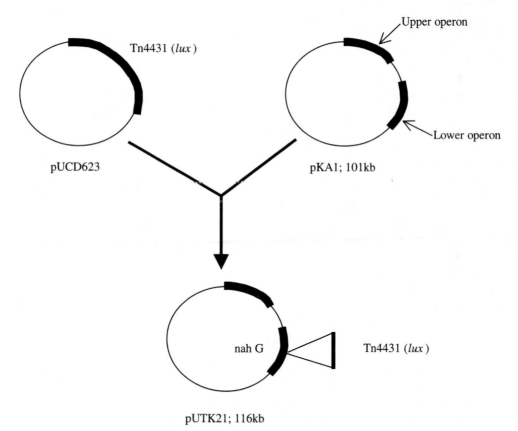

**Fig. 15.2** Construction of bioluminescent reporter plasmid pUTK21. Mid-log phase *E. coli* HB101 (pUCD623) was mated with a spontaneous rifampicin-resistant mutant of *Pseudomonas fluorescens* 5R (pKA1) in a 1:10 ratio and incubated at 28°C for 24 hours on a selective medium to generate the pUTK21 transcojugant. The upper operon encodes for the conversion of naphthalene to salicylate whereas the lower operon encodes for the oxidation of salicylate to acetyaldehyde and pyruvate (Source: King et al., 1987 with permission from the American Association for the Advancement of Science, Washington DC).

the ice nucleation gene, generating Ice-mutants. These GEMs were some of the first to undergo field testing and risk assessment in the United States and subsequent studies showed that some Ice-minus pseudomonads are fairly widespread, reducing the need to rely on GEMs to reduce nucleation (Wrubel et al., 1997). This finding underscores the importance of researchers making an extensive assessment of the cryptic capabilities of the organism under study and taking full advantage of its genomic plasticity rather than fixating only on gene manipulation.

Overall, some obstacles to the release of genetically engineered microorganisms in the environment still exist. These include lack of information about effectiveness in the field and uncertainties about the long-term ecological impact of genetically engineered microorganisms. It is difficult to predict the behavior of these "new" organisms let alone uncertainty about the promiscuity of new genes in the natural environment. Cost and/or risk/benefit analyses as well as ethical issues have also got to be addressed. Nevertheless opportunities in environmental biotechnology do exist, especially for controlling infectious diseases in overcrowded populations, increasing crop yields, reducing the environmental impact of agriculture, providing alternatives to toxic chemical compounds, and in eliminating existing toxic environmental pollutants.

GEMs are perceived to have a reduced level of fitness as a result of the extra genetic material they are provided. Such a scenario is believed to reduce their ability to compete once introduced in the environment (Lenski, 1993). However, generalizations about their survival and fitness have proved hard to come by and remain unpredictable possibly because of the wide range of environmental factors (both biotic and abiotic) that influence survival. Their fitness should be assessed in the context of not just survival, but rather the ability to accomplish the purpose for which they are introduced into a specific environment.

## References

Bloem J.F. and I.J. Law (2001). Determination of competitive abilities of *Bradyrhizobium japonicum* strains in soils from soybean production regions in South Africa. *Biol. Fert. Soils* **33**: 181-189.

Boyd E.M., K. Killham, and A.A. Meharg (2001). Toxicity of mono-, di- and tri-chlorophenols to *lux* marked terrestrial bacteria, *Burkholderia* species *Rasc c2* and *Pseudomonas fluorescens. Chemosphere* **43**: 157-166.

Braun R. (2002). People's concerns about biotechnology: some problems and some solutions. *J. Biotechn.* **98**: 3-8.

Cho J.-C. and J.M. Tiedje (2001). Bacterial species determination from DNA-DNA hybridization by using genome fragments and DNA microarrays. *Appl. Environ. Microbiol.* **67**: 3677-3682.

Crecchio C. and G. Stotzky (1998). Insecticidal activity and biodegradation of the toxins from *Bacillus thuringiensis* subsp. *kurstaki* bound to humic acids from soil. *Soil Biol. Biochem.* **30**: 463-470.

De Leij F.A.A.M., E.J. Sutton, J.M. Whipps, J.S. Fenlon, and J.M. Lynch (1995). Field release of a genetically-modified Pseudomonas fluorescens on wheat-establishment, survival, and dissemination. *Bio/Tech.* **13**: 1488-1492.

Gottweis H. (1998). Governing Molecules: The Discursive Politics of Genetic Engineering in Europe and the United States. MIT Press, Cambridge, MA.

Henderson L. and D.M. Gatewood (1998). Release of genetically engineered microorganisms in the Environment in the United States. *In*: N.S. Subba Rao and Y.R. Dommergues (eds.). Microbial Interactions in Agriculture and Forestry. Science Publ. Inc., Enfield, NH, vol. II, pp. 83-109.

Hirsch P.R. and J.D. Spokes (1994). Survival and dispersal of genetically modified rhizobia in the field and genetic interactions with native strains. *FEMS Microbiol. Ecol.* **15**: 147-160.

Huertas M.-J., E. Duque, S. Marqués, and J.L. Ramos (1998). Survival in soil of different toluene-degrading *Pseudomonas* strains after solvent shock. *Appl. Environ. Microbiol.* **64**: 38-42.

Hurt R.A., X.Qiu, L. Wu, Y. Roh, A.V. Palumbo, J.M. Tiedje, and J. Zhou (2001). Simultaneous recovery of RNA and DNA from soils and sediments. *Appl. Environ. Microbiol.* **67**: 4495-4503.

Jefferson R. A., T.A. and M.W. Bevan (1987). Gus fusions: β-glucurunidase as a sensitive and versatile gene fusion marker in higher plants. *EMBO J.* **6**: 3901-3907.

Jjemba P.K. and M. Alexander (1999). Possible determinants of rhizosphere competence of bacteria. *Soil Biol. Biochem.* **31**: 623-632.

Keasling J.D. and S.W. Bang (1998). Recombinant DNA techniques for bioremediation and environmentally-friendly synthesis. *Curr. Opin. Biotech.* **9**: 135-140.

King J.M.H., P.M. DiGrazia, B. Applegate, R. Burlage, J. Sanseverino, P. Dunbar, F. Larimer, and G.S. Sayler

(1990). Rapid, sensitive bioluminescent reporter technology for naphthalene exposure and biodegradation. *Science* **249:** 778-781.

Koskella J. and G. Stotzky (1997). Microbial utilization of free and clay-bound insecticidal activity toxins from *Bacillus thuringiensis* and their retention of insecticidal activity after inhibition with microbes. *Appl. Environ. Microbiol.* **63:** 3561-3568.

Lenski R.E. (1993). Evaluating the fate of genetically engineered microorganisms in the environment: are they inherently less fit? *Experentia* **49:** 201-209.

Letourneau D.K. and B.E. Burrows (2003). Genetically Engineered Organisms: Assessing Environmental and Human Health Effects. CRC Press, Boca Raton, FL.

Lindow S.E. (1990). Use of genetic altered bacteria to achieve plant frost control. *In:* J.P. Nakas and C.Hagendorn (eds.). Biotechnology of Plant-Microbe Interactions. McGraw Hill, New York, NY, pp. 85-110.

Martin T., R.V. Wohner, S. Hummel, L. Willmitzer, and W.B. Frommer (1992). The GUS reporter system as a tool to study plant gene expression. *In:* S.R. Gallagher (ed.). GUS Protocols: Using the GUS Gene as a Reporter of Gene Expression. Acad. Press, Inc., London, UK, pp. 23-43.

Marx J.L. (1987). Assessing the risks of microbial release. *Science* **237:** 1413-1417.

Moran M.A., V.L. Torsvik, T. Torsvik, and R.E. Hodson (1993). Direct extraction and purification of rRNA for ecological studies. *Appl. Environ. Microbiol.* **59:** 915-918.

Ochman H., J.G. Lawrence, and E.A. Groisman (2000). Lateral gene transfer and the nature of bacterial innovation. *Nature* **405:** 299-304.

Palm C.J., D.L. Schaller, K.K. Donegan, and R.J. Seidler (1996). Persistence in soil of transgenic plant produced *Bacillus thuringiensis* subsp. *kurstaki* ä-endotoxin. *Can. J. Microbiol.* **42:** 1258-1262.

Ramos J.L., A. Wasserfallen, K. Rose, and K.N. Timmis (1987). Redesigning metabolic routes: Manipulation of TOL plasmid pathways for catabolism of alkylbensoates. *Science* **235:** 593-596.

Ripp S., D.E. Nivens, Y. Ahn, C. Werner, J. Jarrell IV, J.P. Easter, C.D. Cox, R.S. Burlage, and G.S. Sayler (2000). Controlled field release of a bioluminescent genetically engineered microorganism for bioremediation process monitoring and control. *Environ. Sci. Tech.* **34:** 846-853.

Rondon M.R., R.M. Goodman, and J .Handelsman (1999). The earth's bounty: Assessing and accessing soil microbial diversity. *Trends Biotech.* **17:** 403-409.

Rondon R.M., P.R. August, A.D. Bettermann, S.F. Brady, T.H. Grossman, M.R. Liles, K.A. Loiacono, B.A. Lynch, I.A. MacNeil, C. Minor, C.L. Tiong, M. Gilman, M.S. Osburne, J. Clardy, J. Handelsman, and R.M. Goodman (2000). Cloning the soil metagenome: A strategy for accessing the genetic and functional diversity of uncultured microorganisms. *Appl. Environ. Microbiol.* **66:** 2541-2547.

Sambrook J and D.W. Russell (2001). Molecular Cloning: A Laboratory Manual. 3rd Edition. Cold Spring Harbor Laboratory Press, Long Island, NY.

Sayler G.S. and S. Ripp (2000). Field applications of genetically engineered microorganisms for bioremediation processes. *Curr. Opin. Biotech.* **11:** 286-289.

Sessitsch A., P.K. Jjemba, G. Hardarson, A.D.L. Akkermans, and K.J. Wilson (1997). Measurement of the competitive index of *Rhizobium tropici* strain CIAT899 derivatives marked with the *gusA* gene. *Soil Biol. Biochem.* **29:** 1099-1100.

Stiner L. and L.J. Halverson (2002). Development and characterization of a green fluorescent protein-based bacterial biosensor for bioavailable toluene and related compounds. *Appl. Environ. Microbiol.* **68:** 1962-1971.

Saxena D. and G. Stotzky (2001). *Bacillus thruingiensis* (Bt) toxin released from root exudates and biomass of Bt corn has no apparent effect on earthworms, nematodes, protozoa, bacteria, and fungi in soil. *Soil Biol. Biochem.* **33:** 1225-1230

Tapp H. and G. Stotzky (1995a). Dot blot enzyme-linked immunosorbent assay for monitoring the fate of insecticidal toxins from *Bacillus thuringiensis* subsp. *kurstaki* in soil. *Appl. Environ. Microbiol.* **61:** 602-609.

Tapp H. and G. Stotzky (1995b). Insecticidal activity of the toxins from *Bacillus thuringiensis* subspecies *kurstaki* and *tenebrionis* adsorbed and bound on pure and soil clays. *Appl. Environ. Microbiol.* **61:** 1786-1790.

Tapp H. and G. Stotzky (1998). Persistence of the insecticidal toxin from *Bacillus thuringiensis* subsp. *kurstaki* in soil. *Soil Biol. Biochem.* **30:** 471-476.

Tappeser B., M. Jäger, and C. Eckelkamp (2002). Survival, persistence, transfer: The fate of genetically modified microorganisms and recombinant DNA in different environments. *In:* D.K. Letourneau and B.E. Burrows (eds) Genetically Engineered Organisms: Assessing Environmental and Human Health Effects. CRC Press, Boca Raton, FL, pp. 223-250.

Vaeck M., A. Reynaerts, H.Höfte, and H. van Mellaert (1988). Transgenic crop varieties resistant to insects. *ACS Symp. Series* **379:** 280-283.

Willardson B.M., J.F. Wilkins, T.A. Rand, J.M. Schupp, K.K. Hill, P. Keim, and P.J. Jackson (1998). Development and testing of a bacterial biosensor for toluene-based environment contaminants. *Appl. Environ. Microbiol.* **64:** 1006-112.

Wilson K.J., A. Sessitsch, J.C. Corbo, K.E. Giller, A.D.L. Akkermans, and R.A. Jefferson (1995). â-glucuronidase (GUS) transposons for ecological and genetic studies of rhizobia and other gram-negative bacteria. *Microbiol.* **141:** 1691-1705.

Wrubel R.P., S. Krimsky, and M.I. Anderson (1997). Regulatory oversight of genetically engineered microorganisms: has regulation inhibited innovation? *Environ. Manage.* **21:** 571-586.

Ye R.W., T. Wang, L. Bedzyk, and K.M. Croker (2001). Applications of DNA microarrays in microbial systems. *J. Microbiol. Meth.* **47:** 257-272.

Yuhashi K.-I., N. Ichikawa, H. Ezura, S. Akao, Y. Minakawa, N. Nukui, T. Yasuta, and K. Minamisawa (2000). Rhizobitoxine production by *Bradyrhizobium elkanii* enhances nodulation and competitiveness on *Macroptilium atropurpureum. Appl. Environ. Microbiol.* **66:** 2658-2663.

Zhou J. and D.K. Thompson (2002). Challenges in applying microarrays to environmental studies. *Curr. Opin. Biotech.* **13:** 204-207.

# Index